The Politics of Privatisation and Trade Union Mobilisation

Trade Unions Past, Present and Future

Edited by Craig Phelan

Volume 3

PETER LANG
Oxford • Bern • Berlin • Bruxelles • Frankfurt am Main • New York • Wien

The Politics of Privatisation and Trade Union Mobilisation

The Electricity Industry in the UK and Argentina

Pablo Ghigliani

PETER LANG
Oxford • Bern • Berlin • Bruxelles • Frankfurt am Main • New York • Wien

Bibliographic information published by Die Deutsche Nationalbibliothek.
Die Deutsche Nationalbibliothek lists this publication in the Deutsche Nationalbibliografie; detailed bibliographic data is available on the Internet at http://dnb.d-nb.de.

A catalogue record for this book is available from the British Library.

Library of Congress Cataloging-in-Publication Data:

Ghigliani, Pablo.
The politics of privatisation and trade union mobilisation : the electricity industry in the UK and Argentina / Pablo Ghigliani.
p. cm. -- (Trade unions, past, present and future ; v. 3)
Includes bibliographical references and index.
ISBN 978-3-03911-961-5 (alk. paper)
1. Electric utilities--Privatization--Great Britain--Case studies. 2. Electric utilities--Privatization--Argentina--Case studies. 3. Electric industry workers--Labor unions--Great Britain--Case studies. 4. Electric industry workers--Labor unions--Argentina--Case studies. I. Title.
HD9685.G72G55 2010
333.793'23--dc22
2010023482

ISSN 1662-7784
ISBN 978-3-03911-961-5

Hochfeldstrasse 32, CH-3012 Bern, Switzerland
info@peterlang.com, www.peterlang.com, www.peterlang.net

Printed in Germany

To Mariana, who knows why;
to Catalina and Teresa, who will;
and to Tecla, who won't.

Contents

List of Tables

List of Abbreviations

AEEU	Amalgamated Engineering and Electrical Union
AEU	Amalgamated Engineering Union
AFJP	Aseguradora de Fondos de Jubilaciones y Pensiones
APEX	Association of Professional, Executive, Clerical and Computer Staff
ART	Aseguradora de Riesgos de Trabajo
BEC	Branch Executive Committees
BIRF	Banco Internacional de Reconstrucción y Fomento
CCT	Convenio Colectivo de Trabajo
CEGB	Central Electricity Generating Board
CGA	Comisión General Administrativa
CGD	Cuerpo General de Delegados
CGT	Central General de Trabajadores
CIAP	Comisión Interna Auto-composición del Personal
CTA	Central de Trabajadores de la Argentina
DEBA	Dirección de Energía de la Provincia de Buenos Aires
DESTUC	Divisional Electricity Supply Trade Union Councils
EDELAP	Empresa Distribuidora y Comercializadora La Plata
EDENOR	Empresa Distribuidora y Comercializadora Norte
EDESUR	Empresa Distribuidora y Comercializadora Sur
EdF	Electricité de France
EETPU	Electrical, Electronic, Telecommunications and Plumbers Union
EMA	Engineers' and Manager's Association
EPEA	Electrical Power Engineers' Association
ESEBA	Empresa Social de la Energía de la Provincia de Buenos Aires
ESI	electricity supply industry
ESOP	Employee Stock Ownership Plan

ESTUC	Electricity Supply Trade Unions Council
FATLyF	Federación Argentina de Trabajadores de Luz y Fuerza
FeNTOS	Federación Nacional de Trabajadores de Obras Sanitarias
FeTERA	Federación de Trabajadores de la Energía de la República Argentina
FTO	Full-time Officer
FUSE	Federation of Unions Supplying Electricity
GMBATU	General, Municipal, Boilermakers and Allied Trade Union
IMF	International Monetary Fund
IPCTT	Internacional de Personal de Correo, Telégrafos y Teléfono
LyF CF	Luz y Fuerza Capital Federal
LyF MDP	Luz y Fuerza Mar del Plata
MNC	Multinational Corporations
NALGO	National and Local Government Officers' Association
NEC	National Executive Council
NGC	National Grid Company
NJB	National Joint Board
NJC	National Joint Council
NJIC	National Joint Industrial Council
NJMC	National Joint Managerial and Higher Executive Grades Committee
NORWEB	North West Electricity Board
PAG	Professional and Administrative Grade
PJ	Partido Justicialista
PPP	Programa de Propiedad Participada
PRP	performance related pay
PSBR	Public Sector Borrowing Requirement
REC	Regional Electricity Company
SEEB	South Eastern Electricity Board
SEGBA	Servicios Eléctricos del Gran Buenos Aires
SOE	state-owned enterprise
STB	single table bargaining
Sutecba	Sindicato Único de Trabajadores del Estado de la Ciudad de Buenos Aires
SWALEC	South Wales Electricity

SWEB	South Western Electricity Board
TEIS	Tarifa Eléctrica de Interés Social
TGWU	Transport and General Workers' Union
TTB	two-table bargaining
TUC	Trade Union Congress
UOCRA	Unión Obrera de la Construcción de la República Argentina
UOM	Unión Obrera Metalúrgica
WB	World Bank
YCF	Yacimientos Carboníferos Fiscales

CHAPTER ONE

Introduction

This book has two main objectives. On the one hand, it aims to explore, empirically, the relationships between privatisation and workers' collectivism. On the other, it is intended as a test of the utility of mobilisation theory for such purposes. As to the former, the research focuses on the privatisation of the electricity supply industry (ESI) in two countries, the United Kingdom (UK) and Argentina, and more specifically, on how privatisation conditioned trade unions' ability to act and how ESI unions have effectively responded to the privatisation challenge. As for the latter, due to differences in the national contexts, the comparative approach compels the analysis to pay attention to intermediate variables that influence the empirical manifestation of the categories of mobilisation theory; in turn, the scope of the study means to incorporate multiple levels of analysis, whereas most works have confined the theory to the micro-level.

The research arose at a point of thematic convergence in the diverse scholarly literature. It comprised studies on the rationale and driving-forces behind the development of privatisation worldwide, and in particular about the privatisation programmes in the UK and in Argentina; the relationships between privatisation and organised labour; and lastly, the use of mobilisation theory in the field of labour studies. The forthcoming chapters review these strands in the literature, where appropriate. Still, a few introductory remarks are necessary to situate the research.

From a mobilisation perspective, the main deficit of the vast mainstream literature on privatisation worldwide is its tendency to approach privatisation as a technical response to an economic problem (Dinerstein 2001). Nevertheless, many scholars have stressed that privatisations were not technically inevitable, but politically driven as part of a wider strategy for overcoming obstacles to capital accumulation (Cook and Murphy 2002;

Dinerstein 2001; Thwaites Rey 1994). In this sense, privatisation is at the heart of a particular phase of capitalist counter-mobilisation. Indeed, for mobilisation theory, privatisation may be viewed as a counter-mobilising force in itself, and hence highly disruptive for organised labour. Chapters Four and Five will discuss this perspective, concomitantly justifying the empirical focus of the book.

As for the relationships between privatisation and labour, scholars concentrate mostly on the impact of privatisation on job loss, industrial relations issues and collective bargaining (Báez-Camargo 2002; Cook and Murphy 2002; Colling and Ferner 1995; Luca 1998; Ferner and Colling 1991; 1993a; 1993b; O'Connell Davidson 1993; Hall 2000, 2005; Murillo 2001; Ogden 1993; Pendleton and Winterton 1993; Pripstein Posusney and Cook 2002; Riethof 2002; Van der Hoeven and Sziraczki 1997; Wallis 2000; among many others). Chapter Three addresses this literature, insofar as it relates to the aims of this research. This book, instead, aims to study how this type of change (and others) conditions workers' collectivism, a topic rather neglected in the literature.

Lastly, mobilisation theory, as promoted by Kelly (1998), has been applied mainly in the field of labour studies to analyse and conceptualise union revitalisation. In this line of research, mobilisation theory plays different roles. In the main, it contributes, together with insights from other theoretical frameworks, to the conceptualisation of partial aspects of the process of union revitalisation. In this vein, it is loosely applied to the analysis of the structure of opportunities afforded by the state and the institutional environment, the politics of coalition building and the key role of union strategies (Baccaro, Harman and Turner 2003; Frege and Kelly 2003, 2004; Heery, Kelly and Waddington 2003; Heery et al. 2003; Kelly 2005). In the cases to which it is strictly applied, the focus shifts to discrete events; for instance, organising campaigns (Kelly and Badigannavar 2003), or struggles around the statutory recognition process (Moore 2004). Besides this strand, the categories offered by mobilisation theory have also prompted a variety of empirical enquiries. This in particular happened with leadership (Darlington 2001; Green et al. 2000; Metochi 2002), injustice (Atzeni 2003; Brown Johnson and Jarley 2004) and mobilisation (Atzeni 2005).

Almost all these works share a preference, particularly when they draw heavily on mobilisation theory, for microanalytical levels. This suggests that, as George Gall (2000) stressed in his review of the book *Rethinking Industrial Relations*, there is still a lack of intellectual engagement with the theoretical framework developed by Kelly at meso- and macroanalytical levels. This research is partly intended as a step toward balancing this deficiency.

The core idea of the research is, thus, that a focus on the consequences of privatisation on organised labour is appropriate to link both levels of the mobilisation theory as outlined by Kelly (1998): the long-run account of the alternating periods of workers' mobilisation and capital counter-mobilisation and the set of explanatory categories for the analysis of collective action in the medium- and short-term: opportunity-to-act (or opportunity structure), mobilisation / counter-mobilisation, organisation, interest definition, and collective action.

The research conceptualises privatisation as not only a part of but also in itself a process of counter-mobilisation, which decreased the opportunity for trade unions to engage in collective action. Additionally it links variability in trade unions' strategies in a context of capital and state counter-mobilisation, not only with differences in the opportunity structure, but also with union organisational capabilities and the dynamics of interest definition.

Outline of the Book

The argument of the book unfolds by gradually building the theoretical and empirical foundations (Chapters Two to Six), which frames the case-study research (Chapters Seven to Ten).

Chapter Two discusses mobilisation theory. While this chapter recognises its potential for the study of workers' collectivism, it states that the micro-level framework adopted by Kelly to account for interest definition is inadequate for analysis of how ESI privatisation conditioned trade unions' ability to act. Thus, when setting up the analytical framework of the research, this chapter puts forward an alternative approach to the category interest, which stresses the collective interactions by which workers process their multiple needs into specific demands and strategies. Concurrently, the discussion underlines the importance of the categories opportunity-to-act and organisation for explaining workers' collectivism, and shows how they are understood and used in the study.

Chapter Three presents the methodology, the design of which combines a case-study approach and the comparative method, and also justifies the selection of the empirical field.

In Chapters Four and Five, the aim is to explain why the process of privatisation was, arguably, one of the most salient aspects of the cycle of capital counter-mobilisation that began in the 1970s. After illustrating the various meanings of the term privatisation, Chapter Four provides a set of arguments to assert its counter-mobilising content against labour. By exploring the driving-forces and rationales of their respective programmes, Chapter Five compares the contextual conditions of privatisation in the UK and in Argentina; in this way, it highlights how far contextual variability contributed to shape the opportunity structures of ESI trade unions in both countries.

Chapter Six analyses, comparatively, how national industrial relations institutions, as intermediate variables, constrained in different ways trade union (re)sources of power, thereby affecting the opportunity structure and the forms taken by both state and capital counter-mobilisation during the process of privatisation.

The remaining chapters of the book explore the relationships between privatisation and workers' collectivism through an empirical analysis of certain counter-mobilising contents of the ESI privatisation, their impact on the mobilising capacity of ESI unions and their strategies.

Chapter Seven focuses on the anti-privatisation campaigns run by ESI unions, paying particular attention to the category opportunity-to-act to which those campaigns were directed. However, as changes in the opportunity structure cannot explain by themselves the strategic choices of the actors, the analysis turns, then, to union organisational features, leadership styles and decision-making processes.

Chapters Eight and Nine address two crucial aspects of the counter-mobilising content of privatisation: firstly, by analysing how monetary incentives were mobilised to prevent workers' collectivism; secondly, by considering how the fragmentation of collective bargaining brought about changes in workers' capacity to engage in collective actions.

Lastly, Chapter Ten examines the impact of privatisation upon the organisational structures and processes of decision-making of ESI unions, pointing to how the growth of sectionalism and the decline of workers' participation in decision-making has threaten the mobilising capacity of organised labour since privatisation.

The book closes with a summary of the main arguments and findings of the research.

CHAPTER TWO

Developing the Analytical Framework

Mobilisation theory, as adapted by John Kelly for the field of industrial relations, appears as a powerful framework for the analysis of workers' collectivism (Kelly 1998) when the insights developed within the social movements and collective action traditions are followed. His formula, which draws upon long wave theory, combines a long-term account of the alternating periods of workers' mobilisation and capital counter-mobilisation with a set of explanatory categories for the analysis of collective action in the medium- and short-term (opportunity-to-act, mobilisation, organisation, interest, collective action). Therefore its application suits, in principle, the empirical aim of the research: analysis of how ESI privatisation conditioned trade unions' ability to act by targeting vital dimensions of workers' collectivism.

However, any theoretical transposition entails a call for prudence, for theories often have blemishes. In addition, the comparative framework in which mobilisation theory is applied adds complexity to the entire theoretical endeavour. So, after explaining why mobilisation theory is more appropriate for this study than standard comparative approaches, the following sections will first address certain weaknesses of mobilisation theory with regard to answering the research questions before arguing that some of these shortcomings underpin the micro-level framework adopted by Kelly to account for the category interest. Therefore the analytical framework of the research is outlined on the basis of the elements of Kelly's model, which are highly relevant to the aims of the book; however an alternative approach to interest definition is followed, which emphasises collective dimensions rather than individual experiences and subjective perceptions.

Mobilisation Theory in Comparative Perspective

Within the practice of sociology, there exists a shared understanding of the advantages of comparison that highlights its role in promoting deprovincialisation, and in revealing both the systematic nature of variations and the broad similarities. According to Crow, these variations and similarities facilitate the identification of what is generally true about a number of cases (Crow 1997). In this shared understanding, it is also stressed, as Durkheim put it, that 'only comparison affords explanation' (Durkheim 1970: 41). Comparative historical analysis has been recently defined as placing the emphasis on three specific points: 'a concern with causal analysis, the exploration of temporal processes, and the use of systematic and contextualised comparison' (Mahoney and Rueschemeyer 2003: 15).

Within the field of industrial relations, Hyman has praised the comparative approach on similar grounds (Hyman 1994a). For him, cross-national comparison forces the observer to address critically what is narrowly accepted as unproblematic within the individual national context, that is, what is otherwise taken for granted is shown to be contingent and, perhaps, exceptional. As a result, and Hyman says always, the researcher is compelled to revise the assumptions about the nature and meaning of the key institutions of industrial relations (companies, trade unions, employers associations, collective bargaining, labour law, and so forth). Thus, comparison would offer a more rigorous test for causal explanations developed in individual countries by forcing the researcher to refine the explanatory propositions in order to make them genuinely applicable as general – rather than single-context – theories.

Hence, it is not surprising that the comparative method had early appeal in the field of industrial relations studies and, more recently, within the research agenda of human resources management.

Originally, conventional comparative approaches in the field tended to focus on the formal institutions and legal structures of either the industrial relations system – *à la* Dunlop (1958), or the collective bargaining system – *à la* Clegg (1976). The aims of the researchers in those studies were, usually, the construction of typologies, the contrast of diverse systems of regulation and the understanding of the basic relations between the actors. Still, during the 1980s, authors like Adams (1981) and Sisson (1987) sought to develop the collective bargaining approach in an international comparative perspective by focusing on the role of employers as a key variable explaining divergence.

The 1990s, in turn, would renew the interest of the debates about the likelihood of convergence or divergence in the evolution of the national industrial relations systems, famously first developed by the book *Industrialism and Industrial Man* (Kerr et al. 1960). But this time, this renewed interest in the evolution of employment relations worldwide arose as a consequence of forces unleashed by globalisation, mainly, the increased international competition and the diffusion new production technologies (Bamber and Lansbury 1993; Bamber, Lansbury and Wiles 2004; Bean 1994; Eaton 2000).

However, despite advocacies of convergence, most comparative studies painted a picture of continuing diversity among the national systems of industrial regulation, while recognising common challenges due to the introduction of new product and informational technologies, industrial restructuring, and managerial pressures towards flexibilisation and decentralisation (Bamber, Lansbury and Wiles 2004; Ferner and Hyman 1998; Locke and Thelen 1995; Ruysseveldt 1995). The MIT project, for instance, was a key force within this trend of research. Scholars like Thomas Kochan and Harry Katz played an important role by stressing the relevance of the strategic choices among the different actors in order to understand how global forces interact with national situations (Katz 1997; Katz and Kochan 2004; Kochan, Katz and McKersie 1987). In addition, the role of nationally specific institutional factors in explaining persistent variation in industrial relations across countries was also typically emphasised by those criticising the convergence theses. Lastly, the literature, which began to pay attention to the international diffusion of employment and human resources management policies due to the global expansion of the Multinational Corporations (MNCs), accentuated the role of not only legal and institutional aspects but also different cultural frameworks (Ferner 1997).

Several scholars commence to feel uneasy about the development of the comparative approach in the field. For instance, Richard Locke and Kathleen Thelen developed an early and convincing critique of the conventional approach, which emphasised that the explanatory power of the conventional model is undermined by three basic assertions. Firstly, conventional comparative analyses often portray external pressures as 'equally pervasive or intense to all national economies' (Locke and Thelen 1995: 340). Secondly, these scholars argued that 'traditional analyses often obscure stark differences in starting-points and hence the significance of the changes for the various national industrial relations systems' (idem). Finally, they underlined that traditional studies assume that same practices have 'the same meaning or valence across the various countries' (idem). By contrast, drawing upon institutionalist and political constructionist analyses, Locke and Thelen put forward an alternative, more contextualised approach, to explain 'why the same international forces have set in motion fundamentally different substantive conflicts in different contexts' (Locke and Thelen 1995: 343). Their main finding was that the specific interactions between institutions and actors' identities, particularly traditional trade union identities, are key to understanding cross-national variations.

Others scholars have attempted to replace the conventional cross-national comparative approach with a more international perspective to account for how globalising forces work and affect the relationships between capital and labour nationally (Bamber and Lansbury 1993; Bamber, Lansbury and Wiles 2004; Bean 1994; Eaton 2000). However, these studies have failed to overcome the conventional approach, in which, characteristically, the institutions and practices of two or more countries are described and systematically analysed. At best, some aspects are isolated and analysed transnationally within this trend – for instance the role of the European Work Councils, the impact of the EU social mandates or the influence of corporate international management – but none of those studies really free themselves from the country-by-country approach at key points of the argument (see Easton 2000, Chapter Five, for example).

In short, this introduction, albeit schematically, helps the reader understand why conventional approaches are inappropriate for a book whose aim is not to construct typologies, contrast systems of regulation or address basic relations between actors, but to focus on the dynamics of unions' responses to change. This is precisely the main focus of mobilisation theory. Indeed, unlike other recent comparative studies on union strategies in which the mobilisation theory is only loosely applied (Frege and Kelly 2003, 2004; Frege, Heery and Turner 2004), a systematic comparison following a mobilising perspective allows for the different stages of mobilisation to be conceptualised and studied within specific contexts.

If it can be argued that this study does partly follow the conventional approach insofar as it examines a similar development, the privatisation of ESI, the explanation of variation in the national responses following mobilisation theory goes beyond the analysis of the alternative institutional arrangements, and hence, surpasses the conventional model. Indeed, like the contextualised comparisons advanced by Locke and Thelen (1995), mobilisation theory also gives a place to the mediating role played by institutions through the categories opportunity-to-act and organisation. Furthermore it puts forward a more sophisticated apparatus to account for trade unions' agency – through the categories organisation and interest definition – which is one of the main objectives of this research. Here mobilisation theory also outperforms Locke and Thelen's emphasis on actors' identities.

In sum, given the focus of mobilisation theory on the dynamics of collective action, it reveals itself as an appropriate framework when accounting for trade unions' responses to change. As a result, it is fruitful to develop the model from a comparative perspective. Whether or not the advantages of comparison are still advocated by scholars from some very different fields of social research, the conventional comparative approach in the field of industrial relations and human resources management has lost its previous appeal for those scholars interested in the analysis of trade unions' agency. Yet the theoretical development of mobilisation theory for a comparative study requires an appropriate methodological design. This latter aspect will be discussed in the next chapter.

Collective Action and Social Movements: Theoretical Insights and Theoretical Shortcomings

Collective action and social movement traditions are rich in theoretical elaborations. A full account of this wealth is not only out of the reach of this chapter, but also meaningless for the purposes of this study. Instead, the examination of some controversial views, sharing these trends in thought, is crucial to any consideration of the prospects for their successful theoretical transposition to the field of labour studies.

The basic questions are: what do these theoretical traditions stand for? Do they focus on identical or dissimilar objects of study? What connotations do the answers to the aforementioned questions have for the study of unionised workers?

In general, modern theories of contentious collective action and social movements have intended to bridge the gap between accounts of structural change and explanations of collective action (Klandermans et al. 1988). By the 1980s, a common research agenda had emerged as the outcome of successive reactions to structuralism (McAdam, Tarrow and Tilly 2001). It comprised four main areas of interest: resource mobilisation, the political processes, the repertoires of contention, and claim framing. Yet, two main concerns permeated this agenda as a whole: why do individuals join contentious collective action and what are the social-psychological processes involved?

The story behind this theoretical development is complex. It includes at least two processes. On the one hand, the social sciences have been colonised by individualistic assumptions taken from neoclassical models of economic thought. As Tarrow complains, 'in the trace of microeconomics, the problem for collective action came to be not how classes struggle and State rules, but how collective action is even possible among individuals guided by narrow self-interest' (Tarrow 1994: 15). Then, having in mind those individual and rationalistic assumptions, Tarrow goes on within this theoretical tradition, 'political scientists and sociologists have begun their analysis of social movements from the puzzle that collective action is difficult to bring about [despite] in so many situations and against so many odds, collective action does occur, often on the part of people with few resources and little permanent power' (Tarrow 1994: 7).

On the other hand, this scientific search comprises at the same time a theoretical endeavour opposing individualistic tendencies, which looks for social-psychological mechanisms, instead of individual rational calculations, but keeps assuming the problem of collective action is one of aggregation. Ironically, in so doing, scholars retrieved during the 1980s the notions traditional within the field until the 1960s and the 1970s, when macro political and structural accounts of social movements, particularly resource mobilisation theory, and also the so-called political process analysis (McAdam, Tarrow and Tilly 2001), came to replace older social-psychological conceptions and push back individualistic assumptions (McAdam 1988).

Indeed, some formulations combined both perspectives. Characteristically, Klandermans (1984: 584) made the explicit case for the renewal of social-psychological approaches to expectancy-value theory, a typical rational-choice framework, which posits the problem in terms of costs and benefits. This renewal, irrespective of rationalistic assumptions, buttressed 'the contention that what is at issue is not merely the presence or absence of grievances, but the manner in which grievances are interpreted and the generation and diffusion of those interpretations' (Snow et al. 1986: 466). This conclusion prompted a cascade of models addressing this problem, in tandem with different terms to describe the shared meanings that inspire collective action (McAdam 1988; Gamson 1992; Klandermans 1984; Snow et al. 1986; Snow and Benford 1988).

What about the object of study of these intellectual schools? Certainly, despite obvious overlap, there is a difference in scope. While some scholars consider, under the heading contentious collective action, any form of action held by a group of individuals based on common purposes against elites, opponents, authorities, and so forth; the social movement tradition, specifically, has narrowed their concerns to the kind of collective challenges that arose during the 1960s and 1970s, mainly in Western Europe and the United States.

In the case of contentious collective action, theory building rests upon a violent abstraction that reduces the importance of the enduring social relationships that structure interaction. This is to isolate behavioural regularities, or to look for similar mechanisms and processes behind different forms of contention (social movements, revolutions, strike waves, nationalism, or otherwise). As for the subjective aspect, different structural constraints tend to be homogenised, reduced and classified under the general term grievance. The truism that people must experience subjectively a grievance first in order to act is reinforced and dominates the research agenda.

In the case of collective challenges that arose during the 1960s and 1970s, the same concern has spread widely as well, but not at the expense of historical perspectives and social structures. However, while collective action traditions do include organised labour, the mainstream of social movement approaches focuses on social movements other than labour, particularly those emerging within the countercultural and emancipatory currents in the 1960s and 1970s (Rucht 1988). Theory building within this trend was grounded in the study of, firstly, the students and civil rights movements, then the environmental and women's movements, and finally the peace, anti-nuclear, neighbourhood movements and the like (Klandermans and Tarrow 1988; Klandermans et al. 1988; Rucht 1988). Organised labour was not the soil in which most social mobilisation frameworks grew. Subsequently, many scholars tended to assimilate 'other forms of contention to prevailing explanations of social movements' (McAdam, Tarrow and Tilly 2001: 14), whereas others explicitly separated trade unions from the new movements due to what they saw as their salient features: formal and bureaucratic organisation, hierarchical structures, and top-down management (Waterman 1998). The institutionalisation of the interactions between workers and employers along with the routinisation of conflict patterns were among the factors calling for demarcation between old and new movements. On the research agenda of this trend, given the type of objects under study, the very processes of building an organisation and engaging in collective action often overlap theoretically, damaging their appeal for those in the field of labour studies. Typically, the starting-point is individual subjectivity (awareness of grievances), and then, through organisation to collective action.

To summarise, individualistic and social-psychological standpoints explaining collective action and either violent abstraction from social structures and historical circumstances, or indifference with regard to the specificities of the institutions and dynamics of the labour movement, are all weak points of social mobilisation theories that pose risks when attempting their transposition to the field of labour studies. It is now time to assess whether any of these theoretical blemishes found their way into Kelly's analysis.

Interest Definition According to Kelly

Kelly states that 'the fulcrum of the model is interests and the ways in which people [...] come to define them' (Kelly 1998: 25); thus it is worth explaining why this research does not follow this crucial part of the model when approaching the problem of interest definition.

Kelly offers a balanced theoretical framework for the study of workers' collectivism by taking insights from different sources. Crude versions of the aforementioned weak points will neither be found in his theoretical presentation nor in his empirical work. In accordance with his own Marxist background, and having as a starting-point Tilly's (1978) aged but exceptional discussion of mobilisation theory, his prime concerns are the institutions and dynamics of the labour movement and the enduring social relationships that structure interaction. Additional proofs of his rejection of individualistic and rational-choice assumptions are his discussion of long wave theory and the work of Mancur Olson (1971).

Still, Kelly's framework to account for interest definition actually rests on individualistic assumptions, and hence is not adequate for the object of study of this book. Although Kelly rebalances the argument by engaging in worthy discussions of some of the social processes involved, a methodological risk remains. This becomes apparent when the origin of collective action is continually portrayed as an aggregation of individuals

(Kelly 1998: 24, 27, 29, 31, 33, 34), and at the same time, socio-psychological mechanisms overshadow other factors in Kelly's theoretical presentation. Gradually, ideational motives and individual dimensions, essentially the notion of injustice, seem to prevail over other levels of analysis. This impression is reinforced by the comparatively little attention Kelly pays to the remaining categories of social mobilisation.

In addition, his book unfolds a linear discussion of the process of mobilisation frequent in social movement studies: that is, from injustice to attribution to collective organisation to mobilisation (see Atzeni 2009 for a critique, though from a different perspective). However, this discussion is more appropriate for an exploration of the rise of new social movements than for the analysis of the dynamics of organised labour. In the former case, this progression parallels, to some extent, the empirical development of the organisation of previously disorganised – or at best, loosely organised – people who share a common concern or grievance, but not a structural antagonism due to capitalist production. That is the reason why underlying that enquiry is, often, a counter comparison, explicit or not, with the realm of class movements. In the latter case, the very existence of a structural antagonism at the root, and the plethora of organisations and institutions in which it is expressed, urge for a starting point other than individual perceptions of injustice. Moreover, in cases like those of this study, in which well-established organisations have been involved in a long history of interactions with employers and governmental institutions, it is recommended to focus strongly upon collectivities and power relations rather than on individual subjectivity.

The point is not that such an approach explains nothing; it may always be possible to manufacture an individualistic explanation of most social events. Indeed, insofar as Kelly focuses on the development of a collective consciousness, which transcends the aggregation of individual interests, this standpoint might be most suitable. However, these assumptions are not appropriate to our object, that is, the effects of privatisation upon trade unions' ability to act, particularly through collective action. It is not what happened to individuals and their perceptions that crucially limited – or not – trade unions' power. Indeed, as ESI unions are very well-established and workers are habituated in this industry to relying upon them, it is fair to say that there is already a strong sense of a collective identity. Therefore, the methodological preference here is to start from the irreducibility of collective action and trade unions' power to an aggregation of individuals' properties, the preponderance of social processes enacted by people but independent of them, and the existence of real workers and institutions acting within discrete social relations governed by a systemic logic (Fracchia and Lewontin 2005). In order to explain workers' collective responses to privatisation, this book will pursue contextual analyses of the impact of the latter upon the opportunity-to-act and trade unions' organisational features and of the social processes by which the interaction of the multiple dimensions of workers' interests unfolds and concrete demands and courses of collective action crystallise.

The Analytical Framework: Power Relations and Organised Labour

Mobilisation theory helps us to understand how workers (and their organisations) deal (or fail to deal) with multiple interests, define collective demands, mobilise organisational and power resources, and translate them into collective actions according to the opportunity-to-act. Its advantage rests upon its flexibility: it identifies factors which are critical to the presence or absence of collective action and, for that reason, makes theoretically informed explanations of specific empirical outcomes possible. The theory's actual outline is contingent on whether one stresses individual and social-psychological variables, or understands it as a theoretical enquiry about power relations. If the latter, scholars working in these traditions approach the analysis of power and power relations through the categories mobilisation and opportunity-to-act, and so does Kelly.

The categories mobilisation and opportunity-to-act were developed through theoretical efforts directed towards outperforming social-psychological explanations of collective action (McAdam 1988). In the case of resource mobilisation theory, most authors underline that grievances are ubiquitous in society, and hence, it is the availability of resources that is the crucial dimension to the explanation of collective action, that is, the degree of mobilisation (Jenkins 1983; McCarthy and Zald 1977). In the political process approach, instead, scholars emphasise that 'the answer should be searched in the variations in political structures and the workings of the political process' (Tarrow 1994: 18), that is, in the political opportunity structure (Eisinger 1973).

Mobilisation, then, refers to the amount and degree of collective control, not only over the resources needed for action (Tilly 1978; Kelly 1998), but also over social processes, interactions and practices that enable it. However, it is, to certain extent, a function and a qualification of the other categories of mobilisation theory. In other words it could be said to be a function of the opportunity to increase the amount of resources, the control over them, or both as well as a qualification of the ability to make them available for collective action. Likewise, the category organisation is of primary importance when analysing a process of mobilisation – particularly, in the field of labour studies – as has been shown by the extensive debates about bureaucratisation (Kelly 1988), for organisational features often shape the workers' ability to exercise collective power. An organisation may be needed to frame the workers' willingness-to-act. Indeed, the ability to exercise collective power is intimately linked to whether collective social processes, which allow the interplay of workers' multi-faceted interests, find a place within a workplace or an organisation.

Etzioni (1968) and Tilly (1978) argued that any process of mobilisation entails, by definition, demobilisation; that is, a decline in either assets or control over resources – and social processes, interactions, practices – by other actors. Kelly shows that these conflicting movements tend to be cyclical, particularly within the field of industrial relations, and that these cycles 'are a normal and familiar feature of capitalist economies [in which] employers and the State typically embark on a wide ranging series of counter-mobilisations against organised labour to restore both their profitability and their control of the labour process' (Kelly 1998: 128). In these phases, 'both the organization and mobilization of workers are eroded and the ideologies of the labour movement are subjected to assault from a variety of sources' (Kelly 1998: 130). Accordingly, a decline of workers' collectivism is to be expected – or at least of its most visible expressions – since counter-mobilisation strategies often target its crucial dimensions: the opportunity-to-act, workers' organisations and the social processes of interest definition. As a result, in such a cycle, workers' mobilisation can only be defensive: initially to oppose the attack; later on, if the assault could not be stopped, to break the state of demobilisation brought about by it. The important point is that in both cases, mobilisation depends upon the aforementioned categories: opportunity-to-act, organisation and interest definition.

As discussed in Chapter Four, privatisation is arguably the most salient aspect of one of those cycles of counter-mobilisation. Hence, this study privileges as its starting point the impact of ESI privatisation upon the opportunity-to-act, workers' organisation and the process of interest definition in order to subsequently evaluate the prospects of mobilisation.

Opportunity-to-act 'concerns the [power] relationship between a group and the world around it' (Tilly 1978: 7), whereas organisation and interest definition refer to the internal structure of a group. Organisation relates to aspects that affect the capacity for collective action; interest definition to subjective and structural causal powers that, through the mediation of certain social processes, compel workers to act (Isaac 1987; Lukes 2005; Sayer 1992).

Accordingly, if one accepts that privatisation has been, not only a part of, but also in itself, a process of counter-mobilisation, then it should have entailed a decline of workers' collectivism, and it should have severely decreased the opportunity for trade unions to engage in collective action. Yet variability in trade unions' strategies within this context, particularly when these strategies included collective action, should be explained by differences not only in the opportunity structure, but also in organisation and the dynamics by which workers process their interests and define their strategies.

So far, the discussion has referred to but not defined the notion of power. It is now time to sketch the meaning of power as applied here, before examining in detail the notions of opportunity-to-act, organisation and interest definition in order to demarcate the empirical enquiry of this research.

A Definition of the Concept of Power

Most categories of mobilisation theory concern power relations (Kelly 1998; Tilly 1978); yet disagreements over the notion of power abound. Lukes, in the re-edition of his classic study, expresses it bluntly: 'There are endless debates [...] which show no sign of imminent resolution, and there is not even agreement about whether all this disagreement matters' (Lukes 2005: 61). So this section avoids engaging in such endless debates; instead, it makes a brief positive case for the use of Jeffrey Isaac's (1987) realist perspective of power and Steven Lukes' (2005) three-dimensional view.

On the one hand, the realist perspective defines social power as 'those capacities to act possessed by social agents in virtue of enduring relations in which they participate' (Isaac 1987: 80). While it locates agency at the heart of power, in its exercise, it crucially 'places power at the center of agency, as a property of human agents that makes their activity possible' (Isaac 1987: 76). Thus, power is defined as a permanent capacity to act which is rooted in the social structure, in other words a necessary property that is independent of its exercise on particular occasions and its contingent effects. It refers to the things an agent might do. This approach is relevant, for it is this understanding of power that underlies the notion of opportunity structure.

On the other hand, the three-dimensional view stresses that power is a capacity and not only in its exercise. This view focuses on power as domination, particularly in its third dimension, incorporating 'into the analysis of power relations the question of the control over the agenda of politics and the ways in which potential issues are kept out of the political process' (Lukes 2005: 25). It examines how willing compliance to domination is secured (Lukes 2005: 7, 10, 110), and engages in a straight defence of power as 'concealment of people's "real interests"' (Lukes 2005: 12). The analysis of power relations when addressing changes in the opportunity structure due to privatisation will be based on this three-dimensional understanding of power.

Opportunity-to-Act

As stated earlier, opportunity-to-act 'concerns the relationship between a group and the world around it' (Tilly 1978: 7). In the field of industrial relations, it is a category that must go beyond the sphere of interactions between workers and management, towards the analysis of power relations in a wider field of enquiry, which comprises the study of governmental policies and capitalists' strategies. This category is therefore a critical dimension when analysing a process of state and capital counter-mobilisation. While at its base lies a realist understanding of power as capacity, opportunity-to-act includes an examination of the contextual conditions that constrain its exercise. It refers to the relations of power as they are historically determined by the general balance of forces between contenders at industrial and political level and the policies and actions performed by employers and the state (Kelly 1998); this includes the consideration of union's sources and resources of power as discussed below.

The empirical analysis of power relations within the field of labour studies has already been based on analytical models developed across the three dimensions discussed by Lukes (2005): a) the capacity of a party in conflict with another to persuade or force the other to adopt a course of action other than the one it originally intended; b) the capacity of a party to control the agenda of interactions such as meetings, and determine which issues are kept on or off the agenda in the face of opposition; and c) the capacity of a party to secure assent to its objectives by another group because of the successful diffusion of a hegemonic ideology.

Those labour scholars who applied this model have stressed its usefulness in evaluating power relations given its relative simplicity, relational character and its focus on behavioural outcomes (Batstone et al. 1978; Frege and Kelly 2004; Kelly and Heery 1994; but see Edwards 2006 for a recent critique).

Literature about union power oscillates between the overlapping notions of sources and resources of power (Batstone 1988; Kelly and Heery 1994; Martin 1992, 1999; Varman and Bhatnagar 1999). It is possible to argue that the notion of sources of power concerns the structural determinant of power (Varman and Bhatnagar 1999); whereas the resources of power are the assets, money, legislation, rules, support, allies or organisational resources an actor may mobilise to achieve an aim (Kelly and Heery 1994). However, this distinction is far from obvious in most cases.

Nevertheless, Batstone underlines three sources of power as analytically important to the field of labour studies: the ability to disrupt production, the state of the labour market and the scope of political influence (Batstone 1988: 223).

While privatisation affected all these sources, its principal and immediate impact was on the scope of trade union's political influence and, therefore, on the political resources of power mobilised by trade unions: statutory requirements regarding the industrial relations in the public industries, legal frameworks, national energy policies, relationships with political parties, frequency of contacts with governments and potential allies, amongst others.

Kelly and Heery are inclined to use the notion of power resources (Kelly and Heery 1994). In their classification, some resources are organisational and these are discussed below. Others resources correspond to unions' external relationships with management, government, other unions and the general public; all aspects which were influenced, or distorted by privatisation. Concerning management, Kelly and Heery highlight the importance for union activities of first, agreement and support from management, and second, procedural and other collective agreements reached between the union and the employer. As for the government, the key aspects are its intervention in disputes and conflicts by means of legislation. Lastly, Kelly and Heery draw attention to two sources of support: other trade unions and their members, and public opinion. Beyond the usefulness of this demarcation, the point to note is the relationship between the impact of privatisation upon trade unions' (re)sources of power, and hence, on their opportunity-to-act collectively.

To conclude, if Tilly (1978) is right, competition decreases workers' chances of gaining and controlling power resources collectively: it reduces the opportunity for collective action to arise. Furthermore, he maintains that competition is detrimental for workers' mobilisation, but would enhance the opportunities for capital's counter-mobilisation by increasing its chances of accumulating and controlling assets. Lebowitz (2003) gives fresh support to this view by identifying competition with capital's logic of accumulation. He argues that when workers' actions are framed under the logic of competition, they express the dynamic force of the political economy of capital, and therefore, are self-defeating for the working class as a whole. In this argument, trade unions are instances of a combination that reduces competition between individual workers and which therefore is an expression of the political economy of the working class. However competition among trade unions may, for instance, recreate capital logic at new levels. As this study illuminates, these insights are important, since privatisation indirectly fostered competition within trade union ranks; on the one hand, privatisation spurred inter-union competition while on the other hand, it promoted changes which favoured the growth of sectionalism.

The Organisation of Workers

While opportunity-to-act alludes to external relationships, the category organisation 'refers to the structure of a group, and in particular those aspects which affect its capacity for collective action' (Kelly 1998: 25). It is closely related to the ability of workers to combine and mobilise resources for the imposition of defensive or offensive sanctions upon the employer. Batstone (1988) identified three main factors which affect the organisational strength and the extent of union influence: membership density, scope of representation (or inclusiveness, see Kelly 1998), and organisational sophistication.

Labour scholars have often seen a certain degree of membership density as a basic aspect, which is a necessary but insufficient condition for mobilisation and collective action. It is taken as a sign of workers' identification with a union and, perhaps, even of potential support. It is assumed to heighten the representative role of unions' officials in negotiations and to make threats of industrial action and sanctions more convincing. Moreover, an increase, or decrease, in membership has direct consequences upon the financial resources of a trade union.

The scope of representation has similar implications although it adds a subtle distinction. While a trade union might cover close to the entire constituency of eligible people, it might still mean a small proportion of the whole workforce of a given industry or workplace. The corollary is that even a high density might not be enough when a key group remains unorganised or organised by others. Therefore, membership density, scope of representation and the dynamics of the labour process combine in diverse ways, increasing or decreasing trade unions' ability to act.

Finally, organisational sophistication is expected to be adequate for a union's scope and inclusiveness. According to Batstone (1988), a proper organisational structure has to allow strategy, co-ordination, representation of sectional interests and organisational resources to plan action. Key to evaluating organisational sophistication would be the balance between centralisation and decentralisation of power within trade unions (Kelly 1998), which is often related to spatial dimensions as well as interactions between full-time officers, shop stewards, lay representatives and activists.

As secondary organisers, trade unions' structures are often constrained by industrial structures, ownership and bargaining arrangements (Muller-Jentsch 1985; Offe and Wiesenthal 1985). For those reasons, privatisation challenged and imposed trenchant pressures upon these three factors, directly or indirectly. Therefore the study needs to explore how trade unions responded to the privatising context in which, for instance, job losses cut not only into membership but also through redundancy packages into the skills, knowledge and expertise of their officialdom; outsourcing and personal contracts reduced the scope of representation; and fragmentation in ownership and bargaining structures seems to have made strategy, co-ordination, representation of sectional interests and distribution of organisational resources more difficult.

Interest Definition

As with the concept of power, controversies around the notion of interest abound, and hence any theoretical and methodological choice will often be contentious. It is what Walter Gallie has called an 'essentially contested concept [which] inevitably involve[s] endless disputes about their proper uses on the part of their users' (quoted by Lukes 2005: 30). Additionally, as Lukes recognises: 'to engage in such disputes is itself to engage in politics' (Lukes 2005: 30). Then a better approach seems to be that the analysis accepts diversity and political connotations, explains the rationality of the choice and, again, avoids engaging in meaningless debates.

In this regard, Kelly criticises the prevalent casual empiricism within labour studies, expressed in the tendency to identify workers' interests with the contents of collective bargaining. He concludes that, in this field, a 'rigorous analytical treatment of interests is quite simply non-existent' (Kelly 1998: 6). In fact, discussions have always oscillated between either taking wants or preferences as the only realm worthy of examination (Armstrong et al. 1981) or deducing objective interests from agents' structural position disregarding their actual pursuit or recognition by the agent (Edwards 1986). Indeed, Edwards suggested years ago that any reference to interests should simply be abandoned, and, instead, reliance should be placed on the analysis of how objective conditions encourage the growth of some preferences and not others among workers (Edwards 1986: 28).

Yet Kelly has insisted on the utility of the category interest, insofar as it is anchored in a theory of exploitation, and in the recognition of multiple levels of analysis along with the complexity of workers' interests under capitalism (Kelly 1998: 8). These latter qualifications are relevant because if they are ignored, it is impossible to sort out the paradoxes alluded to by scholars: if the conceptualisation of the category interest is reduced to what workers can obtain by bargaining or at least, what they actually demand (a highly empiricist way of defining interest), what happens when there are changes in what they demand or obtain (Kelly 1998)? If the focus rests, exclusively, on what workers do rather than what they say, how can the gap between both levels be conceptualised (Edwards 1986)? If it is solely stressed that demands are contingent products of social construction, how can losing sight of the structural antagonism as a crucial determinant of class identities under capitalism be avoided (Isaac 1987)? However, as stated above, in Rethinking Industrial Relations, Kelly overemphasises the link between individual subjectivity – the individual perception of injustice – and interest definition. In this book the emphasis will be on the social processes by which collective demands and ways of action are defined. This standpoint lies in the conviction that Kelly's discussion seems too narrow a platform for the exploration of responses to privatisation policies of well-established organisations of workers.

This is, after all, a rather conventional approach in the field. Labour scholars (Edwards 1986, Hyman 1975, 1989; Kelly 1998, Offe and Wiesenthal 1985, Pizzorno 1978) have traditionally argued that the empirical manifestation of workers' interests is constantly trapped between workers' collectives and particular needs due to the fragmentation, subordination and exploitation of the working-class; and hence that workers' interests can only be advanced insofar as they are to a certain extent collectively reworked. In short, labour scholars have insisted that workers' expressed interests are socially constructed through a set of social practices. This is true, however, with the caveat that under the heading of interest, most labour scholars refer to concrete demands (generally the contents of bargaining) and methods of pursuing them (actions, policies and strategies). The essential point is that it is possible to argue that demands (not only those of collective bargaining) and ways of action are brought about by the social processes with which multiple and sometimes conflicting interests are collectively dealt.

In the main, two social processes have been identified as vital for spurring the collective reworking of conflicting and diverse interests in order to define demands and courses of action: a dialogical democracy (Offe and Wiesenthal 1985; Smith 2001; Hyman 1989) and an appropriate leadership (Batstone et al. 1978; Darlington 2001; Fantasia 1988).

A dialogical democracy implies effective, democratic and participatory channels of communication, debate and collective decision beyond the usual electoral procedures (Gall 2003; Hyman 1989; Kelly 1998; Offe and Wiesenthal 1985). It can be assessed empirically, on the one hand, by the type and frequency of meetings, ballots, assemblies, that form the mechanisms of debate and decision-making of the organisation; and on the other hand, by a qualitative analysis of the flow and type of information used as inputs in this process, and how the latter typically unfolds.

Concerning leadership, several scholars have shown how leaders and activists play a crucial role in the process of collectivisation (Barker et al. 2001; Fosh 1993; Darlington 2001; Gall 2003; Kelly 1998; Smith 2001). As Darlington summarises, these leaders help, first, to construct a sense of grievance amongst workers, attributing blame onto employers and/or the state rather than to uncontrollable economic forces or events. Second, they promote a sense of group or social identity, which encourages workers to become aware to their common interests in opposition of those of employers. Third, they urge workers to engage in collective action, a process of persuasion that is assumed to be essential because of the costs of such action and of the experience of many people with its different forms and consequences. Fourth, they legitimate such collective action in the face of employers' counter-mobilising arguments that it is illegitimate (Darlington 2001: 2).

In sum, the aim of explaining how trade unions came to define concrete demands and strategies in the face of privatisation requires the study of the social processes by which workers mediate the opportunity structure. This includes the analysis of the impact of privatisation upon the decision-making process and workers' participation, the analysis of the mediating role of the different leadership styles, and whether trade unions' choices softened or reinforced the negatives of the opportunity structure.

Conclusion

Mobilisation theory is a powerful instrument for studying how privatisation conditioned workers' collectivism and how trade unions responded to this challenge. The main advantage of mobilisation theory lies in its scope and flexibility. On the one hand, it offers a cyclical and long-term account of conflicting movements of capital and labour. On the other, it identifies certain dimensions, which are critical to the presence or absence of collective action, and for that reason, makes theoretically informed explanations of specific empirical outcomes possible. It provides too, a set of categories for this endeavour: opportunity-to-act, mobilisation, collective action, organisation, interest definition.

This chapter discusses the operationalisation of each of these categories. It advances an understanding of opportunity structure in terms of power relations and identifies crucial aspects of the category organisation. It also explains why the framework developed by Kelly to discuss the category interest seems too narrow a platform for exploration of the responses to privatisation policies of well-established organisations of workers involved in a long history of interactions with governments and managers. Thus, the chapter suggests a different, though quite conventional approach, whose focus is on social processes such as leadership and decision-making.

To conclude, the long-run perspective of mobilisation theory permits the postulation that ESI privatisation programmes were counter-mobilisation events, as shall be discussed in detail in Chapter Four. Hence it may be further postulated that these programmes were processes that affected the dimensions of workers' collectivism as a whole: the opportunity for trade unions to engage in collective action, their organisations and the social interactions by which interests, demands and ways of action are defined. However, in this context of tough external conditions, variability in trade unions' strategies should be explained by mobilisation theory as well, in part, by variability in the categories organisation and interest, as defined above.

CHAPTER THREE

Research Strategies and Methodologies

Although personal choices, opportunities and preferences have an important role in shaping the methodological design of any research, the appropriateness of the strategies and methods adopted has to be evaluated taking into account the object of study, the availability of or the possibility of producing data, and its temporal feasibility (Oyen 1990). Thus, it is possible to claim that research strategies and methods are justified and strengthened if:

a) they are consistent with the aims of the research and the object under study;
b) they are related to a specific methodological tradition in the field of study under consideration;
c) they are feasible in terms of data availability;
d) they are scientific and rigorous in terms of data collection and analysis.

The case-study approach and the comparative method are the two strategies chosen to carry out the research. Bearing in mind the aforementioned criteria, it is necessary to consider whether or not they are pertinent to the research project.

The Aims of the Research and the Object under Study

The empirical aim of this research is to study the relationships between privatisation and workers' collectivism. Why privatisation? Because, as shall be discussed in Chapters Four and Five, the sale of public companies has been a key aspect of the cycle of state and capital counter-mobilisation, which began in the 1970s. This implies the study of how and why ESI privatisation conditioned trade unions' ability to engage in collective action, and also the study of trade unions' concrete responses. The theoretical objective is to test the potential of mobilisation theory for carrying out such a task within a comparative perspective. The case-study method appears appropriate, since the research is driven by explanatory questions with a focus on a small number of cases from the UK and Argentina in order to gather detailed qualitative data.

Why these two countries? From the late 1970s until the mid-1990s, these two countries experienced far-reaching processes of employer and state counter-mobilisation against labour, whose main distinguishing feature was the privatisation of public enterprises, arguably the greatest privatising processes ever witnessed in the Americas and Western Europe. This feature may give us a positive basis for comparison, for both countries experienced the decline of the most visible forms of worker collectivism during the same period. Yet, there are important differences too, which provide bases for testing mobilisation theory in different contexts. These differences not only include these countries' institutional backgrounds, in terms of politics and industrial relations, but also their world market position and the policy forces driving privatisation. While the UK is a G8 economy and is recognised as a privatising pioneer, Argentina has an altogether different position in the world economic order, and has undertaken privatisation later and in different circumstances. In Argentina this policy was largely the outcome of external pressures from international aid donors and banking agencies, such as World Bank (the WB) and the International Monetary Fund (the IMF) (Dinerstein 2001; Thwaites Rey 1994, 2003).

Why electricity? There have been vast differences between privatised enterprises in terms of activity, size, market exposure, and profitability; still, there is a set of activities that shares some basic features which allows a common approach. This set comprises the so-called natural monopolies which are the major public utilities. Following Ernst (1994), the rationale of grouping together utilities (gas, electricity and water) rests upon four defining features: essentialness, inelasticity of demand properties, natural monopoly provision and externalities. Electricity emerges as a good test case because it is a network industry with similar characteristics in both countries.

The ESI involves five vertically related stages of production: supply of energy inputs, generation, transmission, distribution, and retail supply to final customers. Its structure determines a natural monopoly in the transportation activities of transmission and distribution, but not in generation or retail supply, and the need for an especially close vertical coordination between generation and transmission. In both countries, the industry has been divided into several firms producing divergences between companies. This has led to the end of common collective bargaining arrangements. Besides, in both countries, the ESI has been hit by significant workforce reductions, and hence, has passed through an initial tension between pressures towards cost-cutting (clearly visible in the rate of job loss) and the establishment of a high-quality, customer oriented service based around employee skills and commitment (Colling and Ferner 1991, 1993a, 1993b, 1995; Duarte 2001). Also, in both countries, the public sector legacy in the ESI has contributed to a relatively high level of unionisation; so private company managers have had to face strong trade unions. Lastly, new structures of public regulation for the ESI have been put in place in the UK and Argentina as a consequence of privatisation.

As for the potentiality of the case-study approach, even though there are alternative ways to examine the interaction of variables, it is appropriate to explore how complex variables interact with each other (Ferner 1988). This approach seems particularly suitable when researching open systems of social relationships in which variables are to be considered as reacting against each other. As for the comparative dimension of the methodological design, the objective of looking at the relationships between privatisation and worker collectivism, taking as a general framework the explanatory variables developed by mobilisation theory (Kelly 1998; Tilly 1978), matches most of Poole's (1986) criteria for conducting comparative studies in the field of industrial relations. These are: to concentrate on the effects brought about by social structures and processes, to carry out a multidisciplinary approach, to focus on explanatory variables instead of descriptive ones and to introduce a historical perspective. Indeed, most studies applying mobilisation theory have a narrow focus and almost all are based on a single country. Hence the value of comparison lies in extending the generality of mobilisation theory, that is, in enabling the pinning down of a range of intermediate variables and their diverse impacts.

Methodological Traditions in the Field of Industrial Relations

Both strategies, the case-study approach and the comparative method, have been widely used in studies of labour and industrial relations. On the one hand, there is a long tradition of case-study research in these fields. It ranges from the classical ethnographic case-studies (Batstone et al. 1977, 1978; Beynon 1984; Edwards and Scullion 1982) to the common use of this approach in analysing, for instance, the effects of privatisations upon industrial relations and labour conditions (O'Connell Davidson 1993; Pendleton and Winterton 1993) and in testing and advancing general or theoretical propositions (Crouch 1994; Darlington 2001; Gall 2003; Kelly and Willman 2004; Smith 2001). On the other hand, the importance of the role of comparative studies in the development of the discipline of industrial relations is also well-established. Scholars have often underlined the capabilities of comparative studies for the acquisition of a greater insight into one's own country and the development of explanatory theory (Bamber and Lansbury 1993; Bamber, Lansbury and Wiles 2004; Bean 1994; Ferner 1988;

Hyman 1994a), and have usually attempted to combine both approaches (Clegg 1976; Edwards 1986; Hyman and Ferner 1994).

As Hammersley and Gomm (2000) remind us, there is no standard way of applying the label case-study. It is important to specify some of its critical features. These authors consider that a case-study approach broadly distinguishes itself by focusing on a small number of cases, gathering a large amount of detailed qualitative rather than quantitative data for each, and studying preferably ongoing processes. This research matches all these criteria. Regarding theory, while they formally recognise exceptions, these scholars point out a lack of concern with theoretical and empirical generalisations. However this is not inevitably the case. Yin (1994), for instance, offers a different though similar picture. He also understands a case-study to be an empirical enquiry of contemporary phenomena, but one in which context become explanatorily meaningful and in which the researcher has no control over the events. This author stresses the desirability of theory development as the previous step of any case-study in order to guide data collection and analysis. Also according to Yin (1994), there is another factor that may increase the appeal of this approach, and this is the sort of research question the researcher seeks to answer. When explanatory questions such as how and why are the driving forces of the enquiry, case-study emerges as an advantageous strategy.

Comparison, in turn, is a classical approach of socio-historical research that brings into play both convergence and divergence analyses for identification of patterns of similarities and diversity (Hyman 1994b; Mahoney and Rueschemeyer 2003; Oyen 1990; Ragin 1994; Teune 1990). However, it is argued that within the range of qualitative methods 'the distinctiveness of the comparative approach is clearest in studies that focus on diversity' (Ragin 1994: 105). Some authors suggest that this research strategy has become even more important in the current period of increasing internationalisation (Bamber and Lansbury 1993); in this way, the globalisation of problems and processes is identified as a driving force towards comparative studies (Oyen 1990). Privatisation has become a global process which has posed similar problems, although in highly different institutional contexts, for the ability of workers to act collectively.

Contrary to the vacillation referred to above on the theoretical potential of case-study research, the basic features of the comparative method 'make it a good strategy for advancing theory. These features include its use of flexible frames, its explicit focus on the causes of diversity, and its emphasis on the systematic analysis of similarities and differences in the effort to specify how diversity is patterned' (Ragin 1994: 111). Hence, the placement of a set of case-studies within a comparative perspective may help to balance any misgivings about the contribution of the research design to theory development. Additionally, when pursuing qualitative comparisons, this strategy usually demands a limited number of cases in order to achieve familiarity, for knowledge of cases is conceived of as a goal in itself in mainstream academic practice.

Data Availability, Collection and Analysis: The Methodological Design

With regard to the data availability and collection, and bearing in mind that privatisation is a recent phenomenon, access to data benefits from the use of multiple sources of mainly qualitative evidence which are basically documentation, archival records and interviews. Yet, a more fruitful examination of these aspects, availability, collection and analysis of data, demands the presentation of the details of the methodological design.

The methodological design of this book comprises a systematic analysis and comparison of empirical evidence taken from five case-studies: the impact of privatisation upon and the responses to it of five trade unions in two countries (three unions in the UK and two in Argentina; Table 1). Chapters Seven to Ten organise findings according to certain relevant topics with regard to the trade unions' choices, the counter-mobilising factors embedded in the process of privatisation and the impact of the later on organisational and agency type variables.

Table 1: Detail of Case-Studies

Trade unions (case-studies)	Membership before privatisation(1)	Membership 2006
United Kingdom		
EETPU(2) (AEEU – AMICUS)(3)	36,000	20,000 energy & utility sector
EPEA(2) (EMA – PROSPECT)(3)		
NALGO(2) (UNISON)(3)	34,000	14,000 gas/water/electricity
Argentina		
Luz y Fuerza Capital	22,000	4,000 (only ESI)
Luz y Fuerza Mar del Plata	4,000	3,000 (only ESI)

(1) Approximate figures of membership in ESI only
(2) Union names at the time of privatisation
(3) Change in union names due to mergers and amalgamations

This design permitted comparison at two levels: international and national. To anticipate the argument of later chapters, if the impact of privatisation upon the opportunity-to-act and the organisational features of trade unions are relevant to the understanding of variability in unions' strategies, a methodological design should allow for variation in the opportunity structures and the organisational domain. Furthermore it is argued that, in the context of reduced opportunities for trade unions to engage in collective action, the characteristics of the internal processes of decision-making and leadership style may also contribute towards explaining strategic variability. Then it is necessary to take the chance to narrow differences in both the opportunity-to-act and union organisational structures in order to compare the different courses of action taken by trade unions under similar conditions.

With this in mind, it is assumed that variability in the opportunity structure stems, mostly, from differences in two key intermediate variables:

industrial and market factors, and industrial relations institutions. It is also assumed that broader differences in national politics are embedded in the dynamics of the latter.

Following a well-established practice within cross-national comparative studies, the design holds the industry factor constant (Bean 1994; Ferner 1988). Thus, the research focuses on the privatisation of one industry, which, in turn, allows a better appreciation of changes in market and industrial structures due to privatisation. National variability in the degree of fragmentation, restructuring, and boundary redefinitions of the ESI, and in the forms and intensity of capital competition, concentration and integration, are, in this regard, crucial, given that these industrial and market factors have a determining effect on trade unions' (re)sources of power. A comparative study of the process of privatisation of the industry and of the immediate evolutionary aftermath may highlight variations which influence differently the dynamics of counter-mobilisation and the very possibility of workers' collectivism.

Similarly, divergences in the British and the Argentinian Industrial relations systems are to be analysed insofar as they affect patterns of power resource allocation. The research pays particular attention to factors that have determined dissimilar correlations between the political and industrial dimensions of trade unions' (re)sources of power.

As for data collection, this has been completed by holding interviews and by extracting material from archival documents. The principal subjects of the interviews were trade union officials who had experienced the process of privatisation; in particular, national and regional/branch union officials with negotiating and political responsibilities, but also workplace representatives. In Argentina, given the organisational structures of trade unions, there are no regional officials. The choices made have pursued a balance between different levels of officialdom and between different unions. The rationale for the selection of interviewees for the higher levels has been their first-hand knowledge about the process of privatisation. All national and regional/branch union officials have had, at the moment of privatisation, negotiation and political responsibilities. Concerning workplace representatives, interviews have been directed towards obtaining information on the recent evolution of collective bargaining, grievances and conflicts, workplace structures of representation, rank and file participation, and the mechanism of decision-making at shop-floor level.

As the aim of the research is to determine how privatisation impacted upon trade union ability to engage in collective action, the design does not attempt to limit the enquiry to geographical boundaries or particular firms. On the contrary, the survey of diversity has provided positive insight.

A total of fifty-three interviews have been carried out: forty-seven in-depth unstructured and semi-structured interviews, conditioned by the research objectives and the status of the informants, and six focus groups with workplace representatives. The distribution of the fifty-three interviews appears in Table 2:

Table 2: Distribution of Interviews among Trade Unions

2.1. United Kingdom

Respondents	Amicus	Prospect	Unison	Total
National officer	4	5	4	13
Regional/Branch officer	2	3	3	8
Lay reps	3		2	5
Focus groups	2		1	3
Total	11	8	10	29

2.2. Argentina

Respondents	Luz y Fuerza CF	Luz y Fuerza MDP	Total
Union officers	5	6	11
Lay reps	5	5	10
Focus groups	1	2	3
Total	11	13	24

The selection of respondents is intended to balance the representation of different market segments. In the case of the UK, in order to control regional peculiarities, the research design has limited interviews of regional officers and lay representatives to the London region and the Midlands. In both countries, access to lower levels of the officialdom has often been negotiated through contacts with officers at unions' headquarters.

The information collected by this technique has been complemented by data taken from primary and secondary sources. Indeed, the extensive use of historical and written documentation is a distinctive feature of the book. Recent mainstream research on human resources management and industrial relations tends to focus on interviews and questionnaires, and to downplay the historical material built around issues such as privatisation. In this book, by contrast and given the historical perspective adopted by the research, both methods of data collection have been of equal importance and have supported each other. Typically, the interviews have been enriched by the information collected from the primary sources. In mainstream research, data triangulation is usually carried out by holding interviews with people who are expected to hold different views about a given topic. An alternative to this procedure is to use qualitatively different sources. With this in mind, written material has proved to be useful not only in checking data reliability but also in filling information gaps.

The survey, the critique and the analysis of primary sources are very time-consuming tasks. As a result, the organisation of the fieldwork to survey primary sources related to different privatisation processes and five trade unions has been a crucial challenge. The vast array of documentation produced during the period by public authorities, trade unions and other relevant actors compels the researcher to be selective. It follows that the first methodological decision regarding data collection was to prioritise trade unions' sources, for the two axes of the research were how privatisation conditioned trade unions' ability to act and how ESI unions effectively responded to that challenge. However, given the role of the Argentinian government in the process of dismantling the institutions of industrial relations in the public sector before privatisation (see Chapter Six), the survey, in this case, was extended to relevant documentation produced by the Ministry of Economy (Campaño and Caruso 1991; Campaño et al.

1991; Daireaux et al. 1990; Ministerio de Economía 1990a, 1990b, 1990c) and the Ministry of Labour (Ministerio de Trabajo 1990).

Nevertheless, the main sources of the research were trade unions' journals, annual reports (the so-called *Memoria y Balance* in Argentina) and conference reports (see Table 3).

Table 3: Main Primary Sources

Trade union	Main sources
EETPU	– Contact (journal) – Shop Stewards Quarterly Review – Biennial Delegate Conferences – National Conferences for the ESI
AEEU	– AEEU (journal) – Annual Reports
Amicus	– Annual Reports
EPEA	– EPE Electrical Power Engineer (journal) – Privatization news (newsletter) – Agenda of the Delegate Conferences – Annual Reports
EMA	– EMA Newsletter – Annual Reports
Prospect	– Annual Reports
NALGO	– NALGO News (journal) – Focus (journal) – Annual Reports – Electricity Branch Circulars 1989–90
Unison	– Energy Service Group Conferences – Annual Reports
TUC	– Annual Reports
Luz y Fuerza CF	– Dinamis (journal) – Memoria y Balance (Annual Reports)
Luz y Fuerza MDP	– 8 de Octubre (journal) – Memoria y Balance (Annual Reports)

In each case-study, the year taken as the starting-point of the survey was the year in which the privatisation plan was announced. Then, chronologically appropriate reading contributed to the identification of changes in trade unions' policies, critical events and the different phases of trade unions' strategies. In addition, trade unions' journals and reports referred to – and even published – relevant written materials which addressed privatisation issues, thereby orienting the fieldwork towards particular documents. For that reason, during this survey of primary sources a great deal of effort went into locating specific publications and documentation (for instance: EPEA 1986; FATLyF 1992; FUSE 1987; LyF MDP 1994b, 1997, 1998; NALGO 1988c, 1989a, 1989b; NOP 1989, 1991; TUC 1989b, 1989c; Unison 1996a, 1996b; among others).

The accuracy and objectiveness of the primary sources collected during the fieldwork were constant concerns. Documents are often difficult to interpret and demand a high level of scrutiny. Still, primary sources usually offer new input for historical questions and this research was not an exception. While most interviewees were able to give detailed information about certain events and topics, the same respondents tended to ignore other dimensions of the process of trade unions' strategies in the face of privatisation. In this regard, written sources contributed to redressing the balance in the analysis. For instance, aspects like the different forms taken by the politics of money (see Chapter Eight), the contradictory evolution of the structures of collective bargaining (see Chapter Nine) or the scope and timing of unions' organisational changes (see Chapter Ten) could have hardly been studied in absence of written evidence. Interviews proved essential when obtaining information about other facets like inter-unions tensions within single table bargaining (see Chapter Nine) or the rise of sectionalism (see Chapter Ten). Yet, as a stylistic strategy, the narrative of the findings was supported, whenever possible, with quotations taken from interviews. The purpose was to facilitate a more attractive exposition.

A critical problem was where and how to obtain these sources. Two paths were followed. On the one hand, the search began with visits to well-known repositories of primary sources about the history of trade unions. On the other, interviewees were questioned about the type of documentation kept by the organisation they belonged to. In fact, the method of carrying out the fieldwork with regard to the survey of written documents was different in both countries.

In the UK, interviewees manifested that their trade unions did not have policies on documentation storage. Partly due to the process of mergers undergone by ESI unions, past records had been lost or given to public institutions. In this regard, the most important collections of trade unions' documents were the Trades Union Congress Library Collections at the London Metropolitan University and the Modern Records Centre at the University of Warwick. Therefore, it was possible to find in these places, particularly in the former, complete collections of trade unions' journals, annual reports and other relevant documentation.

Instead, in Argentina, both trade unions have collections of the journals published by their organisation and other relevant sources in their headquarters. Thus, access to their archives was negotiated simultaneously with access to carry out interviews with union officials. The libraries of the Ministry of Economy (*Ministerio de Economía*) and Ministry of Labour (*Ministerio de Trabajo*) were two important repositories of official documentation. The former was particularly useful, as primary sources – previously unused by scholars but critical to the study of how the government designed its policies against public unions to ease the process of privatisation – were found during the survey (see Chapters Six, Seven and Nine).

In sum, the written materials collected were crucial to achieving accuracy in the historical reconstruction of the process of privatisation and how trade unions responded to it. They were of primary importance in dating particular changes in trade unions' policies and in identifying certain patterns in, for instance, the evolution of collective bargaining or the counter-mobilisation strategies. In short, these materials were essential for the research, as they provided accuracy where interviews offered diffuse information.

Quantitative data has been collected and processed when appropriate. Some trade union resources affected by privatisation are measurable. For instance, if the main focus of interest is the evolution of organisational resources, variables such as membership, union density, or number of lay representatives and union officers, are quantifiable and apt for analysis along time series. This is also the case with monetary resources. The same can be said about some kinds of collective actions. Strikes, overtime bans, go-slows, works-to-rule, petitions and collective appeals are suitable for quantification. In fact, these types of quantitative data there have formed the basis of the debate on trade unions' power decline. However, if several trade unions' resources may be translated into quantifiable variables, their use and mobilisation may not. Thus, quantitative data was often subordinated to qualitative data. Furthermore, as it was impossible to obtain similar and complete series of data, its use has been very limited and of secondary importance.

Scope and Limits

Any methodological design which takes as a research strategy the case-study approach may face concerns about the generalisation of findings. An author such as Stake (2000) has restricted the scope of case-study research to illuminate what is particularly in-depth and unique. Simons (1996) downplays this aspect by arguing that the tension between what is particular, and what is patterned and regular, underlies any social research. She has pointed out the paradox of case-study research: its strength in generating both unique and universal understanding by use of an in-depth focus and a holistic perspective. Yet, Simons maintains that the advantage of a case-study approach remains its ability to deal with uniqueness and particularities.

By contrast, other scholars stress that a case-study approach does not preclude generalisation; but beyond this elemental agreement, very different positions are taken. Generally, those who support a classic view of social science, in which research aims to generalise findings to diverse populations and times, often insist on the need of a proper design, from sampling in multi-case studies to appropriate thick descriptions for further comparisons. Yin (1994) opposes this view and suggests another understanding based on what he calls analytical generalisation. From this point of view, research designs should not pursue a sampling logic. A case-study approach must avoid statistical generalisation. It is in relation to previous theoretical developments and rival theories that findings should be considered. Finally, some scholars have questioned the traditional meaning of generalisation (Lincoln and Guba 2000; Schofield 2000). They put their efforts into reworking the notion of generalisation in order to enhance the chance of speaking beyond the immediate concerns of their case-studies.

Thus generalisation remains problematic for any methodology built upon a case-study strategy. For this reason, the research neither pursues straightforward generalisations from empirical findings nor pretends to establish universal patterns. Indeed, following the insights of the ontological understanding advanced by critical realism (Ackroyd and Fleetwood 2000; Edwards 2005; Fleetwood 1999; Roberts 1999, 2001; Sayer 1992), our approach starts by identifying serious reservations about the possibility of identifying regular patterns of events in contexts of openness. Yet, what the design does attempt is to test the potential of mobilisation theory to offer comparative and theoretically informed explanations of trade unions' ability to engage in collective action in social contexts characterised by a counter-mobilisation wave. In short, while the enquiry recognises the complexities of generalisation, it also recognises that to ask how far the findings are likely to apply in other circumstances is a task that needs to be undertaken.

Another potential source of concern relates to the rationale for a comparative analysis. Mobilisation theory, as rethought by Kelly for application to organised labour, is still in its infancy. As already mentioned, Gall (2000) stressed the lack of intellectual engagement with the theoretical framework developed by Kelly, especially at meso- and macroanalytical levels. This remains the case, for the theory has often been applied to short-term events (Atzeni 2005; Brown Johnson and Jarley 2004; Darlington 2001; Gall 2000; Kelly and Badigannavar 2003; Kelly and Willman 2004; Moore 2004). This gap, together with the surprising scarcity of studies about how privatisation conditioned trade union strategic capacity, reinforces the case for the investigation of mobilisation theory in a comparative context marked by state and capital counter-mobilisation. Indeed, once meso- and macro-levels are incorporated into the analysis, the diversity in industrial and institutional intermediate variables, which partly determine the opportunity-to-act and the allocation of power resources, proves to be a promising avenue for the evaluation of the potential of mobilisation theory. Likewise, variability in leadership styles, decision-making dynamics and types of organisation serves the purpose of examining how far workers' collectivism depends, in the midst of adverse external conditions on organisational processes and capabilities. In this sense, the virtue of comparing dissimilar cases, as regards the institutional context, is that any patterns of similarity that might be found are likely to exhibit processes with a high degree of generality in relation to workers' (de)mobilisation around privatisation.

Conclusion

For assessment of the adequacy of the methodology, this chapter discusses the research strategies and methodological design, and some areas of potential concern, which stem from these choices. The fundamental conclusion to be drawn from this discussion is that the methodological decisions that underpin this study have the aim to enhance the explanatory potential of mobilisation theory. In this sense, the scope and multiple levels of analysis is thought to be essential for the researcher to pay attention, on the one hand, to the strategic interaction of the contenders, and on the other, to the ability of the actors to modify the opportunity structure through their strategic choices, and in this way, to open or foreclose specific paths of action.

CHAPTER FOUR

Privatisation as Counter-Mobilisation

Chapter Two sets out the analytical framework of the research from a mobilisation perspective. This theoretical starting-point has some advantages. It positions any shift in labour-capital relations in historical perspective, and in this sense, overcomes the ahistorical character of most studies of union organisations (Kelly and Heery 1994: 23). It provides a model for the long-term as well as a related theory of collective action for the short-term (mobilisation theory and its pool of categories). Furthermore, by combining empirical evidence of long waves and the categories of mobilisation, it allows the analysis to transcend the scope of workplace and industrial relations institutions in order to explain workers' collectivism. In this regard, Kelly's proposal matches Hyman's statement: 'the phenomena of industrial relations cannot be understood simply in their own terms. Industrial relations cannot adequately be comprehended as a relatively autonomous "sub-system" of society upon which political and economic forces impinge only exogenously' (Hyman 1994c: 171).

The analytical framework postulates that the process of privatisation is a counter-mobilising force; Chapters Four and Five intend to substantiate this claim. Therefore, after illustrating the various meanings of the term privatisation, this chapter provides diverse arguments asserting its counter-mobilising content. In so doing, it shows, first, that the labour movement had invested in the preceding regime, as trade unions considered public ownership vital to organised labour gains. Then, drawing on Marxist political economy, it develops a theoretical argument about privatisation increasing the scope of operation of private capital. After this, from a mobilisation perspective, it addresses, theoretically and empirically, the relationship between privatisation and appropriation of resources. Finally, there is a review of the literature on the relationships between privatisation and labour, and particularly on the detrimental impact of the former on the latter.

On the Meaning and Origins of Privatisation

Once one becomes immersed in the literature, privatisation reveals itself as an unexpectedly fuzzy term; different sorts of policies have been deployed using this label. As Bishop and Kay pointed out in their approach to the UK's experience, 'since 1979, "privatisation" has become an umbrella term covering a number of British government microeconomic policies. In addition to the sale of publicly owned enterprise with which it is normally associated, privatization has included policies of deregulation, liberalization and franchising' (Bishop and Kay 1989: 643). Such usage focuses on the involvement of market forces in ensuring greater competition, economic efficiency, and the reduction of the role of the state in the regulation, production and provision of goods and services; that is, privatisation is associated with deregulation and liberalisation. This is the normal usage of the term privatisation in most works (Austin et al. 1986; Bishop and Kay 1989; Clarke 1993; Heald and Steel 1981; Marsh 1991; Wiltshire 1987). While this broad definition might be appropriate as a way of depicting a general approach towards economic policy rather than a policy in itself, it might be misleading for certain analytical purposes. Therefore another trend has stressed that the liquidation of publicly owned assets, the sale of minority public-shares of private enterprises, governmental measures towards liberalisation, deregulation of economic activities and the introduction of patterns of behaviour taken from the private sector into public sector enterprises, cannot be considered as instances of privatisation (Bienen and Waterbury 1989; Gupta 2000; Walle 1989).

For authors like Ramanadham, privatisation 'represents marketization of enterprise operations and can be sought through three options – ownership changes, organizational changes and operational changes' (Ramanadham

1993: 2). Such views reduce the concept of privatisation to its practical and political meaning by referring to a wide range of actions taken following the end of the 1970s by governments to reduce the role of the state and to encourage market forces. But, it still comprises many different policies.

In opposition to broad definitions, then, Ramamurti confines the term to the 'divestiture of these enterprises which state owns and operates itself' (Ramamurti 1992a: 1); that is, the 'sale of all or part of a government's equity in state-owned enterprises (SOEs) to the private sector, or the placing of SOEs under private management through leasing and management contracts' (Ramamurti 1992b: 228). Walle expresses the same idea in a slightly different manner: 'the transfer of ownership or control of an enterprise from the government to the private sector, with particular reference to assets sales' (Walle 1989: 601; also Molz 1990). These narrow definitions of privatisation seem to be more useful analytically than broader ones. As Walle (1989) has stressed, they at least allow the distinction between privatisation and liberalisation with regard to one essential feature of any economy: the structure of relative prices. Following Gupta (2000: 31), it would be even more important to make this difference, since the relationship between liberalisation, deregulation and privatisation is diverse and there is no one-way model.

Yet, some scholars claim that in this type of definition an important ambiguity persists: that the percentage of the equity sold by the state to private capital remains unspecified. Beesley and Littlechild (1988) and Weyman-Jones (1993), for instance, have stressed that the sale should represent at least 50 per cent of the shares of the former public enterprises (including mixed enterprises); though they restrict privatisation to just this aspect. On the contrary, concession of control is often considered as an aspect of privatisation in the narrow definitions so far.

In short, following broad formulations, some studies include in privatisation the relaxation or abolition of monopoly powers that prevent private sector firms entering markets previously exclusively supplied by the public sector (Wiltshire 1987). Others, instead, prefer to apply the term market liberalisation (Weyman-Jones 1993), when referring to permission and facilitation of competitive new entrants to the marketplace. Similarly, deregulation – and not privatisation – seems for others to be a better term to use when speaking of the removal or relaxation of restrictions on procedures, pricing, and output and investment decisions of both public and private industries. Empirically, however, these aspects usually appear together, because it is distinctive of privatisation programmes to seek market liberalisation and deregulation in an effort to increase competition, reduce the state activity and free market forces at every level. Yet, this is not always the case as different combinations of privatisation, liberalisation and deregulation are always possible. This is an important distinction in the case of public utilities when new regulatory institutions and policies are put in place. Denationalisation (Gupta 2000) refers to the sale of public assets and shares, regardless of the proportion sold. The term commercialism (Ferner 1988) refers to the action of extending the legal norms that regulate private sector activities to state enterprises working on commercial bases, as an attempt to find a set of objectives and criteria which act as market proxies in the public sector. Now empirically, all these policies have been combined in different ways, and even a strict definition of terms will fail to overcome conceptual and empirical overlaps. It is still worth differentiating between different (or potentially different) processes and policies that have been diversely applied. Moreover, when studying the privatisation of public utilities in a comparative perspective, each of these separate processes might carry consequences for labour that are of a different order and magnitude.

In summary, despite definitional subtleties, all these reforms are intended to reduce the extent of regulated or co-ordinated market activity and, concomitantly, to expand opportunities for accumulation.

As for the origins of privatisation, there are many and diverse explanations, which usually consist of some combination of multiple motives. The few who offer a common and general reason for worldwide privatisation emphasise either the fiscal deficit (Christiansen 1989: 597; Gupta 2000: 19) or the governmental dissatisfaction with the performance of public enterprises (Walle 1989). Beyond general statements, these explanations often add secondary reasons, differentiating between developed countries (DC) and less developed countries (LDC) as shall be shown in the next chapter, or find out particular objectives following specific circumstances.

For the purposes of this book, the main deficit of mainstream approaches to privatisation is not their lack of a common use of the term, or a shared understanding of the origins of privatisation, but their tendency to conceive of privatisation as an administrative measure, that is, as a technical response to an economic problem (Dinerstein 2001). As a result, previous studies seek to evaluate whether privatisation – whatever its operational definition – achieved the declared aims. Whereas it might be important to find gaps between promises and realisations, this approach misses the political and counter-mobilising dimensions of privatisation, for 'as the 1997 [the WB] Report highlights, the current wave of privatization and deregulation is a political program, the result of a strategy designed to serve one set of interests as opposed to another' (Cook and Murphy 2002: 1). Accordingly, many scholars stress that privatisations were not technically inevitable but were part of a strategy in which the role of institutions like the IMF and the WB in designing, promoting and implementing the programmes was critical for their proliferation worldwide (Cook and Murphy 2002; Dinerstein 2001; Thwaites Rey 1994, 2003). Yet, if the picture in the UK is free from the direct interference of the international actors mentioned above, it is no less true that privatisation has also served certain interests as opposed to others in that country. Indeed, diverse accounts make it clear that the agenda for privatisation, especially in the UK but also elsewhere, was partly motivated, overtly and ideologically, towards freeing up markets by rolling back the constraining impact of organised labour on the opportunities for private capital, as shall be discussed in more detail in Chapter Five.

An Historical Argument: Privatisation and the Politics of the Labour Movements (I)

It seems pertinent to begin the substantiation of the counter-mobilising content of privatisation by stressing that privatisation halted a long term and widespread aspiration of the labour movement for the consolidation of state enterprises, at least in key economic sectors (Cole 1953; Ferner 1988; Kelf-Cohen 1958; Weiner 1960). Particularly in the UK, where 'TUC resolutions in favour of nationalisation go back to the 1890s, when the influence of the "New Unionists" first made itself felt' (Cole 1953: 225), state ownership was seen as a victory for the labour movement, which supported the election of a post-war Labour government with a radical mandate to establish social provision of key services. Then, while nationalisation must be understood as a critical aspect of British post-war reconstruction, despite explicit reservations made by Conservative Governments during 1945–79, trade unions' commitment to public ownership and nationalisation has a longer and more independent history (Weiner 1960). In this history, times of war played a crucial role. The First World War completed the conversion of nationalisation in trade unions' policy from a liberal to a socialist conception. State ownership became a political and industrial objective through which the labour movement hoped to defeat exploitation by means of winning a place in industrial management to plan and eliminate wasteful competition, extend democratisation and improve workers' living standards. In this context, there was the first call for the nationalisation of power supply in the Jubilee Congress, which took place two months before the armistice (Weiner 1960). The same year, 1918, witnessed Labour Party's adoption of socialist objectives and the incorporation of the nationalisation policies of trade unions as part of its programme for war reconstruction (Weiner 1960). Moreover, in 1924, nationalisation, in the TUC's resolutions for the previous thirty years, was added for the first time to TUC Standing Orders, the long-range economic and social objectives to be sought by the trade union movement. The Second World War, in turn, opened the opportunity for the labour movement to act as considerable national support developed during this new war for the public ownership of certain industries and services: 'wartime exigencies contributed substantially to win public support for labour's long-standing contention that only by completing the process of national ownership and coordination could electricity supply industry be "tidied up", reorganised, and expanded' (Weiner 1960: 57). This was precisely what the TUC had advocated in 1936. The elected Labour Party then nationalised the electricity industry in 1947. However trade unions had also learnt the lesson of government retreat during the first post-war period, and worked to overcome objections to nationalisation by appealing to wider public opinion, propagandising it as a good for British society. To this end, in 1944, the TUC declared that public corporation boards should be appointed by the Ministers concerned, supporting in this way the Morrisonian solution of no organised interest representation. This was the pattern of nationalisation policy and there was neither workers' control nor trade union representation as such on any body responsible for managing nationalised industries (Cole 1953; Pendleton and Winterton 1993). The Morrisonian Corporation had, by statutory obligation, joint consultation and centralised negotiation procedures with the participation of all representative bodies; that strengthened the trade unions' position and won their support. Although the Morrisonian model was brought into question by some trade unions by the mid-1980s, the commitment to either public ownership or its revitalisation through the notion of social ownership was firm within the ranks of the British labour movement (EETPU 1987; TUC 1984, 1986, 1988).

In Argentina there was no similar long-term commitment to public ownership by the labour movement, at least not explicitly. As the composition of its working-class was an outcome of migratory trends, Argentina's multinational outlook rendered governments and the state alien to workers' aspirations, mainly because foreigners were denied political rights and were often repressed (Falcón and Monserrat 2000; Godio 1987). Distrust of governments increased the appeal within labour ranks of anti-state ideologies such as anarchism and revolutionary syndicalism (Del Campo 1986; Suriano 1990). The labour movement in Argentina never experienced events of exceptional governmental controls and takeovers of industry due to war exigencies, which were crucial in the UK to shaping the evolution of workers' demands regarding nationalisation and to making them a practical issue. Gradually, however, by the end of the 1920s, some key trade unions had begun to negotiate with governments, asking for state intervention through legislation and arbitration in labour disputes, and engaging in a sort of incipient political exchange, but still without including public ownership among their objectives (Falcón and Monserrat 2000). The rise of *Peronism* in the mid-1940s would be the turning point. Its programme of nationalisation and public ownership of key industries as part of a developmental, industrialist and anti-imperialist project found the support of a growing *Peronised* and nationalised labour movement, which appropriated state ownership as their own political and nationalistic objective (Cordone 1993). This trend continued well after the fall of the government at hands of a military coup in 1955, as is shown by the political programmes of 1962 (the so-called *Programa de Huerta Grande*) and 1968 (*Programa del 1º de Mayo de la CGT de los Argentinos*), backed by trade unions, in which amidst acute episodes of open class struggles, radicalised leaderships of the labour movement explicitly asked for wide nationalisations as preliminary steps towards a national version of socialism (James 1988).

The democratic transition in Argentina during the 1980s witnessed strong trade union opposition to any attempt to privatise state enterprises, and this political demand was included in most of the thirteen national strikes launched by the *Central General de Trabajadores* (CGT) between 1984 and 1988, though now completely stripped of any revolutionary connotation. LyF CF, for instance, the electrical trade union of Buenos Aires, demanded, from its origins in 1943, the nationalisation of the industry. This aim was partially fulfilled by the creation of a national enterprise, Agua y Energía Eléctrica in 1947, but the main private companies continued their operations. In 1961, the government planned, as a temporary solution against private monopoly powers, a nearly complete nationalisation of the industry but with the intention of re-privatising it soon after. However, LyF CF aborted this intention and was even able to obtain a short but intense participation in enterprise management between 1973 and 1976, with a view to developing trade unions' management in the public sector. Again, a military coup put a complete and violent stop to this erasing the experience gained. The collective agreement (CCT) negotiated in 1975 by the electrical union was, perhaps, the most advanced ever in Argentina and Latin America from the point of view of social protection to labour, and it was regarded by other public unions as a model to imitate. In short, state control over the industry was the political position of the electrical trade union throughout the period until its re-privatisation in 1993.

Consequently, whether under the ideological wisdom of Fabianism, socialism, anti-imperialism, nationalism or otherwise, both labour movements favoured public ownership, a political aim which was later on reinforced by the good rates of pay and good terms and conditions obtained in the state industries. The argument is not that nationalisations were brought about by the struggles of the labour movement – if this could be a plausible argument for the UK, it is clearly not for Argentina – but simply that nationalisation was amongst, or became a part of, trade unions' long-term political objectives. It is not surprising, then, that privatisation plans were always opposed at the outset by trade unions, though not necessarily by direct industrial action, while the private sector enthusiastically pushed for their implementation. Thus, privatisation came to reverse a well-established preference of organised labour for the extension of public ownership.

A Theoretical Argument: The Enhancement of Market Discipline (II)

Marx's critique of the political economy of capital is well-known (Marx 1973, 1977). He begins with the most simple category to be found in capitalism, the commodity, which is placed in the abstract realm of 'capital in general' (Rosdolsky 1977), in which capital's development is shown in its specific characteristics as a self-valorisation process, and there is no place for surface phenomena such as competition among capitals. Marx considered it methodologically necessary first to grasp the inner nature of capital in order to understand the behaviour and movement of many capitals subsequently (Arthur 2004; Lebowitz 2003). However, Marx often stressed that capital really exists – 'necessarily expresses itself' – as individual capitals, as capitals in competition: 'the immanent laws of capitalist production manifest themselves in the external movement of the individual capitals, assert themselves as the coercive laws of competition' (Marx 1977: 433). For him, competition 'is nothing more than the way in which the many capitals force the inherent determinants of capital upon one another and upon themselves' (Marx 1973: 651). In short, the inner nature of capital development needs the action of many capitals acting in their self-interest by seeking market advantages and the extraction of surplus value, to the extent that left to their own devices, individual capitals will often act to secure monopoly or oligopoly positions against their direct competitors. Hence, 'capital's tendency to increase the workday (extensively and intensively) and to increase productivity is manifested through the efforts of individual capitals to lower their costs of production relative to other individual capitals in the context of competition' (Lebowitz 2003: 82).

Taking into account the role of privatisation in fostering competition in general, and in public utilities in particular, it is theoretically possible to conclude that privatisation plays a major role in enhancing the power of capital *vis-à-vis* labour. It does not matter, at this level of analysis, whether privatisation policies were designed to foster competition or whether strong elements of either monopoly or oligopoly remain (Hall 2005; Hall, Thomas and Bayliss 2002). The crucial point here is that even when competition could have been served better by selling the assets differently, or by different regulatory frameworks, the decision to privatise enhanced competitive forces and the market discipline. Moreover, leading local businessmen were persuaded that services supplied by government such as electricity were too expensive because of inefficiencies and because of the social goals that they pursued, such as equity and employment, and hence that the lack of competitiveness of government providers made private industry uncompetitive as well (Beder 2005).

To put it differently, while profitability is the leading force of private enterprises, this is not exactly the case for state enterprises, whose 'defining characteristic, indeed the rationale for their existence, is that they are subject to some form of political control' (Ferner 1988: 29). Privatisation frees management from political control and interference, and as a result 'the privatized enterprise faces an environment of decision-making very different from the highly politicized public-sector model, with far-reaching implications for the culture of industrial relations' (Colling and Ferner 1995: 493). Public ownership entails a potential diversion from a crude development of capital's dynamic due to political intervening variables and other state functions (Ferner 1988; Saad-Filho 2003). It has been stressed that state enterprises are only partially affected by the discipline of the market and that their multiple and usually contradictory objectives are defined through a political process which combines diverse economic roles – from supporting private industry to implementing macroeconomic policies – with tasks related to political legitimation (Ferner 1988: 30–1; Goldín 1997). Historically, trade unions took advantage of their ability to participate in this political process. At their peak, right-wing thinkers wisely alluded to this aspect when arguing in favour of privatisation and other neo-liberal policies.

A Theoretic Empirical Argument: The Appropriation of Resources (III)

From the point of view of mobilisation theory, privatisation may be conceptualised as constituting in itself a huge process of appropriation of resources, and therefore, as a process of counter-mobilisation as well. Indeed, the main features of the epoch of market liberalisation, the process of capital internationalisation, the expansion of MNCs and the retreat of the state from the direct control of production and services, are closely tied to worldwide privatisations of public property, which entail an enormous transfer at a global scale of material resources from the public sphere to the private domain: in short, an enormous and direct concentration of wealth, power resources and control over key economic sectors into private hands. That is to say, in either Etzioni's (1968) or Tilly's (1978) approach, a process of mobilisation; or, more exactly, within Kelly's amalgamation of mobilisation and long-wave theories, counter-mobilisation. This latter conceptualisation was independent of whether privatisation was a direct consequence of lobbying from capital or the result of a political decision based more on ideological and macroeconomic reasons.

Some figures may help to give a better idea of the dimensions of these transactions. Table 4 offers data about financial proceeds from privatisation between 1990 and 1996 in Latin America, all the LDCs, OECD countries, Argentina and the UK.

Table 4: Proceeds from Privatisation, 1990–6 (in US$ millions)

Year	All LDCs	Latin America and the Caribbean	Argentina(1)	OECD countries(2)	UK
1990	12,658	10,961	891	12,988	4,219
1991	24,243	18,723	1,892	10,168	5,346
1992	26,180	15,560	2,654	18,722	7,923
1993	23,651	10,487	3,823	37,930	8,114
1994	21,704	8,198	746	55,119	4,632
1995	21,802	4,615	954	54,429	5,648
1996	25,175	13,919	272	52,949	2,426
1990–6	155,413	82,417	11,232	242,305	38,308

Source: Manzetti 1999 and *Financial Market Trends: Finance and Investment*, 82, June 2002, Paris, OECD

(1) These are only cash figures. Debt bonds used in transactions totalled US$17 billion in nominal value (as they were considered), but their actual market value was US$8 billion

(2) Mexico is not included

These figures are even more striking if it is taken into account that privatised companies have often made immediate gains, because public offerings were ordinarily underpriced. Although underpricing is also generally found in private equity offerings, research has shown that, for instance, in the UK and France, privatisations involved 'an additional discount of around 10 per cent' (OECD 1995: 9). In Latin America, underpricing was a structural feature of incredible dimensions, to which the use of debt bonds in equity swaps was crucial in a number of cases (Cook and Kirkpatrick 1995). A detailed study of privatisation in Argentina and Brazil has stressed that 'by accepting debt-equity instruments as a form of payment [...] potential investors could make huge savings since the quotation of these equities in the secondary market [during 1990s] was at an all time low' (Manzetti 1999:

304). Privatisation, therefore, opened opportunities for private companies and finance capital. Since the international economic crisis of mid-1970s, they have engaged in an anxious search for profitable areas in which to invest: undervalued businesses, and in the case of public utilities, the appropriation of key undervalued energy resources. It is also important to note that Table 4 misses the timing of the British privatisation process, which began earlier and predated incoming worldwide changes. In this process, the Conservative British Government that took office in 1979 'embarked on a far-reaching effort that lasted through 1980s and during which nearly all state-owned enterprises in the competitive sector were privatised' (OECD 1995). Table 5, then, completes the British picture.

Table 5: Proceeds from Privatisation, 1979–90 (in £ millions)

Country	1979–83	1984–8	1989–90	1979–90
United Kingdom	1,535	18,359	12,180	32,074

Source: Schamis (2002)

Through the appropriation of public resources, privatisation fostered the concentration and integration of key economic sectors into private hands. For example, nine conglomerates in Argentina combined to control oil, telecommunications, electricity, gas, petrochemicals, construction, transportation and banking, aside from other peripheral activities. By this strategy, these capitals have gained economic power, flexibility to respond to economic difficulties, and considerable political influence. If this sort of concentration and capture of the state is alien to the UK, some degree of integration in the utility market is not; utility companies like PowerGen and British Gas (BG), among others, show the increasing importance in Britain of this international economic trend. In short, the appropriation of public assets multiplied the power resources in capital's hands, strengthened their negotiating position vis-à-vis the parallel fragmentation of labour in a context of growing internationalisation of competition, and reduced the power of governments to manage macroeconomic and microeconomic variables.

An Empirical Argument: Evidence of Privatisation as a Counter-Mobilising Force (IV)

The counter-mobilising effects of privatisation on labour cannot be easily isolated from the broader effects of market liberalisation, commercialism, deregulation and productive restructuring which often accompany privatisation programmes (Colling and Ferner 1995; Ferner 1989; Ferner and Colling 1991; 1993a; 1993b; Riethof 2002; Weyman-Jones 1993). However, scholars the world over are generally inclined to associate privatisation with unfavourable developments for industrial relations and organised labour (Báez-Camargo 2002; Cifarelli 1999; Cook and Murphy 2002; Colling and Ferner 1995; Danford, Richardson and Upchurch 2002; Luca 1998; Dinerstein 2001; Duarte 2001; Ferner 1989; Ferner and Colling 1991; 1993a; 1993b; Hall 2000, 2005; O'Connell Davidson 1993; Ogden1993; Petras and Vieux 1999; Pripstein Posusney and Cook 2002; Riethof 2002; Van der Hoeven and Sziraczki 1997; Wallis 2000). This literature sets the empirical foundations for an understanding of privatisation as a counter-mobilising strategy against the social achievements of the labour movement after the Second World War. Only a few, such as Peter Fairbrother (1994, 1996a, 1996b, 2000), have pointed to potentially positive outcomes for labour. Fairbrother advanced the thesis that organisational restructuring due to privatisation, and the decentralisation that this occasions, would be likely to disrupt bureaucratic controls within unions, stimulating mobilisation at the local level. Yet, even this view has not denied that privatisation, in any event, tends to negatively affect pay, terms and conditions of employment.

It is possible to point with relative confidence, therefore, to the likelihood of job loss (and thus membership loss) due to privatisation, whilst being more agnostic about the consequences for management-union relations. Indeed, some scholars have underlined that privatisation damages workers directly through job loss, which often accompanies restructuring, but also indirectly, since changes in the economic role of the state weakened corporatist structures (Cook and Murphy 2002; Palomino 2005; Riethof 2002; Thwaites Rey 2003). Moreover, decentralisation of bargaining over wages – another common outcome of privatisation – terminates any sort of wage indexation by public authorities, forcing trade unions to adapt and shift the focus of their activities from the state to employers. For instance, in her exhaustive study of privatisation in Brazil, Riethof concludes that 'fragmentation of the workforce (dismissal and outsourcing), flexibilization of labour relations and transfer of workers to the private sector tend to weaken trade unions' (Riethof 2002: 232). A similar picture has been drawn for very different countries (Luca 1998; Goldín 1997; Petras and Vieux 1999; Pripstein Posusney and Cook 2002). In addition, state-owned enterprises generally developed a good employer policy that brought better terms and conditions for their workforces (Pendleton and Winterton 1993). Particularly in Latin America, these firms had a policy that made possible generous secondary working conditions, which included pension funds, disability funds and schooling and training for workers and their families (Báez-Camargo 2002; Luca 1998; Margheritis 1999; Pripstein Posusney and Cook 2002; Riethof 2002). Privatisation jeopardises those benefits, to say the least, by shrinking 'the public sector in which workers have achieved the highest levels of organisation and wages, job security, and decent working conditions' (Cook and Murphy 2002: 2). For LDCs, several studies have shown that, in order to make public enterprises attractive to potential investors, revisions to union rights and collective contracts have been standard procedures, and when opposition from combative trade unions arose, governments often embarked on selective and exemplary repression (Báez-Camargo 2002; Luca 1998; Goldín 1997; Palomino 2005; Petras and Vieux 1999; Pripstein Posusney and Cook 2002; Riethof 2002; Thwaites Rey 1999; Tomada and Senén González 1998). In broad terms, this general view applies to both the UK and Argentina.

The privatisation of public utilities in the UK has been associated with a set of work practices such as teamwork, employee participation, directed communication channels, corporate culture policies, team briefings, quality circles, employee stock ownership plans (ESOPs) and joint labour-management partnership, all directed at diluting the power of trade unions through the development of participatory machineries (Arrowsmith 2003; Colling and Ferner 1992, 1995; Danford, Richardson and Upchurch 2002; Ferner and Colling 1991, 1993a, 1993b; Katz 1997; Pendleton and

Winterton 1993). Derecognition, mainly when developing new business units, outsourcing, green field sites and so forth, has been another threat faced by British trade unions as a result of privatisation.

As for the privatised utilities in Argentina, scholars have shown that changes in terms and conditions have adversely affected five main areas: working-time, labour practices, payment structures, employment levels and company-trade union relationships (Brinkmann 1999; Cifarelli 1999; Dinerstein 2001; Duarte 2001; Murillo 2001; Tomada 1999; Tomada and Senén González 1998). In short, empirical research has offered evidence about the counter-mobilising side of the sale of public sector companies from the point of view of labour.

Conclusion

The objective of this chapter is to support the assertion that privatisation was, in itself, a counter-mobilising action against labour; that is, to provide support to the basic foundation upon which this research unfolds. Hence, in order to achieve this aim, different arguments have been deployed.

After addressing the ambivalent meaning of the term privatisation, the chapter embarks on an historical investigation into the politics of labour movements in both countries in order to show that privatisation went against their traditional policies. Yet, since the conceptualisation of privatisation as a counter-mobilising factor cannot rest solely upon historical evidence, further theoretical and empirical arguments are advanced to discuss, from the point of view of the political economy, why privatisation enhances the market discipline; and from the point of view of mobilisation theory, the importance of privatisation to the concentration of power resources in private hands. Finally, the chapter refers to the empirical research that has pointed to the association between the sale of public assets and harmful consequences for organised labour. The essential theoretical point belongs to Marxist political economy and refers to the role of privatisation in fostering competition and the logic of markets on former public industries.

Taking the arguments as a whole, the emerging picture is one in which the relationship between privatisation and workers' collectivism appears as an appealing field for empirical research. However, the study of this relationship in the UK and Argentina demands an increase in depth of the comparative analysis of privatisation and the institutions of industrial relations so as to include surveys of the sort of intermediate variables which contributed towards shaping the opportunity structure confronted by trade unions in the face of ESI privatisation. Chapters Five and Six are devoted to this task. The following chapter compares the contextual conditions of privatisation by exploring the driving forces and rationales of privatisation in the UK and Argentina.

CHAPTER FIVE

The Driving-forces of Privatisation in the UK and Argentina

The preceding chapter mentioned that there is an extended agreement among scholars about the dissimilarity between the driving forces and rationales of privatisation in DCs and LDCs (Bienen and Warterbury 1989; Christiansen 1989; Cook and Kirkpatrick 1995; Cook and Murphy 2002; Dinerstein 2001; Feinberg 1986; Fernández Jilberto and Riethof 2002; Gupta 2000; Herrera 1992; Manzetti 1999; Molz 1990; Nellis and Kikeri 1989; Petras and Vieux 1999; Ramamurti 1992a, 1992b; Riethof 2002; Sánchez Bajo 2002; Walle 1989).

For instance, Christiansen states that even though fiscal deficit (and also inflation) has been a worldwide driving force towards privatisation, in LDCs more serious problems arise due to 'growing current account imbalances, increasing external debt, unfavourable terms of trade, and rising government budgets deficits' (Christiansen 1989: 597). Gupta (2000), in turn, also highlights the disparity in context and objectives of DCs' and LDCs' privatisation processes. In the former, declining growth rates, rising unemployment, falls in investment and rises in inflation led to privatisation in search of faster growth, higher efficiency and wider competition. In the latter, instead, the debt crisis was the crucial turning point towards privatisation, by which was sought debt reduction, creditworthiness and access to foreign capital and investments.

If the analysis lays stress on the political nature of the global spread of privatisation rather than in the dynamic of market forces (Dinerstein 2001; Thwaites Rey 1994), it needs to identify political actors, apart from country specific economic constraints. Undoubtedly, the IMF and the WB were the most important of those actors, two institutions that became intertwined in the 1980s, together with the right-wing governments that backed them, especially the Reagan and the British Conservative administrations (Stiglitz 2002). Indeed, Ramamurti (1992b) concluded – through univariate and multivariate analyses of LDCs' privatisations – that privatisation was likely to be pursued by those LDCs with high budget deficits, high foreign debts, and high dependence on international agencies precisely like the WB and the IMF, which were responsible for the simultaneous shift in favour of privatisation in LDCs, and whose policies were deeply determined by the USA and the UK. Most scholars agree with this view (Cook and Kirkpatrick 1995; Cook and Murphy 2002; Dinerstein 2001; Fernández Jilberto and Riethof 2002; Gupta 2000; Manzetti 1999; Petras and Vieux 1999; Ramamurti 1992a, 1992b; Riethof 2002; Sánchez Bajo 2002; Thwaites Rey 1994; Walle 1989; among many others).

Then, while it is possible to argue that the first privatisations, as in Chile and the UK, 'were experiments driven by business interests and shaped by a mix of neoliberal dogma and, in the case of Britain, pragmatic politics' (Beder 2005), they soon became models for the countries that followed due to the direct involvement of the international donor community in the implementation of the privatisation programmes (particularly in LDCs), not only through the IMF and the WB activities, but also through the Inter-American Development Bank (IDB), the Exim-Bank, the US Agency for International Development (AID) and others.

The privatisation policy was part of, and in itself, a counter-mobilisation wave, that unfolded within the global restructuring pursued by the international financial institutions, hegemonic states from DCs and MNCs in order to enforce market discipline on national economies and public budgets worldwide. To different degrees in each country, even in the pioneers, this wave comprised: an ideological agenda to reduce the role of the state and promote consumer choice; an economic attempt to embrace change in the structure of the economy, facilitate tough labour policies and curb the power of organised labour; a managerial concern about rationalising the internal structures of state-owned organisations; a political interest in achieving support for right-wing administrations and short-term financial objectives (Pendleton and McDonalds 1994; Vickers and Wright 1988).

However, beyond this general understanding, it is worth comparing how specific driving-forces and rationales combined in each of the countries under study (the UK, a member of the G8, privatising pioneer and key international player; Argentina, a peripheral economy, where the external pressures were particularly acute), given that different contexts presumably determined different opportunity structures for organised labour.

The United Kingdom

The British programme was the first large-scale privatisation process and it became a blueprint for other countries (Clark 1993; Marsh 1991); for instance, the Argentinian government explicitly followed the British model for certain privatisations. Scholars coincide in recognising that the reduction of the public sector was probably the most salient feature of the political programme of Thatcherism. Spite of its 1979 election Manifesto, in which denationalisation was supposed to be limited to enterprises recently taken into public ownership, by the end of the successive Conservative terms, the public sector had been reduced dramatically. However, unlike Argentina, the privatisation programme in the UK evolved in a piecemeal, incremental way. Launched in October 1979 with the sale of 5 per cent of public shares in British Petroleum, at the end of the first term (1979–83) twelve SOEs were partially or fully privatised (£1,625 million in revenue). During the second term (1983–7) twenty-four firms were sold in part or in whole (£10,983 million in revenue). The third term (1987–91) involved forty operations, among them ten water suppliers and twelve regional electricity companies (£22,514 million in revenue). Thus: 'as October 1993, about two-thirds of the UK's nationalized industries and about 940,000 jobs had been transferred to the private sector' (Miller 1995: 87). Privatisation continued during the following years.

The intention of privatising the ESI in England and Wales was announced early in 1988. According to Ernst (1994), if the sale of the water supply industry was the most controversial, the sale of the ESI was government's most complex and troublesome. Weyman-Jones (1993) also underlines its complexity regarding the regulatory framework. Unlike the case of BG, a proposal for radical restructuring of its traditional highly integrated structure was made, involving: vertical separation between generation and transmission; horizontal break-up and liberalisation of generation; a regional structure for distribution and retail supply; and phased liberalisation of retail supply. In accordance with these reforms, in 1990, the 12 Regional Electricity Companies (distribution) (REC), which jointly owned the National Grid Company (transmission) (NGC) were sold off. In 1991, 60 per cent of shares of National Power and PowerGen (generators) were sold off as well.

Apologists for British privatisation have tended to confine the analysis to a binary opposition between public sector evil and private sector panacea, and a list of indicators of the alleged success of privatisation in terms of economic efficiency gains. Others have shown a propensity to reduce its driving-force, often critically, to the preponderance of some ideological beliefs (Swann 1988). However, in order to explain the driving-forces of privatisation in the UK, most scholars have pointed to governmental objectives (Pitelis and Clarke 1993; Marsh 1991; Saunders and Harris 1994; Wiltshire 1987). In this regard, scholars have always listed similar reasons: first, reduction of the involvement of the government in the industry to improve efficiency by fostering competition; second, reduction of the public sector borrowing requirement (PSBR) to improve government finances; third, easing of problems of public sector pay determination by weakening public sector unions; and finally, widening of share ownership, in particular through employee share schemes, gaining simultaneously political advantages.

Among those who are inclined to stress the economic aims, only a few unconditional supporters of privatisation credit the governmental claim that such policy was a piece of microeconomics (to foster competition and efficiency) rather than one of macroeconomics (to achieve fiscal objectives and short-term revenues) (Gupta 2000; Wiltshire 1987). Yet, amongst those who agree about the preponderance of macroeconomic motives, differences of emphasis, though slight, persist. For instance, Christiansen (1989) focuses exclusively on the monetarist aim of cutting the fiscal drain and reducing the public sector deficit, whereas Bienen and Waterbury (1989) stress the search for revenues generated from the sale of public assets.

Moreover, other scholars bring into this discussion a political element: privatisation would have been a way of reducing PSBR without unpopular measures (Clarke 1993; Marsh 1991). The appeal of this view is that it re-introduces a political dimension to the origins of privatisation. According to this, at first, it would just mean avoiding losing support and votes. However, as a result of the unexpected public enthusiasm for the sale of British Telecom (BT) shares, and two years later, for the sale of BG shares, the Conservative government ended up developing a positive ideological wisdom, that is, the building up of popular capitalism. In the Conservative Party's election manifesto for 1987 such an objective emerges clearly, depicted as the historic transformation of British society through the widespread of share ownership; in John Moore's words, the birth of a 'property-owning democracy'. Following John Moore in this vein, Miller adds another side of this policy: 'wider share ownership has, of course, made renationalization by future labour governments virtually impossible' (Miller 1995: 92).

Allusions to political and ideological rationales among the driving-forces of privatisation are also common in the literature. Wiltshire (1987), for example, stresses the role played by the ideology of the New Right, and in particular, its strong anti-unionism. He underlines that 'the Thatcher faction and later the Thatcher government have always had a noticeably fierce, antagonistic attitude to trade unions, not just to their actions but to their very right to exist' (Wiltshire 1987: 8). At the beginning of the 1980s, public trade unions, mainly after the winter of discontent of 1978–9, became a main target; hence, Wiltshire emphasises that privatisation was also a powerful anti-union weapon. Many scholars have even argued that anti-unionism was the very driving-force of privatisation in its origins. For instance, Bishop and Kay state that 'the origins of privatization lies in the desire to diminish the power of public sector trade unions' (Bishop and Kay 1989: 649), although, later on, this claim is qualified by stressing that it was not a factor of much relevance in the further development of privatisation. Marsh recognises the same objective with regard to public sector trade unions: 'privatization was seen as a means of reducing their size, bargaining power and influence over policy' (Marsh 1991: 472), and he adds that there is little doubt this was one of 'the chief initial concerns of the Conservative government' (Marsh 1991: 472). However, he concludes that contracting-out and anti-union legislation had more harmful effects on trade unions than the change in ownership. Whilst Marsh is correct on the point of simple changes in ownership, this should not obscure the latterly evident fact that change in industrial relations and human resources in the former public enterprises has been very substantial since privatisation. By contrast, Saunders and Harris (1994) do not reject the point, but argue that to tame the unions was not a major reason for privatising. Instead, liberalisation, not privatisation, would have been the original strategy designed to cope with the union threat. In short, it is relevant that the link between privatisation and anti-unionism in the UK experience is a common reference in the literature (see also Bienen and Waterbury 1989; Clark 1993; Ernst 1994; Ferner and Colling 1991; Gupta 2000; Pendleton and McDonald 1994; Swann 1988).

Among the ideological rationales pursuing privatisation, consumerism has also been a relevant force. This could also be connected to countermobilisation as regards industrial relations, for one of the key criticisms of public ownership was that it tended to privilege producer concerns over those of consumers (Henney 1987). So the insinuation of the language of markets and choice was a deliberate ploy to weaken the claim of internal stakeholders (employees) over external ones (customers). Contrary to the propaganda in favour of popular capitalism, consumerism was part of the privatisation ideological appeal from the very beginning. That consumers should be ridden of the perverse dynamism of nationalised industries, often run in the interest of those who work in them – both managers and workers – was a main argument even before the 1979 election. It remained a principal discursive concern during the first two Conservative governments. Finally, it ended up as an essential component of a more powerful discourse, the aforementioned construction of popular capitalism. As Saunders and Harris (1994) argue, the idea of popular capitalism involved four chief elements: consumer sovereignty, a new popular understanding of capitalism, the breaking of 'them' and 'us' identities through employee identification with the company's aims and fortunes and the undermining of collectivist ideologies. Regarding the latter, for Ernst (1994), privatisation occupies a critical place in the challenge to the legitimacy of collectivist solutions to economic and social problems, and therefore to the appeal of socialism. Ernst states that privatisation also constituted a political and ideological challenge to the very spirit of trade unionism as a channel for defining collective demands by processing, collectively, the multiple interests of workers.

Finally, a matter of debate has been whether privatisation in Britain was part of a master and overall strategy or an element of an incremental and tentative policy, therefore without any particular rationale. Those who subscribe to the second option (Bienen and Waterbury 1989; Bishop and Kay 1989; Clarke 1993; Gupta 2000; Marsh 1991; Saunders and Harris 1994) argue: that there was no commitment to privatisation in the 1979 election manifesto of the Conservative Party; that only during the second and third terms such commitment appeared; that such commitment was due to political rather than economic reasons; that even those reasons changed substantially over time; and lastly, that such evolution imprinted privatisation with inconsistencies and contradictions. However, according to Wiltshire, while it seems absurd to argue in favour of a secret and omniscient plan, not least, because there is no evidence of any such plan at all, 'it should not be thought that the concept is haphazard. Behind it lies a careful strategy guided by a clear and consistent ideology. The destination of the journey is certain even if the route has to be plotted after each leg has been completed' (Wiltshire 1987: 14). After all, denationalisation, deregulation and liberalisation were old commitments of the Conservatives (as exemplified by the Selsdon Declaration, the Ridley Plan and the lobbying activities at that time of think tanks like the Institute of Economic Affairs and its offshoot, the Centre for Policy Studies founded by Margaret Thatcher in 1974). In this regard, privatisation appears as the fastest and most powerful method for achieving such aims. Moreover, the political environment was propitious for the New Right as an embryo, for the 1976 IMF loan to the UK amidst the sterling crisis forced the Labour Government to introduce deep cuts in public expenditure, which affected the nationalised industries by the establishment of tough financial controls (Bertero 2002; Foreman-

Peck and Millward 1994; Heald 1980). This event led to the publication in 1978 of a White Paper on nationalised industries that would then be fully implemented by the Conservatives (Harrison 1988).

Argentina

As for the driving forces and rationales of privatisation in Argentina, most scholars agree that this policy was the outcome of the international pressure to favour external creditors and alleviate the fiscal deficit in the short-term (Azpiazu and Basualdo 1999; Azpiazu and Vispo 1994; Basualdo 2001, 2006; Cifarelli 1999; Dinerstein 2001; Herrera 1992; Margheritis 1999; Rausch 1993; Thwaites Rey 1999, 2003). As Rausch puts it 'the Argentinian privatization policy has been fiscally driven and has also had a significant role in securing external debt as well as increasing its value in the secondary market [...] improving relations with external creditors' (Rausch 1993: 171). It would be difficult to find a discordant voice on this issue within academia. Indeed, in this context, this international constraint is the most critical difference between Argentina and the UK. Consequently, it is impossible to find in the Argentinian literature the diversity of opinions about the driving-forces and rationales of privatisation that are summarised for the British case in the previous section.

Within a context characterised by international conditioning, between 1990 and 1993 the government launched a fast and massive privatisation programme, technically and financially assisted by the IMF, the WB, BIRF (*Banco Internacional de Reconstrucción y Fomento*) and the IDB, in which debt capitalisation mechanisms were widely applied (Alexander and Corti 1993; Cifarelli 1999; Sánchez 1991). Over the implementation, the programme was backed by a World Bank's Public Enterprise Reform Adjustment Loan (PERAL) and a World Bank's Public Enterprise Execution Loan; the former to finance lay-offs, early retirements and enterprise restructuring; the latter to finance technical assistance for the development of tailor-made privatisation plans (usually provided by the WB itself) (Dinerstein 2001; Rausch 1993, 1994; Sánchez Bajo 2002).

In four years, Argentina sold thirty-four companies and let concessions for nineteen services and eighty-six areas for petroleum development. According to Alexander and Corti (1993) Argentina's privatisation program was one of the broadest and most rapid in the Western Hemisphere. By 1993 there had already been 280,509 jobs lost, with a cost to the state of 2,035 million dollars in voluntary redundancy packages, which were financed with public resources and a PERAL. Most SOEs were undervalued and one-third of the amount collected by the government for privatisations, around 18,000 million dollars, corresponded to debt-equity swaps (Azpiazu and Vispo 1994; Dinerstein 2001).

As for the ESI, it underwent significant restructuring prior to privatisation in 1993. It was split into a series of generation and distribution companies. The assets of the former were sold and the latter were given in concession. It was divided into several companies. Generation, considered a competitive activity, was broken up into twenty-five business units that were sold separately to private owners. The transmission, considered a natural monopoly, was given as a concession to only one company in a given area. There are twenty-two main distribution companies; most are under provincial government jurisdiction. Distribution, considered a natural monopoly as well, was given in concession to the three largest distributors: EDENOR (Argentina, France and Spain), EDESUR (Argentina, Chile, USA) and EDELAP (USA and Argentina), which are the main companies in this sector.

Apologists for privatisation, without denying that external conditioning was its crucial driving-force, have often restricted the analysis to either spurious comparisons between the public and the private sector or praiseworthy accounts of the efficiency gains of isolated privatised enterprises. For them, privatisation came to end the chronic incapacity of the state to run economic activities efficiently, enhancing in this manner the competitiveness of the economy as a whole. This is often shown by studies whose starting-point is the inefficiency, inflation and deficit of the public companies, something that they intend to prove by comparing the levels of productivity of the companies before and after privatisation. This type of analysis aims to provide support for the governmental decision and the political recommendations of the IMF and the WB by highlighting the microeconomic rationales of privatisation.

Also Margheritis (1999) points to the need for the government to legitimise the privatisation programme by underlining economic efficiency as a prime motive of privatisation. Nevertheless, she acknowledges that the real motives were gradually incorporated within the official discourse. Consequently, the end of the fiscal deficit, or the cancellation of the foreign debt, ended up as the explicit official rationale of the sale of public companies by means of a simple argument: to transform a source of debt into a source of income. Several scholars underline the fact that the government also claimed, strongly, that given the lack of public funds and the economic crisis, it was impossible for the state to invest in technology to modernise the public services and state industries, and consequently privatisation was presented as the only alternative for improvement of the provision of services and the national economic performance (Margheritis 1999; Thwaites Rey 2003). Thwaites Rey (2003), in turn, has stressed that the privatisation programme was also a strategy for opening new business opportunities to attract foreign investors, which would contribute towards explaining why consumers' interests were so crudely sacrificed for private profitability. Nevertheless, all these arguments are often introduced to complement the main point: that the main driving-force towards privatisation was the external pressure in a context of acute fiscal deficit.

As for public trade unions, they were blamed as in the UK for maximising salaries and benefits for themselves, at the cost of service quality and economic efficiency, and generally at cost to the consumers. Indeed, the need to curb the power of public trade unions was a topic included in the agenda of public debates of that time. However, the economic crisis, which surrounded the implementation of the privatisation programme, downplayed the importance of the anti-union side of privatisation. Rather, the government launched an open attack upon public unions to facilitate the implementation. In this sense, in Argentina, union power was a problem to be tackled prior to privatisation so as to allow the success of the programme. Needless to add that later on, the subsequent and negative impact of this policy on unions' power was welcomed by the government.

Lastly, it may be worth mentioning that, if neither a proper rationale nor exactly a driving-force, the role of the hyperinflation spiral of 1989 is often stressed in the literature as an essential condition, for scholars having diverse perspectives agree that this crisis helped to overcome resistance within civil society, particularly that of trade unions (Dinerstein 2001; Duarte 2001; Hill and Abdala 1993; Rausch 1993; Rofman and Romero 1997). This event brought forward the change of government personnel and paved the way for a wide programme of reforms in which privatisation was decisive. Trade unions found themselves suddenly trapped. The vast majority of trade unions were openly against privatisation, but they did not want to be blamed in front of the population for being responsible for a new hyperinflation crisis, so most unions declined to take industrial action.

Conclusion

By reviewing the literature, this chapter confirms the difference between the driving-forces and rationales of privatisation for DCs and LDCs. The character of Britain as pioneer amplifies the gap between both experiences. While the privatisation programme in the UK evolved in a piecemeal, incremental way, in Argentina privatisation unfolded with astonishing speed. Moreover, while in the former the process was endogenously driven, in the latter, the exogenous constraints posed by international financial institutions were crucial. There were important contextual differences as well, the main one being the deep economic crisis which conditioned the orientations of the political actors in Argentina, including the trade unions.

For organised labour, these differences contributed towards shaping differently their respective opportunity structures. For British ESI unions, privatisation was a foreseeable event, and hence, they had prepared themselves for this possibility to some extent. When privatisation was finally announced the process took enough time to allow unions to engage in campaigns and lobby. In Argentina, privatisation came as a great shock for LyF CF, whereas it was a foreseeable event for LyF MDP. This difference proved to have implications for their respective strategic choices as analysed in future chapters.

Lastly, the opportunities for Argentinian trade unions to oppose the programme of privatisation were limited by the role played by international financial institutions, which exercised all sorts of pressures upon the national government in the midst of an acute economic crisis, and provided resources and technical assistance to implement the recommended policy.

CHAPTER SIX

Comparative Industrial Relations and Mobilisation Theory

Industrial relations institutions are in themselves the shaping forces of the opportunity structure, for they have a determining effect on trade unions' (re)sources of power, in both their political and industrial dimensions. Their comparative study highlights variations, which may influence in different ways the forms of counter-mobilisation and workers' collective action.

The origins and the underpinnings of the Argentinian and the British systems of industrial relations differ deeply. Voluntary regulation instead of legal regulation has been the main historical feature of the British system, whereas the opposite has been the case in Argentina. This basic divergence has had wider effects, especially upon the workings of collective bargaining, the degree of trade unions' political exchange with governments and employers, and obviously, the character of law as a power resource. Consequently, in each system, the political and industrial dimensions of trade unions' (re)sources of power have tended to correlate differently. It is possible to argue that, broadly speaking, industrial (re)sources of power brought political influence to the trade unions in the UK, whereas political (re)sources cemented workplace organisation and industrial strength in Argentina. To put it differently, while in the UK the Industrial mobilisation of workers was the key power resource making employers sit at the negotiating table, in Argentina the political mobilisation of workers was crucial to obtaining bargaining legal rights in order to overcome alienation and repression at workplace level. However, in both countries, the underlying basis of union power in the ESI relates to the political contingency which pervades the public sector, government concern with smooth supply, bureaucratic regulation stimulating micropolitical exchange, statutory responsibilities of the ESI to be a good employer and so forth. In short, the aforementioned difference is less important when studying the specifics of the ESI, though still meaningful as a determinant of the relationships as a whole between the labour movement, the capitalists and the state, that is, the general framework in which ESI trade unions operated during the privatisation.

This chapter aims to analyse similarities and divergences regarding how trade unions have combined industrial and political (re)sources of power in the UK and Argentina, insofar as this combination relates to patterns of allocation of power resources. Five areas will be briefly examined: the basic foundations and features of both systems, their effects upon the dynamics of political exchange, ESI industrial relations institutions, trade union organisation, and finally, the neo-liberal offensive against labour which principally consisted of the legal aspects that framed the context in which privatisation took place. All these areas provide useful insights about the intermediate variables which shape the empirical manifestation of the categories opportunity-to-act, organisation and interest definition. The chapter begins by looking at the British case; after which Argentina is considered.

The United Kingdom

The Voluntary System

In Britain there has never been a body of positive legal rights for trade union activities and collective bargaining. This legal void has constituted the core of the voluntary system. Trade unions from early on enjoyed a range of industrial relations immunities that allowed them to strike and organise workers without the risk of being prosecuted according to common law (Dickens and Hall 1995, 2003; Hyman 2003; Lewis 1986). As there was not any enforceable collective agreement and unions were never agents of their members according to British law, it was workers' deployment of industrial collective strength that forced employers to deal with trade unions; opposite interests were to be reconciled by compromise and accommodation in voluntary agreements (Hyman 2003). Concurrently, employers had no obligation to deal with or to recognise unions apart from a brief and unhappy experience with statutory trade union recognition in the 1970s. It is also often noted that extension mechanisms were weak in the UK, meaning that agreements tended to be observed only in workplaces formally covered by them. In the absence of instruments of this kind, union strength has depended almost exclusively on organisation, that is, union density and the capacity to mobilise workforces. When density and mobilisation dropped, so did the disposition of employers and governments to deal with unions seriously.

Within this framework, while from trade unions' point of view, the role of law was to prevent hostile action by employers through courts, or as a second best alternative when failing to secure basic standards, it has always been a power resource for employers to be mobilised against employees and against unions. The principles of contract that underpin employment law do little to address the fundamental imbalance of power in the employment relationship. The latter was based upon either strong workplace structures or, mainly in the public sector, well developed industrial negotiating machineries, as favoured by the Keynesian orientation of public policies in the period after the Second World War.

In short, labour legislation has had a comparatively much weaker influence than in other countries, and governments usually assumed the role of providing assistance to collective bargainers. Within this context, although the attractions of the voluntary system diminished for trade unions during the 1970s, at the time when this orthodoxy began to break down as manifested in changed strategies around health and safety at work and lobbying for the national minimum wage (Terry 2000), free agreements together with informal systems of workplace negotiations were the bases of social regulation.

This is in striking contrast to other systems like that of Argentina. It entails two important consequences: an ambiguous unions' orientation to political action, and the chance to restrict the right to take industrial action by redefining immunities, and by these means, to distort a crucial source of union power.

Trade Unions, Politics and the Public Sector in the UK

The voluntary system developed in parallel with a shared anti-interventionist understanding of the main actors: employers, trade unions, and governments. Crouch (1979) stresses the general acceptance by the political parties that industrial relations should be taken out from political controversy. Although state ownership was a long term political and industrial objective for unions, they attempted to channel it through the Labour Party, and prevented corporatist biases by supporting the Morrison formula of no organised interest representation on public corporation boards (Cole 1953; Pendleton and Winterton 1993). Moreover, where the trade unions have exercised political influence through the Labour Party, the relationship between unions and Labour has been characterised by a demarcation of functions, which ultimately meant that unions have rarely attempted 'to interfere with the autonomy of the parliamentary leadership in determining general party policy' (Hyman 1994b: 41).

Still, the dynamics of the Keynesian supply policies, adopted by both Conservative and Labour administrations, increasingly triggered trade unions incorporation into decision making, particularly from the late 1960s, through consultation and tripartite agencies (Marsh 1992; McIlroy 1995; Crouch 1994, 2003). There have been controversies surrounding the degree of this corporatist tendency (Middlemas 1991; McIlroy 1995). Yet it seems safe to conclude that even at its height, the main characteristic of industrial relations in Britain as a whole continued to be the low political profile of organised interests, especially on the labour side (Crouch 2003).

In the public sector, in turn, the tradition of voluntarism and Keynesianism united to bring about highly centralised and formal machineries of negotiation and consultation, in accordance with the Whitley philosophy of the post-war Labour government's policy towards industrial relations (Pendleton and Winterton 1993). Thus, public ownership allotted unions a specific political lever as well. Two main aspects should be stressed: first, the weak exposure to market forces of the public sector, and the rigid procedures and regulations, set the conditions for a micro-political exchange at industrial level, which permitted managers to fulfil Ministerial objectives, and trade unions to shape decisions and obtain benefits for workers. Second, at least until 1979, unions took advantage of the state public sector traditionally being a good employer.

So post-war Keynesianism and its commitment to full employment and stable product markets helped workers to strengthen their organisations and collective bargaining overall. Additionally, the general context of compromise in public services allowed workers to benefit from the high level of service provision and public spending stemming from the development of a welfare state, demand-side policies, and good employer policies, these contributed to the development of powerful trade unions and centralised machineries of industrial relations in the nationalised industries (Crouch 2003; Ferner and Colling 1993a).

Then, as a whole, the meaning of the post-war settlement for trade unions was not so much the benefits of a new level of political exchange as the growth and consolidation of their industrial strength. In the particular case of the public sector, it entailed the development of a system of industrial relations in which political contingency and statutory provision meant that unions enjoyed comparatively more power resources (also stemming from micropolitical exchange) than their private sector counterparts.

Industrial Relations in the ESI

In the ESI, trade unions took advantages not only from the peculiarities of the public sector but also from the governmental concern with keeping the smooth provision of electricity supply and balancing public expenditures; these coloured the political context in which industrial relations developed, though with contradictory effects (Ferner and Colling 1993a). In this context, unions profited from the prevalence of securing supply whenever these two trends tended to clash. Indeed, the nature of electricity supply meant an additional source of union power, as conventionally, employee power is seen as enhanced where the product is complex and/or perishable and has strategic importance, as in electricity par excellence.

So as a far cry from the private sector, through the 1947 and 1957 Acts the government placed on the British Electricity Authority, and later, the Central Electricity Generating Board, as it did with other nationalised industries, 'a legal obligation to consult and bargain with trade unions' (Wedderburn 1986: 278); though some negotiating bodies pre-dated nationalisation (Hannah 1979). This implied giving the chance for unions to mobilise statutory resources to secure certain procedural aims. However union density still played a crucial role within the industry.

The negotiating machinery operated at three levels via four specific bodies for manual, white-collar, engineer/technical, and managerial staff. While EPEA monopolised the representation of engineers at the National Joint Board (NJB), several unions disputed the representation of the other bodies; however, before privatisation, EETPU and NALGO were the strongest organisations for, respectively, manual staff (National Joint Industrial Council – NJIC) and white-collar staff (National Joint Council – NJC) (Ferner and Colling 1993a).

The machinery assured the power of the full-time national officers, who carried out detailed and extensive negotiations of terms and conditions at national level, and also assured close relationships between management and unions at all levels (Colling and Ferner 1993a, 1993b). Consultation was kept at three levels too, but in joint bodies uniting all trade unions, and often comprising lengthy and extensive processes. These central negotiations and consultations were complemented, when appropriate, by arrangements outside the formal machinery, mainly in the Area Boards, so far as they did not threat central authority (Hannah 1982). The strong institutionalisation of industrial relations was a vital power resource for trade unions, for it allowed their involvement in the running of the industry through consultation and negotiation, and the formal procedures of remitting problems and disagreements up to the national levels inhibited managerial prerogatives (Ferner and Colling 1993a).

Lastly, both management and trade unions benefited from two essential features brought about by this institutional settlement: the low level of conflict accompanying the modernisation of the industry, and the neutralisation of workplace activism. From the point of view of union leadership, these outcomes were functional in keeping internal threats to a minimum. In brief, not only their industrial strength but also the weak presence of market discipline, the negotiating machinery and the micropolitical exchange at the industrial level, were major power (re)sources for ESI unions.

Trade Union Organisation in the ESI

In the UK several trade unions can operate in one workplace or industry, and this is fundamentally different from the Argentinian system, in which there is a legal monopoly of representation. British multi-unionism is often associated with trends towards fragmentation and inter-union competition (Davidson 1993; Ogden 1991, 1993). This is suggestive, for mobilisation theory finds in competition a negative factor for workers' mobilisation. Additionally, multi-unionism implies likely differentials in bargaining power among the unions, stemming from the role occupied within the labour process by their constituencies. EPEA is the case in point in the ESI. By organising the engineers and technical staff, this organisation benefited from the strategic role of their membership in running the power stations, transmission network, and distribution systems; in their hands lay the capacity to bring the industry to a halt.

Additionally, unions in the ESI organise workers across a wide range of economic activities, which results in heterogeneous constituencies, with the partial exception of EPEA (Blyton and Turnbull 2004; Waddington 2003). This heterogeneity is also important for mobilisation theory as far as interest definition and organisational resources are concerned. Recent mergers have accentuated this feature. All the cases under study, EETPU, NALGO and EPEA, have experienced mergers and amalgamations since privatisation as shown in Table 1 (see pages XXX). In an effort to cope with this organisational fragmentation, ESI unions established an umbrella organisation, ESTUC (Electricity Supply Trade Union Council), which, at the time, led the anti-privatisation campaign, the negotiations to modify the Privatisation Bill and the subsequent contacts with the government. However ESTUC had no constitutional power to direct the organising objectives of individual unions, being primarily a mechanism for consulting autonomous unions about joint responses to employers at the Electricity Association and to the government.

The Neo-Liberal Turn: Thatcherism and the Legal Offensive

The macroeconomic meaning of Thatcherism was the replacement of Keynesianism by a neo-liberal agenda devoted to restoring centrality to market forces. In this view, by definition, organised labour hampers market dynamics by demanding more than the market can afford. Therefore, curbing the power of trade unions and reducing to a minimum the neo-corporatist forms of stability were among conservative political priorities (Crouch 2003).

Macroeconomic political exchange through tripartite bodies was gradually abolished. Additionally, there was a clear change in state attitudes as employer, stopping its good industrial relations commitment. In its new exemplar role, the aim was to encourage employers to recover the managerial initiative and pursue flexibility whether at the labour market or the firm level. Lastly, the government restrained its support for collective bargaining, especially encouraging the dismantling of multi-employer and national arrangements (Crouch 2003; Marsh 1992). Although the coverage of bargaining was not dramatically affected, its scope and content changed: a focus on pay began to prevail, together with productivity or flexibility deals in return for higher wages.

Labour law played a decisive role in this crusade. A piecemeal legislative programme (seven major acts and legislation dealing with specific issues) altered the environment of industrial relations interactions by attacking the voluntary system without replacing it with legal regulation. Its main thrust was to weaken trade unions' power, encourage individualism, and enhance employer prerogatives (Dickens and Hall 1995, 2003; Wedderburn 1986). Two aspects of the legislation are of particular interest: the restriction of industrial action and the introduction of statutory regulation of trade unions' internal affairs.

Concerning the former, Dickens and Hall (1995) pointed out five different strategies: restriction of immunities with regard to particular types of industrial action, redefinition of the notion of a trade dispute, introduction of secret ballots to enable immunity for industrial action, making unions liable for unlawful industrial action, and enlargement of the range of potential litigants so as to allow even union members or citizens (deprived of goods or services) to go to court.

Although the extent of the practical outcomes of the legislation is arguable, as this often operates together with other factors such as the business cycle or changing class composition, there seems to be little disagreement about its overall negative consequences (Gall and McKay 1996; Gennard 1998; Kelly 1998; McIlroy 1999). Generally, the deterrent effects of the employment law were associated with the decline of any kind of solidarity action, the educative role of certain large and well-documented strike defeats in which labour law was applied, and the cautious attitudes of union officials due to 'an atmosphere of self-imposed restraint' (Gall and McKay 1996). Moreover, action against privatisation was prohibited, for it would be deemed politically motivated as, although connected with conditions of employment, it was not wholly or mainly about them (McIlroy 1999). Unions' near universal compliance with legislation when taking industrial action renders the legal prosecutions even more meaningful. They have shown that employers are able, by arguing breaches of technicalities, to 'reiterate the limits of permissible industrial action; keep unions under pressure; engage the energies of officers and officials; divert attention from organising effective action; and, where injunctions are granted, dislocate it' (McIlroy 1999: 528–9). The links between changes in legislation and mobilisation theory, particularly its 'positive relationship with counter mobilization', have been explicitly highlighted by McIlroy (1999: 532): 'it can enhance caution and conservatism and dislocate the processes of mobilization by maximising uncertainty, delay and division'.

With regard to the regulation of trade unions' internal affairs, the aim of the legislation was to discourage collectivist behaviour and solidarity and disseminate a conception of trade unions as mere aggregates of individuals (Dickens and Hall 1995). Members' rights vis-à-vis the trade union organisations were enhanced; but the main area of reform was the gradual implementation of fully postal and independently scrutinised ballots to elect union executive committees, presidents and general secretaries, where they had voting rights in decision-making. These reforms, plus the industrial ballots, favoured the equalisation of democracy with individual balloting, and changed the character of union internal democracy at the expense of participative forms (Dickens and Hall 1995; Waddington 2003). These changes were relevant to the dynamics of leadership and the decision-making of the different organisations. Thatcherism particularly affected the framework of industrial relations within the public sector by implementing financial constraints on the nationalised industries, promoting decentralisation, reforming collective bargaining machineries and encouraging a more confrontational management style (Ferner 1989).

Argentina

The Weight of the Law in Collective Bargaining

The industrial relations institutions of the time in Argentina present a contrasting development. The privatisation programme of the 1990s faced a system of industrial relations laid down in the 1940s, in which the law set the mechanism of collective agreements and unions competed legally to obtain workers' representation.

The juridification of industrial relations began when Perón was in charge of the Secretary of Labour of the Military Government of 1943–6, and continued during his democratic Presidencies (1946–55). Previously, the state had seldom participated as mediator when conflicts disrupted the few voluntary collective agreements which existed at the time between employers and unions. In 1945, the state sanctioned a labour law which regulated trade unions' representation by recognising just one organisation able to collectively negotiate agreements and represent workers, at the workplace or in courts, per industrial sector or economic activity, either nationally or regionally (the so-called *personería gremial*). In most cases, when there was regional fragmentation of the representation, a national federation gathered together the regional or local unions to negotiate agreements on their behalf. Until the mid-1990s, the main level of bargaining had been national, reinforcing in this manner the centralisation of the whole system of industrial relations, and generally, comprising whole industries or economic activities (Bunel 1991; González 2001; Novick 2000, 2001). Though rare, some occupational groups had their own organisation. White-collar workers were usually organised together with blue-collars. Yet, on occasion, upper professional or technical grades might have had a separate organisation. Extension mechanisms secured that unions' operation with low membership levels; although union density was uneven but usually high in key sectors and in the state industries. From the beginning, in every case, the *personería gremial* had been given by the Ministry of Labour. Consequently, a breach of law, or a Ministerial arbitration, often put trade unions at risk of losing their *personería gremial.*

In 1946 a national law (upon a former decree of 1944) established that public authorities had to be present and participate in any collective agreement. It empowered public authorities even further by legislating arbitration procedures. In 1953, a more liberal law formally recognised the autonomy of unions and employers, but two key prerogatives of the Ministry of Labour remained: the final approval of the legal status of any agreement (the so-called *homologación*) and the right to suggest modifications, or reject, an agreement. In this sense, any agreement was always just a project presented by unions and employers to the public authorities. Hence employers and unions usually asked for official advice when they faced an issue that might cause its rejection. In 1958, the Minister of Labour obtained by law the right to force conflicting parts to accept the mediation of public authorities (*conciliación obligatoria*). Despite changes in labour legislation, this legal right is still valid. So during the time that the *conciliación obligatoria* is in force, trade unions can not take industrial action of any type, and employers need to desist from implementing decisions which might lead to conflict. For instance, if the employer dismissed workers, they ought to be reinstated until the official resolution of the conflict. When parties ask the Ministry of Labour for conciliation, the authorities decide whether or not to apply it; thus trade unions have no guarantee of getting a *conciliación obligatoria* to fight back dismissals or other unfavourable situations.

Due to frequent *coups d'état*, the right to bargain was often distorted in the period after 1955. Yet, the collective agreements shaped the field of industrial relations because of a legal disposition by which any agreement remains in place until it is formally replaced or repealed (*ultra-actividad*). In sum, up to the 1990s, the bargaining process was centralised, theoretically periodic, and heavily dependent on the state. The basic foundation was the collective agreement, together with the ability of the union to make employers respect it. Given the features of the system, law and political influence intertwined and became contentious power resources.

Trade Unions, Politics and the Public Sector in Argentina

Several factors explain the comparatively bigger political involvement of trade unions in Argentina, and the development of frequently unstable forms of corporatism and tripartite institutions. To some extent, this involvement was embedded in the origins and characteristics of the system succinctly described. Political influence within the state was rendered a (re)source of power of vital importance by the state's direct participation in collective bargaining, its power to weaken or strengthen trade unions through administrative and legal prerogatives and its role as promoter and guarantor of such influence. This institutional design was a component in a wider political project to industrialise the country by applying an import-substitution model. Most sections of the trade union movement enrolled in this project led by *Peronism* in mid-1940s, obtaining labour laws, benefits and organisational power, unthinkable just a few years earlier, but through political rather than industrial mobilisation (Atzeni and Ghigliani 2009).

Additionally, within this politicised environment, the structure and dynamics of collective bargaining meant that unions, employers and governments agreed not only on industrial and productive matters but also on industrial macroeconomic policies. Indeed, beyond industrial negotiations, unions aimed to influence national economic policy in order to meet their demands concerning wages and levels of employment. Thus, trade unions pressured public authorities as much as they pressured capitalists; capitalists and trade unions frequently pressured governments jointly for industrial policies for the economic sector. In addition, and typical of

Keynesian macroeconomic management, governments needed unions to concert industrial peace and social pacts.

Finally, due to the constant instability of Argentinian democracy and the political proscription of *Peronism* between 1955 and 1973, trade unions, usually through the CGT, fulfilled a political role. Every government, democratic or military, was forced to negotiate with the CGT (or repress it). At the time of writing, it has long been common practice that unions with similar political orientations gathered together in temporary alliances (*nucleamientos sindicales*). These alliances expressed their support or opposition to governments, seeking representation in tripartite bodies, disputing the internal alignments of the *Peronist* Party (*Partido Justicialista* – PJ) and struggling for the leadership of the CGT. Within this framework, national strikes played an important role as manifestations of political strength.

As in Britain, unions in the nationalised industries enjoyed specific (re)sources of power which stemmed from the nature of public ownership: the weak exposure to market forces of the public sector, high levels of employment, political influence, and so forth (Goldín 1997).

In short, different factors explain the political involvement of trade unions; any account ought to incorporate as a vital variable the politicised history and features of the system of industrial relations in Argentina, which goes well beyond the peculiarities of the public sector activities.

Industrial Relations in the ESI

ESI industrial relations, at the time of privatisation, were partially shaped around the basis set by the CCTs of 1975, in themselves, outcomes of trade union political mobilisation. A brief historical detour is needed to put the case into perspective.

From 1964 to 1976, by dealing with all semi-democratic, military and democratic governments, ESI unions gradually increased their participation in the running of the public enterprises. First, workers' directors were appointed, then, LyF CF (the biggest ESI union) obtained co-management rights by law for the enterprise SEGBA, and finally, a former General

Secretary of LyF CF was instated as the president of the company. It was within this highly politicised environment that the CCT 78/75 was negotiated, which act ended up as a symbol of the peak of the trade union's influence in the ESI. In 1976 a new *coup d'état* smashed the participation, suspended the collective agreement, and fired 265 workers, mainly trade union officers, lay representatives and activists (among them the ex-president of the enterprise). Between October 1976 and February 1977, a wave of industrial action launched by LyF CF to force the Military Government to resume the practice of political exchange ended with the physical disappearance of a group of leaders, including the General Secretary of the union, and 570 more dismissals.

Once democracy was restored, the CCT was renegotiated, most terms and conditions of employment recognised, but co-management rights removed. Although LyF CF did not recover the degree of control it had had, it still exerted a great influence in every aspect related to organisation and allocation of work, employment levels and recruitment, training and career development, and managed to maintain virtually complete job stability.

In the case of LyF MDP, the axis of the industrial relations was the CCT 36/75, another highly protected agreement, reached in the same year, 1975, by the national Federation, FATLyF, on behalf of the fourteen unions which organised workers in the company Agua y Energía Eléctrica on a regional basis. The terms and conditions of employment were similar to those negotiated by LyF CF, the leading union of the national Federation, but LyF MDP did not enjoy co-management rights, as the latter were achieved in SEGBA by law (not by a CCT). The CCT 36/75 was also suspended by the Military Government; it was recovered untouched by the union during the 1980s. It was the CCT in place at the time of privatisation of the company.

The institutional structures of the system of industrial relations in the ESI at the time of privatisation were rather simple from the beginning. The cornerstone was the CCT negotiated in a Ministerial commission (the so-called *Comisión Paritaria*), a tripartite body by law integrating public authorities, employers and union officials. Lay representatives (*delegados*) bore the responsibility at workplace level, watching out that the employers did not breach the agreement. If this level failed, lay representatives might have called a union officer, and if still no solution was reached, a formal claim was submitted to the most appropriate of four commissions: terms and conditions of employment, safety, training and fringe benefits. The next step should have been to elevate the claim to an ad-hoc commission (*Comisión Resolutiva de Reclamaciones*), chaired by a Ministerial authority with the power to deliver a final decision. The conflicting parties could appeal to the Ministry of Labour.

These commissions were also the place where ordinary consultation was handled. Important matters often involved officers in face-to-face negotiations, for the system allowed room for regular and informal contacts with managers, which was an essential component of the peaceful resolution of conflicts and an additional mechanism of micropolitical exchange. In fact, these contacts permeated the whole structures. Although the system was highly centralised, it depended on the ability of elected lay representatives to make the agreement work. Mass dismissals and repression during the Military Government explain, in part, why the union could not apply the agreement in its whole extension during the 1980s, which led, in turn, to a deterioration of the industrial relations expressed in an increase, though still within moderate boundaries, of industrial action.

Trade Union Organisation in the ESI

In contrast to the UK, trade union organisation in Argentina was based on a legal monopoly of representation and bargaining rights (*personería gremial*). Any organisation of workers could apply for public recognition providing it presented its internal books and other information to labour authorities. However, to replace a union with *personería gremial*, the competing organisation must have demonstrated that it had had during six months, at least 10 per cent more members than the other union. However, an organisation, without negotiating rights and without health cover institutions (*obras sociales*), had minimal chances of affiliating workers. The *obras sociales* were, perhaps, the most salient feature of Argentinian trade unions. They were created at the end of the 1950s and the beginning of the 1960s by means of collective agreements with employers. Both sides contributed to financing the *obras sociales*, but only the trade union ran the administration. In 1970, a new law compelled employers and unions to create *obras sociales*, not only for union members and their families, but also for workers of the same sector, insofar as they contribute to their financial support. A worker, therefore, could contribute to the *obra social* without being a trade union member. In any case, however, the worker contributed to the financial strength of the trade union. This somehow compensated the union against free riders as collective bargaining covered members and non-members. Thus, *obras sociales*, one of the pillars of the country's health system, became a source of power and an axis of political exchange.

Until privatisation, regional electricity unions (44 organisations in total for the whole country) were part of a FATLyF, a federative body. This body was dominated by LyF CF, the biggest ESI union in the country. FATLyF managed the *obra social* for all electricity workers and offered other social benefits like housing building and tourism. Its members were all organisations of blue-collar and white-collar workers, with *personería gremial* over particular geographical areas. In companies like Agua y Energía Eléctrica, DEBA and later on ESEBA, which extended their operations over huge territories, several trade unions had bargaining rights for certain regions. Thus, FATLyF represented all of them at a single negotiating table thereby securing a united front and keeping homogeneity in terms and conditions. Indeed, the agreement CCT 36/75 closed by FATLyF set terms and conditions nationally, except for the Federal Capital and Greater Buenos Aires where LyF CF signed its own, the aforementioned CCT 78/75 for the public company SEGBA. In contrast to the UK, due to the system of personerìa gremial, there was no room for inter-union competition in the ESI in Argentina, though political disputes between factions within FATLyF were common. Privatisation came to exacerbate the latter.

The Neo-Liberal Turn: Menemism and the Legal Offensive

The neo-liberal offensive was, as in Britain, backed by a legal attack against individual labour rights and trade unions. However, unlike Britain, a group of unions exercised pressure on the *Peronist* Government, which had been elected with unions' support, and engaged in a political exchange, with dubious results for the file-and-rank, but important financial rewards for the organisation.

So during this *Peronist* Government's first term in office (1989–95), only eight out of twenty legislative projects to reform labour laws were passed in the Parliament. Over this period, *Peronist* Members of Parliament (MPs) with union backgrounds blocked in a Labour Law Commission any legislative attempt to decentralise the collective bargaining and make labour contracts flexible. Only after 1994, when a corporative pact was agreed, did these MPs stop blocking the projects. All along, in order to introduce change, the government was forced to give exchange protection and financial support to the *obras sociales*, and permit unions to invest in the new business opportunities brought about by the privatisation and deregulation of the health system, pensions, ESOPs, insurance for labour accidents, and the privatisation of public enterprises. As will be shown later, this form of mobilisation of political resources is critical to understanding of LyF CF's response to privatisation.

However, in the face of opposition, the state also exploited, in key conjunctures, its faculty to issue anti-union decrees and repress unions. Accordingly, the right to strike in public services and utilities was restricted in 1990 in the midst of industrial actions against the privatisation of railways and communications. The strikes were declared illegal, unions' *personería gremial* and arrangements whereby subscriptions were deducted from member's pay by employers (check-off) were suspended, military forces were mobilised to run the services, leaders were threatened with penal prosecution, and more than 400 workers were sacked. In 1991, in the context of the Convertibility Plan (*Plan de Convertibilidad* by which the Argentinian *peso* was pegged to the value of the US dollar by law), the government decided to cut wage increases to prevent a new inflationary crisis, and issued a decree linking rises in wages to productivity growth. Unions opposed this decision arguing that it limited actors' autonomy in collective bargaining but its consequences were much deeper. The decrees impacted on the whole structure of collective bargaining by forcing unions to negotiate wages at firm level and to take into account differentials in productivity between companies. They also constrained corporative strategies by precluding demands for governmental wage polices. Trade unions were obliged to discuss with employers how to increase productivity and concede changes in the labour process that they were previously resisting. Nevertheless, trade unions succeeded in keeping a centralised bargaining process by articulating sectoral and local negotiations. So in 1993, another decree instituted bargaining at enterprise level; during 1995–2000, 90 per cent of collective agreements were of that kind. Consequently, national collective bargaining in the private sector ended.

Finally, there was the specific utilisation of law and decrees to bring change to public industrial relations and break public trade unions' capacity to obstruct the managerial decisions of the future private owners. This move was an essential part of the pre-privatisation period, with full involvement of a horde of consultants paid with the WB loans, who were personally involved in negotiations with managers, trade union officers, and authorities from the Ministry of Labour, the Ministry of Economy and the WB (Daireaux et al. 1990; Campaño and Caruso 1991; Campaño et al. 1991). It entailed the suspension of 718 clauses from collective agreements previously reached by trade unions with thirteen public enterprises (Daireaux et al. 1990). The legal foundations were two laws passed immediately after the election of the *Peronist* Government to deal with the economic emergency signalled by the hyperinflation peak, and to begin the neo-liberal reform of the state. Although this aspect will be analysed in detail elsewhere, for it will help us grasp, on the one hand, the centrality of law in Argentinian industrial relations as a political power (re)source, and on the other, the participation of international agencies in the counter mobilisation process.

Conclusion

The comparison confirms the assumption that constitutes the thread of this chapter: the divergence of the systems of industrial relations of these countries correlates with different patterns of trade union power. Although in both cases industrial and political (re)sources of power are intertwined, the axes, upon which the trade unions movements in Britain and Argentina have traditionally built their respective strength, differ. In the former, industrial resources prevails; in the latter, political resources are paramount. Nevertheless, when turning to the ESI, industry and ownership effects bring similarities: being a crucial economic activity granted industrial power to unions; the political environment of public enterprises assured them a degree of political influence.

Differences remains, however, and from the point of view of mobilisation theory, these had consequences for the opportunity structure, and the forms taken by the process of counter-mobilisation and trade unions' responses to it.

It has been argued that the opportunity for trade unions to engage in political exchange with governments varies. While political influence is likely to be an enduring possibility for ESI trade unions in Argentina in democratic periods, it is less so in the UK, where the chances of mobilising political resources seem to be less. For British trade unions, the fate of the opportunity structure appears to be associated, mostly, with industrial dynamics and managerial policies. Finally, as stressed in the conceptual framework of the research, Tilly (1978) finds in competition a debilitating factor for the opportunity structure of subordinate groups; the British system seems particularly vulnerable to this circumstance.

McIlroy (1999) has remarked on the positive relationship in the UK between legislation and counter-mobilisation. This chapter shows that the Argentinian system of industrial relations amplified this connection. In the UK, though important in framing the context of industrial relations interactions, neutralising threats of industrial action and laying down the general structure of the industry, legislation was still a secondary aspect of the process of counter-mobilisation in the ESI. As shall be shown, the truth of the latter seems to have lain in the industrial restructuring that followed privatisation. In Argentina, given the legal underpinnings of trade union power, counter-mobilisation primarily took a legal form in the run-up to ESI privatisation, which paved the way for further industrial restructurings. Legislation was crucial for privatisation success.

Trade unions' responses depend on contextual conditions, and one of their main determinants is the system of industrial relations. In Argentina, the aforementioned centrality of legislation, which impinged on the counter-mobilisation itself, renders law a contentious resource. Hardly ever is an important event of workers' mobilisation exempt of a legal side. Workers' mobilisation manifests itself in this country, always simultaneously, in the streets, in the workplace and in courts. The system probably gives room for political exchange; therefore, unions often mount political strategies to confront industrial troubles. In the UK, these responses are unusual. Unions prefer to concentrate on strengthening their industrial resources. The structures, contents and strategies of industrial negotiation and consultation, together with the level of membership, dominate union policy thinking. Other variables such as organisation, leadership and decision-making play a determining role on how these resources are combined and to what end. For instance, political influence in Argentina is to be found as an outcome of either union cooperation with privatisation and the reform of labour institutions, or coalitions building to halt the privatisation programme and back industrial action. In the UK, industrial resources are pursued either within the framework of partnership agreements or by organising strategies. If these orientations differ politically, the kinds of resources they aspire to build are of a similar nature; for example, recruitment is paramount in both.

To sum up, this chapter discusses, from the point of view of mobilisation theory, differences and similarities regarding trade unions' (re) sources of power, which stem from the divergent natures of the British and the Argentinian systems of industrial relations, and from the political contingency associated with the management of public industries. In the following chapters, it is necessary to bear these themes in mind, as they recur when analysing the empirical findings of the research.

CHAPTER SEVEN

The Anti-Privatisation Campaigns in the Light of Mobilisation Theory

The aim of this chapter is to analyse the relevance of mobilisation theory to the understanding of the forms taken by trade unions' initial responses to privatisation. According to the analytical sequence laid down in the theoretical framework, the chapter focuses on the opportunity-to-act (or opportunity structure). This category, especially in a comparative perspective, requires detailed examination of the contextual variables that condition the exercise of trade union power. In the realm of industrial relations, the opportunity structure is understood to be empirically determined by the general balance of forces between contenders, not only at political but also at industrial level (Kelly 1998); for by definition, counter-mobilisation by the state and by capital is both a political and an industrial process, and hence it is necessary to approach both dimensions.

Although there is considerable interaction between both levels, the starting-point of a privatisation programme is, by and large, politically determined, and depends on the ability of governments to change power relations at the political level, before they are manifest in the industrial domain. Therefore, though judgments about political priorities may turn on evaluation of the industrial position given that both spheres are mutually reinforcing, it may be argued that the crucial factors shaping the opportunity-to-act in the run-up to a privatisation programme are political. If this is true, the analysis of power relations and the power (re)sources of the contenders should express this predominance.

Objective changes in the opportunity-to-act, however, cannot explain by themselves the strategic choices of the actors (Farinetti 2002). These choices depend on how the opportunity structure is subjectively processed.

In this regard, the combination of certain organisational aspects and certain leadership styles are thought to be vital social mediations for workers to collectively define a set of demands and strategies to face opponents and act back upon the opportunity structure itself. Thus, a brief evaluation of appropriate data regarding workers' organisations and union leaderships will follow. This should contribute towards deepening the understanding of trade unions' defensive responses.

The analysis first approaches the UK, and then, Argentina. In the former, ESI unions launched a single anti-privatisation campaign, which constitutes the focus of the analysis. In the latter, instead, it is necessary to separate the cases of LyF CF and LyF MDP as each union confronted different companies, different branches of the state and their strategic choices differed radically. In every case, a brief sketch of trade unions' initial responses to privatisation is offered first. Then, the category opportunity-to-act is considered by evaluating power relations and trade unions (re)sources of power. This analysis is carried out according to the theoretical framework laid down in Chapter Two. It integrates two different approaches. On the one hand, it follows Lukes (2005) in discussing power in relational terms. It may be worthwhile to present a reminder as this will be applied in the next sections: a) the capacity of a party in conflict with another to persuade or force the other to adopt a course of action other than the one it originally intended; b) the capacity of a party to control the agenda of interactions such as meetings, and determine which issues are kept on or off the agenda in the face of opposition; c) the capacity of a party to secure assent to its objectives by another group because of the successful diffusion of a hegemonic ideology. On the other hand, against this background, the analysis draws on Batstone (1988) and Kelly (2005) for exploration of particular aspects of trade unions (re)sources of power when appropriated. Lastly, organisational and leadership data is incorporated into the analysis to illuminate the interaction between the opportunity-to-act and its reading by organised labour.

The United Kingdom

FUSE Campaign

In March 1987, the joint union forum of the industry, ESTUC, through its campaigning arm, the Federation of Unions Supplying Electricity (FUSE), launched a set of activities opposing the privatisation of the electricity industry. The initial objectives of the campaign were to mobilise public opinion against privatising the ESI, and in particular, to contact MPs as well as candidates at the coming general election, to persuade them not to endorse a manifesto commitment by any political party to privatise the industry (EPEA 1988a). In turn, its primary concern was to prevent the breaking up of the generating boards (TUC 1988a, 1988b). Consequently, using the Divisional Electricity Supply Trade Union Councils (DESTUC), a new, untried local organisation, the collective effort was directed towards lobbying MPs – particularly Conservatives from marginal seats – towards raising public awareness of the issues involved by organising local events, and towards influencing opinion formers (see Electricity News, 8 March 1989). From the outset, the electoral focus swung itself away from industrial action, which was discarded as an option.

The announcement that the general election would be held in June, earlier than expected by ESTUC, meant that ESI unions had to take the campaign to the public before FUSE was mature. The election result marked the failure of the first stage of the campaign, as was openly acknowledged at the time: 'It is a matter of record that the campaign was not successful' (EPEA 1988a: 13); or: 'Let us be honest, the FUSE campaign against privatization of the Electricity Supply Industry actually failed on 11 June 1987. That is when the Conservative Party was re-elected with a majority of 101' (EETPU DJIC no. 4).

Strictly speaking, after the national election, the FUSE campaign faded away under the new secretaryship of NALGO, while the other ESI unions withdrew their active support. The leading unions of ESTUC – EPEA and EETPU, and to a lesser degree NALGO – focused on securing changes over the legislation process. ESTUC established a small working party to monitor the Privatisation Bill's progress, and to consider and submit its own amendments. In contrast to FUSE, this second stage of trade unions' strategy was quite successful insofar as it achieved by legislation their main aims: the continuity of pension schemes, the specification of health and safety procedures, and provisionally, the temporal stability of ESI industrial relations machinery. In particular, the issue of pensions has become the icon of the relative success of this pragmatic strategy as expressed in every interview:

> We secured one very important thing from the government, and that was never ever repeated. That was that we had pension schemes written into the act of Parliament. That meant the pension scheme was protected, and that happened in no other industry [...] It was a collective thing. We, the trade unions, secured an agreement, collectively, from the government. The government wrote into the actual act of Parliament that anyone in the industry prior to the 1 of March 1990, who was in the industry privatised that day, would have the pension protected indefinitely. (National Officer – Amicus)

To summarise, the trade unions' initial response to privatisation was characterised by its pragmatism as it was assumed that the privatisation could not be stopped. It was mainly electoral in the beginning. The possibility of industrial action surfaced in early debates and conferences but union leaderships – particularly those of EPEA and the EETPU, and presumably an important part of NALGO's officials too – were convinced that industrial action would be disastrous for the unions. So trade unions aimed at influencing the would-be private structures of the industry, and in particular, taking advantage of the political environment to secure legislative changes in the Privatisation Bill.

The Opportunity-to-Act

Although a complex exercise, the application of Lukes' model (see pages 31–3) may contribute to exploration of how political and industrial spheres interacted and conditioned the opportunity-to-act of ESI trade unions.

Certainly, their inability to even think about stopping the privatisation – despite opposing the sale of the industry – was an expression of an unfavourable balance of power for organised labour. This unfavourable balance forced ESI unions to actively engage in shaping the future industrial structures, and therefore, to positively collaborate, through know-how and industrial peace, with the privatisation programme. Yet, trade unions' ability to use their industrial latent power, or at least, to make the government fear this possibility, proves the divergence between the political and the industrial facets of power relations in the run-up to privatisation. Political strength gave the government and top managers the lead and the industrial lever; whereas latent industrial power enabled trade unions to bargain defensively and politically in a context in which the overt political influence of the labour movement was at a minimum. This divergence coloured the social interactions by which contenders pushed through their aims causing mobilisation of diverse (re)sources of power or causing contenders to abandon or modify certain objectives.

The scope of political influence was defined in Chapter Two as a source of trade unions' power (Batstone 1988). In turn, social movement theorising evaluates the political strength/weaknesses of subordinate actors by taking into account the openness of the political decision-making of the state, the availability of allies, and the stability of the political alignments and the ruling elite (Tarrow 1994). In applying this model to labour studies, Kelly uses the number of contacts between the TUC and the state as an indirect way of measuring the openness of the political structure to trade unions; as he explains: 'this is far from being an ideal source as it reports only TUC contacts, not those of affiliated unions, and given the nature of the source, there is probably some upward bias in the reporting. But if we assume any such bias is constant over time, then it may provide a reasonably valid measure of trends' (Kelly 2005). Kelly quotes research by Holy Marsh (2002) that shows a steady decline in total contacts between 1983 and 1994, suggesting a reduction of unions' political access, and therefore, as already mentioned in the previous paragraph, of the political influence of the labour movement. Additionally, according to David Marsh (1992), contacts were increasingly initiated by the TUC in search of a new realist approach, and not by the government; these involved a move from face-to-face to weaker contact by writing. Finally, unions' political access was significantly less effective in comparison with those of the 1970s (Marsh 1992). Trade unions' political influence within the Labour Party began to be questioned once the association between trade unions and the Party was identified as part of the explanation of Labour's electoral defeats. Moreover, although civil society in Britain affords the union movement many potential allies in its attempts to influence governments, recent research has shown that coalitions of protest are still rare in the UK (Frege, Heery and Turner 2004). Lastly, the Conservative parliamentary majority stopped all chance of any benefit, at least, at the institutional level, from instability in political alignments or divisions in the ruling elite. In short, as recognised early in the EETPU *Shop Stewards Quarterly*: 'If one examines trade unions' attempts to influence government policies, it is clear that we have had virtually no influence at all. Again, evidence of lack of real power' (EETPU Shop Stewards Quarterly Review, 20 May 1983).

Despite this political environment, ESTUC could break the political exclusion, keep regular contacts with the Secretary of the State, and contribute to policy-making. The key factor explaining why electricity unions were able to secure concessions, whereas other unions elsewhere were unable to, was related to their industrial power: 'Clearly, the Government's tactics are to avoid provoking opposition from the staff in the run up to privatization. They know that on these issues [safety, pensions, and industrial machinery] the membership would be prepared to take industrial action to safeguard their interests' (EETPU 1988).

Thus, ESTUC succeeded in reintroducing the transference of the negotiating machinery by law to the new privatised companies, and the new owner's obligation to give twelve months notice to withdraw from it, against the government's intentions to repeal both guarantees. EPEA commented on this: 'It is a limited protection safeguarding the integrity and continuity of the Industry's existing negotiating machinery, but nevertheless an important one in the political context of today' (EPEA 1988a).

This was still, at the time of writing, the overall framework in which those concessions are analysed by union officers:

Question: Why do you think the government accepted to protect the pension scheme through legislation?
Reply: Well the government didn't want [...] The government gave these concessions because they knew that if they attacked the pension scheme, or if they had compulsory redundancies, they would have given the trade unions the weapon to take industrial action. The objective of the government, considering these things, was to spike the trade union guns [...] They could not encourage our members to challenge them. If they had attacked the pension scheme and threatened to throw people out without given good financial settlements, then we would have balloted for industrial action. And they considered spiking our guns in terms of being able to take industrial action. (National Officer – Amicus)

Question: Why did the government give you those concessions?
Reply: Well, remember that the government didn't want conflicts. The government itself was slightly nervous about how powerful the electricity unions were. And they didn't want industrial action. (National Officer – Prospect)

A complete picture of this relative success should include the privatisation process itself (mainly dealing at that time with one major employer prior to restructuring, rather than with several as in the water supply industry) and the political position of ESI unions (that is, their low profile during the 1984–5 Miners' strike). Yet, it seems reasonable to argue that industrial power opened ESI unions' defensive political influence. Then the analysis of Lukes' first dimension of power should be qualified by incorporating unions' capacity to make the government modify their original preferences with regard to the provision of official guarantees matching specific unions' demands. However, it is necessary to stress the defensive nature of this political influence, for unions' strategy was rooted in an explicit acknowledgement of their own weakness that inclined them to take certain courses of action precluding others. In this sense, the overall influence of the political and legal climate was paramount.

The opportunity-to-act was read in the light of other trade unions' experiences, which as a rule were to be interpreted, unequivocally, as signalling that privatisation could not be defeated. For, as it was graphically put by an EETPU delegate: 'Thatcherism and privatisation is the same thing' (EETPU 1988).

> Remember that the electricity industry was privatised along the line. Gas had been privatised, and so on, OK? We had learned from the experiences of the other unions or we thought we had. Because, in some instances, the other unions just opposed what was happening, blanket opposition. We took a further pragmatic view. We said: 'OK, we don't like what is happening. We will try to make the best of our job and there are certain policies that we want to pursue'. (National Officer – Prospect)

> Question: Did the engineers take any kind of industrial action?
> Reply: No, we never had industrial action of any type [...] Originally, we took the decision not to do that, because we would have been challenging the government, and we all saw what had happened to the miners here [...] We decided what we would do was to work within the system and tried to argue to the best of the system, and also not to challenge the philosophy. (National Officer – Prospect)

> Sometimes you run in two ... It's what we call run in two horses. So you got a public policy but if you know you aren't going to win, at least, you make sure that you protect your members, and that was what we've done. (National Officer – Unison)

From the outset, the whole policy of ESI unions was based upon the recognition of the inevitability of privatisation. Bearing this particular point in mind, and leaving aside the moderate character of ESI unions, it is easy to conclude that the negative development of trade unions' resources with regards to governmental support and intervention in disputes and conflicts – chiefly through legislation – conditioned any chance of their resorting to the offensive mobilisation of industrial resources.

By 1987, when the FUSE campaign was launched, the government had already passed three important pieces of legislation directed towards changing the balance of power between unions and employers. Furthermore, during the run-up to privatisation between 1988 and 1990, three new important Employment Acts were sanctioned. Chapter Six approaches the relevance of these changes in law by which the government crucially restricted, through the 1982 Employment Act, the definition of a permissible trade dispute to the terms and conditions of employment, or the physical conditions under which any workers are required to work. Additionally, picketing and secondary action were restricted by the 1980 Employment Act, unions were declared liable for unlawful acts by the 1982 Employment Act, and ballots were required before industrial action by the Trade Union

Act 1984. It could also be added that the government's attempt via the latter to engage the support of moderate trade union members in dissuading leaders from embarking on political campaigns further narrowed the terrain of legitimate union activity to the industrial sphere, extended the range of activities considered political, and therefore incremented the risks of legal challenges to union activities.

In short, by mobilising their political resources into legal reforms, Conservatives targeted a key industrial source of power: unions' ability to disrupt production (Batstone 1988). As underlined by a member of the NEC of the EETPU in 1988: 'At ESTUC's first weekend conference we seriously considered industrial action as an option to prevent privatisation, but the unanimous view of the council was that action would be illegal' (EEPTU 1988).

The same view is still expressed nowadays when considering the options for trade unions in the mid-1980s:

> You have to remember that at the period of privatisation there was Margaret Thatcher. She made it illegal for people to take industrial action for anything except for trade disputes. I mean, it's quite clear that if we had taken an industrial action against privatisation, it would have been deemed illegal, and therefore, we could have been taken to court. (National Officer – Unison)

> At the end of the day, direct action wasn't an option for us because the legislation introduced by Maggie Thatcher said we would be outlawed because it wasn't an issue which we ... So strike ... Unless it was a trade dispute, and did something to your pay and conditions [...] We couldn't do anything. It would have been deemed to be a political issue. (National Officer – Prospect)

On the other hand, at industrial level, managerial support together with procedural and other collective agreements – which according to Kelly and Heery are crucial power resources too (Kelly and Heery 1994) – did not experience a serious deterioration in the run-up to privatisation, basically, for the national negotiating machinery was in place throughout the process. Changes in managerial styles and policies, the scope of consultation, the range of bargaining issues, and even the structure of the industry did occur before vesting day, but they were not dramatic. It is difficult to know if this was the outcome of ESTUC's success in securing written guarantees from the government that the national bargaining machinery would remain in place until one year after vesting day, and therefore, brought about by union campaigning. It could be the case as a withdrawal from the machinery was submitted by management soon afterwards. In any case, though as signs of the challenges ahead, changes at this level added to the sort of evolutionary change experienced, hitherto, for the industry: 'until privatisation, therefore, the story of industrial relations in the 1980s is essentially one of continuity' (Ferner and Colling 1993: 118).

If one turns the attention to the second of Lukes' indicators, that is, control over the agenda of interactions and its contents, a difference between political and industrial spheres is again noticeable. There were frequent meetings with Cecil Parkinson (Secretary of the State for Energy), submissions to the House of Commons Select Committee on Energy, and contacts with MPs (EPEA 1988b; see also, Privatisation News and Contact, several issues). These permitted unions to maintain some control over the interactions and to keep key issues for trade unions' defensive strategy on the agenda: pensions, safety, the legal underpinning for the negotiating machine and shares schemes. For instance, informal contacts with Cecil Parkinson led EPEA in 1987 to prepare a submission about privatisation and competition (EPEA 1988a). During 1988, ESTUC held meetings on a monthly basis in which issues like research and development, regulation of the industry, the break-up of the CEGB and the processes of consultation were discussed (EPEA 1988b). By the end of 1987, the House of Commons Select Committee on Energy had begun an enquiry into the 'structure, regulation and economic consequences of electricity supply in the private sector' (EPEA 1988a). ESTUC sent written evidence on two occasions, and four ESTUC officials gave oral evidence. Due to ESTUC intervention, the Final Report of the Committee published in July 1988 was critical of government's plans as regards the division of the generation structure and the ownership of the grid by the distribution companies. Lastly, the ESTUC small working party monitored the Privatisation Bill's progress by keeping constant liaison with MPs of all Parties as necessary. Through these interactions ESI unions mobilised political resources to politically influence the process of privatisation.

By contrast, the CEGB and Area Boards refused from the outset to discuss and consult with ESTUC about its future structure; unsurprisingly then, the latter was strongly critical of the CEGB indifference to consultation on fundamental issues such as the break-up of the industry. After strong pressure was applied, the relationships at the industrial level improved and mutual confidence was somehow re-established. It was only by the end of 1988 that informal discussions had begun with the CEGB about privatisation (EPEA 1988b). Things were only slightly better in Area Boards, where ESTUC representatives and a number of Chairmen agreed to facilitate exchanges of views and discussions. In this context, top management also reduced consultation with unions about traditional industrial relations issues, but without pushing meaningful changes.

Therefore while trade unions were able to add industrial relations issues like pensions, health and safety, and the negotiating machinery to the political agenda, ESTUC failed to secure their industrial preferences regarding ESI structure (for instance, their opposition to the split up of the CEGB, the transfer of the obligation to supply to the distribution companies, or the ownership of the national grid by the distribution companies).

Finally, the third level of Lukes' analysis of power relations refers to ideological hegemony. When carrying out this analysis, it is essential to first identify the core ideological leitmotifs of the privatisation of the electricity industry. With this in mind, by analysing the White and Green Papers for privatisation of public utilities, some scholars have suggested that their privatisations were often presented within an ideological package comprising: win-win scenarios, praise of consumer identity, and tributes to the benefits of management and competition (Haslam et al. 1996). At this level, the analytical risks lie in the difficulties in disentangling what is evidence of Thatcherism's ideological domination of debate, and what are tactical choices by unions (for example, it might be argued that emphasis upon the aforementioned issues could have been a conscious choice by unions given the power realities they faced rather than mere acceptance).

In a sense, the incorporation into union discourse of the central tenets of privatisation helps to legitimise the dominant discourse and, hence, reinforces its hegemony, even in the case of just a tactical decision. Although findings suggest that gradual discursive acceptance among ESI trade unions of some aspects of those leitmotifs was a real manifestation of the ideological impact of the changes in the opportunity structure.

The role played by the figure of the consumer in the ESTUC campaign is notably conspicuous. The reference to consumers' interests, instead of those of workers and trade unions, was overwhelming. In the charter in which ESTUC publicised their basic agreements when launching the anti-privatisation campaign, while four out of its ten points were related to consumers' concerns, no single explicit mention of workers' interests can be identified (ESTUC 1987; FUSE 1987). Similarly, in a list of seven points of principle agreed after the Conservative re-election, the first two referred to consumers' concerns; although this time, two points made the case for the continuation of the industrial machinery, and the terms and conditions of their members' employment (EPEA 1987). EPEA's presentations to the TUC and other conferences were deemed to show why privatisation would go against consumers' interests (TUC 1989a, 1989c). After the 1987 election, similar rationale led the ESTUC to consider that there was not an obvious campaign to mount and confined FUSE to lobbying: 'It was agreed that if another campaign is necessary it would be based on a clear issue of concern to the public which might emerge' (EPEA 1987). This approach is still vividly evoked:

> In opposing privatisation, we persistently argued our case on the basis of what was best for the consumer and the nation as a whole. But this didn't stop Conservatives, and they accused us time and again of self-interest. Of course, protecting the interests of our members was a primary concern, but we made clear that not at the expense of the consumer. (National Officer – Amicus)

The water anti-privatisation campaign focused remorselessly on consumer interests, including security and quality of supply and the ability and will of a privatised water industry to comply with European standards. Although it was never sufficient to derail privatisation, at key points it had the government on the defensive (O'Connell Davidson 1993; Ogden 1991, 1993). However water unions spent a lot of effort in building coalitions with consumer groups; these proved to be an effective replacement for low workers' engagement, and also an incentive to workers to become politically active. In the ESI, instead, coalition building was absent, a fact that further demoralised the rank and file. Evidence from unions' conferences suggested that FUSE was experienced by lay workers as a remote event and complaints about its features abounded (EETPU 1988; EPEA 1987; 1988; NALGO 1988b). Certainly, if win-win scenarios and managerial tributes were not endorsed during the campaign, the former matched up ideological developments in the EETPU, while the latter resonated with EPEA idiosyncrasy.

At the time, EETPU was pioneering the politics of a win-win type of social partnership, known by its detractors as strike-free agreements, which primarily proliferated in several Japanese firms (Taylor 1985). The leadership of the union promoted this political line as a modern approach that could collectively add greatly to the potential and prospects of private enterprises, and enhance the role and the involvement at work of the individual employee. Hence, the receptiveness to win-win scenarios among blue-collar workers was likely to be facilitated by the ideological battle of the EETPU leadership against the *us and them* culture: As proudly stressed by a former Officer of the EETPU in an interview:

> We created our own style of trade unionism, a coherent alternative to the political one, which has given game, set and match to the government. We created a new type of union, an effective partner to management. (National Officer – Amicus)

EPEA, an engineering union that organised not only frontline engineering staff but also senior ranks of management, basically grouped together the people responsible for running the industry. First, the nature of their constituencies colours its culture and ideological inclination:

> Our members, if you ask someone, they were proud to be an NJB employee or an NJM, and proud of their status. They were recognised to be elitish sort of people. Our constituencies, technical, professional or managerial staff would prefer to stay out of unions rather than being a small part in big unions representing other types of workers. (National Officer – Prospect)

Second, and most important for this analysis, among the senior members there has always been tension between their worker and managerial souls.

> Plainly, at times we have conflict of interest [...] Let's say to recognise that our members have a duty to do as managers, and that they have to do that first. Most of the time, we don't really have problems with the kind of things we have to achieve on their behalf. But, for instance, they often complain when companies want people redundant, but privately they agree with them. (National Officer – Prospect)

This tension manifested itself in the run-up to privatisation, for many of their members in managerial positions shared the managerial wave of seeing economic and professional advantages in freeing themselves from the rigid structures of the national industry. Many pursued their own managerial agenda in between the government's plans and unions' standpoints. This somehow weakened EPEA endeavours, whose leadership engaged in a subtle ideological battle to prevent senior members from making what the union leadership conceived of as misleading projections about the managerial future in a privatised industry. EPEA's readiness to confront a political issue through a technical discussion could be understood as another hint about this managerial thinking.

To summarise, Lukes' model helps identify how political and industrial variables combined to narrow the opportunity-to-act for ESI unions. Yet, it also contributed to identification of how unions acted defensively upon those variables by mobilising political resources anchored in their latent industrial power. However, the analysis of the opportunity-to-act cannot explain in isolation the forms taken by unions' anti-privatisation campaigns. For a better understanding of the latter, the exploration of the categories organisation and interest definition are needed.

Organisation, Decision-Making, Leadership: Multi-Unionism and Pragmatism

The opportunity-to-act is processed by unions through organisation and the social mediations of interest definition. When applying mobilisation theory to industrial relations, the organisational domain is of prime importance. Moreover, trade unions, as secondary organisations, are highly sensitive to changes in the industrial and bargaining structures, and hence, privatisation in itself targeted trade union organisation as shall be discussed in

Chapter Ten. It is also important to acknowledge that organisations have diverse dynamics of decision-making and are commanded by leaderships with different styles and political outlooks. The combination of certain types of organisation, decision-making processes and styles of leadership is assumed to be crucial in accounting for the existence (or inexistence) of workers' collective actions and unions' campaigns and their particular forms. Then it is important to complete this analysis of the FUSE campaign with a few remarks on relevant data about the category organisation and interest definition, mainly, trade unions' leadership styles.

Chapter Six discusses the meaning of British multi-unionism. By definition, it poses serious challenges to workers' unity, and privatisation was not an exception. However, through a coordinating structure, ESTUC, ESI unions attained a high degree of unity and were able to achieve a remarkable degree of unanimity in ESTUC's response to the privatisation programme. Nevertheless, this unity was overdetermined by the institutional framework within which unions agreed common demands and ways of action, and had different meanings for different unions.

Clearly, for NALGO, compromise at the level of ESTUC was to cause relative detachment from its internal process of decision-making, built upon the values of participatory democracy. Being a strong white-collar union, particularly in the old Area Boards, NALGO organised certain groups with significant disruptive potential (mainly billing staff and customer service functions employees), but without the latent industrial power of the manual and engineering unions who were the leading organisations of ESTUC. While accepting the electoral focus promoted by EETPU and EPEA for the FUSE campaign, NALGO's initial involvement in the latter was characterised by the unsuccessful attempt at prioritising coalition building by working with consumers' organisations and other interested groups as they had done in gas, and were attempting to do in water. NALGO made a list of target consumer groups and organisations, established contacts with the water unions' campaign committee, and held a weekend school for campaigners (NALGO 1988c). After the election, when EPEA consolidated its leading role, while agreeing within the ESTUC to support the Electricity Charter drawn by EPEA (ESTUC 1987), NALGO stressed its commitment to the continuation of FUSE: 'We will be pursuing a number of initiatives of our own and we will seek to ensure that members are fully involved in the campaign' (NALGO News, 299, 9 October 1987). Nevertheless, this campaigning determination was soon simply a discursive device to soften internal dissatisfaction like the sort that arose in the Annual Conferences: 'Conference welcomes the united opposition of the FUSE to privatisation. Whilst urging the NEC to work to preserve this unity, Conference emphasises that NALGO's continued support for public ownership is independent of the views of any other union' (NALGO 1988a).

This warning, however, had no chance of being translated into an independent policy. So although NALGO's lay representatives, who were frequently critical of the campaign, urged the NEC to take a leading role, criticised the DESTUC and raised proposals to set up FUSE committees at every workplace, in the ESI's multi-union structure, coalition building was beyond the individual will of a white-collar organisation.

By contrast, given its constituency, EPEA was a powerful organisation embodying the ethos of meritocracy and moderation which had characterised the ESI. By organising the technical, professional and managerial staff, they easily took a leading role when the opportunity structure narrowed to the extent that a technical approach was the only option to be listened to within the rules of the game, and the only political platform from which to negotiate future industrial relations issues. This leading role within the ESTUC would last in their interactions with the state until the end of the 1990s. It crystallised when, in the run-up to privatisation and after, the NEC of EPEA was in charge of every submission to official spheres endorsed by ESTUC (EPEA 1986). It was reinforced when ESTUC established a small working party co-ordinated by EPEA's leadership to monitor the legislative process and liaise with MPs. Suggestively, the secretaryship of FUSE changed, at this very moment, from EPEA to NALGO. Industrial and organisational features contributed towards incitement of ESI unions to accept EPEA's leadership within ESTUC. Regarding industrial features, not only EPEA's strategic role in running the power stations, transmission network and distribution system, but also its technical expertise and the daily authority its members exercised in the industry, buttressed the campaigning replication of ESI internal seniority, once the negotiation moved to a political sphere but under technical clothing. Turning to the organisational features, while NALGO and the EETPU had membership constituencies outside the industry, those of EPEA were almost entirely within the ESI. Its top leadership was one hundred per cent personally involved throughout the process; its NEC was a general headquarters, which permanently fed the ESTUC with ideological and technical arguments in response to every issue coming up during the process.

The case of EETPU is complex. The union was the main organiser of manual and craft workers, with enough strength to disrupt production alone. This threat was to be ritually used before privatisation to settle the NJIC annual agreement; only at that time, EPEA and NALGO negotiated their own agreements. However the right-wing-populist leadership of the EETPU was a moderate force which contributed significantly to the keeping of industrial peace in the ESI, aborted shop floor activism, and working together with management 'achieved improvements in productivity through technological change, the closure of uneconomic plants and [...] flexibility amongst staff' (EETPU 1990). This approach was undertaken through a political move combining organisational changes towards centralisation with an extensive use of formal democracy under the ideological shell of modernisation. It entailed the abolition of area committees and appeal courts, and the empowerment of the NEC to abolish or amalgamate branches and appoint full-time secretaries and organisers. This power was manifested in disciplinary action against activists and opposition candidates, and in the closing of dissident branches (Hyman 1983). Periodical condemnation of these practices, and debates around participation and branch life, went on from the 1971 to the 1983 Biennial Delegate Conferences, when the EETPU leadership finally took over the opposition by beating their proposals to rigidly bind the Executive Council to conference decisions and to elect the union officials. According to the NEC, the opposition intended to undermine the authority of the elected executive and to reduce the political influence exercised by union members through secret ballots. The privatisation of the ESI found a populist leadership in firm control of a centralised organisation but immersed in a serious conflict within the labour movement, leading to its expulsion from the TUC. In this context, for the EETPU, compromise through ESTUC ensured, first, trade union acquiescence and, second, a common stage from which to pursue the election of a Labour Government to reverse privatisation policies. This was, in fact, the only feasible strategy devised by EETPU, which had consistently opposed industrial action to undermine public policies. Indeed, this compromise was also functional in dissipating misleading dreams of industrial action, as the following quotations show:

> I have yet seen little indication from the EPEA that they are likely to take action. NALGO may like to; NALGO are fond of spending a lot of money on lost causes. (EETPU 1988)

> Does Conference believe that against this background and at this late stage we can go to ESTUC, to our members, asking them to support illegal action? (EETPU 1988)

So while the opportunity structure posed objective constraints to ESI unions, the concrete shape of the FUSE campaign had much to do with the nature of trade unions' compromise through ESTUC, which in turn, was closely related to the organisational features of the individual unions, their respective bargaining powers, their mechanisms of decision-making, and the political orientation of their leaderships.

Argentina

The Case of LyF CF: From Latent Resistance to Active Support

The itinerary of LyF CF's response to privatisation shifts from latent resistance to active support. During the 1980s, the union had openly opposed early calls to privatise the industry (Murillo 2001):

> Question: Did the union oppose privatisation?
> Reply: Of course, there was complete opposition! The same people who later on supported it, during 1980s had been completely against privatisation. 'No to the IMF! No to the financial casino! No to contractors!' We participated at that time in rallies, demonstrations, general strikes. (Lay Representative, LyF CF)

In 1989, the Lay Representatives' Body (*Cuerpo General de Delegados* – CGD) of LyF CF decided to support the *Peronist* candidate for the presidential election, who campaigned for a Keynesian plan to overcome the acute political and economic crisis experienced by the country. However the new administration surprised foes and followers by triggering a programme of neo-liberal market reforms, which included vast privatisations of public enterprises.

As a result, LyF CF engaged in a mute opposition as the union leadership believed that the bitter struggles, which erupted in telecommunications, railways and airlines, would end up preventing the privatising wave. As the government opted to narrow the sources of conflict to a few sectors, isolated threats of direct action were enough for LyF CF to keep the status quo in the industry. Concomitantly, the union opened a political channel to the head of the Ministry of Public Works (*Ministerio de Obras Públicas*), who agreed to engage in negotiations to form a mixed company under public control with worker's participation in reactivation of generation and distribution.

In September 1990, a plan of industrial action in response to delayed payment of wages included anti-privatisation slogans. In response, the government threatened to repress the union, as it had done in railways and telecommunications. As a result, the union leadership, convinced of the inevitability of the privatisation, embarked upon a pragmatic strategic shift. Two months later, the General Secretary of the union suddenly declared his support for privatisation of the industry conditioned by union's active involvement in the process (Clarín, 24 November 1990). According to interviews, despite expressing public support to privatisation, the union still had hope in the Parliamentary opposition:

> We were absolutely convinced that Menem would not win the majority in the Parliamentary elections. His neo-liberal policies went against all traditional ideas of *Peronism*. But people kept loyal to PJ and voted the party. (Union Official – LyF CF)

When, in the election of 1991, the government won the majority in the Parliament, union leadership concluded that the only option left for the organisation was to choose between participation and confrontation, and inclined unambiguously for the former.

Even while opposing government temporary contracts and licences, outsourcing or privatisation of the retail and commercial businesses, the union participated in negotiations and the preparation of the international bidding. In the process, the union attempted to secure the unity of the industrial structure, a single collective agreement, the previous definition of a regulatory framework, the continuity of the *obra social* and the implementation of ESOPs. With such defensive objectives in mind, the unions engaged in novel forms of political exchange by which they obtained economic rewards in exchange for concessions regarding labour flexibility and political support.

The Opportunity-to-Act

Any analysis of the opportunity-to-act against privatisation for Argentinian trade unions should take into account the ideological and political consequences of the hyperinflation processes that devastated the economy in 1989–90. Its disciplining effects upon the population have been stressed by various scholars (Campione 2002; Murillo 2001; Sigal and Kessler 1997; Torre 1998), and even compared to those of a dictatorship or a political repression (Anderson 1996; Bonnet 2007). Indeed, Thwaites Rey (2003) argues that both the political terror implanted in the society by the dictatorship (1976–83) and the economic terror of the hyperinflation explained the popular tolerance of the neo-liberal reforms of the beginnings of the 1990s. The economic crisis prepared the terrain for making the population accept the need for a radical change in economic policy and pass the State Reforms and the Economic Emergency laws which launched the political process of market reforms and privatisation. At the same time, it meant the final surrender of the main political parties to the influence and the privatisation recipes of the IMF and the WB. In the specific case of electricity, this general economic crisis was combined with an energy crisis (1988–9), due to not only climatic contingencies but also to the deterioration of the generation plants as a result of the lack of public investment.

The three-fold model of power referred to at the beginning of the chapter shows the subordination of LyF CF to state imperatives. As for the first dimension discussed by Lukes, the most striking finding is the abrupt conversion of LyF CF to privatisation. The government's determined response to union opposition persuaded the leadership of LyF CF to change reluctance into active support, crudely manifested by LyF CF's joining the officialist faction *Menem Presidente* (Sur, 8 November 1990). Thus, the course of action finally taken by the union was a far cry from its traditional opposition to privatisation. It expressed the unfavourable relations of power underpinning the opportunity structure.

Any analysis of union political resources over the process of privatisation that applies Kelly's (2005) methodology presents difficulties as there are no statistics about the number of contacts between the CGT and the government generally, or between the latter and LyF CF specifically. However, the collection of data from national newspapers and the trade union's journal *Dinamis* gives the firm impression that contacts increased or decreased according to the trade unions' political stance. While LyF CF kept a mute resistance to privatisation, information about official contacts and negotiations were rare in these organs. When LyF CF announced their support for privatisation, news about meetings with the Minister of Economy, the Minister of Labour, the Secretary of Energy, the Secretary of Privatisations, the WB consultants and MPs proliferated in both. Informal channels and direct personal relations with the President, Carlos Menem, were common. The latter was part of a wider process of negotiation and political capture, which included personal favours and concessions as part of a renewed political exchange (González and Bosoer 1999). Regarding trade unions' political influence through the PJ, it underwent a process of de-unionisation during 1980s, which, in turn, eliminated a key source of internal opposition to government's neoliberal reforms. As Levitsky stresses: 'Party reformers dismantled *Peronism's* traditional mechanisms of labour participation, and clientelist networks replaced unions as the primary linkage to the working and lower classes. By the early 1990s, the PJ had been transformed from a labour-dominated party into a machine party in which unions were relatively marginal actors.' (Levitsky 2004: 4) This process is also illustrated by the gradual marginalisation of *Peronist* MPs belonging to trade unions (Table 6). Consequently, trade union political influence within the PJ declined steadily just before the reforms began.

Table 6: *Peronist* MPs with Trade Union Background

Parliamentary period	No. of union MPs	% over total MPs
1983–5	35	13.8
1985–7	28	11.0
1987–9	26	10.2
1988–91	23	9.0
1991–3	18	7.0
1993–5	10	3.9

Source: *Nueva Mayoría*, September 1994 – Boletín no. 83

As for coalition building, interviews and sources showed that LyF CF inclined to pursue traditional union repertoires, without the intention of going beyond its constituency. Finally, any potential instability in political alignments at national level was concealed by the context of economic and political crisis which followed the hyperinflation. It seems safe to claim that trade union capacity to mobilise political resources against privatisation was extremely low; the only room for political exchange would prove to be dependent upon union subordination and cooperation.

In addition to this, trade union power resources due to governmental support and intervention in disputes and conflicts vanished. This is of extreme importance given the prerogatives of the state to intervene and shape the processes and outcomes of industrial relations, as shown in Chapter Six. Concurrently, this explains the dependence of trade unions on government and the role of political action in their strategy (Bunel 1991; Goldín 1997; Atzeni and Ghigliani 2009). The PJ was the historical channel by which the government was accessed and its favour gained. Therefore, confrontation of a *Peronist* Government implied the risk of political isolation and, more importantly, breakage of the alliance that had provided unions with political resources and had served their corporatist interests well so far.

As for legislation, two principal facts conditioned unions' ability to act. First, after trade unions challenged the sale of telecommunications and railways, the government issued an anti-strike decree to avoid industrial action in the public sector in the pre-privatisation process, as demanded by the IMF (Thwaites Rey 1999). So as in the British case, the capacity of unions to disrupt production, a key source of union power, was inhibited. Second, the Decree 1757/90 suspended a number of clauses from the CCTs of the public sector which forced unions to enter into negotiations. In this way, managers and supervisors were excluded from collective bargaining, managerial authority upon the organisation of the labour process was reinforced, outsourcing was increased, and financial resources for the union were cut (Campaño and Caruso 1991; Ministerio de Economía 1990a, 1990b). The aim of the decree was to curtail union power and dismantle obstacles to productivity improvements before privatisation in order to increase the appeal for private capitals (Daireaux et al. 1990; Campaño and Caruso 1991; Campaño et al. 1991; Ministerio de Economía 1990a, 1990b). The whole strategy involved the mobilisation of a variety of legal resources, the involvement of different state branches, and a myriad of well-paid consultants financed and technically supervised by the WB and BIRF (Banco Mundial 1991).

This curtailment of union rights, in turn, changed the balance of forces in the workplace in the run-up to privatisation. The unions witnessed the dilution of agreements and support from management, and with that, how power resources at industrial level withered away. This process will be analysed in some detail in Chapter Nine. For now, it is enough to mention that the process began with the State Reform Law which froze wages and ordered a 10 per cent reduction in the workforce of public companies (Ministerio de Economía 1990c). It continued with the appointment of aggressive company Directors who came from the private sector – partly paid for by the WB (El Cronista Comercial, 17 Augustr 1990) – and, principally, with the suspension of clauses from the CCT that changed the rules of the game (Ministerio de Trabajo 1990). Finally, it was epitomised by the decision taken by an empowered management to stop the functioning of the

Comisión Interna de Reclamaciones, the fulcrum of the negotiating machinery in the workplace. LyF CF experienced negative developments in its political and industrial (re)sources of power in the run-up to privatisation, a situation that saw the subordination of the union to official policy.

The analysis of the second dimension of Lukes' model shows that the union's subordination paved the way to union participation in the process of privatisation, though it did not necessarily open the door to shaping the contents of the agenda. LyF CF took part in the Privatisation Commission (*Comisión de Privatizaciones*), and held several meetings with the Minister of Economy, the Minister of Labour, the Secretary of Energy, the Secretary of Privatisations, the WB consultants, and even with President Menem on two occasions. LyF CF failed to secure any of the major part of its objectives: the integration of the industry, the level of employment, the single collective agreement, and the company responsibility to finance the *obra social*. The union achieved generous voluntary retirement packages (a common policy in the restructuring of public enterprises for privatisation) and the ESOP, which was an essential component of LyF CF's propaganda in legitimising its support for privatisation. At industrial level, the union also kept regular contacts with managers and industrial authorities but these were ineffective because of the gradual change in the balance of forces at the workplace and because decision-making about relevant issues was concentrated in higher political spheres.

When considering the third aspect of the model, the need to disentangle tactical choices from ideological subordination presents itself as a problem again. Be that as it may, the discussion illuminates the power of neo-liberal hegemony. In Argentina, privatisations were framed ideologically by the ethos of the sacrifices demanded from workers by a national crisis. The responsibility of public companies in the crisis was stressed, also the lack of public resources to make investments, the chronic incapacity of the state to manage efficiently economic activities, and the need to repay the foreign debt (Cifarelli 1999).

Findings suggest that the leadership of LyF CF played a substantial role in the diffusion of some components of this ideology, though it rejected some others. Their dilemma lay in justifying their strategic shift so as to neutralise opposition from activists. With this in mind, union leaders consciously assumed and reworked, on their own, the ideological leitmotifs of the neo-liberal discourse.

For instance, while refusing the responsibility of the public sector and the legality of the foreign debt, LyF CF stressed time and again the inevitability of reliance upon private capitals for investment in and revitalisation of the industry. The union reminded workers about the good relationships maintained with the private employers before nationalisation, presented privatisation as a service of electricity workers to the nation in crisis, and stressed the lack of alternatives.

LyF CF did not accept that a public company is inefficient by definition, but took efficiency and competitiveness as leitmotifs:

> It didn't matter for us who would own the company. We didn't care about that. But let me tell you something, we do not think that the state cannot run the industry. Anyway, public or private the important thing was that the company should be efficient. We said to the government and workers: 'Let us work for an efficient company to gain competitiveness'. (Union Official – LyF CF)

LyF CF went further by developing its own discursive devices. It presented the ESOP as the fulfilment of the old participative traditions of the organisation to appeal to a distrustful workforce. NEC's communications to lay representatives strengthened the continuity of aims between the old strategies and the new ones:

> With the opportunity to buy shares, and manage them collectively under worker's organisation leadership, Luz y Fuerza really achieves the implementation of its long-awaited philosophy of responsible participation in order to look after the national interest, consumers' interests, and the present and the future of the electricity workers. (Dinamis, April 1991)

Accordingly, the trade union's journal *Dinamis* devoted important space to presenting win-win scenarios stemming from workers' participation in share schemes:

> We hope that the old 'them and us' model in which interests are antagonistic, that is, where capital opposes workers, becomes a new model, where everybody gains if businesses are successfully run. (Dinamis, December 1990)

Although with less enthusiasm, the same line of argument was reaffirmed in 2006 within the ethos of popular capitalism:

> We fought to get shares. We defended the ESOPs. Why? After 1976 the state was very authoritarian and did not allow us to participate. This situation did not change with the democracy either. That is the reason for our support for privatisation. For us, the ESOP meant a new participative style. A new role for workers in popular capitalism. (Union Official – LyF CF)

Overall, the image, after applying Lukes' model to the case of LyF CF, is one of union subordination with the aim of revitalising the traditional channels of political exchange between unions and the state. The novelty rests in the content of the exchange. The union, as in the past, compromises in order to deliver social peace and collaboration, whereas the government denies unions' influence in the design of public policy and just offers business opportunities as compensation. However, this devaluation of the traditional political resources did not alter LyF CF's inclination to neglect the development and mobilisation of industrial (re)sources.

Organisation, Decision-Making, Leadership: the Birth of a New Style of Unionism

Two organisational developments are relevant at this stage of our enquiry about the relationship between the opportunity-to-act and trade unions' internal capabilities. On the one hand, the workplace structures (*comisiones internas*) grew weaker, and with them, the lively internal democracy capable of imposing limits on a bureaucratic leadership. This fact was the outcome, first and foremost, of the military repression between 1976 and 1983 and of the restructuring of the public enterprises just before privatisation. The latter targeted the union, once more, through the massive retirement of the remaining old and experienced lay representatives. In this way, it stopped the slight recovery of workers' organisations that had continued since the end of the military rule.

On the other hand, this pre-privatisation restructuring also meant financial pressure upon the assets and resources accumulated by the organisation to serve its members. LyF CF defended the so-called multiple trade unionism, a strategy intended to go beyond the horizon of union struggles towards the exercise of political influence in the design of public policy, the participation in the running of the industry, and the satisfaction of workers' social needs. Towards fulfilling the latter, the union devoted energies to the supply of workers with cheap houses, hotels, holidays packages, an institute for secondary studies and training, pharmacies, food cooperatives, personal loans, and so forth. These resources were, historically, important levers of power for union leadership. However job loss, and hence a decrease in union membership, and the suspension of employer contributions to the union threatened the very sustainability of an organisation that, just before privatisation, had employed 1100 workers and had signed as employer five different collective agreements (LyF CF 1989).

These organisational factors help to explain the strategic choices of the union's bureaucratic leadership in the face of an opportunity structure that discouraged forms of collective action based on workers' mobilisation. Pragmatism prevailed before the new contents of the political exchange set down by the government. The bureaucracy of the state rejected the traditional mediating role of *Peronist* unions based on their tactics, which combined doses of workers' mobilisation and negotiation. Instead, business opportunities became the only offer made by the political power in exchange for union political and ideological reliability. The eclipse of the internal life of the organisation, and consequently, of a more participative democracy, and the absence of an articulated opposition, plus a huge institutional structure about to collapse due to lack of money, eased the way for the union leadership to accept the new compromise. In this manner, they inaugurated a new style of unionism in Argentina. Some authors have classified it as a local type of business unionism (Palomino 2005). However, the principles of business unionism were already present in Argentina, mostly in the workings of the *obras sociales*. This new style went beyond that towards a truly entrepreneurial unionism, which assumed directly the function of capital in the running of a diverse portfolio of businesses. In further chapters, its components are analysed in detail.

The Case of LyF MDP: Opposition and Community Alliances

In this case, the anti-privatisation campaign arguably was still an on-going process at the time of interviewing (2005–6), for a new campaign asking for the private concession obtained by EDEA in 1997 to be rescinded was launched by the union in the beginning of 2005. It began in 1987, two years before the accession of Menem, when a combative leadership won the internal election. Since then, the campaign included a vast repertoire of collective actions against the process of privatisation; this process took ten years, partly as the result of workers' opposition. There were three stages of particular conflict during this period.

During the first stage, between July 1987 and August 1990, the union fought the project to divide DEBA – the provincial generation and distribution company covering the Province of Buenos Aires – into three functional firms as the first step towards its privatisation: ESEBA SA, ESEBA Distribución and ESEBA Generación. Over three years, the union launched several campaigns, often based on the democratic vote of the membership through general assemblies, that included widespread propaganda, the exercise of political pressure over the provincial and industrial authorities, and even industrial action. Alongside the campaigns, the union attempted to get support from the CGT and also from FATLyF, which ultimately ended up supporting the privatisation. In addition, the leadership sought to forge alliances with the trade unions operating in other geographical areas covered by the enterprise and with local sections of public unions. Indeed, by the end of this period, LyF MDP had deepened the orientation towards coalition building by making the first call to community organisations, universities and political parties to join the struggle. The union differed from LyF CF by opposing the Emergency and State Reform Laws (8 de Octubre, no. 36, 31 October 1989); the union also differed from LyF CF in opposing the anti-strike decree one year later (see 8 de Octubre, several issues).

After a relative period of peace, the second stage began in May 1992 when the provincial government announced the privatisation of the company. In this context, the union confronted rationalisation and restructuring plans mounting a sustained defence against attacks on three fronts.

The main front was the attack on the collective agreement, which counted on the connivance of FATLyF. The rationale of this assault was the managerial search for control over the labour process and the establishment of flexibility in the workplace. Another front was the downsizing of the workforce through outsourcing, voluntary and early retirements, and compulsory redundancies. The third front was constituted by anti-union policies, and even derecognition. With defence against these attacks in mind, union officials widened their tactical repertoire, which included: all sorts of industrial action, legal submissions, workers' and popular demonstrations, workers' and popular rallies, silent rallies, political alliances with local authorities and local sections of political parties, sitting protest demonstrations in front of the municipality, hunger strikes and fasts, blackouts to show community solidarity, the gathering of 70,000 solidarity signatures, organisation of popular musical concerts, and even the putting up of a tent in front of the company for seven months and eleven days. In all cases, the union forged alliances with community organisations and other local unions. Most crucially, the union promoted during this stage, together with other unions mainly from the public sector, the formation of a new summit organisation called the *Central de Trabajadores de la Argentina* (CTA) to rival the pro-reform CGT (see 8 de Octubre, several issues; LyF MDP 1995b).

The third stage ran between September 1996 – when the union asked the provincial and municipal governments to implement a popular referendum, and July 1997 – when the company was finally privatised. Over this period, the anti-privatisation campaign was again the main focus of LyF MDP, which deepened its political alliances with community organisations in opposition to the sale, and organised a popular assembly when the authorities finally rejected the referendum. Simultaneously, the organisation continued to oppose early and voluntary retirement, and decided to confront compulsory redundancies; every member inclined to accept the offer had to face expulsion by decisions taken in general assemblies. Calls for industrial action completed the picture (8 de Octubre, several issues; LyF MDP 1997a, 1997b).

The Opportunity-to-Act

The political and ideological environment faced by LyF MDP was similar to that confronted by LyF CF; however, there are important peculiarities to note. Argentina is politically structured as a federation of provinces; privatisation developed initially at national level; the provincial government was not in the same hurry to sell off the provincial public enterprises. Similarly, the interference of the IMF and the WB with the design of public policy was less at provincial level, given that the provincial foreign debt was much less. Just as the union succeeded in postponing the sale, although the positive image of trade unions continued to be very low (Nueva Mayoría 1997a), popular support for privatisation began to fade away following other traumatic experiences (Nueva Mayoría 1997b; see Table 7).

Table 7: Percentages of Positive Image of Trade Unions and Privatisation Policy (Opinion Polls)

Year	Trade unions	Privatisation
1987	19	28
1988	16	29
1989	21	35
1990	11	45
1991	8	36
1992	8	36
1993	7	28
1994	7	30
1995	6	28
1996	6	22
1997	7	18

Source: *Nueva Mayoría* 1997a, 1997b

For instance, lack of support persuaded the provincial administration to set down in the privatisation bill a minimum level for employment in the would-be private company in order to avoid massive downsizing. Crucially, the union leadership was not politically involved with *Peronism*, although most union officials belonged to the *Peronist* movement. Together with organisational and agency type variables these factors are relevant to an understanding of the virulence of the anti-privatisation campaigns in this case.

This time, Lukes' model appears as failing to grasp the phenomenon in its whole meaning.

As for the first face of power (see Lukes' model in Chapter One), while the union's opposition to privatisation was unsuccessful, insofar as it did not prevent or reverse it, union campaigns delayed privatisation for years. The argument that the union was either persuaded or forced to adopt a course of action other than the one it originally intended seems insufficient and possibly untrue. The union fought privatisation fiercely all along the way, even after its implementation, by alternating its repertoire of tactics, partly as a result of external pressures stemming from governmental interventions and managerial attacks. It was finally defeated, and therefore the effective privatisation of the industry indicates that power relations were not favourable to workers. However the political direction chosen by the union suffered no alteration, and ESI workers engaged in collective actions until the last minute to stop it. They were still campaigning for the re-nationalisation of the company some ten years after privatisation. Indeed, in 2006 the union launched a national campaign for the re-nationalisation of the energy industry.

As stated earlier, Kelly's model for analysis of the openness of the political decision-making of the state presents difficulties in the Argentinian case. Nevertheless, it is possible to infer some basic tendencies from the collected data about the political dimensions of the (re)sources of power of LyF MDP (Batstone 1988).

Contacts with the provincial government were scarce and often related to situations of conflict, given the complete union opposition to privatisation. There were neither personal channels between union leadership and political authorities nor union affiliation to political parties.

Instead, the union intended to take advantage of differences in provincial and municipal political alignments. The municipal government never assumed an open anti-privatisation stance but local councillors voted for statements of support for the union during its conflict with FATLyF in 1993, and during conflicts with ESEBA related to outsourcing policies in 1994. However, in 1996 the local council voted against the project organising a privatisation referendum which was submitted to the municipal Legislative body by the union. Yet in 1997, just before privatisation, the same body agreed to organise an Open Session in which issues and information about the privatisation of ESEBA were debated, and trade union officials and other representatives of the community had the opportunity to explain the reasons for their opposition. This action was important in popularising union's political views and consolidating community alliances. So LyF MDP's ability to exercise political influence was also small as in the case of LyF CF; but the important thing is that even this small capacity stemmed from a different logic. The union was able to replace the repertoire of traditional political resources that was in decline with a new one: the support and mobilisation of communitarian organisations.

Trade union power resources which depended on governmental support and intervention in disputes and conflicts evaporated. The same adverse conditions outlined above with regard to governmental labour policies and legislation applied to LyF MDP. As for the risk of repression, as time passed, it faded away; in fact, despite acute confrontations the provincial government did not use open violence to defeat the trade union's resistance. Perhaps this was partly due to the richness of the union's tactical repertoire. The union shifted constantly between different types of collective actions, some of them very powerful in terms of symbolism (hunger strikes, for instance), but without damaging production or service levels. In these cases, the political cost of repression could have been enormous for the government. Moreover, the union often reassured the basic provision of minimum standards of power, thus avoiding open defiance of the anti-strike legislation, although the union did not actually comply with the law.

As a result, public authorities and top management attacked procedural and other collective deals in order to undermine the industrial dimensions of union power (re)sources. These counter-mobilising moves included macho-management tactics (authoritarian management tactics), a parallel union and continuous harassment through administrative and legal means. The most striking episode was the recognition by the Ministry of Labour of a collective agreement signed by FATLyF on behalf of the electrical workers in 1992 which replaced the old agreement of mid-1970s with a flexible one preparing the company for privatisation and weakening trade union power in the workplace. This was a hallmark of the connivance between the government and a corrupted faction of the trade union leadership showing a flagrant disregard for the law. Moreover, the spread of political extortion and anti-union tactics was the inevitable corollary. It is possible to list several examples. In 1994, ESEBA replied to union opposition to outsourcing by imposing sanctions on 500 workers and firing twenty-three. After that, the management made a proposal exchanging the compulsory redundancies for trade union voluntary acceptance of the collective agreement signed by FATLyF which replaced the CCT 36/75. In addition, ESEBA used financial penalties to encourage defections from LyF MDP by refusing to pay 120 Argentine pesos in food tickets to those retaining membership. Given that the average wage was 700 pesos per month, and the additional food payment was given to members of the breakaway union, the incentive was significant. Moreover, officials of the breakaway union benefited from paid holidays when carrying out their representative duties, despite the lack of *personería gremial* of the organisation. The story of the policy of early and voluntary retirement is another example of bullying and repression. Threats, home letters, compulsory transfers to other cities, and dismissals likely to be exchanged later for better redundancy packages, were the background of this policy; among the victims were many lay representatives. In 1995, ESEBA failed even to comply with the *conciliación obligatoria* decided by the Ministry of Labour during a labour conflict, confirming 193 dismissals of members of LyF MDP, among them, once more, lay representatives. The company suspended 60 workers soon after, this time attempting to include members of the NEC. At the peak of the confrontation, there was an arson attack on trade union premises, verbal assaults on NEC's members, threats to their relatives, and even a drive-by shooting of the house of a trade union official. Methods of this kind have been very persuasive in deterring social mobilisation in Argentina after the last dictatorship. Shortly, the analysis of power (re)sources at this level will reveal the most negative changes, despite unions' ability to mobilise workers through industrial action. Yet, it is important to keep in mind that the provincial government together with the provincial and the national branches of the Ministry of Labour shared responsibility for this situation, as they tolerated and even encouraged most management decisions despite their illegal character.

Regarding the second analytical variable of Lukes' model, which relates to who is in control of the political agenda and how the latter is defined, union opposition reduced to a minimum the field of interactions. In fact, there was no real agenda of negotiations around privatisation. Negotiations came out during conflicts but not as a part of a concrete agenda, which was absent in the run up to privatisation. By rejecting the idea that privatisation was unavoidable, LyF MDP rejected pragmatism, the common landscape among public unions facing privatisation (Orlansky and Orciani 1994). Thus, the union negotiated neither the future structures of the industry nor compensation for workers. Instead, LyF MDP mobilised industrial action, community alliances or legal resources to confront every decision contrary to their interests.

The analysis of the third dimension of power shows the union's continuous ideological struggle to combat the diffusion of the idea of privatisation among workers. LyF MDP performed an ideological counter-hegemonic role that was very successful among the rank and file in ensuring that union members did not give their assent to the sale of public assets. Union officials combated the association between privatisation and win-win scenarios in the public discourse by pointing to the failure of previous privatisations. They also argued bitterly against the notion that public companies were inefficient, framing it as purposely pursued to justify their sale to private capital. Moreover, union leadership claimed the social role of public utilities and refused to measure their efficiency just in terms of competitiveness. Finally, instead of the abstract praise of the consumer identity, LyF MDP articulated common policies with community groups. The most important achievement of this alliance was the promulgation of the TEIS Law

(*Tarifa de Electricidad de Interés Social* – Electrical Tariff of Social Interest) in 2002 for households that could not afford the provision of electricity. The ideological impact of this type of action was significant.

In this case, a fierce inter-union dispute was a salient factor shaping the opportunity structure faced by the union. While LyF MDP embarked on an intransigent anti-privatisation campaign, the national Federation zigzagged between soft opposition and opportunistic political support. When the latter finally opted for the entrepreneurial strategy, the rupture was unavoidable. The first open conflict had appeared when FATLyF decided in 1989 to support the transformation of DEBA in ESEBA. Then, political differences blew up again when FATLyF refused to consider an alternative negotiation proposal prepared by several unions under the leadership of LyF MDP. However the key event was the process of negotiation initiated by FATLyF and ESEBA in 1991 closing a new collective agreement for the company. LyF MDP first complained to the Ministry of Labour and presented a legal claim before the Courts. Simultaneously, the union leadership organised a political faction within the Federation, which gained the support of nineteen out of forty-four members, but the experience finished in January of 1993, when the leadership of FATLyF expelled LyF MDP from the Federation and dismantled the antagonist faction. Soon after, the Federation took over the control of the *obra social* of LyF MDP, an essential power resource of Argentinian trade unions. This dispute ended in 1994, when FATLyF finally reached a new agreement which replaced the old and protective CCT 36/75; in this way, LyF MDP prerogatives at workplace level were limited. Then the last step was the contribution of FATLyF towards formation of a parallel organisation that challenged the position of LyF MDP, though it failed to gain terrain amongst the workforce. If it never posed a threat to LyF MDP's hegemony, this new organisation together with FATLyF functioned to legitimise several of the illegal actions taken by the company and the Ministry of Labour. By action or omission, the provincial government backed this strategy. This inter-union dispute forced LyF MDP to spend a lot of time and resources on overcoming its negative effects. Essentially, it allowed management to negotiate a new agreement, which undermined union power.

Summarising, the determination of the provincial administration to carry out the privatisation programme, the empowerment of managerial prerogatives, anti-unionism, and inter-union rivalry are variables that speak of an unfavourable opportunity structure for LyF MDP. Yet some other aspects appear to be a bit more ambiguous. Tensions among the municipal, provincial and national levels of the state apparatus let the union get access to local representative bodies. The lesser involvement of the IMF and the WB in the privatisation of the provincial company meant less pressure upon the administration to sell the company quickly. Union opposition was successful in delaying privatisation; concomitantly, popular support for privatisation decreased, easing the field for the development of coalition building as an alternative political resource, which contributed to widening the repertoire of action of the union. Lastly, the union did not form part of the PJ, and consequently, it did not have to demonstrate loyalty to the government. However the capacity to exploit these ambiguities depended, partly, on organisational features and the skills, styles and political objectives of the leadership.

Organisation, Decision-Making, Leadership: Towards Social-Unionism

When considering Lyf MDP, there are four organisational differences between it and LyF CF, which may be of relevance to understanding of the variability of their responses. First, the smaller size of LyF MDP seems to have facilitated wider rank and file participation. Second, the organisation, in contrast to LyF CF, did not suffer the same loss of experienced lay representatives. On the one hand, repression during the last dictatorship was softer in the case of LyF MDP; on the other, the strong opposition of the organisation to voluntary redundancies prevented the exodus of militants during the pre-privatisation process. Third, the organisational resources servicing members were small. In this sense, LyF MDP neither had financial pressures to sustain a huge structure when privatisation policies impacted on its budget nor owned assets likely to be invested in the new businesses. Servicing members was not a key legitimating apparatus for its leadership. For the provision of services to membership the organisation had strongly depended on FATLyF until it left the national Federation. However, most of the unions operating in the same public company (fourteen in total) shared similar features, but chose to accept privatisation and supported ideologically, politically and economically the entrepreneurial unionism of FATLyF.

A relevant difference lies in the style of leadership and in the dynamics of the process of decision-making. The essential features of LyF MDP since 1987 have been the defence of a participatory democracy instead of a formal representative one, the promotion of lay representatives bodies, the development of communication channels for the rank and file, the amount and quality of information to be passed on to members, and strategic coalition building within and beyond the trade union movement. The union leadership devoted time and effort to encouragement of participation, and when the latter occasionally fell, its NEC implemented special policies and propaganda to increase its level. For instance, during 1990, the growing absence of lay representatives from meetings troubled the union, which escalated the problem by different channels and eventually sanctioned some of the absentees. Similarly, when the attendance at assemblies declined, the union applied the rule book in order to discipline members. As a result of these efforts, the internal life of the organisation was livelier than the internal life of the average public union in Argentina. A comparison between the number of general and lay representatives' assemblies held by LyF MDP and LyF CF over the period under study illustrates the latter (see Table 8).

Table 8: Mass and Lay Reps Meetings in LyF MDP and LyF CF

Trade union	Ly F MDP		Ly F CF	
Type of assembly	General assemblies	Lay reps assemblies	General assemblies	Lay reps assemblies
1990	5	5	–	–
1991	10	6	–	–
1992	14	8	–	–
1993	3	6	–	–
1994	14	20	0	1
1995	6(1)	4(1)	0	1
1996	7(2)	5(2)	0	1
1997	5	4	–	–
1998	2	3	0	2
1999	5	–	0	2
2000	4	–	0	3
2001	3	2	–	–
2002	4	6	0	3

(–) No data
(1) Data between 1 January 1995 to 30 April 1995
(2) Data between 1 May 1996 to 31 December 1996

This data reveals the commitment of LyF MDP to a more democratic process of decision-making, which spurred the mobilisation against privatisation. During the interviews, union officials stressed the importance of the interaction between leadership and democratic decisions of rank and file, which has supported every strategic choice along the anti-privatisation campaign and beyond. Information and mutual trust seem to have been key factors:

> The thing is that [...] people were very well-informed; although information is not enough. People saw our dedication to the organisation. We could do things better or worse, but they did not doubt our utter dedication to the cause. And what we anticipated was shown in practice. (Union Official – LyF MDP – Official)

> There is trust [...] Most fellows do not share the political orientation of the leadership, but they know that they will neither betray nor negotiate for their own interest. Let's say that there is mutual trust and that is why you will find discipline among workers. (Lay Representative – LyF MDP)

A basic aim of the leadership of LyF MDP has been the supercession of the traditional model of Argentinian unionism by a new social unionism, autonomous from the political bureaucracy of the PJ and the state, oriented to the community and combative. To achieve this objective the union orchestrated a political strategy, backed by the mobilisation of the rank and file, which comprised a wide repertoire of collective actions, coalition building within and outside the labour movement, the extensive use of legal resources, and the provision of counter-hegemonic information. Those tactics crystallised, eventually, in the formation of the CTA, a new peak organisation to dispute the leadership of the Argentinian labour movement.

Conclusion

By putting forward a conceptual framework in which it is possible to link internal and external conditions that constrain trade unions' strategies and workers' collectivism, mobilisation theory proves to be useful when analysing the forms taken by the anti-privatisation campaigns under study. Furthermore, application of the theory within a comparative research design seems to enrich its explanatory scope.

In general terms, the comparative analysis shows that, during the process of privatisation, adverse changes in opportunity structure due to political and legal factors were not automatically translated into effective changes within the realm of production. This was a heterogeneous development in scope and timing. In this regard, differences between the British and the Argentinian experience suggest, for instance, that the role of governments in changing power relations at the micropolitical level of industrial relations varied according to the degree of external pressures behind the programme of privatisation, and the extent to which the micropolitics of industrial relations are expressed in legislation and formal rules, in which the state plays a central role. Similarly, comparison shows that the mobilisation of political and industrial dimensions of union power resources might vary according to these same institutional variables, but with the internal dynamics of union decision-making and leadership styles.

ESI unions from both countries experienced legal offensives to thwart their ability to disrupt production. Yet, the evolutionary change of the organisation of labour and the industrial relations arrangements in the UK allowed unions to maintain their industrial (re)sources of power, and hence, industrial latent power. This helped British ESI unions to gain defensive political influence within a context of political retreat, though, ironically, this influence failed to shape the agenda of negotiations with regard to the future structures of the industry. On the contrary, in Argentina, public authorities used their industrial relations prerogatives to undermine the industrial power of ESI unions. In this manner, they prevented unions from mobilising industrial resources defensively to oppose the governmental decision to change the scope and content of the characteristic political exchange of the Argentinian system. The latter had guaranteed trade unions the ability to mobilise their influence to doom government's policies to failure, including programmes of privatisation during the 1980s. As privatisation firstly needed to be imposed politically upon the public agenda, this capacity had to be eliminated.

It is stressed in the theoretical framework that trade unions are not only conditioned by their external circumstances but also seek to shape them, according to how they interpret the opportunity structure. This in turn depends on organisational aspects and agency type variables such as leadership styles, workers' participation and decision-making processes which are the social mediations comprised by the category interest. In this regard, the case of LyF MDP is paramount, as a combative leadership was able to confront privatisation through participatory processes of decision-making, which sustained workers' collective action. The union chose not to accept the inevitability of privatisation, and therefore, did not enter into negotiations aimed at achieving the best possible outcome. This union leadership refused to engage in any form of political exchange, either at micro or macro level. A corollary of that choice was that the organisation was forced to replace traditional resources and to find other ideological and political resources to keep members' cohesion over a long and open confrontation. Community alliances were the key substitutes developed by LyF MDP during the process. In the main, this case reveals that different political choices changed the hierarchy and content of union resources. While other organisations neglected coalition building, LyF MDP devoted huge energies towards achieving this aim. Yet, the case of Unison also shows that it is not only a matter of political choice, but also of organisational possibilities. For while coalition building appears to have been among the aims of several union officials of this organisation, who wanted to replicate what trade unions had done in water or gas, ESTUC precluded the evolution of this strategic orientation. The organisational variable also contributes to explaining, partly, the particular choices of LyF CF, as the weight of its accumulated assets and its organisational structures to service members created the material possibilities for the development of a commercial strategy with the favour of the public authorities.

The analysis also shows that political contacts tied unions into a privatising agenda, and hence they reinforced unions' preferences for the mobilisation of political resources whereas they reduced the likelihood of the collective mobilisation of workers. Yet, if power is defined as the ability to achieve outcomes, in the end both alternatives were unsuccessful in preventing privatisation. The first one still permitted British unions, for instance, to secure protection for pensions, whereas the second one delayed privatisation for years in the case of LyF MDP. As for the factors tending to encourage workers' collective actions, the combination of Lukes' model and the comparative approach seems to suggest that an important one was the struggle against the ideological assumptions which backed the privatisation programme.

Finally, the comparative approach gives credit to McIlroy's (1999) insight about how far adverse legislation, and other negative institutional and political developments, may support full-time officials hostile to forms of collective action based upon rank and file mobilisation. EETPU and LyF CF offer the possibility of making the case for this interpretation, as their officials discouraged industrial action by pointing constantly to external limitations. This fact may add to a more subtle analysis of mobilisation and counter-mobilisation by including not only capital and state demobilising strategies, but also the old problem of the role of trade unions in reinforcing capital and state hegemony.

CHAPTER EIGHT

The Forms of Counter-Mobilisation: The Politics of Money

Mobilisation theory has been applied in Chapter Seven to cast light on the anti-privatisation campaigns following in full the sequence set in the analytical framework: first, the analysis of the opportunity-to-act; second, the evaluation of the prospects of mobilisation by assessing the categories organisation and interest. This chapter, instead, analyses a particular aspect of the process of counter-mobilisation: the variety of policies by which workers and union officials were bought off through money incentives. Basically, these policies were intended to counteract discontent during the privatisation process and after, when private companies began to introduce change in the realm of production. It can be argued that the aim was to dampen the sense of injustice (amongst members and negotiators) by providing some incentives for existing employees; these comprised a wide range of initiatives from those with financial consequences for individuals and unions to those with ideological impacts on workers and union leaderships. As stressed by an English union official: 'When the companies were given the power, they wanted to change certain things, and the only way they could change it was by giving people more money'. (National officer – Amicus)

The politics of money was pursued not only by managers after privatisation, but also by public authorities before it. Its most salient aspect, shared by both countries, was the use of generous redundancy packages to downsize the workforce with the minimum of industrial conflict. While in Argentina this policy involved the state and international financial institutions before privatisation, in the UK the bulk of the process began after it. Similarly, ESOPs – the so-called PPP (*Programas de Propiedad Participada*) in Argentina – were implemented in both countries, as sweeteners to reduce union opposition. Yet, sources of variability may be found in the different employment laws, labour market conditions, union structures and governments. These sources also determined country specific phenomena like the use of personal contracts in the UK, or the emergence of an entrepreneurial trade unionism, built around the new business opportunities opened to the union leadership in exchange for social peace and ideological support in Argentina.

Regarding the theoretical and conceptual objective of this chapter, the point is to investigate the multiple consequences of a particular counter-mobilising policy, the mobilisation of money resources, so as to clarify the relationships between counter-mobilisation and change in the categories of opportunity-to-act, organisation and interest definition. The chapter focuses on four main dimensions of the politics of money: voluntary redundancies, shares schemes, personal contracts and business compensations.

The Politics of Voluntary Redundancies

The main consequence of the ESI privatisation in the UK and Argentina was a dramatic job loss. The figures available, though difficult to achieve for the whole period, are revealing. Between 1990–1 and 1995–6, 46,480 employees had left the industry in the UK, that is a reduction of 32.6 per cent (see Table 9).

Table 9: Job Loss in the Electricity Industry between 1990–1 and 1995–6 (United Kingdom)

Company	Workforce						Variation 1990–1/ 1995–6
	1990–1	1991–2	1992–3	1993–4	1994–5	1995–6	
Eastern Electricity	10,001	9,877	8,415	7,003	6,403	6,113	−38.9
East Midlands Electricity	7,382	8,243	8,684	7,914	6,458	5,051	−31.6

Hydro-Electric	3,484	3,480	3,494	3,552	3,584	3,435	−1.4
London Electricity	6,691	6,581	6,258	5,532	4,908	4,404	−34.2
Manweb	5,483	4,623	4,533	4,604	4,582	3,303	−39.8
Midlands Electricity	7,729	7,643	7,370	6,207	5,815	5,114	−33.8
National Power	15,713	13,277	9,934	6,955	5,447	4,848	−69.1
NIE	–	–	3,851	3,536	3,035	2,826	–
NGC	6,550	6,217	5,666	5,127	4,871	4,565	−30.3
Northern Electric	5,528	5,364	4,826	4,714	4,456	3,882	−29.8
NORWEB	8,203	7,917	7,977	7,652	7,617	8,196	−0.1
Nuclear Electricity	13,924	13,300	12,283	10,728	9,426	8,815	−36.7
PowerGen	8,840	7,771	5,715	4,782	4,171	4,148	−53.1
Scottish Nuclear	1,976	2,047	2,172	2,060	1,860	1,704	−13.8
Scottish Power	9,848	9,495	8,724	7,778	8,041	8,036	−18.4
SEEB	6,340	6,257	6,039	5,339	4,680	4,278	−32.5
Southern Electric	8,362	8,340	7,642	7,391	7,091	6,728	−19.5
South Wales Electricity	3,767	3,632	3,166	3,350	3,218	2,979	−20.9
South Western Electricity	5,676	5,553	5,569	5,092	4,656	3,424	−39.7
Yorkshire Electricity	7,126	7,105	6,850	57,643	4,924	4,294	−39.7
Total	142,623	136,722	129,168	115,080	105,243	96,143	−32.6

Source: Centre for the Study of Regulated Industries, United Kingdom

In Argentina, the figures show a similar story. Between 1990 and 1993, 4,499 workers left the industry, that is, a reduction of 28.4 per cent of the workforce (see Table 10). However, this rate is misleading, for the state had already begun a process of rationalisation through voluntary redundancy programmes. At the beginning of 1990, SEGBA employed 22,451 workers; when the company was transferred to the new owners, 6,645 of those had gone by means of an early retirement programme. Thus, if we take the end of 1989 as a reference, the total workforce was almost halved in four years.

Table 10: Job Loss over the First Year of Privatisation (Argentina)

Company	1992 (privatisation)	1993	Variation
Central Puerto	1,115	798	−28.4
Central Costanera	795	661	−16.8
Edenor	6,443	4,164	−35.3
Edesur	6,529	5,051	−22.6
Central Dock Sud	75	60	−22.0
Central Pedro de Mendoza	59	31	−47.5
Edelap	741	542	−26.8
SEGBA residual	59	–	–
Total	15,806	11,307	−28.4

Source: Luca (1998)

The main forces underlying job reductions included rationalisation, technological change (particularly in the case of the UK, as generating companies switched capacity from coal to smaller, gas fired plants, which were more automated), organisational change, and transfer of maintenance functions to external contractors (particularly in Argentina). While some of these aspects might have occurred without privatisation, the latter provided an opportunity for pushing through downsizing measures.

Such job cuts demanded the mobilisation of large financial resources in order to avoid conflict; governments and companies used money to overcome resistance. Only against this background of generous redundancy packages is it possible to understand workers' passivity before the devastation of the employment in the electricity industries.

In the UK, ESI unions overtly opposed compulsory redundancies. Traditionally, electricity workers had never suffered compulsory job losses due to the collective bargaining strength of the unions. Moderate numbers of redundancies in the ESI were usually made voluntarily, as its staff benefited from long service and large pensions at the time of retirement. Yet unions apparently did not foresee the size of the job loss to come owing to the combined pressures of regulatory policies, city expectations and the post-privatisation plans of rationalisation. Therefore, in their eagerness to contribute to a peaceful transition, trade unions considered severance packages to be acceptable whenever workers found them to be generous. In this context, companies used their large money stocks in conjunction with the large profits they were making post-privatisation to fund the pensions of the time and generate selective voluntary severances. In sum, good financial packages coupled with protected pensions – ironically, a key bargaining objective of ESI unions – ensured the companies had possession of a powerful device for downsizing the workforce.

> Question: How many members did the union lose as a consequence of privatisation?
> Reply: Well, as a consequence of privatisation … You are talking about 40,000 … Or maybe more, 50,000 people have left the industry. They have left the industry – I have to say – almost without exception with very good financial packages. There have been no strikes, or very few strikes or confrontations, because the companies continue to offer generous redundancies payments. And also the pension scheme, that meant that you could retire with no loss of benefits at the age of fifty. (National Officer – Amicus)

Moreover, many of those who left one company continued to work elsewhere within the industry, and even within the same firm as in the extreme case of PowerGen, where almost half of 1,000 redundancies returned as short-term contractors by an arrangement included in their severance packages.

The pace of job loss ended up weakening the morale of the workforce, and hence, trade unions' ability to mobilise members into collective actions to stop a policy they themselves had accepted in the beginning. It is possible to argue that the politics of voluntary redundancies led to the development of a redundancy culture among the workforce; that is, a favourable attitude towards voluntary severances as a shortcut to obtaining financial benefits. Qualitative data from interviews supports this assertion:

> The members were demoralised, demoralised. They wanted to get the money and go. Put the past behind them. Leave the industry. (National Officer – Amicus)

> Reply: So, this company immediately announced 2,500 job losses in quite really attractive terms. They were oversubscribed, right? Three times, near everybody in the company.
> Question: Everybody wanted to leave?
> Reply: It is not an exaggeration. There was a massive, massive queue of people wanted out. (National Officer – Unison)

> People came at the time and asked me to get voluntary redundancies. I personally refused to negotiate under such terms. Even active union members [...] just wanting to get good money to leave somewhere else. (National Officer – Prospect)

When, immediately after privatisation, massive redundancies began to hit the financial and organisational strength of trade unions, the officialdom found in this cultural attitude an additional barrier to the organisation of workers against this policy. A former NALGO official depicted the counter-mobilising effects of money as follows:

> Buy [workers'] cooperation. If the company has to announce 3,000 job losses, redundancy terms: 'Well, this is what is on offer. You are going to get between one years' and two years' wages. You get a lump sum, and you can put into a pension scheme [...] You can get an immediate pension now without any loss in benefits'. They actually know that they can take the redundancy, and get a job in a related field, with someone else, the next week. They have no problem. And they got a big lump sum [...] pay the mortgage or whatever. And the management know that even if we want to fight things – because at the time you wanted to fight these things – I knew that the blokes would be split probably fifty-fifty. 'We want you to fight it, we are not going to fight anything, we want severances'. (National Officer – Unison)

The development of a redundancy culture was just one aspect. In addition to subjective consequences, massive job loss depleted trade unions of lay representatives. Workplace structures were seriously damaged, for a lot of experienced union representatives and activists left the industry through severance packages. This trend meant an objective lost in the midst of the growing fragmentation of the industry and the negotiating structures, which, in turn, was putting new pressures upon workplace capabilities:

> Among the people who left the industry, there was a lot of our senior reps. So, overnight, which is lost, it's lots and lots of experienced reps. And at the time, we looked at it, and thought: 'Well, we're dead'. Because, it is a paradox. If you don't get any reps who can organise things to face within the industry all these new structures, and new businesses, and new companies, which began to be set in place, what? (National Officer – Unison)

> But, it was really tough at the time, and we lost a lot of the good people as a result of the redundancies that were taking place. And of course, you know, we can condemn redundancies but on an individual basis a lot of them were men in their fifties, the company offers them a financial package, and they said: 'Yeah, I'm going to take it!' It was a kind of bribery. So we lost a lot of union reps and activists. (National Officer – Amicus)

This policy encouraged in the ESI an extreme manifestation of more general inclinations within the UK: to seek compensation for job loss rather than to contest it. Some critics and union officials have blamed the line of legislation stemming back to the Redundancy Payments Act 1965, explicitly thought to ease industrial restructuring and reduce the scope for mobilisation against job loss by unions (Clegg 1972; Turnbull 1988; Turnbull and Wass 1997).

In the case of Argentina, the research finds the existence of an identical counter-mobilising policy. Still, similarities in management strategies in a different social and institutional context led to differences in implementation and success, for workers very soon confronted it. Since then, the politics of voluntary redundancies was a formal shell that concealed managerial harassment and persecution. It also counted on the benevolence of the Ministry of Labour, which gave legal cover to managerial practices by means of its prerogatives over the employment relationship. As in the previous chapter, it is necessary to differentiate between the two cases: LyF CF and LyF MDP.

The government aimed at restructuring SEGBA to prepare the company for privatisation. In this way, the state shared the social and economic costs with the private companies, which, in turn, followed identical policy later on. As a result, LyF CF had to face voluntary redundancy programmes before privatisation:

> There was a first clear-out before selling the Terms and Conditions documents for the privatisation binding. There was a first wave between 1991 and 1992, when privatisation began. At that time the public company offered 1.4 [monthly] wages per each year of service to workers wanting to leave. Then, there was a second wave from 1992 until 1996. You may have around 14,000 workers that left the industry till today with voluntary retirement. That is, over the years, 14,000 workers from the former SEGBA have gone. (National Officer – LyF CF)

This statement illustrates the general picture, but hides a gradual change in the workers' mood due to the growth of the unemployment rate. This mood is importantly different from that in the UK, where workers had reasonable prospects of re-employment during privatisation. In Argentina, in the beginning, not only was the offer of voluntary redundancy programmes usually oversubscribed but also many workers asked union representatives to arrange severance packages for them:

> In the beginning, all blokes who left the industry thought they were going to do wonderful things with the money: set up small businesses, buy a house or a car, open a kiosk. This debilitated the organisation. We tried to convince the lads, to tell them: 'Please, don't leave. Stay, we shall resist'. Impossible! 'You stay, I go', they said to us. Until 1994, we were like that, losing our voices screaming: 'Don't leave, don't leave, don't leave'. (National Officer – LyF CF)

However, this beginning of a redundancy culture stopped with the rise of unemployment. After 1994, as stressed by scholars studying other privatisations (Ramírez 1999; Goldín 1997), nobody wanted to leave the industry:

> After 1994–5, nobody wanted to leave; everybody had opened their eyes. People saw what had happened with those who left the industry, either they had spent the money or their businesses had failed. And they were again coming to ask for jobs. Uneducated people. Where were they going to find a job? How would they compete? You don't have studies, new technologies, everything changing, they could not compete. They were now unemployed, coming to the union asking for jobs, and there was no work, you know, if companies were still firing people! (Lay Representative – LyF CF)

When workers began to reject the severance packages, companies began raising their amounts. Before privatisation, those made redundant received an average allowance of US$9,912, which was 10 per cent higher than that required by law. The total amount paid out by SEGBA was US$55.5 million. After privatisation, the average received as compensation by employees who were being made redundant was up to three times the pre-privatisation payment. So voluntary retirement packages were up to 300 per cent more than a severance deal that followed the labour law. While in the UK, statutory terms constitute a pretty meagre minimum which is nearly always substantially improved in unionised settings, companies in Argentina pay strictly the amount specified by law. Such severance packages were therefore very attractive. Nevertheless, the politics of money continued to show limits as a counter-mobilising strategy. As a result, management shifted towards harassment and repression to force people into voluntary redundancy programmes.

All interviewees, from national officials to lay representatives, referred to repressive tactics. The reported evidence paralleled evidence collected from other privatised companies by scholars and for legal prosecutions. Indeed, telephone companies, for instance, were denounced to the Parliament by trade unions for such repressive practices (Ramírez 1999; Expte. no. 27,849/02). By the mid-1990s, ESI managers had begun to communicate with those who were no longer in company plans, recommending those affected to join redundancy programmes. If workers refused the offer, management deployed a different range of tactics involving transfers or changes in working times and tasks or both. The latter included downgrading of skilled workers and upgrading of unskilled workers to posts where their skills were insufficient for carrying out the job:

> I took training courses to work as junior programmer, but they sent me out to read meters! What for? Why did they invest in training courses? It was not only me. Everybody got the same, and even more in the administrative sections. They kept my wage up but started to compel me to join a severance package. And I couldn't be out there reading meters! (Former ESI worker – LyF CF)

> The people didn't have training for these tasks ... They had been administrative for years, and suddenly found themselves having to work in the street, meter reading, even maintenance! Then, the company started to ask most of them to leave because they were doing other grades' tasks. Unbelievable! (Lay Representative – LyF CF)

In extreme cases, when workers continued to refuse the invitation to join redundancy programmes, tougher psychological tactics were applied. For instance, reluctant workers were located in empty offices where they had no job to do:

> It was a policy of fear; sow fear and mistreat workers psychologically. And it was not just a matter of swearing at workers. Look, to give you an example, in Central Puerto, workers were brought to an office where they were left without work. Chairs, a couple of desks, a telephone, nothing else. One year! Maybe somebody cannot understand but for a worker to be sat down the whole day with nothing to do ... Every single day! (Lay Representative – LyF CF)

A variation of this psychological pressure was that workers had to face managerial meddling in their private lives. For instance, interviewees manifested that the Human Resources Department used to ring or send letters to families of reluctant workers letting them know that the company had offered him/her an important sum of money; that the trade union had convinced him/her to resist; and that therefore the company would have to fire him/her sooner or later.

This hard line lasted until 1997 when another approach emerged. To cut costs, companies had been outsourcing different services. The soon to be redundant workers began to be encouraged by managers to form cooperatives providing the service to be outsourced to the company as new owners. Hundreds of workers invested their severance packages in the formation of cooperatives, usually under the technical advice of the company and later on, the trade union too, which negotiated with the companies the sale of the service. Most of these cooperatives went bankrupt within few years (LyF CF 2000, 2001, 2002, 2003).

Another peculiarity of the politics of money in Argentina was that it was backed by the WB, the BIRF and the Ministry of Labour. The two former gave loans to the state to finance programmes of retirement before privatisation, and participated in the design of the programmes (Banco Mundial 1991). The state, in turn, funded part of companies' severance programmes once privatised. In addition to this, the Ministry of Labour gave institutional support to the process by authorising the voluntary retirement programmes (*homologación*). This official legitimating practice continued even when the politics of money only worked by putting pressures upon the workforce through repressive tactics. The meaning of this official backing was that workers were unlikely to go to court to denounce unfair practices by employers.

Finally, as in the UK, massive redundancies undermined trade union power. Membership loss obviously implied financial loss. The most relevant thing was that job loss had a direct impact upon the workplace structures of the union:

> I'm talking to you about all those things in a funny manner, but it was demolishing for us. And when a union lay rep left, even more demolishing [...] And they also left, a lot of lay reps left, and also many activists. The companies bribed many activists. Sometimes, they got two times more money than rank and file workers. Exactly the same thing they tried to do with others, and me: companies offered a lot of money to hard line union reps. (Lay Representative – LyF CF)

In order to analyse the case of LyF MDP, it is first necessary to recall the peculiarities of the opportunity structure confronted by this union. In this sense, two facts are relevant regarding the politics of voluntary redundancies. On the one hand, the provincial government neither enjoyed the benefits of the WB's loans nor its technical support. On the other, as the successful opposition of LyF MDP delayed the privatisation, the social costs of earlier sales within the industry, and beyond, became clearer for the workforce reinforcing rank and file decision to confront the sale of the industry. So there was never enthusiasm for voluntary redundancy programmes among electricity workers in Mar del Plata. This attitude was strengthened by the union leadership who framed these programmes as unfair practices and as hidden dismissals of workers:

> Question: Did people join voluntary retirements?
> Reply: Well, yes, there were people, but not many because we rejected it as hidden compulsory redundancies.
> Question: And what did you do? Did you campaign?
> Reply: Yes. We didn't accept it. We were not like other unions that accepted the voluntary retirement. No, we opposed it. (National Officer – LyF MDP)

When by the end of 1992, ESEBA launched a voluntary retirement programme, the company found out that the overwhelming majority of workers rejected it. Company's reaction was to persecute dozens of workers, among them, members of the CGA (*Comisión General Administrativa*, equivalent to the NEC), lay representatives and activists, by threatening them with dismissal if they refused redundancy schemes. The union replied by taking industrial action in the workplaces where voluntary programmes were implemented. Since that time, the union has fought back every attempt to introduce voluntary redundancies, though with variable consequences. The company, in turn, deployed a set of unfair and repressive tactics to force workers to accept special severance packages. Important conflicts around this policy marked every single year from 1992 to 1997.

The description of the practices of the private companies that took over the business in Buenos Aires applies to Mar del Plata. Yet, an important difference is that the company was still under public management when harassment of workers began. Another difference is that, given the opposition of LyF MDP, the company, from the beginning, extorted jobs from workers by placing them in the dilemma of having to choose between either accepting a voluntary retirement package or facing dismissal. Consequently, many workers joined the programmes despite the union's open opposition. Those who left the industry under these circumstances were punished with expulsion from the organisation, a decision often taken by a general assembly.

A critical event illustrates the unfairness and harshness of company's tactics. In 1995, four members of the CGA (including the second person in the union hierarchy), eight lay union representatives and four well-known activists were bribed with generous severance packages – almost three times the usual offer – and left the organisation together:

> They could not privatise the enterprise. We stopped for four years the privatisation of ESEBA with our struggle. So, to break the opposition the company bought off part of our CGA, US$640,000, US$160,000 each one, and twelve union lay reps, all of them members of our workplace organisation [in fact eight union lay reps and four activists (see 8 de Octubre, 126, 11 March 1996)]. The company realised, well, the company and the provincial government realised, that they could not defeat us. We were resisting, and they had already dismissed, unfairly, six of our lay reps. So the company chose to buy off union reps. (Union Officer – LyF MDP)

In spite of the repressive and unfair tactics by which ESEBA attempted to force workers into the voluntary retirement, it is important to underline that between 1992 and 1997, only 36.7 per cent of the total job loss was due to redundancy programmes (LyF MDP *Memoria y Balance*, several issues); this is, in fact, a sign of the partial failure of the politics of money to avoid conflict. The remaining job loss is explained by normal retirement, early retirement programmes, some compulsory redundancies and transferral of employees to other public jobs outside the ESI industry that were mainly in the local councils.

A factor that helps to explain this partial failure is that LyF MDP leadership took full advantage of the opportunities to frame the politics of money under adversarial terms:

> We were warning people about things [...] Honestly, I would say we pioneered many campaigns. For instance, when we talked to fellow lads about voluntary retirements, I'm not going to say that they laughed at us but they looked at you as if they were wondering: 'What is this guy talking to me about?'. And we use to tell them: 'Watch out, it is happening in other places'. We tried to generate antibodies, and there was those who believed and those who didn't, those who prepared themselves and those who didn't. (Lay representative – LyF MDP)

> We have been saying to fellows in every opportunity in the face of every new scheme of voluntary redundancies: 'Don't join the voluntary retirement. Think! Every single job we left meant more work for the rest of us. This let the company introduce precarisation and flexibilisation. Keep with us and fight'. (Union Officer – LyF MDP)

At this point, it is worth directing the attention towards the role of the leadership of LyF CF too. The evidence is mixed, yet it is possible to affirm as a general point that union officials did not embark on an ideological struggle against voluntary retirement programmes. It is also possible to stress the absence of initiatives like those of LyF MDP to punish members who joined the programmes. Nevertheless, the union took industrial action in 1994 and organised a demonstration in front of EDESUR to denounce the persecution from management forcing workers into redundancy programmes. However, while many lay representatives opposed voluntary retirements at workplace level, others encouraged people to leave and negotiate the conditions:

> To be honest, it must be said that in some places, union reps used to tell workers: 'Sign up, ask for five or ten thousand more. They'll give you the money'. And if you asked the company, you got the money. (Lay representative – LyF CF)

> In the case of Costanera, at least, there were two clear-outs, voluntary retirements, often voluntary retirements agreed with the union, it is true. What did the company do? They sent you to an office, every day, without work. And then, union reps came and said to you: 'Come on, guys, don't bother anymore! Get out!'. The company won because workers got tired. It is true. It is painful, but it is true. (Lay Representative – LyF CF)

Besides the data offered by interviews, other sources confirm that LyF CF reached and signed agreements on voluntary retirement with the *homologación* of the Ministry of Labour, hence, legitimising the rationale of the politics of voluntary redundancies (LyF CF 1992, 1993, 1994). LyF MDP, instead, consistently refused to legitimise such practices throughout the process.

In conclusion, the politics of voluntary redundancy had two effects. Not only did it affect power relations as understood in Lukes' model, by cementing ideological understandings and expectations, and hence diminishing the opportunity for trade unions to act, but it also undermined the organisational strength of trade unions. Moreover, as shown by the analysis of Argentina, when a determined leadership opposed it, special severance packages were mobilised to divide this opposition by bribing leaders. All these factors together distorted the social processes of interest formation. Yet when intermediate variables precluded its effectiveness – as in the case of Argentina when unemployment skyrocketed – the mobilisation of money resources through voluntary redundancies covered and intertwined with repression.

The Politics of Shares

Scholars have shown that ESOPs were used to ease acceptance of privatisation (Goldín 1997; Murillo 2001; Pendleton, Wilson and Wright 1998; Saunders and Harris 1994). This literature has illuminated different dimensions of this preventive policy against workers' mobilisation. It points to the financial benefits and also to the intention to refocus a sense of employee identity (around the company rather than the industry) and employee interests (around performance and competitive advantage). It even stated the links between ESOPs and the Conservative ideology of popular capitalism. In any case, while it is difficult to measure their effects, evidence suggests that ESOPs played a counter-mobilising role in the privatisation of the ESI in the UK and Argentina.

In the case of the UK, the issue of shares seems to have impacted differently depending on the union. Overall, the contextual conditions made ESOPs appealing for staff, for companies went into flotation with low valuations in order to ensure full share subscriptions; consequently, share prices usually rose immediately afterwards, offering substantial benefits to those that sold. Qualitative findings from interviews suggest that managerial staff approached the issue as a possible long-term investment, while manual workers opted for quick benefits; but union officials were unable to provide any figures to support this claim. Nevertheless, for EPEA it was an important matter from the beginning which occupied the attention of its membership and union officials: '[we] have little doubt that the effect of free and discounted shares will prove generally popular with our members'

(Privatisation News, no. 2, October 1988). EETPU paid less importance to ESOP schemes in its journal and conferences. Its approach was that the scheme should be 'free from management influence and able to provide membership with effective voice within all sections of the industry [...] according to union long term aim of increasing participation' (EETPU 1988). Hence, EETPU stressed that a collective trust was needed to ensure an independent voice. For Unison, rather surprisingly, the topic went almost unnoticed in the organisation's press and conferences.

As EPEA led the pre-privatisation negotiations, ESTUC lobbied continuously to get the best possible share ownership arrangement for its membership. EPEA channelled its line of argument through ESTUC: that, despite their opposition to privatisation, the unions were protecting members' interests by asking for a share scheme, that an ESOP would not be compensation for any detrimental effects stemming from privatisation, and that the governmental offer should be comparatively better than previous offers to other unions experiencing privatisation (Privatisation News, several issues).

Aside from the pragmatic approach, former EPEA officials have little doubt about the rationale of the politics of shares:

> There have always been in this country share scheme initiatives. As long as I can remember companies used to reward in some instances their employees through the issue of shares. Now, one of the ways the government wanted to gain the consent of the workforce was to give them financial incentives in the success of privatisation. And they followed the path that they had adopted in other privatisations and they gave the staff the opportunity to own shares. (National Officer – Prospect)

In addition to this short-term aim of reducing workers' opposition to privatisation, long-term aims were also part of the rationale of the politics of shares. For instance, it was mentioned earlier that in order to gain public support, a strong ideological campaign was developed around common ideological motives. These were principally that public enterprises were inefficient and ineffective because of their insulation from market forces and bankruptcy, and that public industrial relations needed to be transformed to undermine over-mighty trade unions (Pendleton and Winterton 1993). Yet, together with these criticisms, a positive argument was deployed: the building of a popular capitalism through spreading share ownership (Ernst 1994; Saunders and Harris 1994). There was, in John Moore's words, the birth of a 'property-owning democracy'. In the Conservative Party's 1987 election manifesto such an objective had already emerged clearly as the historic transformation of British society through widespread share ownership. The expected outcome in the realm of production was that this trend would increase workers' commitment to the companies. Whilst it would be imprudent to conclude that from then on, workers' loyalty to the companies effectively increased, it is important to underline this ideological side of the politics of money:

> We were entering into what they called share-owning democracy, and therefore, how you incentivise the employee to be identified with the new owner. So what do you do? You give a share in the company by giving workers shares or enable them to buy shares. And that was the philosophical basis behind the whole privatisation project […] all those incentives for the staff. Well, the idea was that you link, or the philosophy was that you then translate your allegiance as an employee to the financial success of your company. (National Officer – Prospect)

In addition to the effective fulfilment of this ideological aim, what interviewees' testimonies suggest is that, at the very least, the politics of shares sweetened privatisation, because workers invested heavily in ESOPs:

> Reply: So there were further incentives, cash incentives all along the line that conform to the conservative idea. Workers in the electricity industry were given special incentives to buy shares. The government introduced a saving scheme through a building society. And workers were encouraged to deduct or allow deductions from their pay to buy shares on some future day and this was put into a society fund, a saving scheme.
> Question: And was this successful in terms of workers' adherence?
> Reply: Oh, yes, yes, it was. Most workers actually invested and if you bought 500 you got a thousand because you knew the number was double because you got free shares. (National Officer – Prospect)

In the case of Argentina, PPPs were included in the State Reform Law of 1989 to ease privatisation. These involved the chance for workers to have a formal representative (usually a union representative) on the Board of Directors; a situation that did not generally happen as a consequence of

ESOPs in the UK. Again differently from the UK, unions effectively had the means to become institutional investors. These features amplified the influence of PPPs on trade union's and workers' attitudes, at least in the case of LyF CF:

> In our case, when we knew about the PPP, the first thing we did was to fully involve ourselves in getting a PPP, because a lot of unions were unable to get it. We were very busy trying to implement the share scheme. (Union Officer – LyF CF)

LyF CF developed a conscious strategy towards controlling the PPPs in order to gain influence within the companies by accessing business information through worker representatives. Additionally, it was also thought of as a way of financing trade unions' social services:

> You have 10 per cent of each company [...] I believe that this gives you power. The idea was to gather the 10 per cent of shares of each company in our Loan Cooperative, to make a saving stock, a market within the cooperative to gather all workers' shares. Then, every time a worker wanted to get the money, the union could have bought them the shares. And use them to finance services, to build houses, and obviously, to put workers' reps on the board of directors of each company. (Union Officer – LyF CF)

The CGA deployed considerable efforts and resources in publicising the PPPs as an opportunity to achieve the old union objective of workers' participation. As a result, the politics of shares also had an important ideological impact upon union activism. Simultaneously, the rank and file appeared to have simply assumed the PPP was a kind of compensation for the expected effects of privatisation. In any case, over 90 per cent of workers signed up to the programme (see Dinamis, several issues). Once the PPPs were implemented, most workers sought to sell their shares as soon as possible in order to receive cash:

> People wanted to sell, sell, sell ... You know ... It is easy, you see, at worst, one peso. For each worker it can be almost 30,000 pesos. Some may cancel a house mortgage, or buy a car, or sort out a health problem, or whatever. (National Officer – LyF CF)

> What do you think? When are you going to get 45,000 dollars? Where do you get it? That is why workers from both Central Puerto and Costanera sold their shares to the Chilean owners very early. We couldn't stop it. (Lay Representative – LyF CF)

This clash of interests between the organisation and its members illustrates the counter-mobilising success of the policy. As whether to buy or sell shares was an individual decision, the trade union ended up losing control of the acts of its rank and file. In certain cases, the situation led to open opposition between the leadership of the union and the membership:

> In some companies, workers did not want the union managing the PPP. I was lay rep in Costanera. Workers over there didn't allow the union to represent them on the Board of Directors. They just didn't want the union. There were many conflicts. Workers finally chose their authorities by voting between two options: one from the union and an independent one. And the latter won. (Lay Representative – LyF CF)

More importantly, the conflicts around the PPP should be understood as expressing not also workers' lack of confidence in the union leadership, but also the disruption of the internal life of the organisation due to the counter-mobilising effects of money.

On the contrary, the leadership of LyF MDP rejected the PPP, though the organisation attempted to use it, later on, to increase union control upon EDEA. There were again some differences between the national and provincial PPPs, which are of importance for comparative purposes:

> Back to the differences between the Governor of Buenos Aires and the President [...] The former said: 'Gentlemen, we are not going to waste workers' savings as in the national PPPs. We are not going to facilitate the business of the few who can buy workers' shares soon after privatisation'. So the provincial government introduced changes in the scheme. (Union Lawyer – LyF MDP)

The interviewee refers to the Provincial Law 11,771, by which workers received their shares individually only at the time of retirement. They were not allowed to sell the shares in advance. Until the shares were given away, the companies kept the dividends as payment for the shares (as in the case of the privatisation of SEGBA). The implementation of the programme should have taken no longer than nine months after July 1997, and workers would choose their representatives to the Board of Directors.

Union's determined refusal to enter into discussions about different alternatives with regard to privatisation meant, in fact, that LyF MDP did not consider the potential of the programme from the point of view of workers:

> So, what happened? Well, that there is a right, and I shall say, the fellows didn't see this possibility, that there is a workers' right to access company information through their participation on the Board, minority participation of course, but participation. Obviously, the only thing one can do with 10 per cent of shares is to say: 'We don't want this, we don't want that, we don't want this'. But I can also go to the Justice and say: 'I don't want it because this and that, and I want you to make an enquiry'. (Union Lawyer – LyF MDP)

The financial side of the politics of shares failed to attract the union leadership of LyF MDP, and the union, initially, despised the PPP for ideological reasons. Nevertheless, EDEA took a preventive decision securing the absence of LyF MDP's representatives from the Board of Directors. The company colluded with FATLyF to appoint pro-company union representatives. Since then, LyF MDP campaigned against this usurpation by publicising this situation to the community, by taking the case to court and by mobilising workers in order to increase the pressure upon the company and the public authorities.

It is possible to detect a clear counter-mobilising rationale behind this particular policy, in this case, sweetening the acceptance of privatisation through money incentives. Again, the same policy shows multiple sides with potential effects, in particular on power relations (due to the ideological message attached to it), and also on the social processes of interest definition and the strategic orientation of union leaders (as shown by the cases of LyF CF and LyF MDP).

Peculiarities of the UK: the Politics of Personal Contracts

Once more, differences in legal frameworks are sources of variability. The notion of a personal contract is more powerful in a legal system based substantially on common law as in the case of Britain, where the individual contract of employment acts as the conduit for most employment rights. In Argentina, instead, unions were able to set stricter legal limits on individual contracts. The diffusion of personal contracts to eschew collective bargaining and rights to pre-determine pay rises in favour of performance management systems was a relevant dimension of the politics of money in the UK (Evans and Hudson 1993). Companies were able to take out of collective bargaining powerful groups of staff with consequences not only for them of course, but also for those that continued to negotiate. Unsurprisingly at this stage of the analysis, the counter-mobilising effects of this policy have been multiple.

To begin with, emphasis has been commonly placed on the role of personal contracts in furthering loyalty to the company by shifting managerial identification away from trade unions and in sending a message to managers as to their new position as employers. This has been borne out by interviewees:

> The idea was to divorce senior and middle managers from strong support to the trade union to strong support to the company. And the way they did it was by getting them away from relying on the union to negotiate their salaries, to deal on the one to one basis. (National Officer – Prospect)

In fact, it constituted a determined strategy towards the building of a true private management:

> In the public sector, the companies didn't need certain commercial skills, certain accounting and financial skills, in short, certain managerial skills. Once in the private sector they need all those skills. And also there was a sense that they wanted to change culture. You've grown up in a public sector management culture. And they want you to change your mind. That was definitely the case to introduce personal contracts. (National Officer – Prospect)

Their impact went beyond these cultural boundaries because they posed a concrete threat to the organisational strength of EPEA.

In 1987, EPEA reported the growth of the initially covert but finally admitted policy of the Company Boards, developed with the least consultation, of inducing certain NJMC staff to accept non-negotiable posts. The report presented the policy as part of a process of 'softening up' of managers prior to privatisation. This policy was facilitated by holding down management pay until personal contracts had been made. The introduction of personal contracts continued in the following years despite complaints by EPEA. By 1990, all ESI companies had managed to hire the majority of their managers on the basis of an individual contract. According to a managerial survey organised by EPEA, while in 1989 only 17 per cent of respondents admitted to having personal contracts, in 1990 this figure reached 72 per cent with half of them said to have been transferred in the first half of the year (EPEA 1990; EPE, 73, 2 February 1991). Thus, when companies began to give notice of withdrawal from the negotiating machinery, most of them had already placed their managers on personal contracts. Though the same survey showed that 77 per cent of those on personal contracts manifested that collective representation would still be relevant for issues like pensions, safety organisation, professional responsibility and the like, the danger to EPEA was evident. When interrogated about whether the process of privatisation impacted immediately on trade unions, former officials of EPEA often mentioned the introduction of personal contracts:

> The impacts on trade unions were several. First and foremost, from our point of view, because we represented the engineers and the higher managers, the first the companies did was to take the higher managers and engineers out of the collective bargaining. (National Officer – Prospect)

This quotation points to the reason why personal contracts have been perceived by trade union officials as a force undermining collective bargaining:

> Our legal framework was very permissive in this respect in allowing employers to undermine collective bargaining through personal contracts. (National Officer – Prospect)
>
> So having got the senior managers out of the collective bargaining, they then sat down and discussed with the trade unions a revised method of collective bargaining. Looking away from the collective agreements, bringing in a company-based bargaining, a single table, where all the unions in the industry sat down and talked to the employer. (National Officer – Prospect)

This policy is also explicitly associated with an anti-union approach through the use of money:

> There was a trend in the UK at that time: union derecognition. The electricity employers were subtler about it. Instead of getting into confrontations with the unions, they bribed individuals away and said: 'Look you could stay in the collective agreement, or if you like, there is an extra 2,000 pounds. What do you want? And you can have free medical care and we'll give you a better car'. So there were inducements, personal inducements. (Regional Officer – Prospect)

Additionally, personal contracts deprived unions of vital information for collective bargaining. It has been stressed in Chapters Six and Seven that the power of EPEA rested, partly, upon the role occupied in the labour process by its technical, professional and managerial constituencies. By organising the employees in charge of running the industry, EPEA exercised great influence and gathered worthwhile technical and financial information for bargaining within the industry from a position of strength. In the case of EPEA, more than in the case of other ESI unions, membership loss implied information loss. This side of the policy was reinforced by the early introduction of confidentiality clauses into the letter of the contracts (EPEA 1989a).

After privatisation, most companies, though not all – PowerGen maintained all employees in collective bargaining for instance – continued with this policy depending on their own individual approaches. Some companies began to take out technical and some administrative grades from collective bargaining, peeling off the top levels layer by layer and restructuring the company grading. In National Power, for example, the policy began to affect not only EPEA but also NALGO, which had opposed personal contracts throughout 1990 and 1991, because it was thought of as an attempt to break workers' solidarity and to introduce multi-skilling without remuneration (NALGO 1990b; NALGO News, no. 494, 20 September 1991). Again during 1997, National Power intended to persuade the overwhelming majority of the professional and technical employees to leave the collective bargaining machinery. However, a significant number of EPEA's members rejected the offer this time, according to the data collected in interviews, perhaps an early indication of a changing environment:

> Also, I think, our members' experiences of personal contracts have not taken them in the way they were portrayed by the employer, because personal contracts are less secure. People who are in personal contracts are now much easier to be dismissed than people in collective agreements. So people have got an experience now and personal contracts haven't been a positive experience. To some extent, this makes personal contracts more difficult to initiate, and less attractive, because the ongoing government is more or less supportive of collective bargaining, at least, more than the conservatives during 1990s. (National Officer – Prospect)

> What actually happened, in practice, is that in the early years of that process people was seduced to take a personal contract, and then, the favourable terms were actually narrowed. So, when the company car was not renewed, they suddenly discovered that the car they had previously was no longer available to them. So, the company then cut back the costs by reducing the bonus and just moderately increased the salaries. (National Officer – Prospect)

Furthermore, some companies began to bring people into collective bargaining after 2002:

> Question: What is the reason for this trend?
> Reply: the reason is that it is too much time-consuming to deal with individuals. It appears to be the reason. And you got a political climate that is changed, a political climate that is more encouraging towards collectivisation, you got that the climate within the country is shifting with the Labour government. Not dramatically, as some people had expected, but it is there. Overall, it is easier to deal with five hundred people sitting with the trade unions, two or three trade unions around the table, than it is to deal with three hundred people in collective bargaining, and the other two hundred on the individual basis. And there was no real individual contracts. The only individual thing was the salary. It was the same contract, all the terms are identical but with different pay. The only thing there was pay. (Regional Officer – Prospect)

However, the administrative argument does not seem to provide the full answer to this, because the multiplied administration involved should have been noticed before, and in fact, as recognised by the interviewee, individual terms did not vary much in practice (Evans and Hudson 1993). It might be worth mentioning that offering inducements to eschew collective bargaining became more difficult after unions began challenging UK law in the European Court of Human Rights; the law subsequently prohibited such activity where the main objective was to undermine membership levels or union organisation (Collins 2006; Wilson and Others v. United Kingdom 2002).

To conclude, a political process whose objective was to build a private management was embedded in the diffusion of personal contracts. This process comprised ideological elements on the one hand and the partial decollectivisation of managers and other high-ranking employees on the other. In this latter sense, it undermined the collective processes of interest definition. This policy also targeted the organisational power of EPEA by encouraging senior and middle managers out of collective bargaining. In so doing, the process weakened the engineers, who were more reluctant to enter into these arrangements and who, in fact, had disputed managerial prerogatives since 1970. Once in full control of the right to manage, some companies began to reverse this policy somewhat due to practical and legal pressures.

Peculiarities of Argentina: The Politics of Compensation

This aspect of the form taken by the politics of money in Argentina needs to be put in a wider context. As stressed by scholars (Etchemendy 2001; Etchemendy and Palermo 1998; Murillo 1997, 2001; Palomino 2005), the CGT accepted market reforms in return for business concessions. Apart from generous severance packages and employee-owned stock programmes managed by unions, the latter obtained important compensations during

1992 when the government attempted to reform the pension system and union welfare funds, the so-called *obras sociales*. The CGT negotiated its participation as a service provider in both reforms. The government explicitly included a provision for unions to create pension funds (AFJPs) and restricted competition in the social security system to the existing *obras sociales* by excluding new private health insurance companies during an undefined transitional period starting in 1993. Later on, in 1994, unions also secured from the government the right to establish insurance firms for work accidents (ART) along with subsidies for restructuring *obras sociales* and a bailing out of their debts in relation to social security provisions.

It is within this context that LyF CF adopted the outlook of aggressively taking business opportunities as compensation for the financial loss brought about by a declining membership. This position was channelled through FATLyF, actually a national arm of LyF CF, that even participated in the privatisation of less profitable segments of the electricity industry through a body created to that end. As a result, FATLyF bought five of the privatised units, which were three utility groups (Patagonia, Litoral and Northwest) and two transportation companies (Transnea and Transnoa). It also obtained special conditions for the concession of the state-owned coalmining zone (Yacimientos Carboníferos Fiscales – YCF) providing inputs for electricity utilities. This policy orientation arose with the union strategic shift from latent resistance to active support and crystallised in the Conference of FATLyF of 1992 (FATLyF 1992). In this conference, LyF CF imposed a motion on smaller, reluctant unions by which FATLyF was authorised to adopt whatever measures necessary to take advantage of the business opportunities coming along with the market reforms. In a fifteen point document, FATLyF tackled how to seize opportunities opened not only by the reform of the social security and pension systems but also by the privatisation of less profitable electricity industries, administration of the PPPs, and other independent areas such as tourism and banking (FATLyF 1992). At the request of LyF CF, this orientation was ratified in the 1993 VI International Conference of Trade Unions and the 1994 LyF CF Annual Conference (IPCTT 1993; LyF CF 1994a).

Together with other unions (FeNTOS, SUTECBA and Sindicato del Seguro that together and in total owned just 12 per cent of the business),

FATLyF created an AFJP, Futura, by far the biggest business controlled and run by the Federation. The pension fund, by 2006, owned a portfolio of more than $267 millions (Argentincan *pesos*), including $1.5 million shares in privatised utilities (AFJP Futura website). Findings revealed that, apart from the five electricity businesses, YCF and Futura, FATLyF owned at least ten more companies (Contacto, October 2000).

Beyond the activities developed from within FATLyF, LyF CF became involved in the politics of microentrepreneurship, which followed the failure of the politics of voluntary redundancies to achieve the expected downsizing of the industry. As mentioned above, the union organised cooperatives of former employees who invested their severance packages in providing services for the new privatised companies. In 1995, three years after privatisation, 290 workers grouped into twenty-four cooperatives worked providing outsourced services to three companies: EDESUR (twelve), EDENOR (six) and Central Puerto (six). Due to this microentrepreneurship, the union received in 1995, as subscriptions deducted from member's pay by employers, the annual amount of US$29,648 (LyF CF 1995).

The union also created a company to provide food tickets. The rationale underlying the next statement is paradigmatic of union justification for its embarking upon economic activities:

> As we saw the food tickets coming to replace part of our wages, in fact, a plain attack against our institutions, the *obras sociales* and the union itself [...] Our policy was to say: 'Why don't we organise a tickets' company?'. We, as a union. We were the only union, I think, in the whole of Latin America, that created its own enterprise. We called it Lyfchek. (Union Official – LyF CF)

The enterprise survived just a few years before succumbing to competition. Yet, this has been often the rationale accompanying the investments: as the union became debilitated because of declining membership and employer contributions, it was necessary to counterbalance this trend by developing alternative sources of income.

The politics of compensation rested upon specific institutional features of Argentinian unions, that is, their traditional involvement in providing social services through institutional bodies, independently from union administration, mainly through the *obras sociales*. This policy worked as a significant counter-mobilising force, for it served to buy the union leadership off and helped to moderate union politics. Moreover, it also increased the tensions between negotiating and combative trends within FATLyF. From the beginning, the counter-mobilising nature of engaging in businesses was clear for LyF MDP and other organisations that opposed the politics of compensation within the national Federation and beyond. The conflict between both orientations ended up with the expulsion of LyF MDP from FATLyF. This occurred when the Federation, pushed by LyF CF, made changes to the rule books of the organisation adopting a pro-business orientation.

Although LyF MDP refused any form of political exchange based on business compensations, the politics of microentrepreneurship impacted upon the organisation. Many workers left the industry seduced by the opportunity to become independent microentrepreneurs. Once more, this happened under ESEBA, which was still a public company when the policy began. In 1995, the union launched an important campaign against this tactic, in which the CGA framed the issue as a labour fraud, given the conditions of the contract that workers had to sign up to provide the service (LyF MDP 1995a, 1995c, 1995d). This campaign was crucial in reducing the success of this aspect of the politics of compensation, the only one that truly affected LyF MDP.

Conclusion

This chapter focuses on the counter-mobilising side of the politics of money, and how it acted to prevent conflicts over the process of privatisation and after, in the UK and Argentina. This does not necessarily mean that the counter-mobilising content was the only rationale behind these policies, which may have been implemented for other aims as well (cost reduction, organisational efficiency and the like). Yet findings support the view that all these initiatives shared common features, that is to say, that all of them implied the mobilisation of money resources, had counter-mobilising objectives and affected workers' collectivism.

If the previous chapter shows that the conceptual framework laid down by mobilisation theory proved useful when analysing the forms taken by workers' initial responses to privatisation, this chapter exemplifies its potential to study how governments and managers deployed resources that target crucial dimensions of workers' collectivism, in this specific case, financial resources. However, its usefulness seems to be limited to mapping the main dimensions that a counter-mobilising force ought to target in order to undermine workers' collectivism. The comparative dimension helps in turn to identify variability in what the theory defines as the opportunity-to-act.

The four dimensions of the politics of money studied in this chapter manifest a two-fold purpose: to break up resistance and to further loyalty to the new private firms. Regarding the former, findings show that public authorities and managers had to conceive of policies that broke resistance to change at individual and collective levels. For instance, while the politics of voluntary redundancies was a device principally designed to buy off workers individually, the politics of compensation in Argentina was clearly a strategy directed at establishing a tacit pact with trade unions – a new type of political exchange – whose content consisted of business opportunities. All four dimensions nevertheless ended up affecting both levels: for instance, voluntary redundancies targeted workforce morale, and therefore, arguably, made it more difficult for unions to mobilise workers through collective action. It also depleted unions of experienced union officials and activists. The politics of shares was an incentive for individual workers but in Argentina, it was also an important component of the politics of compensation. Personal contracts in the UK debilitated the organisational strength of EPEA by taking individuals voluntarily out of collective bargaining arrangements.

Regarding the aim of furthering loyalty to the companies among the workforce, findings are not conclusive as to whether companies succeeded. It seems possible to argue that the above was a secondary objective, a desire of the private management often clashing against the rough reality of seeking profits by cutting labour costs and increasing productivity.

The comparative dimension of the study is also crucial in highlighting factors that may explain variability. For instance, differences in the employment law may contribute towards explaining the important role played by personal contracts in the UK, a strategy unlikely to be successful in Argentina. Differences in legal frameworks also seem to have affected the chances of unions representing workers as share owners and mobilising them around this issue. Similarly, the fate of the politics of voluntary redundancies appears to have been closely tied to the state of labour markets. While the study points to the development of a redundancy culture among the British workforce, a rising rate of unemployment thwarted similar beginnings in Argentina. Soon after privatisation, Argentinian workers began to reject voluntary retirement programmes. Consequently, not only private companies but also the provincial government deployed a wide range of repressive tactics and psychological persecution to force people to join the redundancy programmes. From then onwards, what appears formally as a process of voluntary downsizing hid what was, in fact, a process of compulsory redundancies. In the language of the mobilisation theory, the opportunity structure for companies and the government changed, and hence the chance of using money to downsize the industry faded away. Given this scenario, the state not only encouraged harassment and repression, but also devoted more efforts and resources to bribing unions with business opportunities in order to fragment workers' resistance, and keep discontent as an individual expression.

At this point it is necessary to refer once more to the case of LyF MDP, which shows how far counter-mobilisation strategies are mediated by workers' organisational capabilities. So union structures and government proved to be relevant sources of variability too. This is not to say that LyF MDP was immune to the politics of money; but to stress that union leadership is relevant when explaining variability in workers' responses. First, the politics of voluntary redundancies and compensation succeeded at the individual level, and then in this way it impacted on the organisation. The politics of money seduced individual workers, though fewer than the industrial average. It corrupted members of the CGA and other union officers, yet it did not pervade union officialdom. In sum, union leadership could mobilise the collective power of workers throughout the process against different manifestations of the politics of money.

While the analytical sequence laid down in the conceptual framework – opportunity-to-act, organisation, interest definition – proved to be useful to the study of the forms of workers' defensive mobilisation, this chapter shows that a focus on the effects of counter-mobilisation upon workers' collectivism may require a more flexible and empirical approach. However, it is essential to emphasise that it was mobilisation theory which signalled the direction of the analysis.

Finally, the chapter also shows that changes in the intermediate variables may influence very quickly the field of social interactions and subjective processes. This was the case in Argentina when a sudden rise in unemployment impacted upon the effectiveness of the politics of voluntary redundancies. Workers began to oppose it, and concomitantly, managers began to back redundancy programmes with repression. In this new context, even an extremely moderate union like LyF CF mobilised the rank and file through industrial action and rallies. This reminds us of Tilly's self-criticism: 'the model has no time in it. Collective action does. The most obvious defect of the model is that it makes no allowance for the ways a contender's collective action affects its opportunities and its power. The model provides no place for strategic interactions and no place for the conquest or loss of power. Collective action affects a group's power, but that effect takes time' (1978: 58). This chapter suggests that changes in external variables also affect contenders' opportunities and power.

CHAPTER NINE

Privatisation and Collective Bargaining

The focus of this chapter is the vital counter-mobilising forces unleashed by privatisation; in this case, the fragmentation of collective bargaining and the changes of its procedures. The empirical analysis is guided by mobilisation theory which sheds light on how and why these phenomena brought about changes in workers' capacity to mobilise. With this in mind, particular attention is to be paid to factors which diminished trade unions' ability to confront management, or which impacted upon their organisational structures. At theoretical level, the chapter provides room for the study of the strategic interactions of the contenders and the conquest or loss of respective power, which Tilly demanded in the quotation closing Chapter Eight. The comparative perspective shows, in turn, important differences between the UK and Argentina, within a common horizon of growing obstacles to collective mobilisation. Its basic aim having been to increase managerial prerogatives, the assault upon collective bargaining happened in national contexts, which differed regarding the system of industrial relations, the extent of decentralisation and restructuring, and the evolution of market structures. So these intermediate variables influenced the final outcome of the process; at the same time, these differences set limits to the symmetry of the comparison. Lastly, the analysis also points to meaningful differences between unions.

The United Kingdom

The Process of Counter-Mobilisation against Collective Bargaining

By lobbying the government, ESI trade unions ensured through ESTUC the continuation of the national negotiating machinery throughout the process of privatisation, and obliged the new private companies to announce, twelve months in advance, any intention to withdraw from it. This achievement was, in fact, a tacit recognition by ESTUC of the inevitability of the end of the traditional arrangements as well as a reassuring guarantee of time to negotiate the future institutional framework of industrial relations. There had been, in the run up to privatisation, abundant signals of enthusiasm among management for the opportunity to put an end to national bargaining, preventing any optimistic hope for the establishment of a common negotiating machinery across companies. In several of the would-be private companies, middle managers began to let unions know about their distaste for national and industrial institutions in the new environment. The message took a diversity of forms, from the rhetoric of corporate culture building (as in PowerGen for instance) to the crude economic language of regional differentials in the cost of living (as in Norweb). As managerial pressures mounted, the EETPU had, for example, to alert shop stewards to resist the undermining of the bargaining machinery until new negotiating structures had been agreed by the union. In brief, as stated by an interviewee: 'Most of the companies, if not all of them, saw privatisation as an opportunity to liberate themselves from national bargaining'. (National Officer – Prospect)

It is possible to argue that the decentralisation and fragmentation of collective bargaining was a key aspect of capital counter-mobilisation: partly intended, partly the unconscious by-product of other requirements. Still, as Lukes (2005) might put it, the end of centralised procedures showed the capacity of the new management to force trade unions to accept a course of action other than the one they originally pursued. In marked contrast to their early calls to defend the national machinery, trade unions ended up negotiating company agreements and procedures with stoic pragmatism in search of the best possible arrangements. Decentralisation and fragmentation were clear signals of unfavourable changes in power relations. In the face of this situation, unions of craftsmen were generally more pragmatic than the predominantly public sector ones.

In December 1991, East Midlands and National Power gave notice of withdrawal from the machinery. Soon after, the remaining private companies followed their example. By June 1992, unions reached the first agreement with PowerGen, which included single table bargaining (STB), flexible working, and the harmonisation of terms and conditions for all employees. The bulk of company agreements, however, were balloted during 1993 (Table 11).

Table 11: Introduction of Company Agreements

1992	1993	1994
PowerGen / June Scottish Power / December	National Power / January Southern Electric / February Yorkshire Electricity / February East Midlands Electricity / April NGC / April Midlands Electricity / May SWEB Connect / May Manweb / July SEEB / July Northern Electric / November	Eastern Electricity London Electricity Norweb Swale SWEB

According to a common view among current national officers, there was no widespread radical change at that stage, apart from the new bargaining structures:

> So [companies] were not all necessarily radical themselves; though they had, certainly, their own ambitions. So, different sources of bargaining formats came out of this move from public to private. (National Officer – Prospect)

> In the immediate post-privatisation aftermath, there was not any visible change in bargaining strength. So, trade unions were able to achieve at company level, in company agreements, in company bargaining, terms and conditions that were still very much acceptable in terms of the economy as a whole. (National Officer – Amicus)

> Early on, the employers were more concerned about stabilising their businesses; in stabilising employee relations in their companies than they were about attacking in any dramatic sense the terms and conditions. (National Officer – NALGO)

The interviewees stressed that, initially, management was satisfied just with the breakup of the national machinery, which allowed managers to take responsibility and authority for their own employment conditions. There is an extended belief that the managerial agenda was principally dominated by how change might be achieved in collective bargaining structures and also, if possible, in payment structures. In 1993, at least five companies succeeded in introducing performance related pay (PRP), and by 1995, most if not all of them had PRP for certain categories of workers. Still, in the main, employers restrained their full agenda of desired changes while establishing companies' frameworks for negotiation. It follows that over the period of transition from national to company bargaining (1991–2), when the new firms inherited the national agreement, the companies amplified the rigidities of the old arrangement under market competition by compromising a common line of negotiations composed of minimum offers. The strategic aim was to discourage unions' attachment to national machinery whilst avoiding radical changes and confrontation. This negotiating context pushed unions into seeking mandates from their constituencies for exploration of alternatives at company level. For instance, EPEA had convinced themselves by 1991 that either unions succeeded in backing up a unified claim across the industry, or they had better contemplate separate company arrangements. By the same period, the EETPU had found comparative advantages over other unions. On the one hand, the negotiating experience of the union in the private sector could translate into a relative organisational strength at local level. On the other, company agreements opened a window of opportunity for their union members to move up through additional skills and career progression. NALGO, in turn, while keeping up opposition to the breakup of the machinery, joined manual and engineers' unions and asked its membership to mandate the definition of new institutional frameworks at company level through ESTUC.

It took months of painstaking negotiations to finalise the company agreements. Although trade unions were overworked throughout a process that put a lot of strain on their human resources, five companies were still negotiating at the end of 1993. The final outcome was a variety of bargaining formats, though STB prevailed in the main. Some companies set up multi-table bargaining structures, either by strict replication of three agreements, or by two-table bargaining (TTB) that generally maintained the split between, on the one side, technical and administrative staff, and on the other, manual and craft workers. However, this was just the starting-point of a process of decentralisation in which, initially, pay bargaining and, later on, terms and conditions too, tended to be split into smaller units, as the electricity companies themselves sub-divided the bargaining structures in their different business units:

> Over a very rapid period, two or three years, the companies themselves began to subdivide their own company's agreements because they rapidly realised that they, at the company, they faced more than one labour market, more than one commercial circumstance, and actually began to shape with us new agreements, which were at the company level. (National Officer – Prospect)

Additionally, the external influence of the regulator in requiring the unbundling of distribution businesses accelerated the process even more by providing companies with a justification for pushing through the decentralisation of bargaining. From 1995 onwards pressures towards devolution of bargaining within companies began to grow. On the union side, there was no uniform policy. When, at the end of 1995, SEEB indicated the desire to move away from company-wide agreements to agreements based on the separate businesses; trade unions agreed to enter into discussions despite not having received any concrete proposal from any business unit. In Northern Electric, instead, the same intention faced mass meetings and strong protests. However, once again, companies forced the acceptance of a course of action, which unions disliked and even opposed, though in a lukewarm way. By the end of the 1990s, most companies had subdivided their negotiating structures into smaller units.

The next sections address the counter-mobilising effects of the process of radical decentralisation of bargaining, particularly the growth of inter-union competition and latent tensions, and the appearance of differences in bargaining power.

The Growth of Inter-Union Competition

For mobilisation theory, the growth of inter-union competition is a debilitating factor for workers and their chance to gain and control power resources collectively in order to increase the opportunity for collective action to arise (Tilly 1978). Decentralisation, the introduction of STB, and finally, the devolution of bargaining to business units furthered competitive trends.

Early in the pre-privatisation period, trade union leaders foresaw an uncertain future likely to encourage inter-union competition for members. They arrived at this conclusion for various reasons. It is necessary to recall the aggressive recruiting policies pursued by the EETPU during the 1980s by means of single union agreements. Inter-union disputes led to the expulsion of the EETPU from the TUC and increased the negative expectations. Later on, the imminent end of the law enforcement for the carrying out of the national negotiating machinery fed fears that companies would give notice to leave the national agreement whatever one's own trade union did or said, and that other unions could then try to negotiate separately with the companies concerned. The pace of job loss and plant closure announcements also showed the hard times to come regarding union memberships. Unions consequently diverted their resources and energies from other uses in the run-up to privatisation to the launching of recruitment policies, or at least, their reinforcement. The first symptoms were a set of mandates on the topic of recruitment from union conferences, branch motions, official policies and campaigns. This emphasis continued during the early days of privatisation.

For instance, EPEA decided in December 1988 to offer single union agreements to new entrants to the industry to counteract the threat embedded in the bargaining policies of the EETPU and the three large manual workers' unions: TGWU, AEU and GMB (EPEA 1988). In November

1990, again in the midst of a debate about conflicts with other unions in competition for recognition in new companies, EPEA's NEC officially established the promotion of recruitment as a prime union task (EPEA 1990). NALGO declared 1989 as a recruitment year for the ESI (NALGO 1989a). The central tenet was a programme of action aimed at maintaining NALGO's position as the main union for NJC staffs, and equipping it for negotiating in the private sector (NALGO 1988b; NALGO News, no. 347, 30 September 1988). During 1990, NALGO's reports on privatisation often stressed the danger of the growth of inter-union competition as the background for the extension of the 1989 recruiting campaign (NALGO 1989b). NALGO's insistence seemed justified at the time by the moves by EETPU throughout 1989 and 1990 to poach NALGO's members in some branches and to begin recruiting NJC employees, and by the attempts by APEX, GMBATU's arm for the recruitment of white-collar workers, to increase its influence and numbers of seats within the different negotiating bodies (TUC 1990). EETPU, in turn, did not need special recruiting policies to be implemented. The union had already been involved in bitter inter-union disputes for years for that cause; as the head of the union, Hammond, graphically stated: 'We are up to our necks in the struggle for membership. We make no apologies for that' (Contact, 17, 4, August 1987).

During 1991–3, inter-union competition decreased as workers confronted the establishment of private companies and jointly negotiated the agreements anew. However, the introduction of STB, and to a lesser degree TTB, ended up blurring the traditional frontiers of recruitment. This gave unions greater incentive to poach members and revive the competition for new starters:

> Question: How have union relationships been in the sector?
> Reply: You have effectively, Unison – I use the present names – and Prospect, representing white collar. Amicus, GMB and TGWU representing blue-collar workers. That is not so clear now, because a number of the industrial negotiating forums allow all unions to sit at the same table [...] We now represent a small number of engineers as well, you see, whereas before we wouldn't. Relations between the unions at that time were very good, and now, well, things changed. There is more competition for members. (Regional Officer – Amicus)

> I really don't like them [Amicus and Prospect] at all, because they are aggressive poaching unions. Well, we are not pure on this either. But we shouldn't actually poach each other's members. There is no point in that. There are large pockets of non-members out there, we should go and recruit those members. But the easiest members to recruit are existing trade unions members, for the very reason that they show an interest in joining a trade union. (Regional Officer – Unison)

These practices brought about disputes, which were usually sorted out on-site. Only rarely did inter-union conflicts reach the TUC Disputes Committee, as in the case of Southern Electric in 1996, when Prospect was found guilty of poaching members from Unison and the AEEU (EPE, July/August 1996; TUC 1996). While disputes have not helped to encourage trust and unity among unions, it is necessary to recognise that unions gradually began to comply with agreed procedures for inter-union transfers. This happened particularly once it was generally recognised that there were huge pockets of non-unionised workers in the industry, and hence that efforts should clearly be directed towards their recruitment; for instance, one in five of Southern Electric staff were non-unionised at the time of the dispute. For some interviewees, though important, the tendency towards inter-union competition should not be exaggerated:

> Question: Did job loss provoke inter-union competition?
> Reply: Yeah, I mean, perhaps not as much as you expect because, you know the English system, there is always a certain amount of competition between unions. And I suppose, after privatisation, all unions were looking to survive [...] We were all quite ruthless in trying to recruit new members and hold on to our existing members. (National Officer – Unison)

The Introduction of STB

The implementation of STB was pushed by companies, whereas trade unions shared a common distrust about it. Instead the latter attempted to favour, when possible, the replication of the traditional negotiating bodies or, at least, TTB arrangements. There were exceptions, however, among companies. Paradoxically in Yorkshire Electricity, due to managerial fears about the prospects of an unintended empowerment of unions by STB, unions succeeded in keeping their traditional separations. SEEB, in turn, inclined to TTB to guard supervisory and professional terms against being indirectly determined by those they supervised, and to maintain different levels of call-out payments for manual and non-manual staff (Gall 1994). Eastern Electricity set up TTB machinery, which comprised a Professional Group covering ex-NJB employees, PAG grades and ex-NJIC foremen, and a Staff Group covering industrial and clerical support employees. This division was thought to facilitate the introduction of PRP schemes in the Professional Group Agreement, and the end of annual incremental progressions. Yet time would increase the number of companies with STB. However time also finished with trade unions' uniform and contrary viewpoint about it as craft groups perceived it as an opportunity to crack the interface problem and the administrative staff as an opportunity to press their claims for equal pay.

Two central arguments were offered by employers to explain the early introduction of STB: the reduction of the amount of time and resources spent on union negotiations and the establishment of intra-company consistency with regard to pay and conditions (Gall 1994). However findings from the ESI do not support any of these claims. They show that the adoption of STB by electricity companies was just a first step towards devolution of bargaining to business units. This evolution was already announced by the devolution of decision-making down the managerial hierarchy before companies gave notice of withdrawal from the negotiating machinery. It manifested itself on the one hand in the proliferation of profit centre policies and, on the other, in the procedural agreements established by a number of RECs to discuss organisational change and work flexibility. Although senior managers were freed up, the time and resources taken up by negotiations multiplied with the devolution of bargaining involving a mass of junior managers; companies attempted to counterbalance this factor by signing agreements for two years, and even three years. A more important consequence was the tendency towards equalisation of terms and conditions across different grades of workers within each bargaining unit together with growing inconsistencies across the company. Repeated mergers and take-overs only added to such inconsistencies. Then it comes as no surprise to find in some companies the same rationale, the saving of time and resources, for current programmes to harmonise terms and conditions, and to have one pay negotiation for all business units. In short, STB rather seems to have been the favourite means to further devolution, the main counter-mobilising tool for debilitating unions' ability to resist change in payment structures, work practices and terms and conditions of work. Its subsequent subdivision seems to have been a strategic move to match more easily the conditions of local labour markets with regard to both regional wage differentials and skills.

The engineers organised by EPEA were the main losers within STB, for its implementation downgraded their representational rights. Gall refers to this type of phenomenon as a form of derecognition or decollectivisation (Gall 1994). According to a Prospect official:

> As you gave unions seats according to their size, the manual unions, because of the largest numbers of staff, of course, took more seats than the others. So the unions were forced to work together. And you no longer have individual policies being dealt with at national level. So, the NJB's, which are mine union's policies, were always different from the manual workers' policies. Well what, of course, is done is weaken our union because we had exclusive rights for the technical staff. (National Officer – Prospect)

By contrast, even when STB somehow threatened the preponderance of the EETPU's representational rights among former NJIC members as well as the preponderance of NALGO among former NJC members, their situation was different from that of EPEA. In May 1992, EETPU amalgamated with AEU to form the AEEU, strengthening its position with regard to other manual unions. EETPU was already used to reaching compromise policies within the NJIC body, with TGWU – despite major political differences between them – and GMBATU. More importantly, it opened an opportunity for industrial craft unions to sort out favourably the dispute with EPEA about task demarcations as the interface question was addressed. Lastly, Amicus ended up dominating most STBs until their subdivision, and then it continued to dominate STBs in the leading business units of ESI companies. The latter, in part, compensated manual workers for the damages brought about to their bargaining power by the end of the national machinery.

NALGO – Unison from 1993 when companies' agreements proliferated – was the main victim of the subdivision of the STB into business units' agreements, which revealed the disparity of strength among different groups of workers but not of the STB arrangements in themselves. This was firstly because white-collar workers had often built their negotiating power on the shoulders of manual and engineering unions, and the STBs did not put such a strategy at serious risk initially. Secondly, and despite their utter opposition to the breakup of the negotiating machinery, this was because Unison found out that STB might give administrative and clerical staff the chance to seek harmonisation of pay, terms and conditions. This hope ended with the subdivision of bargaining into business units, but it has enjoyed a rebirth ever since, as some companies expressed the wish to harmonise terms and conditions by reunifying the company STB.

For EPEA, instead, there were no similar compensating factors. EPEA had built up its power by maintaining the exclusive rights to representation of the engineers in an independent body within a highly centralised and formalised negotiating machinery, by keeping strict demarcation of tasks to prevent craft workers diluting from their control upon the labour process and by establishing a pay link between NJB and NJIC pay. All these aspects were targeted through STB, in which EPEA was outnumbered by AEEU-Amicus, a factor that contributed to the erosion of their leading position within the industry. Unsurprisingly, EPEA's initial reaction was to oppose any arrangement which involved the possibility of their being outvoted concerning their own members' pay and conditions; yet, this position proved to be untenable. The union was left with no choice but to accept sharing the negotiating table with other unions. In the years that followed the establishment of the new arrangements, concern with STB was common amongst EPEA's members from different companies. To give just a few examples, in SWALEC's STB, EMA obtained three places out of fifteen; in SWEB's just two out of fifteen, whereas in the past there had been a committee of nine members who regularly met the company at high level to discuss NJB issues. Early complaints arose not only in companies from the South West but also in NGC, Midlands Electricity and Southern Electric. These differences awakened latent tensions and brought about new ones.

The Old Sources of Inter-Union Frictions under the New Arrangements

As privatisation opened dissimilar opportunities for manual workers and engineers with regard to the inter-union rivalry for the demarcation of technical and craft skills, management exploited this traditional source of friction to its own ends. The company-based bargaining agreements negotiated throughout 1993–5 opened up the work traditionally carried out by engineers to ex-NJIC employees to a greater or lesser degree. Craftsmen took on extra work and responsibilities for little or no additional remuneration; the carrot was career development. SWEB, for example, launched a job redesign programme, which included a new team structure to dispense with engineers being involved within the teams, confining them to project work and reducing their numbers. For this exercise, the company targeted the necessity for formal authorisations by engineers to do specific works thus pushing down levels of responsibility. Operational engineers were replaced by operational technicians who had received a concentrated period of training. The strategy adopted by Southern Electric, instead, was to devolve what had been engineers' duties, particularly safety documentation and work planning, to craftsmen but without seeking radical change in the organisation of labour. This process continued all the time. In all cases, these managerial policies benefited from the enthusiasm of AEEU-Amicus, whereas they were resisted by EPEA:

> Because our guys go into the job, and then they say: 'All right, sit down for two hour's, wait for that bloke coming out'; 'put in a piece of paper to do the next job'; or 'switch out, switch in', and that was it. We can do that, but we were not authorised to do that at that time. (Lay Representative – Amicus)

> But the other people [engineers] are saying, well you are taking our job. Oh, yeah, we know, we are comfortable with that, because we are the AEEU union; we had a major conflict with them, a major conflict. (Regional Officer – Amicus)

Another old source of tension that has provoked several skirmishes between Amicus and Prospect under the new arrangements was the antagonistic position held by their constituencies within the industry: Amicus organised manual workers, whereas Prospect organised people who occupied managerial roles and exercise authority over the former. Interviewees referred to this note of discord between the organisations:

> Question: How do you agree common lines of negotiation with a trade union that organises mostly engineers and managers?
> Reply: Well, the answer to your question is that this situation does cause tensions. There is no doubt about this. It causes tensions. And sometimes there has been a break up and the trade unions went in different directions. Because the industrial staff feels that guy is my manager, he disciplines me. Why should we deal with them? And there are tensions, very real tensions. (National Officer – Amicus)

> In fact, I've been in the middle of an argument developed between the Prospect's and the Amicus' reps because Amicus feels that Prospect ... Well, you know: 'They are all managers, they defend the company, they are fighting the workers'. So they [Amicus' stewards] started an argument about a recommendation and decided not to go with the recommendation. So the other man, from Prospect, got accepted. And he said: 'I'm not speaking with them because I don't agree'. And one of the stewards started to say: 'That's because you are a manager too!' [...] And I'm the one saying: 'OK, let's cool down'. And that quite often happens. So, there are tensions, obvious tensions between trade unions, and that's because one sees the other as the management's union, and the managers' union thinks the other is a left-wing, militant union. (Regional Officer – Unison)

Conflict between Unison and other unions in the industry was mainly related to differentials in pay and labour conditions. Although there had always been frictions due to this issue, the national machinery reserved to NALGO exclusivity as the body discussing white-collar needs, thereby lessening inter-union disputes. Rather, it encouraged NALGO to press upon management the enhancement of the status of white-collar workers. While STB, on the one hand, opened new space in which to pursue equalisation, on the other, it confronted Unison with the agendas of engineers and manual workers. As stated by a trade union official from Unison, which organised most low-paid workers in the ESI:

> I mean, there were tensions between Unison and the other unions because the reality is that other unions, when we are actually in the meeting, they say: 'We are not going to have our members lose any money'. So, on the one hand, they say: 'Yes, we agree with this, we agree with doing this because it is equal pay issues, and the law,

> and so on'. On the other hand, they try to make sure that people get their money. Of course, that created tensions. (Regional Officer – Unison)

The frame of reference for this quotation is the harmonisation policy pursued by Unison to improve the terms and conditions of their members:

> We tend to support what we call harmonisation. So, we do away with the distinctions between manual workers on the one hand, and staff workers on the other. We don't like to have second class citizens so we like to have everybody treated the same. Obviously, there are going to be different pay scales, but you know, it shouldn't be on poor terms and conditions. That is not always the case with the other unions, because sometimes their membership is stratified, and they only recruit within a certain band of workers so they like to keep those separated. (National Officer – Unison)

The growth of tensions within STB due to differentials is confirmed by officials from manual unions:

> Sometimes there are clashes between unions, because, sometimes, trade unions' constituencies are people who are on lower pay, who prefer a fixed money increase; where people on higher pay resists that on the basis that they will prefer a percentage increase. This is all to do with differentials [...] So there are tensions in STB. On most occasions unions find an accommodation to go forward; but in some occasions, they don't. (National Officer – Amicus)

All those tensions have militated against the opportunity to articulate unified fronts and collective demands. One official from Prospect, while confirming the existence of conflicts between unions, introduced a subtle viewpoint on the problem:

> What is interesting is that while full-time officers work well together, there are certainly tensions at the level of the members, but not between the full-time officers. I think, what happens is, in fact, some people tend to think that, well, if you are working in this area you can belong to this union but you can't belong to another union. Some of that is there; there are tensions there between the staff. (Regional Officer – Prospect)

The important point to stress seems to be that, given that one consequence of decentralisation was that more negotiating tended to be done by lay representatives, by overburdening the full-time officer cadre, changes in bargaining structures amplified inter-union tensions at firm level.

The Irruption of Differences in Bargaining Power

The counter-mobilising effects of devolution were widened by the subdivision of STBs into business based agreements, as this policy showed the disparities in the bargaining power of different groups of workers. Broadly, differences emerged between, on the one side, the people employed in generation and distribution, and on the other, those employed in commercial and retail activities. The causes of this diversity have to be understood in relation to the characteristics of the labour process, the labour market and historical traditions:

> Initially, it was more a company agreement but along with time the companies began to segment their own agreements. This began to reveal the disparity in power strength between different groups within the same company. So, the engineering staff and craft staff still get good deals because they have, literally, the key to the power. The retail and the commercial people were more exposed, and also, it began to expose that the companies face more than one labour market. The labour market in which they obtained the engineer and craft staff was different, than the labour market in which they obtained the retail staff, for example. (National Officer – Amicus)

> Over the time certain parts of the company businesses revealed themselves as weaker. You have to make a distinction here, between the engineering activity and income, and the retail and commercial activities. In the engineering, the trade unions were stronger. There, you have an environment where engineers and craftsmen went to work with a different ethos. And they had a very strong strategy independence [...] Union density in these areas was much more substantial. In some other parts of the companies there was less commitment to trade unions. In the past, the terms and conditions in the retail and the commercial areas had been built upon the strength of the industrial areas because it was one national agreement. (National Officer – Prospect)

This difference in bargaining strength impinged upon the respective powers of the various trade unions, for it mainly targeted Unison, whose members belonged mostly to powerless sectors (Unison 1994, 1996a, 1996b). Sales and marketing employees, though reasonably well unionised, were never as willing to take collective action as manual workers, and their collective bargaining power diminished too. Later on, trade unions faced similar problems in Customer Services as companies consolidated call centres, in metering as outsourcing developed, and even in the newer small generating plants, where neither Amicus nor Prospect were always strong. For some interviewees, contracting was able to keep reasonable bargaining strength, similar to that of distribution. Other interviewees, instead, argued that contracting suffered a continual attack on terms and conditions by employers, similar to that of Retail, mainly in subsidiary companies and green-field sites.

Tactics to lower pay, and weaken terms and conditions in powerless sectors, varied. It was usually the result of gradual changes introduced in pay bargaining settlements. Yet, on several occasions when faced with proposals to close certain business, unions negotiated concessions. For instance, by this means Southern Electric obtained important changes in Energy Sales and Marketing. In Retail, these manoeuvres comprised effective closures, mass redundancies, and the reopening of businesses as franchises, employing new staff with lower pay and conditions. In SEEB, the same strategy included tens of workers who had left the company with redundancy packages and who were employed the next week by the franchiser to do the same job but with lower terms and conditions. According to interviews, even Prospect members doing technical work were treated in exactly the same way on a number of occasions.

Union officials from the ESI and engineering feared that, under STB, the bargaining position of manual workers would be 'undermined by having joint negotiations with white-collar workers, as the former were more readily unionized than their white-collar counterparts' (Gall 1994: 70). Even so, neither STB nor bargaining at business units seems to have seriously undermined the power of manual workers for this reason. In 2006, however, after the long detour of bargaining devolution, some companies were attempting (like Eon) or considering (like EdF) unification of pay bargaining and harmonisation of terms and conditions. While workers in call centres and retail activities welcomed this policy, workers in generation did not like the idea. The former saw harmonisation as an opportunity to overcome the difficulties they had to face in obtaining money from their business units; the latter, instead, considered that companies attempted to lower their wages and erode their terms and conditions. A pattern seems to emerge: in the recent past, separation provided the opportunity for employers to exploit the lack of bargaining power of clerical employees; at the time of writing, the weakened position of white-collar workers was used to undermine the power strength of craft and engineering staff.

To summarise, fragmentation and devolution of bargaining in the ESI have had important counter-mobilising effects, as was accordingly feared by ESI officials: 'There was a common fear among trade unions that the break of a national bargaining would undermine our bargaining strength' (National Officer – Prospect).

Argentina

The Process of Counter-Mobilisation against Collective Bargaining

The process of counter-mobilisation against collective bargaining was different in Argentina. While in the UK the majority of changes occurred after privatisation, without a clear and concerted agenda, and were pursued by private companies, in Argentina, a set of relevant changes happened before privatisation as a result of a detailed plan, and these were carried out by public authorities and consultants employed by the Ministry of Economy. The reason was that, from the point of view of the government and the international financial institutions which backed the sale of public companies, the power of trade unions in the public sector expressed itself in the contents and procedures of collective bargaining, so that public collective agreements were deemed to be an obstacle for the success of privatisation. The government decided to dismantle the agreements in order to attract private investors. This entailed a different starting-point for the counter-mobilisation process against collective bargaining in Argentina. The involvement of the government, however, should also be understood as being determined partly by the institutional features outlined in Chapter Six and particularly by the legal underpinning of the Argentinian system of industrial relations, for this narrows management's opportunities to push change without legal support.

As the analysis will also show, there are differences in Argentina between the case of LyF CF and that of LyF MDP, which relate to the peculiarities of the opportunity structures faced by these organisations and their distinctive processes of decision-making and leadership styles. Yet, despite their diverse characteristics, the process of change of collective bargaining underwent two phases in both LyF CF and LyF MDP. The first phase, before privatisation, was characterised by the offensive of public and industrial authorities forcing unions to accept radical modifications in the content of CCTs. The second one was after privatisation, when the electricity companies either refused to continue with joint arrangements bringing about the fragmentation of negotiations as in the case of LyF CF, or simply targeted workers' collectivism but without fragmenting bargaining as in the case of LyF MDP. During this second phase, two additional variables help to differentiate between the Argentinian cases and the British ones: a different structure of union representation and the absence of mergers and takeovers affecting the boundaries of the industry.

The Case of LyF CF: Two Steps towards the End of the Centralised Bargaining Machinery

The decision of the national government to curtail union rights stemming from public CCTs was part of a bigger programme of reforms of the public sector, which counted on financial resources provided by the WB. It involved a number of branches of the state under the supervision of the Ministry of Economy and the participation of dozens of experts in industrial relations, public sector, law and economics, who carried out a diversity of studies and projects (Banco Mundial 1991; Daireaux et al. 1990). In the case in point, a team of consultants was responsible for the design and implementation of a strategy, whose aim was to change the conditions, which had allowed public trade unions to defend rather successfully their pay and terms of employment. The explicit objectives were to remove obstacles to productivity growth and to regain managerial control. It is important to bear in mind that the clause of the so-called *ultra-actividad* included in the CCTs permitted unions to keep past achievements by boycotting negotiations, while they simultaneously negotiated wage increases by means of ad hoc acts. In the private sector, this tactic has also been important for trade unions; but employers have had the power to introduce change despite the restrictions of collective agreements. In the public sector, this possibility has been always smaller. The dilemma for the government resided in how to force unions to negotiate collective agreements anew avoiding open conflicts and resistance through legal manoeuvres.

After the evaluation of alternatives, the solution was to be found in a strategy whose core was founded on a complex legal apparatus. It was composed of two steps: first, the suspension of a set of clauses deemed to damage productivity and managerial prerogatives, and second, the opening of negotiations between the enterprise and the public unions. In carrying out this task, a team of experts in industrial relations together with personnel managers analysed more than 6,000 clauses, which belonged to fifty collective agreements of thirteen national enterprises. In total, the team suggested the suspension of 718 clauses. In the case of SEGBA, fourteen clauses and one act were suspended through an administrative decision of the Ministry of Labour, which on 14 December 1990 rejected the formal claim of the LyF CF. In a context characterised by the massive mobilisation of legal resources to push through change, the court appeal attempted by LyF CF was condemned to failure; the Supreme Court of Justice threw it out.

The suspension of clauses affected five main areas. First, public authorities targeted clauses which warranted subsidies to the union above those required by labour legislation. Until this decision, LyF CF had enjoyed an amount, equal to four per cent of total wages, deposited in the trade union's account by companies for social ends. The union distributed this money between its Housing Fund, the Children's Holiday Camp, and its Cultural, Educational and Sports Fund. In addition to this, the union received a fixed amount of money for a programme of Life Insurance organised by its Loan Cooperative. This cutting of funds hit the union's services severely. Second, automatic increases in line with inflation were also stopped. The policy behind this measure was to tie wage increases to productivity growth, anticipating a specific decree on the matter that would be issued by the government one year later. Although the suspension of index-linked wages was particularly unpopular among public workers, it was backed by a vast percentage of the population. The government had partly attributed economic problems and inflationary pressures to wage increases in the public sector. Third, the Ministry of Labour suspended the group of clauses that guaranteed trade union participation in recruitment, promotions and levels of employment. Fourth, a clause included in the collective agreement in 1975, by which the enterprise committed to job stability, was removed. Lastly, higher levels of management were taken out of collective bargaining.

The governmental strategy was thought up to take advantage of the opportunities opened by the economic and institutional crisis. It would prove to be successful: public trade unions opted for negotiations and refused to take industrial action. Chapter Seven already mentioned the disciplining role of the crisis, which immobilised civil society. The consultants who were in charge of designing the plan congratulated themselves on the extent of the transformation; for them, the suspension of clauses had created favourable pre-negotiating conditions for public enterprises for the first time in more than forty years (Campaño and Caruso 1991; Daireaux et al. 1990).

In launching the second stage, the same team of consultants devoted their efforts to training negotiators and drafting directives about the issues to be achieved in bargaining (Ministerio de Economía 1990a, 1990b). They wanted to change managerial culture adapting the behaviour of management to the dynamics of the private sector. The idea was to prepare public managers to face negotiations with trade unions from a position of strength, and in particular, to help managers draft their own proposals. Historically, union officials had had the initiative. They used to set the agenda of negotiations by putting on the table their own draft agreement; managers limited themselves to accepting or objecting to the content of union documents. For carrying out the negotiations, the government set up a single table for the electricity enterprises affected by the decree: SEGBA, Agua y Energía, and Hidronor. The trade union side was taken by LyF CF and FATLyF, the trade unions that had negotiated the previous agreements. However, partly because of changes in the authorities of the Ministry of Economy, and partly because of delays due to trade unions' opposition to the employers' project, negotiations became derailed. The new collective agreements of the electricity sector would be finally signed by the new private owners from 1992 onwards.

Despite the failure of negotiations, the government delineated the agenda of change, for the list of aims of the industrial authorities, as far as contents were concerned, would be replicated by the new companies. The bottom line of the list was to reduce provisions in the public agreement as to the minima required by labour law; thus, forty years of workers' achievements over the minimum recognised by law was put at stake. The official directives emphasised the need to maintain management's exclusive right to decide over recruitment, career development, discipline, transfer of employees and others; to establish links between increases in wages and productivity, or better, productivity and actual profits; to create new procedures and bodies to prevent and sort out collective conflicts; and to oblige unions by concerted agreements to maintain essential services in case of industrial action. Perhaps the most contentious directive to negotiators was to introduce the so-called *cláusula de blanqueo*, a clause annulling every single right or obligation emerging from previous agreements, acts or company resolutions. This clause targeted the notion of *ultra-actividad*, for if accepted by trade unions, this clause meant that the negotiating parties had to negotiate every single issue anew. The clause was softened by the official policy, according to which, whenever a controversial topic came up, the negotiating parties could extend the validity of a clause belonging to an older agreement for six more months; but after this period, its validity expired.

Fragmentation of Bargaining

As the government dismantled the protective collective agreement of 1975, the new private owners found themselves empowered, for LyF CF had been momentarily deprived not only of the right to participate in managerial decisions, but also of important sources of income. In fact, by declaring most daily issues as non-negotiable until the finalisation of a new agreement, the decree froze the ability of the union to mobilise workers on the shop-floor, as proved by the evolution of trade union claims through the CIAPs (Table 12). The number of claims formally dealt with at company level only recovered from 2001 onwards.

Table 12: Evolution of Trade Unions Claims Via CIAPs

Type of claim	1989	1992	1995	1998	1999	2001	2003
Claims from last year	596	588	112	22	23	80	n/d
New claims	100	32	22	1	–	74	136
Total claims	696	620	134	23	23	154	n/d
Favourable	95	180	2	–	8	58	102
Quit from procedure	40	118	7	–	5	48	n/d
No favourable	1	–	34	–	–	–	n/d
Waiting for resolution	560	322	47	23	10	48	n/d

Source: LyF CF Memoria y Balance (several years)

As a result, the privatised companies began running the business in a context in which the risk of interference from trade unions was very low. Thus, companies faced the negotiations from a position of strength. Trade union officials were convinced that the CCT 78/75 would not be recovered:

> We could do nothing. It was clear we would neither stop the government nor recover the CCT 78/75. We decided not to expose members to the consequences of industrial action. What for? How do you fight back the government? And our guys would not strike ... None of the public unions took industrial action [in fact, workers from telecommunications, railways, public health, public schools and many others did take industrial action]. There was one thing clear for us: the CCT 78/75 was lost. (Union Official – LyF CF)

Given the extent of the counter-mobilisation, LyF CF was on the defensive. Its aims were, on the one hand, to avoid the fragmentation of collective bargaining and, on the other hand, to defend firstly pay and secondly terms and conditions of employment. The union found out that companies were ready to take full advantage of the situation and were therefore unwilling to negotiate. Then it was the union which, at the time, had to insist on renegotiating the suspended clauses to renew the agreement and union officials were ready to make concessions in exchange for money. However, the union failed to achieve its main objectives.

While LyF CF was initially able to avoid the fragmentation of bargaining on the distribution side, company agreements proliferated on the generation side. In 1993 the union closed an agreement with the distribution companies EDELAP, EDESUR and EDENOR (CCT E 225/93). In contrast, it took time to finalise the company agreements in generation, apart from Central Puerto where agreements were concluded in 1993. In Central Dock Sud, the agreement was closed in 1994, and in Central Costanera the negotiations ended in 1995. The contents of the agreements were similar to those of the CCT E 225/93 with regard to terms and conditions; but in generation workers received better wages and additional cash benefits. Hence, as in the UK, fragmentation highlighted disparities in the power of different groups of workers. Still, the absence of multi-unionism, business agreements within the same firm, and company mergers and takeovers softened the effects of this disparity upon organised labour.

Although the distribution companies succeeded in introducing change, they refused to continue with joint bargaining. As soon as the CCT was closed, the companies announced that they would not embark on joint negotiations in the future. LyF CF complained but was unable to counter the decision; consequently, fragmentation developed as well in distribution. Fragmentation nevertheless stopped at this stage, for the law forbids negotiation of CCTs below company level. Moreover, the market structure of the electricity industry in Argentina did not pass through the process of mergers and takeovers experienced by the industry in the UK, both factors that benefited the stabilisation of the bargaining institutions. However, as will be shown in Chapter Ten, fragmentation constrained the ability of LyF CF to engage in collective action.

Distribution companies, particularly EDESUR, were reluctant to honour the agreements, which brought about demoralisation and conflict with union lay representatives. In 1994, the union threatened industrial action if EDESUR did not abide by the agreement or if it continued to neglect health and safety measures or persecute lay representatives (Dinamis, no. 48, Julio 1994; LyF CF 1994b). In this context, the negotiations in reaching the company agreement were tortuous. Bargaining began in 1995 and lasted almost three years. Between 1995 and 1998, when EDESUR finally acceded to closing the CCT 316/98, management reinforced direct communications with employees to minimise the influence of the union on the shop-floor and union lay representatives were ignored whenever possible. In EDENOR, despite following a similar path, the company agreement was implemented in 1995. However, it was only after painful negotiations including threats of industrial action that EDENOR sat at the negotiating table in 1998.

The reluctance of EDESUR and EDENOR to negotiate and their intention to sidestep the union cannot be explained by their failure to introduce change in the agreements. The CCT 223/93 was a crude expression of the counter-mobilisation launched by the government in 1990. It manifested the new power relationships within the industry as it incorporated almost every recommendation given in the directives to the public negotiators, now pushed forwards by the private managers. It is necessary to recall once more that the government had issued a decree (1334/91) forbidding wage increases without productivity growth. Accordingly, a special commission was set up by the Ministry of Labour to evaluate whether increments in wages were accompanied by changes spurring productivity. This was a weapon enabling employers to push through changes deemed by the commission as proofs of expected rises in productivity before authorising (*homologación*) the CCT. In short, the distribution companies took full advantage of the opportunities opened by legal means. The CCT E 225/93 was limited to employees and workers: middle and higher levels of management were not included; their clerical support employees were also outside the agreement. More importantly, workers employed in contracted companies were explicitly excluded too; this would later become an important source of inter-union competition as outsourcing grew. Exclusive managerial prerogatives over recruitment, staff levels, job design and the labour process were reaffirmed. Flexible working practices were also conceded by the union by means of general statements of intention, thereby increasing the opportunity for managers to introduce change as they no longer had any obligation to consult or negotiate with unions over organisational issues. This objective was reinforced by the reduction of workers' categories from fifteen to six. Lastly, the working day was extended from seven hours to eight hours and twelve minutes. LyF CF, in exchange, recovered their sources of income; the union also obtained money for workers as compensation for signing the new agreement and, at the end of the year, in the form of a productivity bonus. The union was also able to keep certain clauses without alteration despite managerial insistence on the contrary.

The Irruption of the Unknown: Inter-Union Competition

Apart from fragmentation, outsourcing was another vital dimension of the process of counter-mobilisation against collective bargaining. After privatisation, private companies began to outsource marketing and customer services, maintenance and repair, construction, wiring, cleaning and security. These contractors, in turn, subcontracted work to smaller companies: cooperatives formed by workers who had been dismissed by the core companies, or independent workers, often under worse conditions of employment and with low security levels. These contractors promoted unionisation of their workforces with other unions to avoid abiding by the CCT of the LyF CF:

> The companies, instead of choosing our CCT, join the CCT of UOCRA [*Unión Obrera de la Construcción de la República Argentina*, a union of building workers], or UOM [*Unión Obrera Metalúrgica*, a union of metalworkers], or whichever CCT from whichever trade union as far as it is lower than ours. And we are talking about legal workers, for they have only 200, 300 guys within the CCT. When the contractors need more workers, they themselves contracted a labour force too! (Union Official – LyF CF)

This created a new situation for LyF CF, which had never experienced this type of competitive challenge. For example, the power station Central Costanera had 795 workers at the time of privatisation (Luca 1998), and over 95 per cent of them were members of LyF CF. By the year 2000, it had only around 200 workers and some subcontracted companies which, between them, employed a further 200 workers under different CCTs, including that of LyF CF. Masa, for example, the most important contractor of Central Costanera, employed seventy workers under the CCT of LyF CF; work for all of them was outsourced straight to the contracted company and they were transferred without losing their benefits. However when the workforce was insufficient for the tasks required by Central Constanera, Masa hired additional workers under the CCT of the UOCRA. In 2002, the company had to employ 100 extra workers, who earned two pesos per hour, while the company was paying six pesos per hour to its core workers (LyF CF 2002).

This type of situation created tensions between LyF CF and competing unions. LyF CF presented formal complaints to the Ministry of Labour in which the union requested the recognition of its right to represent workers (*personería gremial*) who were carrying out tasks proper to the ESI activities.

As this trend developed, outsourcing became a key issue on the agenda of LyF CF, particularly after 1996 (LyF 1996). The union began to pressure contracted companies to organise their workers, while making core companies responsible for the conditions in outsourced companies. At the same time, LyF CF looked for official recognition from the Ministry of Labour of its right to monopolise the representation of workers of contracted companies. In 1998, for instance, LyF CF formally complained to the Ministry of Labour that EDESUR had 700 contracted workers outside the electricity agreement and without health coverage (LyF CF 1999). By the end of the 1990s, union policy had begun to pay off, although often with the concession of lower wages and worse terms and conditions in the CCTs agreed with contracted companies. During 2003, LyF CF closed twelve such agreements, seven of them with companies providing services to EDESUR (LyF CF 2003). If these employees benefited in comparison to workers covered by CCTs of other unions, differences with core electricity workers persisted. So not only did the amount of CCTs bargained by the LyF CF grow, but also the union developed a kind of second-class CCT, which was applied to contracted companies.

The truth of the matter is that the extension of the phenomenon concerned union officials:

> The private owners began to form contracted companies. Who did those companies start to employ? The workers who had gone from the industry; the company needs them. It is not true that you managed all this with 4,000 workers. The reality is that we should be 9,000, 10,000 blokes according to circumstances; we are 4,000: where are the remaining 5,000? They are working. They are working in contracted companies. (National Officer – LyF CF)

> Although approximate, these figures illustrate the magnitude of the counter-mobilising effect of outsourcing. As stated by a shop steward when discussing the problem: 'We may have, moving around, not always as permanent workers ... Let's say 3,000, 3,500 people working for contractors. And another 1,500 coming and going; today they are working, tomorrow they aren't. They are swarming around the contracted companies. And the thing is that we don't represent most of these companies'. (Lay Representative – LyF CF)

As they shared the same place of work, shop stewards attempted to recruit them, at least as volunteers, when contracted workers had no union representation at all. However this tactic proved to be a failure, for while workers feared dismissal, a volunteered affiliation meant just a precarious access to trade union services, but not to the CCT, which applied just to the core company. In short, trade unions' strategy has been threefold: negotiation with contracted companies obtaining their consent to facilitate the official recognition; making core companies responsible for the situation of outsourced companies; and lobbying the Ministry of Labour for recognition as the only union with the right to represent and negotiate collective agreements within the contracted companies.

The Case of LyF MDP: Struggles around Collective Bargaining

When explaining the distinctive character of the counter-mobilisation process in the case of LyF MDP, as compared to that of LyF CF, it is necessary to bear in mind two factors. Firstly, the Decree 1757/90 could be applied to curtail trade unions' rights by national companies like SEGBA, but not by the provincial company ESEBA. Consequently, the CCT 36/75 remained untouched in ESEBA until its replacement in 1994 by the CCT

E 1052/94, the product of a political manoeuvre planned by the public authorities, which were running the company.

Secondly, the process of fragmentation of ESEBA had an outcome different to that of SEGBA. While in SEGBA there was only one union representing the entire workforce, in ESEBA there were fourteen. Yet in ESEBA, multi-unionism was completely different from that of the UK. There was no inter-union competition for membership, for unions organised the workforce on a regional basis and delegated their bargaining rights to FATLyF, which negotiated a single agreement for the whole company. By this delegation of power, the weaker unions benefited from the aggregate strength of the national Federation.

The division of ESEBA into five companies multiplied the instances of negotiation from the point of view of FATLyF. However, it simultaneously meant the reduction of the number of regional trade unions in each of the would-be private companies. The division additionally implied that LyF MDP would become the only union organising the workforce of the most important of these new companies, and hence, the traditional reasons for delegating the bargaining power to the Federation evaporated. The consequent fragmentation of the company was an opportunity for LyF MDP to take over the process of bargaining and fight back the counter-mobilisation wave and its outcomes, including the illegitimate company agreement imposed on the union by ESEBA with the connivance of FATLyF.

As in the previous case, counter-mobilisation against collective bargaining unfolded in two phases: first, the replacement of the CCT 36/75, and second, the managerial attempt to undermine workers' collectivism through anti-union practices. But in this case there is an additional element which coloured the context in which counter-mobilisation developed: the intransigent character of the union leadership with regard to privatisation. Faced by a determined opposition, the provincial government, ESEBA and, since 1997, the private company EDEA besieged the union using multiple strategies.

The First Step in the Counter-Mobilisation against Collective Bargaining: The Replacement of the CCT 36/75

Similar to the privatisation of SEGBA, the sale of ESEBA required annulment of the collective agreement to attract the attention of private capitals. Unlike SEGBA, managers in ESEBA could not count on the national Decree 1757/90 to dismantle the CCT. According to an interviewee, there were also financial reasons behind this assault:

> The only chance to privatise ESEBA, I repeat, was by giving absolute power to management through a new CCT. According to the confession of an interested party, ESEBA with the historic CCT, the 36/75, had 30 per cent less value than the final price at which it was finally sold off. It was the key to selling ESEBA. There is no other explanation. If not, why was it privatised in 1996–7 when the idea was already in place in 1993? (Union Official – LyF MDP)

The way to achieve the objective was to obtain the support of FATLyF in bargaining a flexible CCT. FATLyF and LyF MDP had had continuous clashes ever since privatisation invaded the public agenda. By 1990, they had already dissented when the ownership status of ESEBA's precursor company DEBA, was restructured to prepare for its privatisation. This time, FATLyF supported the official project and ignored the alternative one elaborated by the regional trade unions under the leadership of LyF MDP. In the 1991 Federation Conference, their antagonistic standpoints regarding privatisation blew up into a row. LyF MDP accused FATLyF of having given consent to the sale of ESEBA; the latter accused the former of political extremism (FATLyF 1991). In 1992, LyF MDP was excluded temporarily from the Federation; in 1997 its expulsion was finally formalised (FATLyF 1997). In between, FATLyF deployed a battery of resources to vanquish the leadership of LyF MDP.

ESEBA found in FATLyF an ally to displace LyF MDP from the bargaining process. This was legally possible because FATLyF had represented the workforce in the negotiation of the CCT 36/75 with DEBA, on behalf of the fourteen trade unions operating in the company. The Federation had the legal right to decide about future changes to that agreement. The separation of LyF MDP from FATLyF therefore coloured the process of illegitimacy, in view of the fact that the local union called a halt to the traditional delegation of bargaining rights. However, as ESEBA was still undivided in 1994, it was easy for FATLyF to legally justify its right to sign a company agreement replacing the CCT 36/75 despite the opposition of LyF MDP. Nine out of the other thirteen unions voted in favour of the agreement; four voted against. The Federation thus obtained the majority necessary to close the CCT E 1052/94 in 1994.

LyF MDP went to court to present a formal complaint against the deal, whereas the Ministry of Labour gave formal recognition to CCT E 1052/94 replacing CCT 36/75 in ESEBA. The leadership of LyF MDP refused to accept this, and denounced its illegitimate and legally debatable character; however, the union was forced to abide by the agreement:

> So, to publicly refer to the agreements closed by FATLyF as just an Act, to call it 'Act' instead of CCT, is to put it beneath the whole legal edifice of the collective bargaining process. And I understand why Rigane [the General Secretary of the Union], politically, does not want to accept the agreement as a CCT and will never accept it, and talks about it as a simple Act. Now, I, as a lawyer, when I'm writing to a Jury, have no choice but to accept the agreement. (Union Lawyer – LyF MDP)

When ESEBA was finally privatised the union sought to recover the old CCT. The argument was that, given that the new CCT was a company agreement, it could not extend its coverage beyond ESEBA. So according to the union, the fragmentation of ESEBA meant that the CCT 36/75, which covered the electricity industry nationally, had to be applied in the new companies. Once more, the Ministry of Labour rejected the argument of LyF MDP. Far from acceptance of defeat, the union continued looking for the recognition of the CCT 36/75 whenever it could. In 2006, for instance, the union argued that this was the agreement that should have been applied to contractors which provided services to EDEA. The argument was the same as before, as the CCT E 1052/94 was a company agreement, it could not have been applied to other companies, therefore the national CCT should have been automatically applied to new entrants to the industry. Public authorities were left to decide.

The process of counter-mobilisation was based on a tacit alliance between managers, FATLyF's officials and Labour authorities which, initially, succeeded in stopping LyF MDP from taking full responsibility for collective bargaining.

By opening the door to labour flexibility, the CCT E 1052/94 was crucial to the introduction of the type of changes in terms and conditions demanded by private businesses:

> This [the CCT E 1052/94] was the main weapon. It launched flexibilisation; it legalised flexibilisation. In ESI the agreement of 1994 was the inauguration of a set of policies; it was not just the confirmation of changes already achieved in practice, as would be the case in other industries from the private sector. (Union Official – LyF MDP)

The agreement meant loss of job stability; shortened maternity leaves, sickness leaves and holidays; elimination of overtime and other cash benefits; introduction of PRP, multi-skilling and the unilateral redesign of tasks; flattened categories damaging promotions; amongst other things. According to a comparative study carried out for LyF MDP, the agreement implied the loss of 60 per cent of the benefits of the CCT 36/75 (LyF MDP 1994a).

The Second Step: Deepening Counter-Mobilisation against Workers' Collectivism

Not satisfied with the removal of the CCT 36/75, public and private managers assailed workers' collectivism to weaken further the joint regulation of industrial relations. Chapter Eight points to the combination of repression and money persuading workers to join voluntary redundancy programmes; in this analytical context, it should be said that ESEBA and EDEA were also paradigmatic cases of the combination of anti-unionism and the politics of money to avoid bargaining and achieve decollectivisation.

For instance, ESEBA discriminated between union members and non-union members, favouring the latter with the payment of food tickets during 1994–6. The workers who left the union received automatically 120 pesos monthly in food tickets, when the average wage at the time was 700 pesos.

EDEA, in turn, pursued a policy of marginalisation of LyF MDP, whose core elements were the restriction of issues under negotiation. As a rule, the company attempted to sidestep the union and to negotiate face to face with the workforce. The aim of this tactic has been to ease the exchange of money for benefits in order to narrow the field of trade union's interference. This is the reason why the company was often ready to pay workers to drop collective benefits. In reply to this policy, the union persisted with campaigns to warn workers against the risks of losing spaces of negotiation and representation.

The individualisation of wages has been another common practice followed by EDEA's management, offering different pay rates to the same category of workers or workers doing the same job, most of the time discriminating against LyF MDP's members.

With the basic idea being to break solidarity, these practices were accompanied by others, often with the aim of undermining collective bargaining. After 1997, EDEA established direct communication channels with employees at the workplace, even sending letters to their homes in which the company criticised union policies; it implemented credit lines for workers to compete with those of the union while refusing to deposit in trade union accounts the amount of money deducted from wages corresponding to union mortgage instalments; it disciplined workers who took industrial action and discounted their wages; and it frequently violated the CCT in its weakened sections. Overall, the company put the workforce, particularly union activists, under the threat of redundancy, which was used politically in every labour conflict.

Workers' collectivism had already been challenged from within the labour ranks as well. In 1994, FATLyF not only closed an agreement against the will of LyF MDP and its members to replace the CCT 36/75 as explained above, but also provided support and resources to a breakaway group of workers who created a parallel union. It was initially born as a sister organisation of LyF Mercedes, the biggest regional union to operate in ESEBA before privatisation. Later on, it became an independent organisation, named LyF Pueyrredón. FATLyF incorporated the union into the Federation, despite its lack of *personería gremial*. It also offered to its affiliates health coverage while cutting out any type of service to members of LyF MDP. EDEA worked upon this division. Management acceded to paying time off to the lay representatives of the parallel organisation, whom nobody had elected, while it did not treat in the same way the democratically elected representatives of LyF MDP. To fulfil the requirements stemming from the PPP, the company gave a place on the Board to an official of the parallel union, a member of FATLyF who was not elected by EDEA's workers. Anyway, LyF Pueyrredón was still an empty shell at the time of writing.

In brief, counter-mobilisation did not stop at replacing the CCT but targeted workers' collectivism in undermining the joint regulation of labour issues. However, if harassment was important, the main weapon against collective bargaining was the development of outsourcing as in SEGBA.

The Irruption of the Unknown: Inter-Union Competition

In Argentina, where the system of industrial relations precluded competition, outsourcing became the prevailing strategy to weaken collective bargaining institutions. Outsourcing permitted companies to open the door to other unions; this practice was paramount in privatised companies. As in the previous case, this was an entirely new challenge for LyF MDP.

In 2006, LyF MDP denounced EDEA for subcontracting 357 workers out of 922 whose tasks corresponded to those covered by the CCT E 1052/94. According to the law under which ESEBA was privatised, EDEA should have had a minimum of 683 of this type of worker. In fact, the company had many more, but only 570 workers were covered by the CCT negotiated by LyF MDP due to outsourcing. LyF MDP demanded the right to represent all these workers.

Outsourcing affected mainly metering, meter readers, installation, wiring, and construction. Most of the outsourced workers were under the CCT of UOCRA, which organised construction workers. Despite their being concentrated in some particular areas, members of LyF MDP and outsourced workers, unionised or not, who carried out equal or similar tasks but with different pay were always working together on the shop-floor. Workers under the CCT of UOCRA earned significantly less than those under the CCT of LyF MDP.

> You have to compare both groups of workers over one year. Because a LyF worker earns approximately eighteen month wages a year against thirteen, which is the average in other unions [...] If I compare a LyF craft worker to a UOCRA craft worker, say, the higher UOCRA category, the worker from UOCRA gets eight, nine thousand a year; the worker from LyF gets twenty or twenty-two thousand a year. There is one more detail: it is more difficult to dismiss a unionised worker in ESI. Instead, workers under the CCT of UOCRA have a fund to cover unemployment, and this makes it much easier to get rid of them. (Union Officer – LyF MDP)

This was a source of clashes between workers and unions which hurt collectivism. Electricity workers saw in outsourcing a threat to their terms and conditions of employment; union officials considered that the attitude of UOCRA was disloyal to LyF MDP. This feeling was reinforced by the fact that UOCRA initially lent their premises for the establishment of a headquarters for the parallel union. The point was that industrial action in certain areas faced difficulties, for the LyF MDP officials certainly knew that outsourced workers would not take action. Lastly, as in the case of LyF CF, the counter-mobilising meaning of outsourcing should be evaluated by its success in diminishing the scope of representation of the union, which undermined the process of collective bargaining. In the case of EDEA, while before privatisation LyF MDP organised over 90 per cent of workers, by 2006, the union organised 62 per cent of them.

Conclusion

The comparative approach illuminates the association in both countries between privatisation and changes in collective bargaining, fragmentation, devolution and other changes in the negotiating structures manifested as important counter-mobilising forces. Even so, comparison reveals a certain ambiguity as to whether these initiatives were part of a deliberate counter-mobilising strategy. The suspension of clauses of the LyF CF's CCT as part of the pre-privatisation legal changes, and the attack upon the CCT of LyF MDP, before and after privatisation, were clear components of a counter-mobilisation strategy against trade unions in Argentina. In the UK, the counter-mobilising component of several managerial decisions seems to have been the effect of actions taken for other motives, for instance, cost reduction or organisational efficiency. Yet the effective achievement of those aims has partly rested in the managerial capacity to prevent trade unions' opposition to change. In this sense, fragmentation, devolution and changes in bargaining procedures, even when they may not have had any inevitable consequences by themselves, contributed to the weakening of unions' aggregate strength. Indeed, the fact that changes in bargaining procedures were required to further industrial restructuring meant that management needed to weaken the position of ESI unions. Devolution of bargaining was key to this. Finally, comparison reveals diversity, at the same time, in the forms taken by these processes due to the intervention of country specific variables.

While the British Government left the introduction of change in collective bargaining to private capitals, the Argentinian counterpart assumed this task as a condition for the success of privatisation. The features of their respective systems of industrial relations seem to explain this initial difference. The particular place of law in the Argentinian system of industrial relations forced public authorities to act in order to shape the legal dimensions of the opportunity structure. As in other initiatives, they counted on the support of international actors who provided the necessary financial and human resources to work through the privatising agenda. LyF CF was thereby a victim of a national counter-mobilisation strategy which also affected employees, including sailors, of national public companies in the gas, railway, communications, water, postal services and oil industries. At provincial level, the combination of a different opportunity structure with an intransigent union leadership determined forms of counter-mobilisation which involved, to a greater degree, illegal and unfair procedures for dismantling previous agreements. LyF MDP was victim of a tacit alliance of employers, trade union leaders and public authorities, who connived at forcing change despite workers' resistance. In contrast, public authorities in the UK did not occupy a similar role in the process of counter-mobilisation; the initiative in this country was left to the industrial actors.

Despite the involvement of the political power in the Argentinian case, the process of counter-mobilisation against collective bargaining did not bring about the extent of changes that hit the institutional framework of collective bargaining in the UK. In fact, there occurred less fragmentation of negotiating structures, and fewer changes in the procedures, than in the UK. British trade union structure and the dynamic of the electricity markets seem responsible for this diversity.

British multi-unionism emerges as a key variable explaining the possibility of inter-union competition and tensions as decentralisation ended the bargaining structure that had entrenched the respective spheres of influence of ESI unions. Even then, there had always been some friction between the AEEU and EPEA/EMA over the interface question, and also between NALGO and the industrial unions due to pay and labour conditions differentials, though much more muted. The fragmentation of the bargaining machinery post-privatisation left management free to transform those frictions into opportunities to introduce change and to undermine unions' ability to mobilise their constituencies. Although the growing power of engineers was the main target of managerial counter-mobilisation, the trend has had consequences for the whole range of unions representing ESI workers.

In Argentina, where multi-unionism coexists with clear recruiting demarcations, competition and tensions emerged as well, but taking different forms. Its general form was spurred by outsourcing, in which competition expressed itself as inter-union conflict for official recognition at the administrative level of the Ministry of Labour. However there was also a political dimension to inter-union conflict, particularly in one of the cases under study. Privatisation divided the labour movement leading to internal confrontations, which debilitated the intransigent leadership. The latter was prevented from exercising a fruitful defence of the bargaining institutions due to a pro-company union leadership which supported managerial policies. This episode opens the door to consideration of the counter-mobilising role of certain union officialdoms. LyF MDP kept, however, the ability to mobilise the rank and file throughout the period but these divisions undoubtedly reduced the union's effectiveness.

The process of mergers and takeovers that took place in the UK might contribute towards explaining the eagerness of private managers to bargain pay, terms and conditions at the lowest possible level to counteract competitive pressures. This feature was absent from the industry in Argentina, where the risk of a takeover was almost non-existent, and companies neither bought nor sold business units, neither did they integrate industrial structures after privatisation.

Particularly in this chapter, the combination of mobilisation theory at micro-, meso- and macroanalytical levels, with the comparative method enhance the need to incorporate intermediate variables in order to account for the particular forms adopted by the process of counter-mobilisation and its effects. This need seems to be proportional to the analytical scope. The broader the scope, the more compelling seem to be the demands for taking into account industrial, institutional and political intermediate variables.

Additionally, the sequence of changes under analysis in every case highlights the importance of approaching counter-mobilisation from a historical perspective. Changes in bargaining formats and procedures gradually modified the opportunity structure and organisational capabilities of trade unions; trade unions evaluated the new conditions and essayed various organisational and political responses which, in turn, compelled managers to act accordingly when needed. However the organisational and strategic responses of trade unions cannot be isolated from their internal capabilities, in themselves constrained by the continuing process of change in the UK or by the adoption of hard line managerial tactics in Argentina.

What appears as a common feature in both experiences is the growing diversity in institutional arrangements, terms, conditions and wages that trade unions have had to face as a result of privatisation and the concomitant counter-mobilising forces analysed in this chapter. It is possible to argue that this diversity has impacted upon the organisational structures of trade unions. And in so doing, that it has also affected the organisational process by which workers shape their interests, define concrete demands, and choose actions to achieve their aims. From the point of view of mobilisation theory, these phenomena are related to the categories organisation and interest, which are vital aspects of the analysis of the fate of workers' collectivism. Hence, the next and last chapter is devoted to exploration of those dimensions.

CHAPTER TEN

Organisational Change, Leadership Styles and Decision-Making

According to mobilisation theory, workers' collectivism depends on a conglomerate of factors, which include workers' organisation and interest definition. Indeed, scattered references have already been made about the impact on those variables of the counter-mobilising forces embedded in the politics of money and the fragmentation of collective bargaining. This chapter addresses qualitative findings related to organisational change, workers' participation in decision-making and leadership styles.

The comparison illuminates two scenarios regarding the category organisation: profuse change in the UK and relative stability in Argentina, within a shared field of membership loss. This is explained by variability in the evolution of the industrial structures and the bargaining arrangements of each country. Concerning interest definition, the comparison points to how agency variables such as decision-making, union strategy and leadership style intertwined with structural and institutional variables to condition, differently, workers' mobilisation.

The United Kingdom

Organisational Change

Interviewees share the view that both membership and union density have developed negatively in the ESI ever since privatisation. Indirectly, the TUC's statistics broadly support this impression. Only fourteen trade unions had membership totals in 1996 above their 1979 levels; none of the ESI unions appeared among them (TUC 1980; 1997). Scattered figures also support this view. The EETPU was the biggest ESI union at the time of privatisation with 40,000 members, its union density being over 90 per cent. At the time of writing, Amicus had slightly over 20,000 members in the energy sector (Amicus 2005). During the 1980s, NALGO's density oscillated around 80 per cent. By the beginning of 1992, the union had 33,000 members in the ESI, that is a loss of 6,712 members since 1981. Restructuring following privatisation affected the union severely; membership barely reached 20,000 in 1996 (Unison 1996c). According to interviews, ten years later they had no more than 15,000 members 'across the whole lot, water, gas and electricity'. (National Officer – Unison) Union density in the case of EPEA, was 95 per cent for engineers and 80 per cent for managers when privatisation was announced (EPE, May 1991). Data from the certification office shows that the organisation counted on 33,127 members in 1991 (Certification Office 1992). Prospect officials confirmed they had approximately 17,000 members in the utility sector in 2006. Yet lack of data prevents the analysis going beyond tentative descriptions.

By way of contrast, qualitative findings point to important changes in the organisational sophistication of ESI unions (Batstone 1988). As intermediary organisations (Muller-Jentsch 1985; Offe and Wiesenthal 1985), unions' representative and negotiating bodies could isolate themselves neither from changes in ownership and industrial structures nor from changes in bargaining structures. Thus, trade unions anticipated privatisation by embarking on their own internal restructuring, which experienced continuous pressures as private managers were ready to exploit the counter-mobilising sides of the never-ending process of change in the ESI. On the whole, findings show trade union resilience on the one hand, and their failure to counteract workers' demobilisation and sectionalism through organisational adaptation on the other.

In all cases, trade unions' reorganisation was constrained by the same factors: organisational mergers and amalgamations, massive redundancies, the devolution of bargaining, and company mergers and takeovers. Previous chapters already discussed the counter-mobilising meaning of most of these phenomena, but their organisational consequences were seldom referred to. Yet the latter is relevant to the study of privatisation as counter-mobilisation.

Overall, the response of trade unions to the organisational challenges posed by privatisation was led by the pragmatic ideology of adaptation, which might be considered as an expression of trade union resilience in the face of an unfavourable opportunity structure and negative power relations:

> We had to accept the way companies merge and employers change. We couldn't stop that, although our members would have liked to do it. So, we had to adapt our role and, basically, be ready to bargain with the new employer, or new management team. We change much more rapidly now. (Regional Officer – Prospect)

> What we had to accept was that privatisation was something we were not going to stop, so we had, therefore, to be prepared to adapt. (National Officer – NALGO)

> We don't have a national structure that was broken upon a geographical basis. We have to change that to a national structure that was broken up on a company basis. And we have to adapt over the years because companies have merged, and changed again, and changed again, so we have just been able to control them by adapting to the change. It was difficult. (National Officer – Amicus)

The core of the adaptation was the devolution of power to lower organisational levels.

The EPEA 1989 Annual Conference instructed the NEC to review the organisation of the union on a regional basis and to make possible a mixture of members from distribution, generation, and transmission companies in each section (EPEA 1989a). The conference also concluded that the negotiating functions of the EPEA had to be organised on a basis matching the new management structures. The plan included the incorporation within the NEC of representatives from each of the future private companies (EPEA 1989b, 1989c). Additionally, in 1990, the NEC organised a seminar to consider the unification with EMA anticipating the financial, staffing and organisational challenges which privatisation would pose. EPEA expected a loss in membership after privatisation (EPEA 1989b). As union density among NJB staff was over 95 per cent, the NEC concluded that the area of expansion to counteract this trend had to be in the EMA field with the committed cooperation of EPEA (EPEA 1990, 1991). There were also acute financial reasons. A single organisation would yield economies in administration, structure and the utilisation of staff, for it would cut out the unnecessary duplication of effort and resources involved in the existence of separate NECs, FTOs, Annual Conferences, and so forth. Hence, the unification was approved in the 1991 EPEA Annual Conference (EPEA 1991). EPEA foresaw a gradual demise of centralised negotiations, which would shift the focus to the regions; thus, a reorganisation of the regional offices was carried out. After that, branch structures were also revised to ensure the devolution of authority to local levels.

By 1994, thirty-four new company branches and fifteen new section boundaries had been established, together with the election of Branch Executive Committees (BEC) for every company and the creation of annual branch conferences to debate pay and conditions. Since then, the National Conference adopted a biennial frequency. Then seven regional teams were established, composed of a national officer called 'national secretary' as leader, and one or two negotiator officers. The national secretary was responsible for the larger companies, while the negotiator officers were responsible for the smaller companies. In every case, the BEC determined the policies to be pursued at company level. In the national companies, there were usually side committees which fitted into the BEC. The basic idea underlying this reorganisation was to have a representative structure at each industrial and bargaining level. The decentralisation of authority was accompanied by the provision of additional resources to counterbalance the increased workload of regional officers. As explained by a FTO:

> Whereas before we had one central pay bargaining in the NJB that was dealt by one person at head office, and there would be some executive members present as well, with twenty-two separate companies you can't expect one person to do that. So, you have to give your authority to officers, locally, in the regions, to deal with those companies. What we did was, if we use the football analogy, man for man marking. We have identified each company with one officer. And that officer, then, has responsibility for all the pay and conditions, which was previously part of central bargaining. (National Officer – Prospect)

In the case of NALGO, the 1988 Annual Group Meeting concluded that fragmentation, particularly of the CEGB, would render ineffective their branch organisation, and that the likely devolution of bargaining would demand strengthened local skills and the provision of new resources by full-time officers (NALGO 1988b). The large number of small branches covering the industry was recognised as a weakness, only hidden by the nature of the national machinery; so it was that many branches relied on just one or two over-worked individuals. It was finally decided in 1989 that branch restructuring would maintain an employer base, in agreement with NALGO's general rule, and with the ultimate aim of establishing one-company branches (NALGO News, no. 374, 28 April 1989; no. 402, 10 October 1989; no. 409, 12 January 1990). Between 1989 and 1991 the number of branches fell from around 100 to approximate seventy. As mergers and takeovers developed, by the mid-1990s company-based branches were replaced by multi-employer branches on a regional basis supported by regional FTOs. Ironically, by 2006, Unison was considering reversing its branch structure:

> Now, in the electricity industry, companies tend to be on a bigger scale. And the interesting thing is, because the electricity market is being dominated again by the three or four big companies, we are thinking about whether it makes more sense to organise our branches on a company basis than on an area basis, or a geographical basis, but that's to be still going on. It is a nightmare! (National Officer – Unison)

As did EPEA, NALGO decided in 1990 to reorganise the National Electricity Committee adapting its structure to a company-based system of representation (NALGO 1990a). It was also decided to abolish the generation and transmission committees, which were replaced by specific advisory committees for National Power, PowerGen, Nuclear Electric and NGC. The system of representation was changed again with the creation of Unison, as electricity and gas workers formed the Energy Service Group. Then, as company mergers and takeovers blurred business boundaries, Unison ended up gathering the energy, water and transport workers under a single forum: the Business and Environment Service Group.

In the case of the EETPU, the organisational response to privatisation and the amalgamation with the AEU ran almost simultaneously. The line adopted was to devolve power to shop stewards within a horizon of minimum trade union reorganisation (EETPU 1989; AEEU 1993). In 1988, the EETPU had already set three working parties to provide information on the needs of distribution, generation and transmission, within a context of privatisation (EETPU 1988). The conclusion was to emphasise the need for stronger workplace organisations, a topic neglected for years by the Executive Council, but raised in the 1989 Annual Conference (EETPU 1989). By 1990, this preoccupation had continued with proposals to reconstitute any works committees into shop stewards' committees to cope with more localised discussions as the union had predicted, by that time, a move towards company rather than national bargaining after privatisation (see Contact, several issues).

The official decision to reinforce shop steward structures had to face up to many years of branch life deterioration. From 1971 to 1983, every biennial delegate conference witnessed hot contests around the problem of the gradual demise of EETPU branches (EETPU 1971, 1977, 1979, 1981, 1983). Since 1977, complaints had increased focus on the closure of branches, compulsory amalgamations and the appointment of full-time officials to replace elected ones in branch management. At that time, the Executive Council refused the claims, emphasising that massive industrial conferences, divisional shop stewards' meetings, the appointment of area officials to service individuals, systems whereby union subscriptions are deducted from member's pay by employers systems and the like had rendered the existence of branches a thing of the past. Yet a Policy and Rules Revision Conference held in 1983 ended the disputes with the triumph of the leadership over the opposition, which was demanding changes binding the Executive Council to Conference decisions and to elect officials (Contact, 13, 4, December 1983). In brief, the EETPU's policy together with the dynamic of the bargaining machinery weakened workplace organisations in the industry. The AEEU EETPU section therefore needed to reinforce the new orientation by launching, in the 1993 Policy Conference, a campaign which included a working party with the aim of restoring active branch life (AEEU 1993). The conference urged the organisation of special and occasional conferences for the shop stewards of newly formed companies which emerged from the privatisation of the ESI to deal with issues peculiar to them.

It was only after the amalgamation with the AEU that proposals for limited structural change arose, first, to ensure that the two electricity sections worked in unison, second, to match developments in the ESI brought about by the eventual end of the national machinery. In the main, the changes were to make principle FTOs responsible for PowerGen, National Power, Nuclear Electric and NGC, and to put the twelve RECs under the responsibility of the Executive Councillors in their respective geographical areas. The coordination of the activities was left to an EETPU Section Committee, chaired by an Executive Councillor member. The creation of Amicus did not change the basic tenet: the development of shop stewards' skills and structures towards establishing a narrow working relationship between elected senior shop stewards and FTOs responsible for the companies. Regarding representative bodies within the union, energy and utilities companies were gathered into a single unit, the Energy and Utilities Sector. Energy and utilities workers elected an Energy National Sector Committee, composed of twenty-eight people, which organised its own annual conferences (Amicus 2005).

To conclude, privatisation forced unions to look inwards. Trade union introspection in all cases began as soon as privatisation was announced. Overall, the shared line was the devolution of power, although the emphasis varied. Restructuring due to privatisation mixed with restructuring due to union mergers and amalgamations, both led to continuous organisational changes. New systems of workers' representation emerged as a result, in which ESI constituencies lost prominence compared to previous arrangements. Still, variability among trade unions would demand some qualification. For instance, in relation to EPEA and NALGO, the structural reorganisation within EETPU was small. Moreover, it was often linked to union mergers. In general, change was located chiefly in the devolution of responsibility and authority throughout unchanged hierarchical bodies. Another example: ESI engineers and managers kept a more prominent representative position in Prospect than that of ESI blue-collar workers in Amicus, or white-collars in Unison.

Organisational Readjustment and Union Resources

Union reorganisation brought about uneven results as regards the opportunity for workers to mobilise collective resources of power over a period of readjustment. In this respect, the case of NALGO would be paradigmatic. This organisation approached the privatisation as an opportunity to empower workplace and branch structures; however, union resources were stretched due to redundancies and fragmentation, slowing the build-up of local capacity. Indeed, many branches were left without their most experienced individuals; some of its branches were even extinguished. The lack of voluntary people, though not an entirely new phenomenon, became a critical point as interviewees agree on how hard it had been to persuade members to take on branch and steward positions ever since privatisation. For instance, the West Midlands Region Branch of PowerGen was seriously stretched, even though this was the area where the majority of NALGO's membership was located at the time. Similarly, to find lay representatives in certain activities, for example, in call centres, was often hazardous according to interviewees.

Scarce resources were critical too, due to cultural features. ESI white-collar workers, as compared to other white-collars organised by Unison, appear to have had a greater reliance on FTOs. While some branches had many activists and keep FTOs at a distance, most were heavily dependent on the full-time officials: 'You know, too dependant than anything that comes up, they are on the phone and say: "Look can you come in and sort this out?"'. (National Officer – Unison)

So deeply rooted was this attitude among white-collar workers that changes in the negotiating dynamics brought tensions between branch members and FTOs, as the latter were not able to deliver to the expectations of the former. By 2006, Unison's FTOs had agreed to train and develop lay representatives to take on the first line work. They encouraged membership involvement:

> I get involved up to a certain level. The things that we [FTOs] used to do, we now say to them: 'No, you need to do it. You've got the ability'. And the distinction I make is that if there are issues to do with redundancies or changes in pay and conditions

> in a significant way, I will get involved in those negotiations. Other than that, no more. (National Officer – Unison)

> Well, I can't do everything. They [branch reps] expect to be able to demand from you, without actually taking any responsibility. They expect you to do it all. You know: if it is a big company, its got locations all over the country. If you get your full-time officer to be in all of them ... It's just impossible! But, in fact, with the other full-time officer, we are actually trying to change that. We are moving towards, to try to rely on people to do a lot of things that we can't do efficiently. (Regional Officer – Unison)

In addition to cultural attitudes, employers' preferences and tactics put ever greater pressures upon scarce union resources:

> In energy, companies prefer to deal with union officials, full-time officials rather than lay workers' representatives. So, it is much more difficult to get lay membership activity in the energy side, as compared with the water side. (National Officer – Unison)

> The employer encourages that [members reliance in FTOs], I think, the employer wants that. The employers will be quite happy if they just meet me all the time, and don't meet members, so I get to make sure that, you know, if you want to meet me, fine, but I'm going to be bringing with me, you know, the representatives. (Regional Officer – Unison)

In the case of Amicus, officials have stressed that, initially, there were frictions within the union, as FTOs, who were accustomed to operate in full control of the negotiating agenda and union-management interactions, were reluctant to change. In their new role, FTOs should have been supporting shop stewards, whereas before, they had kept them at arm's length. As privatisation unfolded, FTOs had to learn to operate in an environment where things changed more quickly. Moreover, FTOs incorporated new responsibilities. For instance, they had to write their own reports about what happened in their companies for communication to members, when, prior to privatisation, all communication was channelled through one full-time structure at national level. FTOs additionally had to establish closer relationships with the rank and file through meetings and discussions on how to run company bargaining together with senior shop stewards. All that generated resistance:

> Before, the bureaucratic system operated in the industry uniformly, and slowly, and pedestrian. Well, FTOs had to adapt. So, they had to change in many ways. Before, the officers, under the old set up, had had complete control. They had to delegate that control to the stewards [...] And a number of people did not want to give up that control. (National Officer – Amicus)

Moreover, the change of the organisational focus towards the workplace structures demanded by privatisation increased union logistical problems. While Amicus had too many shop stewards in some companies, it did not have enough in others. FTOs stressed in the interviews that changes in working practices were used by managers as counter-mobilising devices, for certain working practices can make it more difficult for the union to gather members. In many distribution companies where the workforce is scattered the union needed more shop stewards to sort out communication and logistical difficulties. In most power stations, where it was usually easier to identify the constituency, shop stewards' structures were stronger; however, the union remained weak in a number of power stations too.

Variability appears as the rule:

> So, it varies, it varies. It depends on each situation, how many stewards you have and how many you need. Some companies support you, and some don't. And then, it is hard to organise structures that work [...] And you need good shop stewards to make you strong. (Regional Officer – Amicus)

The Rise of Sectionalism

The most critical finding is that union reorganisation has spurred the growth of sectionalism. The cause of this unintended outcome seems to be clear: while the national union remained relevant for members on issues like pensions, energy policy and regulation matters, it was at lower levels where negotiations on pay, terms and conditions are conducted.

For instance, in Prospect, each BEC determined the bargaining agenda; whereas, previously, the NEC had decided the priorities and policies to be negotiated, which were applied uniformly to everybody in the industry:

> The branch executive committee decided the policies that they would follow. So then, the PowerGen branch committee may have a different philosophy, and a different aim to the company next door, because the company next door may want to go in a different direction, may have a different philosophy, may have a different concern, so, that is the way we deal with it. (National Officer – Prospect)

Still, the NEC could exercise control if there was a particular branch going in the opposite direction to union policy. However, as this principle was often loosely interpreted, only extreme and unlikely cases of lay representatives agreeing on compulsory or massive redundancies would have pushed the NEC into intervention. Prospect members consequently had, by 2006, much greater identification with their BECs than with the national union.

NALGO's FTOs pointed to similar problems. The problem was not only that company bargaining agendas were quite different, but also that shop stewards lacked more comprehensive views:

> Some shop stewards can look beyond, but most, in my experience, only look at their own company. And in fact, you'll probably find that you've got someone who works for that bit of the company, and he can't even think beyond that bit, he can't even see his own company. (National Officer – Unison)

Consequently, the exacerbation of sectionalism attached workers' immediate interests to the business unit, spoiling branch life:

> What happens is that the branch secretaries concentrate in the unit that they come from, and don't always appreciate such a fact, that you are actually representing everybody across the whole of the company. (Regional Office – Unison)

In this sense, privatisation changed the role of the national officer, who had to assume the coordination of policy so as to avoid the profusion of local bargaining agendas serving managers playing different groups of workers off against each other. As sectionalism grew, national FTOs appear to have met the challenge with difficulty:

> Well, one of the things that had to change was the role of the national officer because you have to co-ordinate policy. But, it has been very difficult to avoid an FTO, or a group of workers, considering something that then is imposed on another company. (National Officer – Amicus)

Parallel to the growth of sectionalism, there was a debilitation of national focus, which increased the need for coordination. Prospect attempted to overcome these problems in two ways: by setting up coordinating committees for distribution and generation, and by reinforcing the coordinating role of the NEC. While useful for gathering and sharing information, these initiatives seem to have failed to promote effective actions at the time:

> We set up coordinating committees for various distribution companies, for generation and so on, to make sure there is a cross-flow of information and exchange of experiences. (Regional Officer – Prospect)

> We try to counter the trend towards sectionalism through coordination. And the other thing to remember is that we have the NEC, who receives monthly reports, and the NEC was restructured to make sure that it was representative, as far as we can make it, of the whole industry. In that way, the reports were channelled into the NEC, and they can see what is going on. So, we have this exchange of information. (National Officer – Prospect)

NALGO's officers manifested identical worries:

> The reality is that it's a struggle to cope with so many different companies, all with their own identity and priorities. It isn't easy. What we are trying to do from the centre, here, is to co-ordinate activities. So, we will co-ordinate our full-time officers working in regions, so we keep them informed. We keep them advised of developments, we bring them together, so they can learn from each other's experiences, generally trying to maintain a national focus. (National Officer – Unison)

This shift in bargaining responsibilities changed the role of the NEC which, by 2006, was essentially about administering policy and managing the union. This brought about a new emphasis on union services, strong in Prospect, overlapped with an organising discourse in Unison, and changing to a more organising profile in Amicus slightly later.

Paradoxically, then, while the national machinery was thought of as a barrier to the militant outlook associated with workplace unionism, the latter showed a greater readiness to compromise in the ESI within the privatising context. This inclination to compromise manifested itself in the growth of sectionalism. This was the case because privatisation increased, as compared to the past, the dependence of the staff upon the fate of the particular companies where they were employed. The new structure brought about a deeper understanding among the staff of unfavourable market circumstances, and hence this fact eventually ended up softening workers' demands.

Leadership Styles and Decision-Making

Many scholars emphasise the role of mediating factors such as the style of leadership, or the dynamics of decision-making, in the definition of workers' interests, demands and actions. So does the analytical framework applied in this book, built upon mobilisation theory, which pays special attention to the knotty relationship between leadership styles and participation, as the latter is deemed to be essential for workers engagement in the type of dialogical interactions which enable workers' collectivism.

Chapter Seven broadly defines the various leadership styles predominant among ESI unions at the time of privatisation. Although, they varied in a number of aspects, a common feature underpinned all of them: a dose of paternalism. As summarised by a FTO from Unison:

> The older paternalistic style, acting according to the idea that they [FTOs] come in, and do everything [...] The male full-time officer, who is the one who does everything for everybody, and all members, willingly or not, subordinate to his decision. I think this is just the way most unions used to be, and that is slowly changing right now. (Regional Officer – Unison)

By replacing membership involvement with FTO expertise, paternalism tended to be substituted for workers' participation. Nevertheless, neither is paternalism a function of FTO's values, nor is participation simply an outcome of the leadership style. On the one hand, paternalism emanated partly, at least in the ESI, from the centralised dynamics of bargaining; so by breaking down the national machinery, privatisation challenged its rationality. On the other hand, patterns of participation often fluctuate following the impact of events such as privatisation which disturb the context of industrial relations through changes in management-union relations, the wider political situation, or the union itself (Fosh 1993).

For instance, the right-wing orientation of the leadership of the EETPU (Hyman 1983; Kelly and Heery 1994) is an established fact. Indeed, most of its FTOs and shop stewards also expressed centre-right political views, a feature explained by the tight central control of appointments within the union (Kelly and Heery 1994). Concrete manifestations of this centre-right standpoint were the policy towards branches and the promotion of the principles of liberal democracy by which participation was limited to the right to vote. Moreover, the EETPU leadership downplayed conflicting interests, at least in the ESI, and avoided industrial action whenever possible. However, the EETPU used to mobilise workers to put pressure on reluctant managers when the opportunity structure was favourable as far as the demands were deemed achievable, this being the crucial defining criterion for action of EETPU's FTOs and shop stewards in the industry (Kelly and Heery 1994).

Yet privatisation distorted this landscape. After the announcement of the sale of the industry, there were signs of an increment in workers' participation in meetings related to privatisation as workers were eager for information (EETPU 1988; Flashlight, no. 47, September 1988). Within the same context, demands on the NEC to improve communication were raised (Contact, 17, 4, August 1987; EETPU 1988). Lastly, there were, if not pervasive, at least repeated and noticeable calls for industrial action to defend the national machinery of negotiations (EETPU 1987; Contact, 17, 6, December 1987). This incipient movement was channelled by union leadership through the formal procedures of the ESTUC, the establishment of working parties and the organisation of FUSE (the anti-privatisation campaign). The remoteness of the activities of union leaders and the lack of enthusiasm for the characteristics of the campaign, translated into demoralisation and apathy amongst the rank and file (see Chapter Seven).

Nevertheless, the end of the national machinery obliged the union to develop workplace structures to match the gradual devolution of power to first line managers. The union began promoting rank and file participation and shop steward involvement in negotiations. Although this move was resisted among FTOs, new structures ended up encouraging new practices. The likelihood of a revitalisation of workplace activity was, however, severely limited by a set of countertendencies: tradition, restrictive laws, increased market competition, an unfavourable opportunity structure, and a type of leadership that preferred the stability and predictability of interactions with management.

Kelly and Heery (1994) state that NALGO's FTOs had a greater disposition than those of EETPU to attend to member wishes and rank and file support. Additionally, during the 1980s, NALGO advocated the values of participatory democracy embedded in the workplace voting system, while fighting back the introduction of postal ballots (NALGO 1986; NALGO News, no. 271, 27 March 1987; no. 277, 8 May 1987). Unison FTOs stressed the particularities of ESI members as compared with members from water or local government throughout the interviews, especially their lower levels of participation and their stronger dependence on them. Additionally, as workplace organisation had been established only recently in NALGO, these cultural attitudes remained strong by the time of privatisation. So the latter opened a window of opportunity for FTOs to deepen workplace structures, but lack of local leaders, cultural attitudes and massive redundancies conspired against participatory policies. Findings show that even when FTOs expressed an organising concept of unionism, that is, that members should be making the decisions rather than their representatives, in practice, they have failed to motivate the rank and file. The problem has manifested itself, most clearly, in the lack of volunteers to run union activities:

> Very few of our branches have elections, because we don't get more than one person standing. Actually, in a lot of cases, people are elected unopposed. So some branches have elections but some don't, because you don't have enough voluntary people. (National Officer – Unison)

> It is a problem. I've got six branches and we are really short of representatives in almost all of them, it is a real problem. In some areas there are no representatives. (Regional Officer – Unison)

Moreover, problems of communication aggravated the lack of participation locally, thereby preventing the extension of collective discussions. FTOs complained that branches did not communicate each other, neither did members from different business units and neither did branches with their own members. Hence, information did not circulate as FTOs believed it should:

> So, even if all members in Unison are at one branch, they don't have any contacts. There are limited contacts with other business units, with other members. (Lay Representative – Unison)

> What we've got is a branch structure, but I think that where branches are poor is in communicating across. (National Officer – Unison)

> Branch reps must communicate with their members [...] You should put information down to your members. But, also, how do you know what needs to be done? What needs to be negotiated? How do you know there is a problem if the branch doesn't talk with the members? How do I find out that there is a problem in that area, and negotiate with the employers to make it better? You can't. If you don't communicate with your members, you may never know about it. (Regional Officer – Unison)

Consequently, low levels of participation and poor communication conspired against extended democratic interactions. This situation promoted workers' attachment to their immediate environment. Astonishing manifestations of this problem emerged, for instance, over the process of integration launched by Eon Company. Most Unison members from the retail unit, which was still operating under the brand name PowerGen at the time of writing, did not even recognise that they were working for Eon. Similar misunderstandings came out in other units, although less acutely. The union had to make strenuous efforts to get members to understand they were all part of one company.

The distinctiveness of EPEA among ESI unions was also underlined in Chapter Seven. It stemmed from the features of its constituency, whose role in the industry boosted a pragmatic ideology which embodied the ethos of meritocracy and moderation. In EPEA, the distance between leaders and members seems to have been narrower than that of other ESI unions. Concurrently, findings point to a strong identification between members and leaders. Both factors would have contributed to moderate tendencies towards the divergence between the objectives of members and union leaders. As a result, the leadership of EPEA corresponded to the representative outlook as understood by Batstone et al. (1977), that is, a leadership that is ready to take independent initiatives as well as to execute policies according to membership's wishes. Given those features, workers' participation was often subordinated to pragmatic considerations and seniority, but the union could be fairly rated as member-led. A similar picture continued to characterise the union after privatisation.

If members initially expressed in the EPEA Annual Conferences their disposition to fight back privatisation and, possibly, its effects, as soon as the consequences of privatisation became apparent, members showed signs of demoralisation and so participation declined. However, the reorganisation carried out by EPEA in 1994, with regards to their branch structure, had ambiguous consequences. If it contributed to consolidating the trend towards sectionalism, it had positive effects upon participation, as the new structure increased rank and file's involvement in the decision-making over matters related to their immediate environment. EMA's own evaluation of members' participation concluded that the new company branch organisation – the new basic unit of organisation formed around members working for a specific company or, where appropriate, for a group of companies – increased the number of people involved in the decision-making process (EMA 1994).

Apart from breaking down collective bargaining, privatisation distorted EPEA's internal life through the introduction of new communicational strategies for technical and managerial ranks and the spread of personal contracts. The response of EPEA was, then, to improve internal procedures and consultation. Additionally, the union took special care to improve the communication channels to members:

> One thing the employer did was to improve their communication with the staff. They started introducing leaflets, magazines, newsletters, whereas before they had done nothing. They started to tell employees about terms of employment; previously, it was left to the unions. They took on board all these measures and we had to match it. So, we have improved our communications: not only the content but the style, their appeal, and the breadth of the appeal. (National Officer – Prospect)

As with personal contracts, they segmented the immediate interests of EPEA's members. Those members who were taken out of collective bargaining began to appreciate the union mainly for its ability to represent them when tackling issues like pensions, safety organisation and professional responsibility (NOP 1991). This appreciation also extended to the quality of union services. Thus, the NEC sought to enhance its union servicing profile.

In short, the previous analysis shows that, in each case, structural and institutional (fragmentation of the industry and collective bargaining), organisational and agency variables (mainly, strategic choices and leadership styles) intertwined to shape trade unions' organisational responses to privatisation.

Argentina

Organisational Stability

As with previous topics, findings point to relevant differences between LyF CF and LyF MDP, within a common context of scant organisational restructuring.

In the case of LyF CF, the first thing to be noted is that its membership declined dramatically, whereas union density remained within acceptable boundaries. It would be wrong, however, to conclude that the latter contributed to concealing the meaning of the former. On the contrary, the decline in membership from 22,000 to 4,000 workers was the main structural determinant of the union's strategic shift towards entrepreneurial initiatives. Additionally, it offered union leadership a convincing rationale for this policy.

In turn, the union's scope of representation received impacts from two different sources. On the one hand, the government decided to take out managerial ranks from collective bargaining, a policy that was followed later by private investors. If not numerically relevant, this situation was important qualitatively, as those were the ones who had access to company information. FTOs have complained during the interviews about the chronic lack of bargaining information since privatisation. On the other hand, outsourcing impacted upon the ability of the union to organise ESI workers. Due to this challenge, union leadership adopted policies organising outsourced workers with varying results.

In the main, however, findings show that organisational change limited the downsizing of existing structures, namely, that privatisation brought adjustment but not proper restructuring. By the beginning of the 1990s, the union owned a large infrastructure with which it serviced more than 20,000 members in areas like health, personal loans, culture, education and tourism. However massive redundancies caused massive losses in membership, which translated directly into financial hardship for the union.

At the time of privatisation, the union employed 1,114 people, whose employment relationships were regulated by five different CCTs. As a consequence of the organisational adjustment during the first year after the privatisation, the reduction of trade union's employees was 23 per cent. By 1998, downsizing had affected 53 per cent of the union's own workforce, the number of employees who remained employed by LyF CF being 497. To achieve this, the union implemented their own programmes of voluntary redundancy; for instance, just between 1997 and 1998, a total of 120 people left the organisation this way (LyF CF 1997, 1998, 1999). In short, the union confronted the challenges posed by privatisation with a structure that was leaner, but not adapted in any meaningful way.

Financial hardship boosted not only adjustment but also the union's entrepreneurial policy. As stated by a shop steward, while speaking of the commercial exploitation of organisational assets by the union:

> As we lost members, we had to privatise our own capital, which was built up over forty years with the effort of every one of us. You see, on the one hand, we criticised privatisation; on the other, we also privatised. (Lay Representative – LyF CF)

To give an example, LyF CF opened their hotels to members of other unions; in 1993, 35 per cent of the accommodation capacity was booked by non-LyF CF members. Additionally, the adjustment affected the union's services for members: the *obra social* began charging fees for certain medical practices. Union assets were also affected: the organisation, for instance, sold off all their cars.

Whether these figures relate mainly to the impact of the financial losses upon union's social services, the most important sign of the deterioration of resources regarding the union's ability to defend members was the decline of *permisos gremiales*. The latter are formal authorisations by which union representatives are paid their wages by the company for carrying out full-time union activities. Since privatisation, their number fell from 200 to ninety-eight. Private owners first downsized their workforces, and then they made unions accept the reduction of the *permisos gremiales*. Moreover, by 2006 the union was paying seventeen out of these ninety-eight. Still, the crucial decision taken by LyF CF was to modify the union's rule book in order to reduce from thirty-two to twenty-one the number of members of the CGA (*Comisión Directiva*), whose wages were paid by the union as well.

If the composition of the CGA was traditionally a source of tension among different groups of workers within the industry, the reduction of the number of CGA's members, together with the fragmentation of the industry, increased the likelihood of conflicts:

> I'm speaking of conflicts between companies and sections. Because each section has lay reps, and when there are elections, every section chooses, often among the shop stewards, candidates for CGA's posts. If you reduce them, then, there are more disputes among sections around which section gets people to be elected for the CGA. So, as companies multiplied, this has been the great battle since privatisation. (Union Official – LyF CF)

The union did not impose a quota of representation to prevent disputes by agreed procedures. Indeed, the problem was sorted out through political struggles over electoral periods. As there were unopposed general elections since privatisation, the hegemony of the union leadership was consolidated through political clientelism, while the opposition was displaced by various methods, sometimes in connivance with the employers.

Regarding LyF MDP, union membership also declined due to the job loss caused by privatisation. Yet, it should be stressed that privatisation affected only 30 per cent of LyF MDP's constituency. The rate of the decline was not equivalent to that of the redundancies; and then, as a whole, the proportion of the impact was less than for other unions. In this case, outsourcing has been the main factor undermining union's density and scope of representation. It is debatable whether outsourcing meant a reduction of the former or a narrowing of the latter; but the concrete outcome was that LyF MDP organised less company workers in 2006 than it did before privatisation. So the union organised 62 per cent of company workforce, whereas before it had covered around 90 per cent. This damaged the effectiveness of the union's collective actions.

As with LyF CF, the organisational sophistication of the union remained essentially unchanged. Yet the union did have, however, organisational problems due to the decision of FATLyF to expel LyF MDP from the national Federation, and therefore, to halt the provision of social services like health coverage and tourism. Although these problems were not in the area of collective bargaining, they are worth mentioning because the provision of social services in Argentina continued being crucial to union leaders maintaining their legitimacy into the future. However, difficulties were not directly produced by privatisation policies in themselves, but by inter-union conflicts brought about by privatisation and the strategies of the soon-to-be private company to get rid of LyF MDP. To tackle this problem, the union leadership requested workers to pay a special and voluntary contribution for the creation of their own services through a political discourse based on the notion of class solidarity against individualism. An astonishing 99 per cent of union's members answered the call.

Although the union did not change its internal structures, there were very important innovations at higher organisational levels. These innovations related particularly to the union's global response to privatisation, and more generally to market reforms. Briefly, LyF MDP founded, together with other unions, mainly public ones, a new national Federation: the CTA. This was one of the reasons why it was expelled from FATLyF, which was enrolled in the rival CGT. After that, once inside the CTA, LyF MDP created FeTERA, an umbrella organisation for energy workers based on principles different to those of the traditional unions in Argentina. This began to be organised around 1995, and represented around 12,000 workers at the time of writing. From this platform a huge arch of international relations with diverse organisations was also built. Every time the union engaged in bitter conflicts, this organisational umbrella was used to ask international solidarity to put pressure upon the company and the provincial government. Also, LyF MDP joined the International Federation of Chemical, Energy, Mine and General Workers' Unions and founded the *Coordinadora de Sindicatos del Sector Energético del MERCOSUR* (*Mercado Común del Sur*) with ESI unions from Paraguay, Brazil and Uruguay. International solidarity has to be understood as a replacement for missing power resources at national level since LyF MDP was expelled from FATLyF.

The Rise of Sectionalism

As in the UK, fragmentation of the industry brought about the rise of sectionalism. In LyF CF, rivalry between lay representatives from the distribution companies EDENOR and EDESUR had been noticeable as they competed for declining union resources. Additionally, while generation workers were better off negotiating acceptable terms and conditions, distribution workers engaged in bitter conflicts, which led on occasion to industrial action. However, the union never coordinated common actions, even though EDENOR and EDESUR underwent simultaneous periods of conflict. Findings show that, overall, differences in working conditions caused tensions within the union. Workforces experiencing worse conditions of employment tended to attribute blame to FTOs:

> The division of the public company broke workers' solidarity. People know nothing about what is going on in other companies. Yet, if lads hear about better conditions in other companies, they complain to us, as if we were responsible for their lack of ability, or strength, or whatever, to back the negotiation of better agreements. (Union Officials – LyF CF)

Generation workers were inclined to isolate themselves from the wider picture every time they achieved favourable terms and conditions; this was particularly marked between 1993 and 1998, a period of major conflicts in the distribution area:

> The companies divided everything and everybody; they advanced as far as they could. They broke down companies' structures. Before privatisation, we were a family. Things improved somewhat between 1996 and 1998, but anyway, what they got is a change in mentality. Workers arrange something with a company; others arrange a different thing in a different way in another one. Nobody cares for fellow workers next door, especially, generation workers. (Lay Representatives – LyF CF)

However, attempts to co-ordinate policies were often circumstantial. Indeed, union leaders seem to have assumed sectionalism to be an inevitable outcome of industry fragmentation, a new feature of the landscape which they should be accommodating.

Instead, as explained in previous chapters, the fragmentation of ESEBA benefited LyF MDP, as the union bargained with a single private company. This partly explains why sectionalism, a common finding so far in the analysis of the organisational impact of privatisation, was not a problem. Still, pressures to devolve managerial decisions to lower levels were present. This obviously impacted upon the workplace as shop stewards had to deal with issues previously bargained at higher levels. Nevertheless, most crucial bargaining decisions, pay for instance, remained still centralised in CGA's hands.

Leadership Styles and Decision-Making

Chapter Seven sketches the contours of LyF CF's style of leadership, whose ideological pillars have been the notions of multiple trade unionism and participation. Based on these, a highly concentrated and pragmatic bureaucracy, historically sustained, on the one hand, the trade union's accumulation of assets, and on the other, the disposition towards establishing channels of dialogue with public authorities opening the door to political exchange. During the 1970s, these notions eventually backed union involvement in co-management, various forms of political exchange, disposition to negotiate with democratic as well as military governments, and even direct participation in the apparatus of the state (Palomino 2005; Pozzi and Schneider 1994). Traditionally, policy making has been concentrated in just a few people, who have controlled the process of decision-making through a vertical and bureaucratic leadership:

> Our union was always ready to take a seat at the negotiating table. We have a different sort of outlook than other organisations. For us, industrial action is a sign of failure. We have always tried to deliver the goods by negotiation and participation in industrial and political spheres. And look, the union has always trained people, and created small elites – five, ten blokes – who run the organisation this way. (Union Officer – LyF CF)

However, workers' participation in the *comisiones internas* had often balanced this tendency somewhat, at least until the last dictatorship. After the latter, the internal life of the union never recovered fully:

> I'm not going to lie to you, the deals are often, more or less, what the CGA wants them to be. It is very difficult for a lay rep to reverse something that you have negotiated with the company. If not, you wouldn't have leadership. It is sad, but it's the way. The union is a vertical organisation, head, body, and a political line goes down, and that's it. (Union Official – LyF CF)

> OK, lay reps may influence to certain extent the policy, they can disagree with something and oppose it; but to be honest, the CGA has a mandate. It defines general policies, and communicates them upside-down. Otherwise, there is no authority. (Union Officer – LyF CF)

Representative democracy predominates over participatory democracy. Currently in 2006, lay representatives are elected every two years by fellow workers, while CGA posts and the General Secretary of the union are elected every four years in general elections. Despite the lack of opposition, the high level of workers' turnout legitimised the authority of the leadership. Indeed, the exercise of the representative democracy actually served the leadership by hiding the absence of true collective definitions about crucial decisions. In 1992, when the companies had already begun to be transferred to the new owners, the union leadership brought forward the general election to back, ex post, their decision to support privatisation.

The union kept the requirement for lay representatives to approve every CCT agreed by the union, as they had to endorse annually the Memoria y Balance, which are the reports of the annual activities of each union department, and the official report detailing the union expenses. However, given the absence of true workers' participation beyond the elections, these meetings became a kind of automatic approval of the leadership. Additionally, the CGA maintained, when necessary, informative meetings with the shop stewards of each firm about specific company problems. Also, FTOs usually participated in the sectional meetings of the *Comités de Lugar* (Site Committees), where problems relating to a particular area of production were discussed. Lastly, there were informal communication channels activated by FTOs, often embedding profuse doses of paternalism. All these instances were mostly informative or, at best, informally consultative. Thus, workers' involvement in the process of decision-making seems to have been constrained by the extended prerogatives of union leadership:

> Shop stewards meetings have no regularity. The union calls an assembly only when a big problem comes out. Otherwise, most meetings are sectional meetings, just informative ones, for instance, meetings of the *Comité de Lugar*, or meetings with company's shop stewards, but often without a mandate. It is the leadership who truly decides. (Lay Representative – LyF CF)

This panorama was, partly, an outcome of privatisation, which came to renew the obstacles to participatory democracy. Shop stewards emphasised, time and again, the low level of workers' participation as a result of fear, demoralisation and lack of belief, mentioning as the main evidence the lack of volunteers assuming responsibilities as lay representatives. During the worst period (1991–7), the union hardly covered shop stewards positions; indeed, the union at that time had neither second shop stewards, nor militant representatives, that is, the old union posts to back activities of the principal shop stewards at firm level.

Fear of retaliation rated high among the reasons that would explain this development:

> A lot of people refused to be shop steward. You know, the managers saying: 'Be careful my friend, you are shop steward today, but tomorrow …' (Lay Representative – LyF CF)

> Nobody wanted to be shop steward. There was fear because the shop stewards are always in the eye of the storm. Managers look for them first when there are problems. (Union Officer – LyF CF)

Workers' demoralisation and lack of belief, in turn, had a two-fold reason. On the one hand, shop stewards recognised that most lay representatives did not oppose voluntary redundancy programmes; quite to the contrary, they accompanied the process. This is not surprising as the trade union agreed some of these programmes with the company and the Ministry of Labour (*homologación*) (LyF CF 1992, 1993, 1994). Even worse, a few shop stewards made blacklists of combative activists and negotiated voluntary redundancies for themselves above the normal rate.

> Let us be honest, lads, we all learnt about colleagues who made blacklists at the time. And even worse, they are still reps. (Lay Representative – LyF CF)

> It is true. There were people in the union who connived with the bosses. Just a few; don't think, please, that this was common. Most were good fellows. But, you know, for money, I guess. Although, I wouldn't be able to prove that they got money from the companies. But, everybody knows they did. (Lay Representative – LyF CF)

On the other hand, many others joined redundancy programmes due to their own demoralisation, as the union lacked initiatives opposing the privatisation. In fact, union discourse explicitly assumed a demobilising content, which contributed to feeding the exodus. By 1992, unions' slogans stressed that the role of leaders was to preserve the organisation and to avoid pointless struggles which might dishearten militants and activists.

Three other aspects are significant in this respect. First, the lack of new entrants to the industry made it difficult for the union to train new lay representatives. According to interviewees, this conspired against the renewal of the union leadership. Second, lay representatives complained that their workload increased and was more difficult to perform as a consequence. Companies neither consulted nor communicated with them about change; hence, shop stewards had to be continuously dealing with managers and fighting back their unilateral decisions. As an FTO put it: 'Today, to be a shop steward is much more costly in terms of time, money, and hours of rest'. (National Officer – LyF CF) Third, the divide between union officials and the rank and file widened after privatisation. The reason for this separation is broadly attribute to the general factors.

It was within this context of declining participation that union leadership could develop an entrepreneurial unionism. This style was characterised by union leaders as a two-fold strategy: the reinforcement of workers' solidarity around the organisation and an entrepreneurial project to produce new resources to increase and strengthen union social services. Findings have shown that the union neglected the former, developing only the latter aspect.

In the case of LyF MDP, instead, the leadership framed union actions through political discourse, which stressed workers' democracy, autonomy and activism in opposition to the exercise of formal democracy. Indeed, the latter is to blame for being empty of any social content.

The promotion of workers' participation was deemed by union leaders to be essential to development of a powerful organisation. From the beginning, the union attempted to root the confrontation to privatisation in mass meetings, shop stewards assemblies and rank and file gatherings at the workplace. Workers' participation then grew over the long-lasting anti-privatisation campaign, despite obstacles. Only the determination of the leadership to punish the lack of commitment by applying union's rules made participation grow. For instance, shop stewards who failed to take on their responsibilities were sanctioned, and occasionally replaced. Also, in the face of insufficient attendance at meetings, suspensions of those who were absent were applied through members' ballots, a prerogative rarely used by other trade unions. So privatisation faced a mobilised union used to engaging in collective actions often ratified collectively. Union's efforts to communicate with members were crucial for workers' mobilisation:

> People, firstly, were very well-informed, though it has been shown that it is not enough to give information. It is necessary to discuss it with the lads. Also, people saw that we have full dedication to the organisation. Things can be done better or worse, but workers did not doubt about our dedication to the workers' cause. The other thing is that everything we said about privatisation was gradually proved by facts. (Union Official – LyF MDP)

The capacity of the union leadership to foresee privatisation policies and explain to members their meaning and their consequences was vital to creation of a platform of trust between themselves and the rank and file.

> There is trust. People do not necessarily share the ideology of the CGA [...] I would say that the vast majority does not have a Leftist orientation. But, there is mutual respect and trust. The CGA does not go beyond where the rank and file is ready to go. And the latter know that the CGA will not betray or negotiate for themselves [...] Mutual respect and trust translate into discipline. Neither is the CGA for wild strikes, nor do workers cross picket lines. (Union Lawyer – LyF MDP)

Trust, in turn, allowed the union to overcome critical events. For instance, when a group of CGA members and shop stewards joined a redundancy programme in 1995, the leadership called mass meetings to discuss the situation politically. When individual members began to accept food tickets against a general assembly decision, the union leadership decided, once more, to discuss the situation openly in mass meetings, reversing earlier decisions by the mass vote of workers.

Despite the profusion of mass mechanisms of decision-making, the style of LyF MDP's leadership corresponds to that defined by Batstone as representative, that is, a leadership with capacity to take decisions independently (Batstone et al. 1977). Furthermore, general assemblies are, mostly, called by union leaders to debate and vote on concrete tactics and forms of action. The absence of an organised opposition after the breakaway facilitated the hegemony of the current leadership. A fact deemed by several interviewees to be damaging for the leadership as it contributes to relaxing self-discipline and increased risk of bureaucratisation:

> We need opposition [...] It would help us to have better reflexes against our own bureaucratic tendencies. (Lay Representatives – LyF MDP)

> Actually, it was harmful for our internal life the disappearance of the opposition. If this were not the case, who knows, we wouldn't be still managing the union. And I'm not speaking of the opposition, but of fellow workers who have lost the enthusiasm to participate as they have complete confidence in the CGA. But the renewal of leaders would be good for the organisation. (Union Official – LyF MDP)

This appears as a weakness in common with LyF CF. There was no renewal of leaders, though for different reasons. In the case of LyF MDP, the demarcation between active and passive groups of workers seems to have gradually crystallised after the breakaway. Besides this, a degree of demoralisation after privatisation, due to aggressive company human resources policies debilitated participation. So whether or not workers still responded positively to calls for industrial action by union leaders, their engagement in the process of decision-making seems to have decreased. The latter would explain why the union gave up the strict application of the rule book to punish the lack of participation in mass meetings.

The key aspect to be underlined is the ability of the leadership to complement, and even replace, traditional union resources by new ones based on external solidarity and the politics of social unionism. The former was manifested by its engagement in organisational building at higher national and international levels. Also the union understood from early on the power of communitarian alliances to fight back privatisation. Chapter Seven already describes how this orientation translated into a widening of the union's repertoire of collective actions. In this way, the union widened its representation by linking its own demands with communitarian demands. It also forged relationships with organisations of neighbours, students, small business owners and other social movements for formulation of common policies. So, for instance, while workers' participation decreased, the union was able to mobilise the unemployed and neighbourhood organisations in demonstrations and rallies against the firm. An outstanding outcome of this strategy was the campaign for the TEIS. After a long campaign led by LyF MDP, an agreement reached between the government, EDEA and a coalition of diverse organisations secured, in 2000, the cheap provision of electricity for poor neighbourhoods, thanks to a 40 per cent reduction in the price afforded by the company, and a 15 per cent and 6 per cent reduction of taxes afforded by the provincial and municipal administrations respectively. For the consumers, this meant a final reduction of 50 per cent. Needless to say, this achievement increased the appeal of the union in the community. In 2006, the same alliance, again under the leadership of the union, campaigned for the re-nationalisation of the company; and this just after the union had pushed FeTERA into campaigning for the re-nationalisation of the oil and energy industry.

Conclusion

This chapter attempts to shed some light on the relationships between privatisation and changes in the organisational sophistication of ESI unions and the dynamics of decision-making. The aim is to add new elements to the judgement of trade unions' ability to mobilise workers in the context of privatisation. As with previous chapters, the comparative analysis makes it possible to identify similarities and differences, both nationally and internationally.

The first thing to stress is that, to some extent, union mergers and amalgamations in the UK obscured the impact of privatisation on trade union resources. By contrast, in LyF CF, the steady decline of union's resources due to dramatic losses in membership appears as the chief organisational aspect underlined by the analysis. While neither union density nor the scope of representation were seriously affected by job loss, the extent of this obviously distorted the organisational life of trade unions. Chapter Nine shows that its main manifestation in the UK was the proliferation of recruiting campaigns. Occasionally in this country, union leaders also expressed misgivings about the financial impact of downsizing. Yet unions' mergers and amalgamations ended up softening the financial consequences of job loss. In Argentina, instead, the whole policy drive of LyF CF is analysed taking into account the huge financial implications of membership loss. It reinforced workers' demobilisation as the strategic focus was to safeguard the union against the consequences of labour conflicts as the opportunity-to-act was deemed to be unfavourable. LyF MDP, in turn, benefited from an organisational structure which depended only partially on the privatised company, so the impact was less. In this case, the leadership chose from the beginning to oppose privatisation through a diverse repertoire of actions based upon the wide-ranging mobilisation not only of the rank and file but also of the affected community.

Certainly, variability in the extent of the organisational restructuring of British and Argentinian trade unions proved to be salient. The former engaged in a never-ending process of change, adapting their structures to the development of the industry. Inevitably, the gap between the decision to devolve power to lower organisational levels and the effective settlement of the new structures weakened workers' collective capacities. The lack of strong workplace organisations was a crucial deficit. Companies took advantage of the situation by pushing through change before the eyes of a disconcerted workforce, who expected to negotiate concessions for job security. However, unions did their best to resource shop stewards through training programmes and FTOs' support, and readapted their organisations to counteract changes in bargaining and industrial structures. On the contrary, Argentinian unions kept their organisational structures unchanged, although LyF CF launched a severe programme of adjustment to equate action to resources. This variance is explained by the combination of the particularities of the institutions of industrial relations and the evolution of the electricity industry as privatisation unfolded. Both factors contributed in Argentina to the maintenance of a centralised control of bargaining. Although lay representatives gained prominence, the unions were able to form a sole negotiating team to bargain with ESI companies. Obviously, company's shop stewards have played an outstanding role in the definition of the contents of the dealings. In this respect, LyF MDP stands out due to its relative stability as the company was not divided and the structure did not suffer as much as the other unions from job loss. Thus, a different structural context facilitated a different strategic choice by union leadership, who followed traditional patterns of mobilisation, though enriched by the adoption of the orientation of the social unionism.

It is worth mentioning that both trade unions analysed in Argentina had to face acute problems in the maintenance of their social services (*obras sociales*) to members, though for different reasons; this was the only area where change has been considerable. Then different strategic choices led to different outcomes. In the case of LyF MDP, they served to reinforce the notion of workers' collective solidarity, whereas in LyF CF, they provided the platform from which an entrepreneurial style of unionism developed. Still, problems regarding the provision of social services became a potential threat to the legitimacy of both union leaderships.

From the point of view of mobilisation theory, the hallmarks of privatisation regarding organisational developments were the growth of sectionalism and the decline of workers' participation in decision-making. Since privatisation, sectional interests have spread across companies thereby undermining solidarity. Additionally, the uncertainties provoked in the UK by the hectic evolution of the industry have conspired against inclusive forms of collectivism. Yet workers' mobilisation has not been absent, as short localised actions have taken place throughout the industry. In Argentina, sectionalism affected LyF CF too, due to industry fragmentation. Differences in bargaining power between generation and distribution militated against unified policies. Distribution workers failed to mobilise in support of their fellow workers employed in generation; in fact, employees from distribution even failed to coordinate policies, despite sharing similar problems. LyF MDP is, once more, the exception for the reason that the division of ESEBA did not affect the representational reach of the union; on the contrary, it consolidated its autonomy.

Additionally, in both countries, most respondents have stressed that workers' participation was sooner or later affected by the demoralisation brought about by the process of privatisation. Massive job losses and voluntary severance packages, the deterioration of terms and conditions of employment, aggressive human resources policies and so forth – and findings show that it would be necessary to add to that list trade unions' own strategic choices – would have produced a general debilitation of workers' disposition to participate, actively, in decision-making. Certain qualifications are necessary, however. Whereas it is true that British unions have faced problems in recruiting lay representatives and organising workplaces, it is also true that there have been plenty of success stories. Before privatisation, the traditional bargaining arrangements tended to hide workers' lack of participation as centralisation rendered it superfluous; it was enough for union leaders at the time to count on workers' support when needed. The general picture, then, stands out for its variability. In Argentina, in turn, brutal evidence of a declining participation dates back to the mid-1970s when there was massive repression of working-class politics. If the 1980s witnessed a recovery in workers' mobilisation, the costs of the dictatorship were strongly felt. In this period, a bureaucratic leadership consolidated itself in LyF CF. It faced privatisation through a policy which explicitly praised workers' demobilisation in search of the preservation of the organisation. LyF MDP, instead, was able to mobilise not only ESI workers, but also community interests and external solidarity behind a policy of total confrontation. Nevertheless, after privatisation, workers' participation also decreased in LyF MDP, thereby putting at risk the lively dynamic of the process of decision-making, and damaging the commitment of many shop stewards. If this phenomenon had developed so as to affect the ability of the union to engage in collective actions, the decisions would have been concentrated more than ever in union leaders, whose combative outlook ensured the survival of workers' mobilisation. Additionally, communitarian alliances and external solidarity are important resources post-privatisation, which partly compensate for the deterioration in union internal dynamics.

CHAPTER ELEVEN

Concluding Remarks

This book opened stressing two principal objectives: the exploration of the relationships between privatisation and workers' collectivism and the insight provided by mobilisation theory for such a task. In this sense, the most elementary and general conclusions to be drawn are that ESI privatisation did make workers' collective action more difficult and that the research did prove the potential of mobilisation theory to analyse the effects of counter-mobilisation on labour and how trade unions respond to that challenge.

Chapter One specifically posed a set of empirical and theoretical research questions, which are worth restating. A broad empirical question concerned the reconstruction of the counter-mobilising content of privatisation and trade unions' defensive actions. In this regard, the research was intended, on the one hand, to identify sources of variability in the forms taken by the counter-mobilisation and workers' strategic choices as well as the type of resources mobilised by the actors. On the other hand, it aimed to illuminate how the forces unleashed by the process of privatisation targeted workers' collectivism. At the theoretical level, the main questions related to the contribution of a cross-national comparison at meso- and macroanalytical levels towards the development of mobilisation theory, and the place of agency type variables in the conceptual framework.

Thus, this book was grounded on a discussion of mobilisation theory which laid the conceptual foundations of the research: the postulate about the counter-mobilising character of privatisation; the expectation of privatisation undermining the mobilising capacity of trade unions; and the conclusion that the categories opportunity-to-act, organisation and interest were appropriate tools to use when carrying out the empirical investigation into that problematic. The methodological discussion, in turn, justified the adoption of a comparative approach to identification and evaluation of the intermediate variables which impinged on the categories of mobilisation theory, as far as the comparison might be properly contextualised (Locke and Thelen 1995). Mobilisation theory allows for the different stages of mobilisation to be understood and studied within a specific context; so this was an additional reason for adopting this theoretical framework.

Then, after exploring the diverse meanings and complexities of the term privatisation, Chapter Four developed four sorts of arguments in order to substantiate the postulate about the counter-mobilising content of privatisation against labour. It began by providing historical evidence about the long-term orientation of organised labour towards public ownership in both Argentina and the UK, and finished by reviewing empirical evidence showing the positive association between privatisation and unfavourable developments for labour. In between, it offered theoretical arguments taken from the perspectives of Marxist political economy and mobilisation. Yet the essential point of the chapter was that privatisation enhances market discipline.

By making a cross-national comparison, Chapters Five and Six intended to identify various mediating variables which intervened to shape the opportunity structure and the forms and prospects of the counter-mobilising forces. These chapters surveyed similarities and differences regarding international pressures over the process of privatisation, the speed of the programmes, the evolution of the industrial structures, the role of the labour law in the respective systems of industrial relations, and so forth. Bearing in mind the two main objectives of the book, while public ownership and the type of industry pointed towards similarities between case-studies, differences in the national systems of industrial relations and in the evolution of the industrial structures after privatisation called attention to variability in trade unions' (re)sources of power, and hence, in the opportunity structure, and in the timing and forms of the process of counter-mobilisation. Basically, it was argued that in Argentina, given the legal underpinnings of the system of industrial relations, and concomitantly, of trade union power, the role of the government was crucial in undermining workers' collectivism and paving the way to privatisation. In the UK, instead, that task was left to private capitals and industrial restructuring after privatisation.

Chapters Seven to Ten testified to that by providing an empirical reconstruction of key aspects of both, the counter-mobilisation process associated with the privatisation programme and the trade unions' defensive responses. Together, these four chapters constituted a sort of spotlighting approach, which illuminated different sides of the counter-mobilising wave and their impact upon workers' organisational strength and interest definition. Their different starting-points allowed the research to inquire about the various conceptual connections between the categories. In this sense, these chapters also gave evidence about the importance of union internal capabilities and agency to mediate the opportunity structure and the process of counter-mobilisation. Moreover, they showed that the organisational features and developments as well as the agency factors and social interactions, which facilitate the collective definition of interests, may impinge back positively on the opportunity structure.

Let us summarise the main arguments and findings offered by these chapters as they serve as a useful platform for the presentation of some theoretical conclusions. The empirical focus of Chapter Seven was the reconstruction of the anti-privatisation campaigns of ESI unions in both countries; the theoretical aim was to test the explanatory power of the conceptual sequence laid down in Chapter Two. The chapter showed that in the UK, during the run-up to privatisation, ESI unions maintained their industrial (re)sources of power, and hence, industrial latent power quite unchanged. This would help them to obtain defensive political influence within a context of political retreat. This defensive influence was mobilised to obtain concessions regarding pension schemes, health and safety procedures, and provisionally, the industrial relations machinery. In Argentina, instead, the industrial power of national public unions was legislatively undermined by the government in order to prevent workers from mobilising industrial resources against privatisation. In this context, LyF CF inclined to reformulate its strategy at the expense of increasing affiliation and organising workers' collectivism, and towards an entrepreneurial unionism to underpin the financial strength of the organisation. The case of LyF MDP, in turn, expressed the significance of agency type variables to accounting for workers' collectivism. A determined anti-privatisation leadership was able to delay privatisation for years by combining the mobilisation of political, industrial and legal resources through collective and participatory mechanisms. Theoretically, the chapter demonstrated the utility of the sequence opportunity-to-act, organisation and interest definition in accounting for the strategic choices and forms of collective action of well-established labour organisations in the face of a counter-mobilising event.

Chapter Eight addressed an obvious but often neglected aspect of counter-mobilisation, that is, the mobilisation of money resources to prevent opposition, at both individual and collective levels, and to further loyalty to new private firms. Concomitantly, the chapter explored the connections between counter-mobilisation and the categories offered by mobilisation theory for the analysis of collective action in the medium- and short-term (opportunity-to-act, organisation, interest), and in particular, the effects of specific counter-mobilising forces upon the aforementioned categories. The chapter initially pointed to the influence of various intermediate variables on the fate of the politics of money. The active involvement of the government before privatisation was prominent in Argentina, backed and resourced by international financial institutions, mainly, the IMF and the WB. In the UK, the bulk of the process took place after privatisation. Employment law too, explained diversity in the form and outcomes of the politics of money, for instance, the importance of personal contracts in the UK or the chance for Argentinian unions to assume the representation of workers as shareholders on Company Boards. The most important factor explaining country specific outcomes, however, was the growing rate of unemployment, which paralleled in Argentina the implementation of voluntary redundancy programmes, the axis of the politics of money. It had two consequences: it precluded the indulgent attitude toward voluntary redundancies policies found among the British ESI workforce and it caused workers' resistance. Since then, in Argentina voluntary severance packages hid the development of managerial harassment and repression to achieve downsizing. Finally, agency type variables come to the fore, once more, in the explanation of variability between unions in Argentina. From the beginning, LyF MDP leadership framed voluntary programmes as hidden dismissals, took industrial action against them, and expelled from the organisation those members who accepted severance packages. The union was not immune from the politics of money: dozens of workers were seduced and, even worse, four CGA officers and eight union representatives were corrupted by the company ESEBA. However the leadership could mobilise the collective power of workers throughout the process against the different manifestations of the politics of money, and reduce its consequences. At the level of theory, the diffuse effects of the various forms adopted by the politics of money seemed to warn against assuming any simple relationship between counter-mobilisation and its consequences on the empirical manifestation of the categories of mobilisation theory. Nevertheless, the analysis provided evidence that the organisational domain is often a main target.

Chapter Nine, in turn, researched another side of the counter-mobilising wave associated with privatisation: the process of fragmentation of collective bargaining and devolution. It substantiated the central role of the Argentinian government backed by the international financial institutions, in this case, in dismantling collective agreements before privatisation. It differentiated this from the case of the UK, where the prospects of competition compelled managers from privatised firms to push the devolution of bargaining structures so as to achieve change in working practices and pay structures. In this case, it is also necessary to refer to the major industrial variables. These are: the hectic process of mergers and takeovers experienced by the ESI ever since denationalisation, which explains the profuse change in collective arrangements in the UK; and multi-unionism, which explains why this change furthered inter-union competition. In Argentina, instead, where industrial structures remained almost untouched after privatisation, bargaining structures stabilised quickly; concurrently, a different union structure also limited competition between ESI unions, appearing as it did for the first time in the landscape of the industry. In turn, the chapter related fragmentation and devolution to workers' demobilisation by addressing tensions and conflicts within trade union ranks which weakened unions' aggregate strength, and by exposing differences in bargaining power that managers exploited in their favour. Again, the particular side of the process of counter-mobilisation under analysis proved to have wider effects upon the categories opportunity-to-act, interest and organisation, and specifically upon the latter.

Finally, Chapter Ten turned the theoretical spotlights onto the analysis of organisational and agency type variables such as organisational sophistication, union leadership, participation and decision-making. The aim was to shed light on the ways by which unions mediated the opportunity-to-act and the state and capital's counter-mobilising policies they faced during the process of privatisation, and after. Findings revealed the impact of privatisation upon these dimensions of workers' collectivism, too. In this realm, the evolution of ESI structures was again the chief intermediate variable when understanding unions' organisational changes today. The picture presented was one of never ending organisational change in the UK, and relative stability in Argentina as far as workplace structures and other representative levels are concerned; although Argentinian trade unions have reduced their size since privatisation, mainly due to job loss and dramatically so in the case of LyF CF. Additionally, in the UK, trade union mergers added complexity to this picture. This process of union reorganisation, which ran parallel to privatisation, at times intertwined and often overlapped with it. The financial consequences of membership loss were softened as a whole by union mergers, although, from the point of view of ESI lay representatives, mergers meant at the same time fewer resources. To some extent too, the new systems of workers' representation that emerged with the creation of the new unions implied that, compared to the past, ESI constituencies lost prominence. Nonetheless, from the point of view of mobilisation theory, the main findings regarding the relationship between privatisation and workers' collectivism were the growth of sectionalism and the decline of workers' participation.

After this summary, the first general conclusion to be drawn regarding theory, is that the cross-national comparison and the scope of the analysis contributed to presenting as evidence the relevance of various mediating factors in the explanation of variability in the opportunity-to-act, trade union's organisational change and the empirical forms taken by the counter-mobilisation wave.

Concerning the role of mediating factors in the understanding of variability between countries, the use of mobilisation theory confirmed the relevance of alternative institutional arrangements as has been widely shown by conventional approaches and other competing models (Clegg 1976;

Ferner 1997; Ferner and Hyman 1998; Katz and Kochan 2004; Kochan, Katz and McKersie 1987; Locke and Thelen 1995). It also showed that political economy variables, which were in part determined by different international trends in capital investment, played an important mediating role too, as they contributed to explanations of differences in industrial restructuring in the UK and Argentina, and consequently, variability in trade union's organisational change. This is an important finding, for it evidences the capability of mobilisation theory to illuminate aspects which were rather neglected by conventional approaches mostly focused on institutional factors.

Also, mobilisation theory, being different from the conventional approaches, has proved to be a powerful device with which to study the kind of political and agency factors that set in motion the various counter-mobilising forces unleashed by privatisation and the dynamics of trade union responses to them. Moreover, it has shown its particular appeal to those researching the strategic interactions between capital and labour over processes of counter-mobilisation.

This is the case partly because the basic understandings underpinning Kelly's conceptual framework direct the analyses towards the detail of the demobilising dimensions of processes, which might otherwise go unnoticed. In this latter case, and put simply, the theory proved to be empirically productive.

Taking a view from a mobilisation standpoint, as illustrated in Chapter Eight, allows the researcher to gather together a diverse set of policies, whose negative effects for workers' collectivism are usually taken for granted, but hardly ever explicitly incorporated into academic explanations of workers' demobilisation. Indeed, this encouraged the researcher to approach well-established facts, like the fragmentation of industrial structures and collective bargaining, which scholars had often listed among the threats posed by privatisation for organised labour, but had not scrutinised in detail in order to establish why and how fragmentation and decentralisation brought about division, competition and sectionalism.

Nevertheless, the distinctive character of mobilisation theory must be assessed by its sensitivity to variability in the (re)sources of power mobilised by unions, and hence, by its power to question and explain how trade unions face all those counter-mobilising challenges posed by privatisation. In this regard, and more specifically, the second theoretical conclusion is that agency type factors were paramount in explaining variability between unions facing similar opportunity structures and counter-mobilising policies at national level.

The research substantiates this conclusion, mainly, through the comparison of the cases of LyF MDP and LyF CF. In the former, a determined leadership engaged in fierce fighting against privatisation for years. A case that underlined the relevance of participatory and dialogical processes of interest definition to sustaining workers' collectivism and mobilisation, particularly in a context of an increasingly negative opportunity structure. Acting contrarily, the leadership of LyF CF, ready to compromise without involving the rank and file in the decision-making, feinted to oppose the sale of the ESI, and then, supported the privatisation process in a desperate attempt to maintain its organisational power, though at the expense of the terms, conditions and levels of employment. This latter case points to the problem of the demobilising role of trade union officials in the face of a counter-mobilisation wave, an issue not addressed in this book. Similarly, the English case-studies also illustrate, although less dramatically, how different responses and inclinations in the face of similar opportunity structures depend not only on agency type variables, but also on organisational developments, a topic to which this concluding chapter returns later. To give just one example, Unison's inclination towards coalition building seems to have been seriously restricted by multi-unionism and the characteristics of ESTUC, features that ended up leading the union towards compromise with blue-collar and technical unions.

Regarding the placement of agency factors in the explanation of variability, the methodological design was the key. By complementing the cross-national comparison with the comparative study of union responses within the same country, the study was able to make a better evaluation of the role played by agency type variables in those cases in which unions faced similar counter-mobilising dynamics and similar opportunity structures. Conventional approaches usually set the comparison within a country-by-country research design, and therefore have difficulties in accounting for union internal variables, as different union responses are usually understood to be the outcomes of the alternative institutional arrangements in which workers make their strategic choices. Then, in such conceptual and methodological models, the specific role of agency type variables is more difficult to address.

This second general conclusion points towards the need for mobilisation theory to be even more sensitive to questions of union internal politics, structure and leadership. Both the discussion of the analytical framework developed in Chapter Two and the final findings of the study suggest that an approach focusing on the social mediations of interest definition might be fruitful. Few studies using mobilisation theory, however, have dealt with these important aspects systematically (Atzeni 2005; Baccaro, Harman and Turner 2003; Frege and Kelly 2003, 2004; Heery, Kelly and Waddington 2003; Heery et al. 2003; Kelly and Badigannavar 2003; Moore 2004).

In fact, the investigation into the styles of leadership, the level of workers' participation and the process of decision-making proved essential in explaining how trade unions reacted defensively against the opportunity structure. Yet findings also show that these processes of social mediation cannot be analysed in isolation from the category organisation. Indeed, they are inextricably intertwined with organisational features. Perhaps this is the most relevant conclusion to be drawn as far as the interaction among the analytical variables is concerned, though this was an unexpected finding of the research, and therefore the issue could not be explored in its full extension in this book. So this is a dimension that needs further research, as this empirical finding has important theoretical consequences for the future development of mobilisation theory.

To continue with the metaphor, the category organisation seems to occupy centre-stage under the conceptual spotlights of mobilisation theory. This category pertains to the workers' main armour facing a given opportunity structure. Moreover, it can be argued that the organisational domain is, in itself, a component of such structure. This explains why the organisation of workers is often the main target of any counter-mobilising move as shown by Chapters Eight and Nine. Indeed, given its dependent status (Muller-Jentsch 1985; Offe and Wiesenthal 1985), it is a category constrained by the type of industrial and institutional variables analysed in Chapter Six. However, and this is crucial, it is also both an enabler and a limiter of dialogical leaderships, workers' participation and democratic decision-making as explored in Chapters Seven and Ten. Additionally, organisational change is the outcome of workers' strategic choices, usually, in the face of a changing environment – as in this study, for instance – but often according to a certain set of political aims. Therefore it is an analytical arena, which serves as a liaison within mobilisation theory between structural and agency type categories.

As already suggested, the theoretical perspective advanced by Kelly is useful when studying the strategic confrontations of the various actors involved in the struggles around counter-mobilisation events. This positive aspect was again enhanced by the methodological design. In this regard, it was the scope and multiple levels of analysis that permitted the researcher to pay attention, on the one hand, to the strategic interaction of the contenders, and on the other, to the aforementioned ability of the actors to modify the opportunity structure through their strategic choices, and in this way, to open or foreclose specific paths of action. For instance, the sequence of changes in collective bargaining was illuminating. In both countries, although by different means, the fragmentation of the bargaining machinery transformed the field of interactions, and forced unions to accommodate their organisational structures and mobilise their power resources accordingly. This mobilisation, in turn, brought about new scenarios for the contenders. Similar dynamics were typical of other counter-mobilising events addressed in this book; for instance, the fate of the politics of voluntary redundancies in Argentina when unemployment rose. The latter discouraged workers from signing up to redundancy agreements, and then changed the social context in which the policy had been accepted by trade unions. After that, opposition from workers brought about repression and harassment, and the latter, brought about industrial action. From that point, the content of the politics of money changed. Another example: the analysis of the anti-privatisation campaigns made clear that the initial responses of workers faded away with the evolution of the events and the changes in the socio-political context. In the UK, the FUSE campaign stopped with the announcement of the Parliamentary election. In Argentina, in the case of LyF CF, latent resistance metamorphosed into active support as the government threatened to resort to violence; in the case of LyF MDP, the repertoire of collective action evolved according to not only the type of policies implemented by managers and public authorities, but also the changes in the (re)sources of power of the union.

To put it theoretically, the development of mobilisation theory within a comparative framework proved to match the methodological requirements of comparative historical researchers, that is, 'to explicitly analyze historical sequences and take seriously the unfolding processes over time' (Mahoney and Rueschemeyer 2003: 12). This means that, keeping a mobilising perspective, the comparison of trade union responses to change should take into account the temporal structures of processes and events in the explanation. To put it mildly: history matters.

As a consequence, this implies, empirically, that the study was able to address not only how trade unions mobilised but also how their strategies evolved over time. Furthermore, the book explains both dimensions according to a contextualised study of the changes in the opportunity-to-act, the timing of those changes and the dynamics of interest definition by which the opportunity structure was interpreted by workers and certain ways of action decided by their organisations.

So the potential of the theory has again been enhanced by methodological decisions: the scope and the multi-analytical levels of the research. So the methodology proves to be essential for mobilisation theory to show its explanatory potential, and this in itself constitutes another meaningful conclusion of this study. In support of this conclusion, it is important to add that a methodological warning is also necessary in relation to the number of case-studies: the bigger the number, the more difficult it is to engage in systematic and contextualised comparisons following mobilisation theory. This is so because this conceptual framework, as discussed in Chapter Two, entails the exploration of the connections between several variables, which may have different causal effects across heterogeneous contexts. A close inspection of particular cases requires detailed historical and contextual knowledge, which is only possible to achieve when the number of cases is limited. Hence, the application of mobilisation theory to a large number of cases seems to be only possible as the outcome of a collective academic work.

So far, the examination of the pros and cons of mobilisation theory has underlined the positives; what, then, about its weaknesses? In this regard, it is important to stress that the process of research uncovered weak spots in mobilisation theory too. These are that the causal relationship among its categories appears, at times, ambiguous and unstable; and that the categories, on occasion, even overlap. Therefore, the success of mobilisation theory seems to depend to a large extent on its concrete operationalisation, as shown by the previous studies, which have drawn on Kelly's framework (Atzeni 2005; Brown Johnson and Jarley 2004; Darlington 2001; Kelly and Badigannavar 2003; Metochi 2002; Moore 2004). At this point, it is essential to reiterate that the operationalisation should be sensitive to questions of union internal politics, structure and leadership in order to account for the social processes of interest definition, for this contributes to a better understanding of union choices.

This remark may disturb those who conceive of theory as a rigid model of causation. However, this is also acknowledged by the champions of mobilisation theory (McAdam, Tarrow and Tilly 2001). For them, mobilisation studies should not look for explanations that rest upon low-level empirical generalisations, in turn, subsumed under higher-level empirical generalisations, which refer at the end to covering laws. That is, the type of explanations which better fit the construction of rigid theoretical models. Instead, for these scholars, explanations following mobilisation theory are to be conceived of as the identification of causal chains consisting of mechanisms and variables that reappear in a variety of social contexts in diverse combinations and sequences (McAdam, Tarrow and Tilly 2001).

From the view that underpins this book, this identification is the most fruitful understanding of mobilisation theory, for it is a perspective that posits a flexible framework, in which there is a set of interacting factors and categories critical to the presence or absence of collective action. Mobilisation theory, then, should be conceptualised as a sort of middle-range theory for the empirical analysis of workers' collectivism (or its absence); a middle-range theory which needs to be complemented by other theoretical insights, the definition of intermediate variables, the establishment of empirical causations, and fundamentally, a historical perspective. In this way, the comparative approach facilitates all those methodological and theoretical definitions.

Still, mobilisation theory, at least as developed here, does posit analytical hierarchies, which are perceived as appropriate for approaching workers' collectivism. For the study of the strategic choices and collective actions taken by organised labour, it prioritises the analysis of the opportunity structure as the starting-point of given power relations and the structural variables upon which the latter rest. In short, it gives a certain privilege to structural determinants over agency type variables like interest definition, and obviously over (secondary) variables like (union) organisation (as in Chapter Seven). For the analysis of the effects of counter-mobilisation upon organised labour, it emphasises its diffuse character but prioritises effects upon organisational variables, which, in turn, distort the process of decision-making (as in Chapters Eight and Nine).

At first sight, it might appear as contradictory that the emphasis is placed by this concluding chapter on agency type factors. However, there is no contradiction. On the contrary, this accent explains why a mobilising perspective is able to bring to the fore trade union agency and strategic choices thereby giving a firm grasp on the meaning of the concrete responses of trade unions to events of counter-mobilisation (like privatisations, for instance), but through a contextualised and comprehensive analysis that is able to avoid both deterministic structural causation and over-politicised and voluntaristic accounts of workers' collective action.

Needless to say, however, this research is far from exhausting the empirical and theoretical issues involved in the application of mobilisation theory to a process of counter-mobilisation, in this case, to the privatisation of the ESI. To mention just a single important one: there seems to be plenty of room for a closer examination of the ideological processes which intervened in shaping the opportunity-to-act during the privatisation of the ESI. Several references have been made to the ideological consequences of the counter-mobilising forces at play, and the role of union leaderships in disputing meaning. Indeed, Lukes' (2005) framework proved to be useful in treating them in their more general aspects (see Edwards 2006, however, for a recent critique). Another instance is that the Gramscian notion of hegemony could have been of prime importance when carrying out this task. This is another pending matter for further research. For the time being, if this book helps us to a better understanding of why privatisation conspired against the power of organised labour and sheds light on the perspectives opened by Kelly's theoretical proposal, its most basic aims will have been fulfilled.

Bibliography

Ackroyd, S. and Fleetwood, S. (2000) *Realist Perspectives in Management and Organisation*, London: Routledge.

Adams, R. (1981) 'A Theory of Employer Attitudes and Behaviour Towards Trade Unions in Europe and North America', in K. Weiermair and G. Dlugos (eds) *Management Under Differing Value Systems*, Berlin: De Gruyter, 275–93.

Alexander, M. and Corti, C. (1993) *Argentina's Privatization Program. Experience, Issues and Lessons. Development in Practice*, Washington DC: The World Bank.

Anderson, P. (1996) 'Balance del Neoliberalismo: Lecciones para la Izquierda', *La invención y la herencia. Cuadernos Arcis – LOM, Santiago*, 4, noviembre-diciembre, 5–28.

Armstrong, P., Goodman, J. F. B., and Hyman, J. (1981) *Ideology and Shop-floor Industrial Relations*, London: Croom Helm.

Arrowsmith, J. (2003) 'Post-privatization industrial relations in the UK rail and electricity industries', *Industrial Relations Journal*, 34 (2), 150–63.

Arthur, C. (2004) 'Subject and Counter-Subject', *Historical Materialism*, 12 (3), 93–102.

Atzeni, M. (2003) 'Argentinian Workers Facing Flexibilisation: between Mobilization, Socio-political Consciousness and Trade Unionism', *Royal Geographical Society Annual Conference, session on 'Globalisation and Labour Relations'*, 5 September 2003, London.

—— (2005) *Mobilization Theory, Workers' Solidarity and the Evolution of Conflict: Collective Action in Multinational Companies in Argentina*, PhD Thesis, University of Warwick.

—— (2009) 'Searching for Injustice and Finding Solidarity? A Contribution to the Mobilisation Theory Debate', *Industrial Relations Journal*, 40 (1), 5–16.

—— and Ghigliani, P. (2009) 'Trade Unionism in Argentina since 1945: the Limits of Trade Union Reformism', in C. Phelan, *Trade Unionism since 1945: Towards a Global History*. Volume 2: The Americas, Asia and Australia, Dusseldorf: Peter Lang, 223–48.

Austin, J. et al. (1986) 'Privatizing State-Owned Enterprises: Hopes and Realities', *The Columbia Journal of World Business*, 21 (3), 51–60.

Azpiazu, D. and Basualdo, E. (1999) 'El Papel de las Privatizaciones en el Proceso de Concentración y Centralización Económica', *Documento de Trabajo*, 6, Flacso.

—— and Vispo, A. (1994) 'Some Lessons of the Argentine Privatization Process', *Cepal Review*, 54, 129–47.

Baccaro, L., Harman, K. and Turner, L. (2003) 'The Politics of Labour Movement Revitalization: The Need for a Revitalized Perspective', *European Journal of Industrial Relations*, 9 (1), 119–33.

Báez-Camargo, C. (2002) 'Union Responses to Privatization: Lessons from the Mexican Case', in A. Fernández Jilberto and M. Riethof (eds) *Labour Relations in Development*, London: Routledge.

Bamber, G. and Lansbury, R. (1993) *International and Comparative Industrial Relations: a Study of Industrialised Market Economies* (2nd edition), London: Routledge.

—— and Wailes, N. (2004) *International and Comparative Employment Relations. Globalisation and the Developed Market Economies* (4th edition), London: Sage.

Barker, C., Johnson, A. and Lavalette, M. (eds) (2001) *Leadership and Social Movements*, Manchester: Manchester University Press.

Basualdo, E. (2001) *Concentración y Centralización del Capital en la Argentina durante la Década del Noventa*, Buenos Aires: Flacso.

—— (2006) *Estudios de Historia Económica Argentina. Desde mediados del siglo XX a la actualidad*, Buenos Aires: Siglo XXI.

Batstone, Eric (1988) 'The Frontier of Control', in D. Gallie (ed) *Employment in Britain*, Oxford: Blackwell, 217–48.

Batstone, E., Boraston, I. and Frenkel, S. (1977) *Shop Stewards in Action: the Organization of Workplace Conflict and Accommodation*, Oxford: Blackwell.

—— (1978) *The Social Organization of Strikes*, Oxford: Blackwell.

Bean, R. (1994) *Comparative Industrial Relations: an Introduction to Cross-National Perspectives* (2nd edition), London: International Thomson Business.

Beder, S. (2005) 'Critique of the Global Project to Privatize and Marketize Energy', *Envisioning a Renewable Public Energy System, International Symposium* organised by the Korean Labor Social Network on Energy (KLSNE), Seoul, South Korea, June.

Beesley, M. and Littlechild, S. (1988) 'Privatization: Principles, Problems and Priorities', in Beesley, M. (ed) *Privatization, regulation and deregulation*, London: Routledge, 27–42.

Bertero, E. (2002) 'Does a Change in the Ownership of Firms, from Public to Private, Make a Difference?', *Departmental Working Papers*, 2006–03, Department of Economics, University of Milan.

Beynon, H. (1984) *Working for Ford*, Harmondsworth: Penguin.

Bienen, H. and Waterbury, J. (1989) 'The Political Economy of Privatization in Developing Countries', *World Development*, 17 (5), 617–32.

Bishop, M. and Kay, J. (1989) 'Privatization in the United Kingdom: Lessons from Experience', *World Development*, 17 (5), 643–57.

Blyton, P. and Turnbull, P. (2004) *The Dynamics of Employee Relations* (3rd edition), Basingstoke: Palgrave – Macmillan.

Bonnet, A. (2007) *La hegemonía menemista. El neoconservadorismo en Argentina, 1989–2001*, Buenos Aires: Prometeo.

Brinkmann, M. E. (1999) 'Las Relaciones Laborales en el Sector Telefónico', in A. Fernández and R. Bisio (eds) *Política y Relaciones Laborales en la Transición Democrática Argentina*, Buenos Aires: Humanitas, 257–90.

Brown Johnson, N. and Jarley, P. (2004) 'Justice and Union Participation: An Extension and Test of Mobilization Theory', *British Journal of Industrial Relations*, 42 (3), 543–62.

Bunel, J. (1991) *Pactos y Agresiones. El Sindicalismo Argentino ante el Desafío Neoliberal*, Buenos Aires: FCE.

Campione, D. (2002) 'Estado, Dirigencia Sindical y Clase Obrera', available at <http://fisyp.rcc.com.ar/Dirigencia%20sindical%20y%20clase%20obrera.pdf> (last access: 12 December 2005).

Christiansen, R. (1989) 'Editor's Introduction', *World Development*, 17 (5), 597–99.

Cifarelli, V. (1999) 'Las Privatizaciones en la Argentina', *Seminario Internacional Impactos de la Privatización Eléctrica a Nivel Mundial*, México.

Clarke, T. (1993) 'The Political Economy of the UK Privatization Programme: a Blueprint for other Countries?', in C. Pitelis and T. Clarke (eds) *The Political Economy of Privatization*, London: Routledge, 205–33.

Clegg, H. (1972) *The System of Industrial Relations in Great Britain*, Oxford: Blackwell.

——(1976) *Trade Unionism under Collective Bargaining: a Theory Based on Comparisons of Six Countries*, Oxford: Blackwell.

Cole, G. D. H. (1953) *An Introduction to Trade Unions*, London: George Allen and Unwin Ltd.

Colling, T. and Ferner, A. (1992) 'The Limits of Autonomy: Devolution, Line Managers and Industrial Relations in Privatized Companies', *Journal of Management Studies*, 29 (2), 209–27.

——(1995) 'Privatization and Marketization', in P. Edwards (ed) *Industrial Relations. Theory and Practice in Britain*, Oxford: Blackwell, 491–514.

Collins, H. (2006) 'The Protection of Civil Liberties in the Workplace', *Modern Law Review*, 69 (4), 619–31.

Cook, L. and Murphy, C. (2002) 'Introduction', in L. Cook and M. P. Posusney (eds) *Privatization and Labor*, London: Edward Elgar, 1–15.

Cook, P. and Kirkpatrick, C. (1995) 'Privatization Policy and Performance', in P. Cook and C. Kirkpatrick (eds) *Privatization Policy and Performance. International Perspectives*, New York NY: Prentice Hall/Harvester Wheatsheaf, 1–27.

Cordone, H. (1993) 'El Sindicalismo Bajo la Hegemonía Peronista: Emergencia, Consolidación y Evolución Histórica (1943–1973)', in O. Moreno (ed) *Desafíos para el Sindicalismo en la Argentina*, Buenos Aires: Fundación Ebert, Legasa, 49–80.

Crouch, C. (1979) *The Politics of Industrial Relations*, Glasgow: Fontana/Collins.

—— (1994) *Industrial Relations and European State Traditions*, Clarendon Press: Oxford.

—— (2003) 'The State: Economic Management and Incomes Policy', in P. Edwards (ed) *Industrial Relations. Theory and Practice* (2nd edition), Oxford: Blackwell, 105–23.

Crow, G. (1997) *Comparative Sociology and Social Theory: Beyond the Three Worlds*, Basingstoke: Macmillan.

Danford, A., Richardson, M. and Upchurch, M. (2002) *New Unions, New Workplaces. Strategies for Union Revival*, London: Routledge.

Darlington, R. (2001) 'Union Militancy and Left-Wing Leadership in London Underground', *Industrial Relations Journal*, 32 (1), 2–21.

Del Campo, H. (1986) *El Sindicalismo Revolucionario*, Buenos Aires: CEAL.

Dickens, L. and Hall, M. (1995) 'The State: Labour Law and Industrial Relations', in P. Edwards (ed) *Industrial Relations. Theory and Practice in Britain*, Oxford: Blackwell, 255–303.

—— (2003) 'Labour Law and Industrial Relations: A New Settlement', in P. Edwards (ed) *Industrial Relations. Theory and Practice* (2nd edition), Oxford: Blackwell, 124–56.

Dinerstein, A. (2001) *The Violence of Stability. An investigation of the Subjectivity of Labour in Argentina*, PhD Thesis, University of Warwick.

Duarte, M. (2001) 'Los Efectos de las Privatizaciones sobre la Ocupación en las Empresas de Servicios Públicos', *Realidad Económica*, 182, agosto–septiembre, 32–60.

Dunlop, J. (1958) *Industrial Relations Systems*, New York: Holt.

Durkheim, E. (1970) *Suicide: A Study in Sociology* (1st edition), London: Routledge & Kegan Paul Ltd.

Eaton, J. (2000) *Comparative Employment Relations. An Introduction*, Cambridge: Polity Press.

Edwards, P. (1986) *Conflict at Work*, Oxford: Blackwell.

—— (2005) 'Critical Realism: Progress and Challenges', available at <http://www.mngt.waikato.ac.nz/ejrot/cmsconference/2005/proceedings/criticalrealism/

Edwards.pdf#search=%22Critical%20Realism%3A%20Progress%20and%20Challenges%20edwards%22> (last access: 30 April 2010).

——(2006) 'Power and Ideology in the Workplace: Going Beyond even the Second Version of the Three-Dimensional View', *Work, Employment and Society*, 20 (3), 571–81.

—— and Scullion, H. (1982) *The Social Organization of Industrial Conflict: Control and Resistance in the Workplace*, Oxford: Blackwell.

Eisinger, P. (1973) 'The Conditions of Protest Behaviour in American Cities', *American Political Science Review*, 67 (1), 11–28.

Ernst, J. (1994) *Whose Utility?*, Buckingham and Philadelphia: Open University Press.

Etchemendy, S. (2001) 'Construir Coaliciones Reformistas: la Política de las Compensaciones en el Camino Argentino hacia la Liberalización Económica', *Desarrollo Económico*, 40 (160), 675–703.

—— and Palermo, V. (1998) 'Conflicto y Concertación. Gobierno, Congreso y Organizaciones de Interés en la Reforma Laboral del Primer Gobierno de Menem (1989–1995)', *Desarrollo Económico*, 37 (148), 559–90.

Etzioni, A. (1968) *The Active Society*, New York: Free Press.

Evans, S. and Hudson, M. (1993) 'Standardised Packages Individually Wrapped? A Study of the Introduction and Operation of Personal Contracts in the Port Transport and Electricity Supply Industries', *Warwick Papers in Industrial Relations*, 44, University of Warwick.

Fairbrother, P. (1994) 'Privatization and Local Trade Unionism', *Work, Employment, and Society*, 8 (3), 339–56.

——(1996a) 'Trade Unions and Human Resource Management in Britain', *Employee Relations*, 18 (6), 8–25.

——(1996b) 'Workplace Trade Unionism in the State Sector', in P. Ackers, C. Smith and P. Smith (eds) *The New Workplace and Trade Unionism*, London: Routledge, 110–49.

——(2000) *Unions at the Crossroads*, London: Mansell.

Falcón, R. and Monserrat, A. (2000), 'Estado, Empresas, Trabajadores y Sindicatos', in R. Falcón (ed) *Nueva Historia Argentina. Democracia, Conflicto Social y Renovación de Ideas (1916–1930)*, Buenos Aires: Sudamericana, 151–94.

Fantasia, R. (1988) *Cultures of Solidarity. Consciousness, Action and Contemporary American Workers*, Berkeley, CA and London: University of California Press.

Farinetti, M. (2002) 'La Conflictividad Social después del Movimiento Obrero', *Nueva Sociedad*, 182, 60–75.

Feinberg, R. (1986) 'Latin American Debt: Renegotiating the Adjustment Burden', *The Columbia Journal of World Business*, 21 (3), 19–30.

Fernández Jilberto, A. and Riethof, M. (2002) 'Labour Relations in the Era of Globalisation and Neo-Liberal Reforms', in A. Fernández Jilberto and M. Riethof (eds) *Labour Relations in Development*, London: Routledge, 1–25.

Ferner, A. (1988) *Governments, Managers and Industrial Relations: Public Enterprises and their Political Environment*, Oxford: Basil Blackwell.

——(1989) 'Ten Years of Thatcherism: Changing Industrial Relations in British Public Enterprises', *IRRU Discussion Paper*, 11, Warwick University.

——(1997) 'Country of Origin Effects and HRM in Multinational Companies', *Human Resource Management Journal*, 7 (1), 19–37.

——and Colling, T. (1991) 'Privatization, Regulation and Industrial Relations', *British Journal of Industrial Relations*, 29 (3), 391–409.

——and Colling, T. (1993a) 'Electricity Supply', in A. Pendleton and J. Winterton (eds) *Public Enterprise in Transition. Industrial Relations in State and Privatized Corporations*, London: Routledge, 100–33.

——and Colling, T. (1993b) 'Privatization of the British Utilities: Regulation, Decentralization and Industrial Relations', in C. Pitelis and T, Clarke (eds) *The Political Economy of Privatization*, London: Routledge, 125–41.

——and Hyman, R. (1998) 'Introduction: Towards European Industrial Relations?', in A. Ferner and R. Hyman (eds) *Changing Industrial Relations in Europe*, Oxford: Blackwell, xi–xxvi.

Fleetwood, S. (1999), 'The Inadequacy of Mainstream Theories of Trade Union Behaviour', *Labour*, 13, 445–80.

Foreman-Peck, J. and Millward, R. (1994) *Public and Private Ownership of British Industry, 1820–1990*, Oxford and New York: Oxford University Press.

Fosh, P. (1993) 'Membership Participation in Workplace Unionism: The Possibility of Union Renewal', *British Journal of Industrial Relations*, 31 (4), 577–92.

Fracchia, J. and Lewontin, R. (2005) 'The Price of Metaphor', *History and Theory*, 44 (1), 14–29.

Frege, C., Heery, E. and Turner, L. (2004) 'The New Solidarity? Coalition Building in Five Countries', in C. Frege and J. Kelly (eds) *Varieties of Unionism: Struggles for Union Revitalization in a Globalizing Economy*, Oxford: Oxford University Press, 137–58.

Frege, C. and Kelly, J. (2003) 'Union Revitalization Strategies in Comparative Perspective', *European Journal of Industrial Relations*, 9 (1), 7–24.

——(2004) *Varieties of Unionism: Struggles for Union Revitalization in a Globalizing Economy*, Oxford: Oxford University Press.

Gall, G. (1994) 'The Rise of Single Table Bargaining in Britain', *Employee Relations*, 16 (4), 62–71.

—— (2000) 'Debating Mobilization, Class Struggle and the Left: a Response to a Reply', *Historical Materialism*, 7 (1), 175–80.
—— (2003) *The Meaning of Militancy? Postal Workers and Industrial Relations*, Aldershot and Burlington: Ashgate.
——and McKay, S. (1996) 'Research Note: Injunctions as a Legal Weapon in Industrial Disputes', *British Journal of Industrial Relations*, 34 (4), 567–82.
Gamson, W. (1992) *Talking Politics*, Cambridge: Cambridge University Press.
Gennard, J. (1998) 'Labour Government: Change in Employment Law', *Employee Relations*, 20 (1/2), 12–26.
Godio, J. (1987) *El Movimiento Obrero Argentino (1870–1910). Socialismo, Anarquismo y Sindicalismo*, Buenos Aires: Legasa.
Goldín, A. (1997) *Privatizaciones y Relaciones Laborales. El Caso Argentino*, ILO Action Programme on Privatization 1996–7, Bern.
González, M. (2001) 'Transformaciones de Hecho y de Derecho en las Relaciones Laborales en los Noventa', *AEPA, Jornadas Argentinas de Estudios de Población*, Buenos Aires.
Green, A., Black, J. and Ackers, P. (2000) 'The Union Makes Us Strong? A Study of the Dynamics of Workplace Union Leadership at Two UK Manufacturing Plants', *British Journal of Industrial Relations*, 38 (1), 75–93.
Gupta, A. (2000) *Beyond Privatization*, London and New York: Macmillan.
Hall, D. (2000) *Impact of electricity privatisation on industrial relations, – lessons from the UK (and Hungary)*, London: PSIRU, University of Greenwich.
——(2005) *Electricity privatisation and restructuring in Latin America and the impact on workers*, London: PSIRU, University of Greenwich.
——Thomas, S. and Bayliss, K. (2002) *Resistance and alternatives to energy privatisation*, London: PSIRU, University of Greenwich.
Hammersley, M. and Gomm, R. (2000) 'Introduction', in R. Gomm, M. Hammersley and P. Foster (eds) *Case Study Method*, Thousand Oaks: SAGE, 1–16.
Hannah, L. (1979) *Electricity before Nationalisation: A Study of the Developement of the Electricity Supply Industry in Britain to 1948*, London: Johns Hopkins University Press.
——(1982) *Engineers, Managers and Politicians: The First Fifteen Years of Nationalised Electricity Supply in Britain 1948–1962*, London: Johns Hopkins University Press.
Harrison, A. (1988) 'The framework of control', in C. Whitehead (ed.) *Reshaping the Nationalised Industries*, Newbery: Policy Journals, 38–40.
Haslam, C. et al. (1996) 'Stakeholder Economy?: From Utility Privatization to New Labour', *Capital and Class*, 60, 119–34.

Heald, D. (1980) 'The Economic and Financial Control of UK Nationalized Industries', *Economic Journal*, 90 (358), 243–65.

——and Steel, D. (1981) 'The Privatization of UK Public Enterprises', *Annals of Public and Co-operative Economy*, 52, 351–68.

Heery, E. et al. (2003) 'Trade Union Recruitment Policy in Britain. Form and Effects', in G. Gall (ed) *Union Organising: Campaigning for Trade Union Recognition*, London: Routledge, 56–78.

——Kelly, J. and Waddington, J. (2003) 'Union Revitalization in Britain', *European Journal of Industrial Relations*, 9 (1), 79–97.

Henney, A. (1987) *Privatise Power – Restructuring the Electricity Supply Industry*, Policy Study no. 83, London: Centre for Policy Studies.

Herrera, C. (1992) 'The Privatization of the Argentine Telephone System', *Cepal Review*, 47, 149–61.

Hill, A. and Abdala, M. (1993) 'Regulation, Institutions, and Commitment. Privatization and Regulation in the Argentina Telecommunications Sector', *Policy Research Working Paper*, 1216, World Bank.

Hyman, R. (1975) *Industrial Relations: a Marxist Introduction*, London: Macmillan.

——(1983) 'Trade Unions: Structure, Policies, and Politics', in G. Bain (ed) *Industrial Relations in Britain*, Oxford: Blackwell, 35–66.

——(1989) *The Political Economy of Industrial Relations: Theory and Practice in a Cold Climate*, Basingstoke: Macmillan.

——(1994a) 'Changing Trade Union Identities and Strategies', in R. Hyman and A. Ferner (eds) *New Frontiers in European Industrial Relations*, Oxford: Blackwell, 108–39.

——(1994b) 'Introduction: Economic Restructuring, Market Liberalism and the Future of National Industrial Relations Systems', in R. Hyman and A. Ferner (eds) *New frontiers in European industrial relations*, Oxford: Blackwell, 1–14.

——(1994c) 'Theory and Industrial Relations', *British Journal of Industrial Relations*, 32 (2), 165–80.

——(2003) 'The Historical Evolution of British Industrial Relations', in P. Edwards (ed) *Industrial Relations. Theory and Practice* (2nd edition), Oxford: Blackwell, 37–57.

——and A. Ferner (eds) (1994) *New Frontiers in European Industrial Relations*, Oxford: Blackwell.

Isaac, J. (1987) *Power and Marxist Theory. A Realist View*, Ithaca and London: Cornell University Press.

James, D. (1988) *Resistencia e Integración. El peronismo y la clase trabajadora argentina. 1946–1976*, Buenos Aires: Sudamericana.

Jenkins, J. C. (1983) 'Resource Mobilization Theory and the Study of Social Movements', *Annual Review of Sociology*, 9, 527–53.

Katz, H. (1997) 'Introduction and Comparative Overview', in H. Katz *Telecommunications. Restructuring Work and Employment Relations Worldwide*, Ithaca: ILR Press, 1–28.

—— and Kochan, T. (2004) *An Introduction to Collective Bargaining and Industrial Relations* (3rd edition), New York: McGraw-Hill.

Kelf-Cohen, R. (1958) *Nationalisation in Britain: the End of a Dogma*, London: Macmillan.

Kelly, J. (1988) *Trade Unions and Socialist Politics*, London: Verso.

—— (1998) *Rethinking Industrial Relations*, London: Routledge.

—— (2005) 'Social Movement Theory and Union Revitalization in Britain', in S. Fernie and D. Metcalf (eds) *Trade Unions: Resurgence or Demise?*, London: Routledge, 62–82.

—— and Badigannavar, V. (2004) 'Union Organizing', in J. Kelly and P. William (eds) *Union Organization and Activity*, London: Routledge, 32–50.

—— and Heery, E. (1994) *Working for the Union*, Cambridge: Cambridge University Press.

—— and Willman, P. (2004) *Union Organization and Activity*, London: Routledge.

Kerr, C. et al. (1960) *Industrialism and Industrial Man*, New York: Oxford University Press.

Klandermans, B. (1984) 'Mobilization and Participation: Social-Psychological Expansions of Resource Mobilization Theory', *American Sociological Review*, 49 (5), 583–600.

—— et al. (1988) *From Structure to Action: Comparing Social Movement Research across Cultures*, International Social Movement Research, vol. 1, Greenwich, CT: JAI Press.

—— and Tarrow, S. (1988) 'Mobilization into Social Movements: Synthesizing European and American Approaches', in B. Klandermans et al. *From Structure to Action: Comparing Social Movement Research Across Cultures*, International Social Movement Research, vol. 1, Greenwich, CT: JAI Press, 1–38.

Kochan, T., Katz, H. and McKersie, R. (1987) *The Transformation of American Industrial Relations*, New York: Basic Books.

Lebowitz, M. (2003) *Beyond Capital. Marx's Political Economy of the Working Class* (2nd edition), London: Palgrave.

Levitsky, S. (2004) 'Del Sindicalismo al Clientelismo: la Transformación de los Vínculos Partido-Sindicatos en el Peronismo, 1983–1999', *Desarrollo Económico*, 44 (173), 3–32.

Lewis, R. (1986) 'The Role of the Law in Employment Relations', in R. Lewis (ed) *Labour Law in Britain*, Oxford: Blackwell, 3–43.

Lincoln, Y. and Guba, E. (2000) 'The Only Generalization is: There is No Generalization', in R. Gomm, M. Hammersley and P. Foster (eds) *Case Study Method*, Thousand Oaks: SAGE, 27–44.

Locke, R. and Thelen, K. (1995) 'Apples and Oranges Revisited: Contextualized Comparisons and the Study of Comparative Labor Politics', *Politics and Society*, 23 (3), 337–67.

Luca, L. (1998) *Labour and Social Dimensions of Privatization and Restructuring – Public Utilities Water, Gas, Electricity*, Geneva: ILO.

Lukes, S. (2005) *Power: a Radical View* (2nd edition), Basingstoke: Palgrave – Macmillan.

Mahoney, J. and Rueschemeyer, D. (2003) 'Comparative Historical Analysis. Achievements and Agendas', in J. Mahoney and D. Rueschemeyer (eds) *Comparative Historical Analysis in the Social Sciences*, London: University of California Press, 3–40.

Manzetti, L. (1999) *Privatization. South American Style*, Oxford: Oxford University Press.

Margheritis, A. (1999) *Ajuste y Reforma en la Argentina (1989–1995). La Economía de las Privatizaciones*, Santiago de Chile: Nuevohacer, Grupo Editor Latinoamericano.

Marsh, D. (1991) 'Privatization under Mrs. Thatcher: A Review of the Literature', *Public Administration*, 69 (4), 459–80.

—— (1992) *The New Politics of British Trade Unionism*, London: Macmillan.

Marsh, H. (2002) 'Changing Pressure-Group Politics: the Case of the TUC 1994–2000', *Politics*, 22 (3), 143–51.

Martin, R. (1992) *Bargaining Power*, Oxford: Clarendon Press.

—— (1999) 'Mobilization Theory: A New Paradigm for Industrial Relations?', *Human Relations*, 52, 1205–16.

Marx, K. (1973) *Grundrisse*, New York: Vintage Books.

—— (1977) *Capital*, vol. 1, New York: Vintage Books.

McAdam, D. (1988) *Political Process and the Development of Black Insurgency, 1930–1970*, Chicago: University of Chicago Press.

—— Tarrow, S., and Tilly, C. (2001) *Dynamics of Contention*, Cambridge: Cambridge University Press.

McCarthy, J. and Zald, M. (1977) 'Resource Mobilization and Social Movements: a Partial Theory', *American Journal of Sociology*, 82, 1212–41.

McIlroy, J. (1995) *Trade Unions in Britain Today* (2nd edition), Manchester: Manchester University Press.

—— (1999) 'Unfinished Business – The Reform of Strike Legislation in Britain', *Employee Relations*, 21 (6), 521–40.

Metochi, M. (2002) 'The Influence of Leadership and Member Attitudes in Understanding the Nature of Union Participation', *British Journal of Industrial Relations*, 40 (1), 87–111.

Middlemas, K. (1991) *Power, Competition and the State*, vol. 3: 'The End of the Postwar Era', Basingstoke: Macmillan.

Miller, A. (1995) 'British Privatization: Evaluating the Results', *The Columbia Journal of World Business*, 30 (4), 82–98.

Molz, R. (1990) 'Privatization in Developing Countries', *The Columbia Journal of World Business*, 25 (2), 17–24.

Moore, S. (2004) 'Union Mobilization and Employer Counter-Mobilization in the Statutory Recognition Process', in J. Kelly and P. William (eds) *Union Organization and Activity*, London: Routledge, 7–31.

Muller-Jentsch, W. (1985) 'Trade Unions as Intermediary Organizations', *Economic and Industrial Democracy*, 6 (1), 3–33.

Murillo, V. (1997) 'La Adaptación del Sindicalismo Argentino a las Reformas de Mercado en la Primera Presidencia de Menem', *Desarrollo Económico*, 37 (147), 419–46.

—— (2001) *Labor Unions, Partisan Coalitions, and Market Reforms in Latin America*, Cambridge: Cambridge University Press.

Nellis, J. and Kiker, S. (1989) 'Public Enterprise Reform: Privatization and the World Bank', *World Development*, 17 (5), 659–72.

Novick, M. (2000) 'Reconversión Segmentada en la Argentina: Empresas, Mercado de Trabajo y Relaciones Laborales a Fines de los '90', in E. De la Garza Toledo (ed) *Reestructuración Productiva, Mercado de Trabajo y Sindicatos en América Latina*, Buenos Aires: CLACSO, 49–72.

—— (2001) 'Nuevas Reglas de Juego en la Argentina: Competitividad y Actores Sindicales', in E. Garza Toledo (ed), *Los Sindicatos frente a los Procesos de Transición Política*, Buenos Aires: CLACSO, 25–46.

O'Connell Davidson, J. (1993) *Privatization and Employment Relations. The Case of the Water Industry*, London: Mansell.

Offe, C. and Wiesenthal, H. (1985) 'Two Logics of Collective Action', in C. Offe *Disorganised Capitalism*, Oxford: Polity Press, 291–332.

Ogden, S. (1991) 'The Trade Union Campaign against Water Privatization', *Industrial Relations Journal*, 22 (1), 20–34.

—— (1993) 'Water', in A. Pendleton and J. Winterton (eds) *Public Enterprise in Transition. Industrial Relations in State and Privatized Corporations*, London: Routledge, 134–65.

Olson, M. (1971) *The Logic of Collective Action: Public Goods and the Theory of Groups* (revised edition), Cambridge: Harvard University.

Orlansky, D. and Orciani, M. (1994) *La Reconversión Sindical después de la Reforma del Estado*, Universidad de Buenos Aires, Facultad de Ciencias Sociales, Instituto de Investigaciones.

Oyen, E. (1990) 'The Imperfection of Comparisons', in E. Oyen (ed) *Comparative Methodology. Theory and Practice in International Social Research*, London: SAGE, 1–18.

Palomino, H. (2005) 'Los Cambios en el Mundo del Trabajo y los Dilemas Sindicales', in J. Suriano (ed) *Nueva Historia Argentina, Dictadura y Democracia*, Tomo X, Buenos Aires: Sudamericana, 377–442.

Pendleton, A. and McDonald, J. (1994) 'Which Affects Industrial Relations Most – Privatization or Competition? A Study of the UK Bus Industry', *Paper to the Employment Research Unit Annual Conference*, Cardiff Business School, University of Wales.

—— Wilson, N. and Wright, M. (1998) 'The Perception and Effects of Share Ownership: Empirical Evidence from Employee Buy-outs', *British Journal of Industrial Relations*, 36 (1), 99–123.

——and Winterton, J. (1993) 'Introduction: Public Enterprise Industrial Relations in Context', in A. Pendleton and J. Winterton (eds) *Public Enterprise in Transition. Industrial Relations in State and Privatized Corporations*, London: Routledge, 1–21.

Petras, J. and Vieux, S. (1999) 'Las Trampas de la Privatización', *Seminario Internacional Impactos de la Privatización Eléctrica a Nivel Mundial*, México.

Pitelis, C. and T. Clarke (eds) (1993) *The Political Economy of Privatization*, London: Routledge.

Pizzorno, A. (1978) 'Political Exchange and Collective Identity in Industrial Conflict', in C. Crouch and A. Pizzorno (eds) *The Resurgence of Class Conflict in Western Europe since 1968*, New York: Holmes and Meier, 277–98.

Poole, M. (1986) *Industrial Relations: Origins and Patterns of National Diversity*, London: Routledge and Kegan Paul Ltd.

Pozzi, P. and Schneider, A. (1994) *Combatiendo el Capital*, Buenos Aires: El Bloque.

Pripstein Posusney, M. and Cook, L. (eds) (2002) *Privatization and Labor: Responses and Consequences in Global Perspective*, Cheltenham: Edward Elgar.

Ragin, C. (1994) *Constructing Social Research: the Unity and Diversity of Method*, London: Thousand Oaks.

Ramamurti, R. (1992a) 'The Impact of Privatization on the Latin American Debt Problem', *Journal of Interamerican Studies and World Affairs*, 34 (2), 93–125.

——(1992b) 'Why Are Developing Countries Privatizing?', *Journal of International Business Studies*, 23 (2), 225–49.

Ramanadham, V. (1993) 'Privatization: Constraints and Impacts', in V. Ramanadham (ed) *Constraints and Impacts of Privatization*, London: Routledge, 1–63.

Ramírez, L. (1999) 'Terrorismo Laboral: el Retiro (In)Voluntario en las Empresas Privatizadas', en C. Lozano (ed) *El Trabajo y la Política en la Argentina de Fin de Siglo*, Buenos Aires: Eudeba, 346–61.

Rausch, A. (1993) 'Privatization in Argentina', in V. Ramanadham (ed) *Constraints and Impacts of Privatization*, London: Routledge, 166–96.

——(1994) 'Monitoring and Regulatory Aspects of Privatization in Argentina', in V. Ramanadham (ed) *Privatization and After. Monitoring and Regulation*, London: Routledge, 337–63.

Riethof, M. (2002) 'Responses of the Brazilian Labour Movement to Privatization', in A. Fernández Jilberto and M. Riethof (eds) *Labour Relations in Development*, London: Routledge, 224–46.

Roberts, J. (1999) 'Marxism and Critical Realism: The Same, Similar, and Just Plain Different?', *Capital and Class*, 68, 21–49.

——(2001) 'Critical Realism and the Dialectic', *British Journal of Sociology*, 52 (4), 667–85.

Rofman, A. and Romero, L. A. (1997) *Sistema Socioeconómico y Estructura Regional en la Argentina*, Buenos Ares: Amorrortu.

Rosdolsky, R. (1977) *The Making of Marx's Capital*, London: Pluto Press.

Rucht, D. (1988) 'Logics, and Arenas of Social Movements. A Structuralist Approach, in International Social Movements', in B. Klandermans et al. (eds) *From Structure to Action: Comparing Social Movement Research across Cultures*, International Social Movement Research, vol. 1, Greenwich, CT: JAI Press, 305–28.

Ruysseveldt, J. (1995) 'Growing Cross-National Diversity or Diversity Tout Court? An Introduction to Comparative Industrial and Employment Relations', in J. Ruysseveldt et al. (eds) *Comparative Industrial and Employment Relations*, London: Sage, 1–15.

Saad-Filho, A. (2003) 'New Dawn or False Start in Brazil? The Political Economy of Lula's Election', *Historical Materialism*, 11 (1), 3–21.

Sánchez Bajo, C. (2002) 'Institutionalising Labour Throughout Argentina's "Second Great Transformation"', in A. Fernández Jilberto and M. Riethof (eds) *Labour Relations in Development*, London: Routledge, 194–223.

Sánchez, M. A. (1991) 'El FMI, el BM y el BID Financian la Reestructuración del Estado y las Privatizaciones en Argentina', in A. Bocco and N. Minsburg (eds) *Privatizaciones. Reestructuración del Estado y la Sociedad. Del Plan Pinedo a 'los Alzogaray'*, Buenos Aires: Ediciones Letra Buena, 125–53.

Saunders, P. and Harris, C. (1994) *Privatization and Popular Capitalism*, Buckingham and Philadelphia: Open University Press.

Sayer, A. (1992) *Method in Social Science: a Realist Approach* (2nd edition), London: Routledge.

Schamis, H. (2002) *Re-Forming the State. The Politics of Privatization in Latin America and Europe*, Michigan: The University of Michigan Press.

Schofield, J. (2000) 'Increasing the Generalizability of Qualitative Research', in R. Gomm, M. Hammersley and P. Foster (eds) *Case Study Method*, Thousand Oaks: SAGE, 181–203.

Senén González, S. and Bosoer, F. (1999) *El Sindicalismo en Tiempos de Menem*, Buenos Aires: Editorial Corregidora.

Sigal, S. and Kessler, G. (1997) 'La Hiperinflación en Argentina: Comportamientos y Representaciones Sociales', in D. Cantón and J. Jorrat *La Investigación Social Hoy*, Instituto de Investigaciones Gino Germani, Buenos Aires: UBA, 155–88.

Simons, H. (1996) 'The Paradox of Case Study', *Cambridge Journal of Education*, 26 (2), 225–40.

Sisson, K. (1987) *The Management of Collective Bargaining: An International Comparison*, Oxford: Blackwell.

Smith, P. (2001) *Unionization and Union Leadership: the Road Haulage Industry*, London and New York: Continuum.

Snow, D. et al. (1986) 'Frame Alignment Processes, Micromobilization, and Movement Participation', *American Sociological Review*, 51 (4), 464–81.

——and Benford, R. (1988) 'Ideology, Frame Resonance, and Participant Mobilization', *International Social Movement Research*, 1, 197–217.

Stake, R. (2000) 'The Case Study Method in Social Inquiry', in R. Gomm, M. Hammersley and P. Foster (eds) *Case Study Method*, Thousand Oaks: SAGE, 20–6.

Stiglitz, J. (2002) *Globalization and Its Discontents*, New York: W.W. Norton and Company.

Suriano, J. (1990) 'El Estado Argentino Frente a los Trabajadores Urbanos: Política Social y Represión, 1880–1916', *Anuario 14*, Segunda Época, UNR, Rosario, 109–36.

Swann, D. (1988) *The Retreat of the State. Deregulation and Privatization in the United Kingdom and United States*, Ann Arbor: University of Michigan Press.

Tarrow, S. (1994) *Power in Movement: Social Movements, Collective Action and Politics*, Cambridge: Cambridge University Press.

Taylor, R. (1985) 'The Union No-Strike Force', *Management Today*, April, 81–4.

Terry, M. (ed) (2000) *Redefining Public Sector Unionism: UNISON and the Future of Trade Unions*, London: Routledge.

Teune, H. (1990) 'Comparing Countries: Lessons Learned', in E. Oyen (ed) *Comparative Methodology. Theory and Practice in International Social Research*, London: SAGE, 38–62.

Thwaites Rey, M. (1994) 'Privatizaciones. La Lógica del Capital', *Oikos*, 2 (3), 68–73.

—— (1999) 'Ajuste Estructural y Reforma del Estado en la Argentina de los '90', *Realidad Económica*, 160, 76–109.

—— (2003) *La Desilusión Privatista. El Experimento Neoliberal en la Argentina*, Buenos Aires: Eudeba.

Tilly, C. (1978) *From Mobilization to Revolution*, Reading MA: Addison-Wesley Publishing Company.

Tomada, C. (1999) 'Inversión Extranjera y Privatizaciones ¿Otras Relaciones Laborales?, in A. Fernández and R. Bisio (eds) *Política y Relaciones Laborales en la Transición Democrática Argentina*, Buenos Aires: Humanitas, 180–97.

—— and Senén González, C. (1998) 'Relaciones Laborales y Sindicatos en Empresas Privatizadas en Argentina (1990–1997)', *Trabajo*, 1 (2), 121–54.

Torre, J. C. (1998) *El Proceso Político de las Reformas Económicas en América Latina*, Buenos Aires: Paidós.

Turnbull, P. (1988) 'Leaner and Possibly Fitter: the Management of Redundancy in Britain', *Industrial Relations Journal*, 19, 201–13.

—— and Wass, V. (1997) 'Job Insecurity and Labour Market Lemons: the (Mis) management of Redundancy in Steel-making, Coal Mines and Port Transport', *Journal of Management Studies*, 34, 27–51.

Van der Hoeven, R. and Sziraczki, G. (eds) (1997) *Lessons from Privatization: Labour Issues in Developing and Transitional Countries*, Geneva: ILO.

Varman, R. and Bhatnagar, D. (1999) 'Power and Politics in Grievance Resolution: Managing Meaning of Due Process in an Organization', *Human Relations*, 52, 349–82.

Vickers, J. and Wright, V. (1988) 'The Politics of Industrial Privatization in Western Europe: an Overview', *Western European Politics*, 11 (4), 1–30.

Waddington, J. (2003) 'Trade Union Organization', in P. Edwards (ed) *Industrial Relations. Theory and Practice* (2nd edition), Oxford: Blackwell, 214–56.

Walle, N. (1989) 'Privatization in Developing Countries: A Review of the Issues', *World Development*, 17 (5), 601–15.

Wallis, E. (2000) *Industrial Relations in the Privatised Coal Industry: Continuity, Change and Contradictions*, Aldershot: Ashgate.

Waterman, P. (1998) *Globalisation, Social Movements and the New Internationalisms*, London and Washington: Continuum.

Wedderburn, K. (1986) *The Worker and the Law* (3rd edition), Harmondsworth: Penguin.

Weiner, H. (1960) *British Labour and Public Ownership*, Washington: Public Affairs Press.

Weyman-Jones, T. (1993) 'Regulating the Privatized Electricity Utilities in the UK', in C. Pitelis and T. Clarke (eds) *The Political Economy of Privatization*, London: Routledge, 53–61.

Wiltshire, K. (1987) *Privatization. The British Experience*, Melbourne: CEDA.

Yin, R. (1994) *Case Study Research. Design and Methods* (2nd edition), Thousand Oaks, CA: Sage.

Documentary Sources

Trade Unions' Journals

8 de Octubre, Revista de LyF MDP (LyF MDP's Magazine).
Contact, EETPU Journal.
Contacto, Revista de FATLyF (FATLyF's Magazine)
Dinamis, Revista de LyF CF (LyF CF's Magazine).
Electricity News, NALGO.
EMA Newsletter, EMA.
EPE, Electrical Power Engineer, EMA Journal.
Flashlight, A paper for EETPU members. The other voice of the EETPU.
NALGO News, NALGO Journal.
Privatization News, EPEA.
Privatization News, NALGO.
Shop Stewards Quarterly Review, EETPU.
Unison Focus, Unison.

National Newspapers

Clarin
El Cronista Comercial
Sur

Other Sources

AEEU (1993) *Policy Agenda Biennial Delegate Conference* (EETPU section), Blackpool.
Amicus (2005) *Energy Report.*
Banco Mundial (1991) *Informe Anual*, Buenos Aires: Banco Mundial.
Campaño, A. and Caruso, J. (1991) *Los convenios colectivos de trabajo y otros aspectos de la actividad laboral en la Empresas Públicas. Análisis de la Legislación*, Argentina. Ministerio de Economía Obras y Servicios Públicos (MEOYSP)., 11 p.; Buenos Aires, Jun. 1991. Proyecto: Programa de Asistencia Técnica para la Gestión del Sector Público Argentino. Préstamo BIRF 2712-AR PNUD/OSP:ARG/86/R01.
——et al. (1991) *Estrategias Laborales y productividad en las Empresas Públicas. Informe Final.* Argentina. Subsecretaría de Coordinación Administrativa y Técnica., 46 p. en 3 informes; Buenos Aires, Jul. 1991. Proyecto: Programa de Asistencia Técnica para la Gestión del Sector Público Argentino. Préstamo BIRF 2712-AR. PNUD/OSP:ARG/86/R01.
Certification Office (1992) *Papers of the Certification Office for Trade Unions and Employers' Associations* (MSS, 388), Modern Records Centre, University of Warwick.
Daireaux, Emilio et al. (1990) *Relaciones Laborales en las Empresas Públicas*, Argentina. Ministerio de Economía Obras y Servicios Públicos (MEOYSP), 266 p.; Buenos Aires, Ene. 1991. Proyecto: Programa de Asistencia Técnica para la Gestión del Sector Público Argentino. Préstamo BIRF 2712-AR PNUD/OSP:ARG/86/R01.
EETPU (1971) *Third Biennial Delegate Conference*, Blackpool.
——(1977) *Fourth Biennial Delegate Conference*, Blackpool.
——(1979) *Fifth Biennial Delegate Conference*, Blackpool.
——(1981) *Report of the 7th Biennial Delegate Conference*, Blackpool.
——(1987) *Conference Report*, Scarborough.
——(1988) *Abridged Report of the National Conference for the Electricity Supply Industry*, Scarborough.
——(1989) *Annual Conference.*
——(1990) *Centenary Industrial RelationsYearbook.*
EMA (1993) *Annual Report.*
——(1994) *Annual Report: 'Your support achieves our aims'.*
——(1995) *Annual Report: 'Coping with Change'.*
——(1996) *Annual Review 1996: 'Security and Service'.*
——(1997) *Annual Review 1997.*
——(1998) *Annual Review 1998.*
EPEA (1986) *Public or Private Electricity? The American Experience.*

——(1987) *Annual Report.*
——(1988) *Annual Report.*
——(1989a) *Annual Conference Report.*
——(1989b) 'The Association's Future', *NEC Seminar.*
——(1989c) *Final Agenda. 15th annual delegate conference of the Electrical Power Engineers' Association*, University of York, Heslington, York, April.
——(1989d) 'Review of the Association's Organisation. Proposals by the National Executive Committee for the Consideration of Membership', August.
——(1990) *Final Agenda. 16th annual delegate conference of the Electrical Power Engineers' Association*, The Pavilion Bournemouth, April.
——(1991) *Final Agenda. 17th annual delegate conference of the Electrical Power Engineers' Association*, University of York Heslington, York, April.
ESTUC (1987) *Press Conference.* Charter on the agreed points of ESTUC, March.
Expte. no. 27.849/02, García José Antonio y Vázquez Aníbal Félix Contra Telefónica de Argentina S.A. Sentencia Definitiva, no. 3006. Poder Judicial de la Nación.
FATLyF (1991) *Declaración Pública.*
——(1992) *Resolución del XXXIV Congreso General Ordinario de F.A.T.L.Y.F*, Mar del Plata.
——(1997) *Informe del Congreso*, Buenos Aires.
FeTERA (2000) 'Los trabajadores de la energía, proponemos cambios políticos y económicos de fondo', Federación de Trabajadores de la Energía de la República Argentina, 20 de Septiembre, Buenos Aires.
FUSE (1987) *Campaigning Charter*, leaflet, April.
IPCTT (1993) *Declaración de Santiago*, Santiago de Chile, Chile. 10 de Noviembre.
LyF CF (1989) *Memoria y Balance. 1988–89*, no. 46, Buenos Aires.
——(1992) *Memoria y Balance. 1991–92*, no. 49, Buenos Aires.
——(1993) *Memoria y Balance. 1992–93*, no. 50, Buenos Aires.
——(1994a) *Memoria y Balance. 1993–94*, no. 51, Buenos Aires.
——(1994b) 'Nuestro Conflicto Gremial', *Solicitada*, febrero.
——(1995) *Memoria y Balance. 1994–95*, no. 52, Buenos Aires.
——(1997) *Memoria y Balance. 1996–97*, no. 54, Buenos Aires.
——(1998) *Memoria y Balance. 1997–98*, no. 55, Buenos Aires.
——(1999) *Memoria y Balance. 1998–99*, no. 56, Buenos Aires.
——(2001) *Memoria y Balance. 2000–01*, no. 58, Buenos Aires.
——(2002) *Memoria y Balance. 2001–02*, no. 59, Buenos Aires.
——(2003) *Memoria y Balance. 2002–03*, no. 60, Buenos Aires.
LyF MDP (1989) *Memoria y Balance. 1989*, Mar del Plata.
——(1990) *Memoria y Balance. 1990*, Mar del Plata.
——(1992) *Memoria y Balance. 1991–92*, Mar del Plata.

——(1993) *Memoria y Balance. 1992–93*, Mar del Plata.
——(1994a) *Estudio Comparativo del Convenio Colectivo de Trabajo*, Mar del Plata, febrero, Sindicato Luz y Fuerza Mar del Plata.
——(1994b) *Memoria y Balance. 1993–94*, Mar del Plata.
——(1995a) 'La impunidad de ESEBA SA', Sindicato de Luz y Fuerza Mar del Plata – *Solicitada* – 7 de agosto.
——(1995b) *Marcha Provincial. Por la reincorporación de todos los cesantes de ESEBA S.A.*, Mar del Plata.
——(1995c) *Memoria y Balance. 1994–95*, Mar del Plata.
——(1995d) 'La propuesta de conformar una "empresa obrera" configura un "fraude laboral"', Sindicato de Luz y Fuerza Mar del Plata – *Solicitada* – 7 de agosto.
——(1996) *Memoria y Balance. 1995–96*, Mar del Plata.
——(1997a) *Conflicto con EDEA S.A.*, Mar del Plata, febrero, Sindicato Luz y Fuerza Mar del Plata.
——(1997b) *Memoria y Balance. 1996–97*, Mar del Plata.
——(1998a) *Denuncias de acciones judiciales persecutorias promovidas por EDEA S.A. con el sindicato de Luz y Fuerza Mar del Plata y sus dirigentes José Jorge Rigane, Oscar Alberto Escalante y Néstor Daniel Cuenca*, 28 de enero.
——(1998b) *Memoria y Balance. 1997–98*, Mar del Plata.
——(1999) *Memoria y Balance. 1998–99*, Mar del Plata.
——(2000) *Memoria y Balance. 1999–2000*, Mar del Plata.
——(2003) *Memoria y Balance. 2002–03*, Mar del Plata.
Lyons, John (1989) *The Picture Unfolds: What the Consequences of Privatization are likely to be*, Speech made by John Lyons, General Secretary of the Engineers' And Managers' Association to the Economist Conference, 7th November.
Ministerio de Economía (1990a) *Directivas a las Paritarias de las Empresas Públicas. Para la renovación de los convenios colectivos de trabajo*, Ministerio de Economía, Octubre, Buenos Aires.
——(1990b) *Instructivo impartido a los Señores Interventores sobre la aplicación del Decreto no. 1757/90*, Ministerio de Economía, Buenos Aires.
——(1990c) *Plan Financiero para reducir aproximadamente en 5 per cent las plantas de personal en las empresas públicas* – diciembre.
Ministerio de Trabajo (1990) *Laudo suspendiendo cláusulas del CCT 78/75 celebrado entre SEGNA / Luz y Fuerza Capital Federal*, Ministerio de Trabajo.
NALGO (1986) *Annual Conference.*
——(1988a) *Annual Conference.*
——(1988b) *Conference, Annual Group Meetings – Electricity.*
——(1988c) *Privatization Special Report*, NALGO.
——(1989a) *Electricity Recruitment news*, NALGO.

——(1989b) *Future priorities in the Electricity Supply Industry*, Report of the National Executive Committee, June 1989.
——(1990a) *Annual Conference.*
——(1990b) *Conference. Annual Group Meetings – Electricity.*
——(1992) *Conference. Annual Group Meetings – Electricity.*
——(n/d) *Facing the Future.*
NOP (1989) *A Report for the Electrical and Power Engineers' Association*, National Opinion Polls Market Research Ltd., May.
——(1991) *Survey of Managers' Opinions,* National Opinion Polls Market Research Ltd., December 1990.
Nueva Mayoría (1997a) *Imagen Positiva Factores de Poder*, Informe, Buenos Aires.
——(1997b) *Imagen Positiva de las políticas de Privatizaciones*, Informe, Buenos Aires.
TUC (1980) *TUC report*, London: Trades Union Congress.
——(1984) *The Future of Nationalised Industries*, TUC Discussion Paper. London: Trades Union Congress.
——(1986) *Public Ownership*, Brighton Resolution. London: Trades Union Congress.
——(1988a) *TUC Information Note. 'The EETPU suspension'.* London: Trades Union Congress.
——(1988b) *TUC report*, London: Trades Union Congress.
——(1989a) *Power Failure Conference*, London: Trades Union Congress.
——(1989b) *TUC report*, London: Trades Union Congress.
——(1989c) *The Ultimate Power Failure, Report on the Consequences of Electricity Privatization*, March.
——(1990) *TUC report*, London: Trades Union Congress.
——(1991) *TUC report*, London: Trades Union Congress.
——(1992) *TUC report*, London: Trades Union Congress.
——(1996) *TUC report*, London: Trades Union Congress.
——(1997) *TUC Directory*, London: Trades Union Congress.
Unison (1994) *Service Sector Conference*, London: Unison.
——(1996a) *Annual Report 1995–96*, London: Unison.
——(1996b) *The Privatised Utilities*, Policy Document, London: Unison.
——(1996c) *Unison. The Trade Union*, London: Unison.
——(1997) *Annual Report 1996–97*, London: Unison.
——(2005) *Energy Service Group Executive. Annual Report 2004/05*, Unison, Business and Environment.
Wilson and others (2002) *Wilson and the National Union of Journalists; Palmer, Wyeth and the National Union of Rail, Maritime, and Transport Workers; Doolan and Others v United Kingdom [2002]*, IRLR 568 ECHR.

Index

Trade Unions Past, Present and Future

EDITED BY CRAIG PHELAN

This series publishes monographs and edited collections on the history, present condition and possible future role of organised labour around the world. Multidisciplinary in approach, geographically and chronologically diverse, this series is dedicated to the study of trade unionism and the undeniably significant role it has played in modern society. Topics include the historical development of organised labour in a variety of national and regional settings; the political, economic and legal contexts in which trade unionism functions; trade union internationalism past and present; comparative and cross-border studies; trade unions' role in promoting economic equality and social justice; and trade union revitalisation and future prospects. The aims of the series are to promote an appreciation of the diversity of trade union experience worldwide and to provide an international forum for lively debate on all aspects of the subject.

Volume 1 Craig Phelan (ed.): Trade Unionism since 1945: Towards a Global History. Volume 1: Western Europe, Eastern Europe, Africa and the Middle East.
467 pages. 2009. ISBN 978-3-03911-410-8

Volume 2 Craig Phelan (ed.): Trade Unionism since 1945: Towards a Global History. Volume 2: The Americas, Asia and Australia.
364 pages. 2009. ISBN 978-3-03911-950-9

Volume 3 Pablo Ghigliani: The Politics of Privatisation and Trade Union Mobilisation: The Electricity Industry in the UK and Argentina.
293 pages. 2010. ISBN 978-3-03911-961-5

Volume 4 Heather Connolly: Renewal in the French Trade Union Movement: A Grassroots Perspective.
Forthcoming. ISBN 978-3-0343-0101-5

Volume 5 Magaly Rodríguez García: Liberal Workers of the World, Unite? The ICFTU and the Defence of Labour Liberalism in Europe and Latin America (1949–1969).
354 pages. 2010. ISBN 978-3-0343-0112-1

Volume 6 Jeffrey Sluyter-Beltrão: Rise and Decline of Brazil's New Unionism: The Politics of the Central Única dos Trabalhadores.
481 pages. 2010. ISBN 978-3-0343-0114-5

1. Introduction

This study originates from the practical need that is common to all those who have to face the problem of creating reading-comprehension language tests: the choice of the texts that are suited to a given level of linguistic competence and the construction of equally suitable tasks to assess the candidates' comprehension of the texts. The background and purpose of this study will be illustrated in section 1.1, the theoretical and methodological framework adopted will be sketched out briefly in section 1.2 and an outline of the contents will be given in section 1.3.

1.1 Background and Purpose of this Study

This study brings together two areas of research in an investigation of text complexity and English-language test difficulty. One aspect of this study is more strictly linguistic and is related to attempts in linguistics – most importantly text-linguistics, functional linguistics and cognitive linguistics – to define and characterise the notion of 'text complexity'. The other aspect is related to research into the notions of 'test difficulty' and 'task difficulty' carried out in the field normally called applied linguistics and particularly in research on language testing. In order to investigate these phenomena, a corpus made up of twenty-five reading comprehension tests was designed and analysed. The tests in this 10,613-word long corpus offer the material for analyses of text complexity and of test difficulty. Specifically, they consist of an input text and of one or more accompanying tasks and have been elaborated by their authors to assess the reading skills of students of English as a second or foreign language at different levels of linguistic proficiency. Statistical methods were made use of to quantify, wherever possible, both the inherent complexity and the (receiver-oriented) difficulty of the corpus.

The distinction made in this study between 'complexity' and 'difficulty' derives from Merlini Barbaresi (2003: 23), who distinguishes between 'text inherent complexity' on the one hand and 'the (receiver-oriented) notion of text difficulty' on the other. She defines text complexity as:

> a multifaceted quality of the text starting with the speaker's locutionary and illocutionary planning, but adjustable and negotiable during the text development.

Difficulty, on the other hand, is defined as "a predictable and explainable perlocutionary effect, subject to situational variables". She then goes on to say that:

> Text complexity is a dynamic configuration resulting from the contributions of complex phenomena, as they occur at the various text levels. Text is here viewed as a system, whose components are interacting to achieve a global effect.

Merlini Barbaresi (2003: 29) also hypothesises that 'markedness':

> can be measured by means of a grid of criteria, that complexity is the result of the amount and values of marked phenomena in the specific text and that difficulty derives from the value of complexity, but it is relativised by contextual variables.

For Merlini Barbaresi (2003: 33), text complexity is an 'emergent cumulative quality', due to the interplay of the various linguistic subsystems a text is made up of. She underlines how, in order to effectively, economically, and 'optimally' convey the intended meanings, compromise solutions between the various textual sub-systems are necessary, and marked choices among the options available within the sub-systems might have to be made. The theoretical framework that Merlini Barbaresi proposes to account for the interplay of marked choices at different discourse and textual levels is the theory of 'naturalness/markedness', which she elaborated beginning in Merlini Barbaresi (1988), on the basis of the works by, for example, Peirce (1965), Dressler (1985), and Dressler, Mayerthaler, Panagl, and

Wurzel (1987)[1]. In Merlini Barbaresi's model, text complexity is, in particular,

> a global assessment obtained by evaluating the number and mutual impact of the various marked phenomena. Its degree is proportionally dependent on the percentage of marked vs. unmarked phenomena, and on their markedness values. A large proportion of marked phenomena in a text will account for a high degree of global complexity, but it will not allow a conception of global markedness. From the receiver's perspective, complexity is evaluated in terms of processing difficulty, but this value is situationally biased, i.e. depending on many variables, whereas markedness values are not. Such factors as text type, micro- and macro-goals, addressee and his/her approach to the text are powerful variables, which come in to regulate complexity and relativise processing difficulty. (Merlini Barbaresi 2003: 34)

However, Merlini Barbaresi (2003: 29) admits that their research project on markedness as complexity in texts has not yet achieved "a definition of graded values of markedness, complexity and difficulty and of their mutual significance," although "this remains a prominent goal in the research project". Indeed, although the great potential that a markedness framework might offer for this research was clearly recognized, it was decided that an attempt to use it as the basis of this study would be premature. The approach, methods and procedures described in 1.2 below were, therefore, chosen and applied.

1.2 The Theoretical and Methodological Framework

The theoretical and methodological framework for this study was, as said above, selected and designed with in mind the investigation of both the text-inherent complexity and the processing (and test) difficulty of the twenty-five tests of the corpus. The investigation into their inherent complexity was carried out by applying both quantita-

---

1 The notion of markedness has been used by linguists belonging to different schools and traditions. For instance, some structural linguists of the Prague School applied it to the fields of phonology and grammar (Givón: 1990: 946), while Givón (1990: 945-986) applied it to syntax.

tive and qualitative text- and corpus-analyses. Specifically, on the one hand methods and findings of systemic functional linguistics, corpus linguistics, text linguistics, and psycho-linguistics were drawn upon for the qualitative analysis, and on the other some calculations already available in the literature were made use of to obtain quantitative data: the 'type/token ratio', 'lexical density', 'lexical variation', 'vocabulary difficulty', 'readability formulae', as well as 'grammatical intricacy' and the calculation of 'lexical density' following Halliday's (1989) method. These last two calculations, in particular, require hand tagging, but they are also amply described by various researchers, including the present author (Castello 2002 and 2004).

The phenomena investigated were captured by using thirteen measures in all:

- The type/token ratio;
- The standardised type/token ratio;
- Lexical density calculated following Ure's method;
- Lexical density calculated following Halliday's method;
- Lexical variation;
- Lexical variation calculated on the basis of the number of word families in a text;
- Grammatical intricacy;
- The ratio of secondary clauses in hypotactic structure to independent, paratactically related clauses, and primary clauses in hypotactic structures;
- The Flesch Reading Ease index;
- The Flesch-Kinkaid Grade Level index;
- The Gunning-Fog index;
- The percentage of academic words;
- The percentage of words that do not belong to any frequency list.

The type/token ratio is the ratio of the number of different words (types) to the number of running words (tokens) in a given text or corpus. This index indicates how often, on average, a new 'word-form' appears in the text (McEnery and Wilson, 1996: 158), and can be used as a way of measuring vocabulary and lexical diversity in a text or corpus. For the comparison of texts of different length the standardized type/token ratio is more informative (Scott 2006).

Lexical density (Halliday 1987: 60) is "the proportion of lexical items (content words) to the total discourse." It can be measured either as the ratio of lexical items to the total running words or as the ratio of the lexical items to the clauses in a text. Grammatical intricacy is the ratio of the total number of ranking clauses to the total number of clause complexes. Halliday (2004: 654) considers lexical density and grammatical intricacy two different kinds of complexity, the former associated typically with written language and the latter with spoken language.

Lexical variation is a measure of the lexical repetition or lexical diversity in a text (Gibson 1993: 157). This will be measured in two ways. The first is the ratio of the total number of different lexical words in a text to the total number of lexical words. The second, on the other hand, is the ratio of the total number of word families in the text to the total number of lexical words. The type/token ratio, lexical density, grammatical intricacy, and lexical variation can all be considered measures of text complexity.

The so-called readability formulae represent a method of assigning a numerical estimate of 'readability' to a text. Bruce and Rubin (1988: 5) define readability as 'ease of reading', 'interest' or 'ease of understanding', which are receiver-oriented criteria for difficulty or lack of difficulty. However, the readability formulae used in this study – i.e. the Flesch Reading Ease formula, the Flesch-Kinkaid Grade Level formula, and the Gunning Fog Index – although experimented on readers, only take into consideration text-inherent factors (e.g. word length, sentence length), and, therefore, they can be included in this overall attempt to quantify the complexity of the present corpus.

It was also decided to concentrate on the area of text cohesion and, in particular, the problem of 'participant identification' and lexical cohesion. *Participant identification* is the term used by Martin (1992: 93-157) to indicate the textual resources that allow readers and listeners to track the identities of participants, that is, to introduce people and things into a text and keep track of them once there. Participant identification is, in particular, concerned with the 'phoric' relationships between the items which allow readers or listeners to recover the correct referent. In the present study, the analysis of participant identification was carried out to identify the tracking resources used in the tests of the corpus, and not just in the texts. The focus was, therefore, on the resources used to identify the relevant participants which occur in both the input texts and in the accompanying tasks. The analysis of cohesion in chapter 4 is qualitative rather than quantitative, and takes into consideration various phenomena which might be considered elements of complexity. Although the findings on text cohesion have not been quantified, indications have emerged of marked and complex types of cohesion, where it was foreseeable that there could be difficulty in properly identifying a participant retrieved through cohesive devices. These indications were confronted with the actual test results and questionnaire answers.

The investigation into the difficulty of the language tests in the corpus was carried out by applying language testing methodologies. Specifically, some language testing procedures were made use of, firstly, to properly administer the tests to a sample population of 379 readers/test takers, and analyse their test results, and, secondly, to devise an evaluative feedback questionnaire about the difficulty of the texts and tasks under investigation and analyse and interpret the data collected.

As stated above, in language testing one of the main areas of research concerns test and task difficulty. Test and task difficulty, in particular, is not only related to the characteristics of the task itself (e.g. answering a multiple-choice question), but also to the characteristics of the readers and to those of the texts to be read. Text, task (and test) and readers are, therefore, conceived of as interrelated and interdependent variables. In this respect, Alderson (2000: 32-33) maintains that the 'reader' and the 'text' are the two main 'constellations' of variables involved in the process of reading, and the two of them are seen as interacting with each other. As he points out, the readers and their characteristics have important effects on the reading 'process' – i.e. the interaction between the reader and the text – as well as on the 'product' of reading – i.e. the 'understandings' which readers end up with – because:

> different readers will develop somewhat different understandings of what a text 'means'. This is at least in part because a text does not 'contain' meaning which is waiting to be discovered by an able reader. Rather, meaning is created in the interaction between a reader and a text: the text has what Halliday (1979) and Widdowson (1979) call meaning potential, and the potential is realised – in the product of understanding – only by readers reading. Since

> [...] readers' knowledge and experiences influence the realisation of this meaning potential, and since readers may differ in their knowledge and experiences, then the products of reading will also necessarily differ. (Alderson 2000: 6)

Similarly, according to Urquhart and Weir (1998: 112-113):

> Readings may differ in the case of readers from different cultures, either ethnic or professional, or in the case of the same reader at different times, with different knowledge or different preoccupations. Such differing readings, which are generally not under the control of the readers, Urquhart terms 'interpretations'. The reading product may also vary according to a dimension controlled by the reader's purposes. As Candlin points out, the reader 'may decide to glimpse at the text, extracting 'gist', or work conscientiously through it, satisfying himself that he has made sense of all of it'. In Urquhart (1987) such variations are labelled 'comprehensions'.

From this perspective then the different understandings or comprehensions of a text – i.e. the readers' realisations of the text's meaning potential – can be thought of as the perlocutionary effects of a given text on different readers, and, following Merlini Barbaresi's definition of text difficulty given above, as related to the difficulty of a text. Furthermore, since text, readers, and task are interdependent variables, text difficulty and task difficulty also depend on each other and on the characteristics of the readers.

Alderson (2000: 33-60) reviews some of the main contributions and findings concerning the readers' characteristics which are relevant to language testing. According to his classification, the main variables that have been investigated in this area are: the reader's knowledge, e.g. background knowledge, knowledge of language, schemata[2]; the

---

2 Alderson (2000: 34) writes that a distinction is often made between 'formal schemata' and 'content schemata'. Quoting Carrell (1983), he writes that formal schemata include "knowledge of language and linguistic conventions, including knowledge of how texts are organised, and what the main features of particular genres are." Content schemata, by contrast, include "knowledge of the world, including the subject matter of the text." He also writes that "content schemata can be divided into background knowledge – i.e. knowledge which may or may not be relevant to the content of a particular text – and subject-matter knowledge, which is directly relevant to text content and topic. Moreover, some researchers have focused upon certain aspects of

reader's motivation; the strategies that readers use when processing texts; the reader's skills and abilities; the reader's purpose and motivation in reading; the reader's affect, e.g. the emotional state induced by a testing situation; and other stable characteristics, such as sex, social class, occupation, and whether the readers are beginning or fluent. As far as text variables related to text complexity are concerned, Alderson (2000: 60-79) lists: 'text topic and content', 'text type and genre', 'literary and non-literary texts', 'text organisation', 'text readability', 'typographical features', 'verbal and non-verbal information', the 'medium of text presentation', and 'traditional linguistic variables'. These include, for example, 'syntactic complexity', "the opacity and heaviness of the constituent structure of sentences which make it difficult for readers to parse sentences", and 'vocabulary difficulty.' Some of these variables are investigated in chapter 3, where an attempt is made to quantify and capture some aspects of the complexity of the texts which form the corpus for this study (see chapter 2). However, Alderson warns us that inherent text complexity alone is not a sufficient concern for those investigating test difficulty and says that:

> The language of texts would seem, prima facie, highly relevant to the testing and assessment of reading. The interesting thing about much of the research is that a common-sense assumption proves too simplistic, and that identifying text variables which consistently cause difficulty is a complex task. Clearly at some level the syntax and lexis of texts will contribute to text and thus to task difficulty, but the interaction among syntactic, lexical, discourse and topic variables is such that no one variable can be shown to be paramount. Moreover, even the ability to guess words from the context has to be seen in context: the context of the reader, and other variables in the text. (Alderson 2000: 70-71)

Alderson's concerns are shared, and in fact the choice of asking different groups of Italian university students to read some texts suited to their respective levels of language proficiency and answer related test questions (see chapter 5) was made partly to take into account their characteristics as readers.

According to Alderson (2000: 85), the difficulty of a task, or item, or test question is "very simply measured by the proportion of

---

background knowledge, in particular that knowledge which is common to a particular culture, or culture knowledge". (Alderson 2000: 34) candidates getting the answer correct compared with those getting it wrong". The causes of task difficulty are manifold and depend mainly on the characteristics of the tasks. As the testing 'techniques' or 'test methods' used in the tests of the corpus are mainly 'multiple-choice', 'multiple-matching', 'true/false', and 'true/false/not given' (see chapter 2), the two main types of task characteristics that will be taken into account and investigated in this study are the so called "item variables that may influence the reading item difficulty measure" and the "text-by-item overlap variables that may influence the reading item difficulty measure" (Freedle 1997: 402). The former includes such item characteristics as correct choice length, incorrect choice plausibility, 'stem[3] length', and 'stem-unique content words[4]' (Alderson 2000: 88), which are specific to multiple-choice questions. The latter includes the so called "lexical overlap between the necessary information and the correct option" (Buck, Tatsuoka and Kostin 1997: 450), where necessary information means "the information in the text which the reader must understand to be certain of the correct answer" (Buck, Tatsuoka and Kostin 1997: 437). An investigation into these task characteristics, which, however, are also to a certain extent text characteristics, is carried out in chapter 4. As said above, this chapter analyses the ways participants are identified in the input texts of the corpus and in the accompanying tasks, that is, in the overall tests. This analysis, complemented by an analysis of the cohesive resources used in these tests, was carried out to gain insights into the complexity of a given test dealt with as a text.

Another important characteristic of tasks which might influence text difficulty has to do with the type of reading that is required. There are, in fact, different 'purposes' of reading that readers/test takers consciously 'adopt' to carry out the specific tasks in a test. It can be claimed, in particular, that different purposes of reading need different 'types of reading' of a given text, that different 'readings' in turn require one to process the text in quite different ways, and that these differences can cause variations in processing or text difficulty. In this

3 'Stem' is the first part of multiple choice question, appearing before the optional choices.

4 Unique content words are those "nouns, verbs or modifiers that appear in the question but not in the passage" (Alderson 2000: 88).

respect, Urquhart and Weir (1998: 115) propose to break down reading into reading types according to the skills and strategies[5] that need to be deployed to reach specific goals. To this end, they distinguish between 'careful' and 'expeditious' reading and between 'global' and 'local' reading. By combining these reading typologies, they further distinguish between the four reading types represented in Figure 1.1.

	Global	*Local*
Expeditious	*A.* Skimming quickly to establish discourse topic and main ideas. Search reading to locate quickly and understand information relevant to predetermined needs.	*B.* Scanning to locate specific information; symbol or group of symbols; names, dates, figures or words.
Careful	*C.* Reading carefully to establish accurate comprehension of the explicitly stated main ideas the author wishes to convey; propositional inferencing.	*D.* Understanding syntactic structures of sentence and clause. Understanding lexical and/or grammatical cohesion. Understanding lexis/deducing meaning of lexical items from morphology and context.

Figure 1.1. Matrix of reading types (adapted from Urquhart and Weir 1998: 123).

As can be seen, by reading a text expeditiously and globally (type A) one seeks to either establish the discourse topic and the main ideas in a text (i.e. skimming) or to locate and understand information relevant to predetermined needs (i.e. search reading). Expeditious and local reading (type B), by contrast, aims to locate specific information such as symbols or names, whilst the purpose of careful and global reading (type C) is to accurately comprehend explicitly stated main ideas and to infer 'propositional' meaning. Finally, by reading carefully and locally (type D), one processes the immediate co-text and occasionally

5 According to Urquhart and Weir (1998: 96-98), there is confusion in the literature as to what distinguishes a skill from a strategy. Some of the differences they list are the following ones: strategies are reader-oriented, whilst skills are text-oriented; strategies represent conscious decisions taken by the reader, skills are deployed unconsciously; strategies, unlike skills, represent a response to a problem.

the wider context in order to deduce the meaning of unknown lexical items, the reference encoded by grammatically cohesive items, or the syntactic structures of sentences or clauses. In chapter 2 information about the specific skills and/or strategies assessed by each item or group of items in the tests of the corpus is given. It must be said, however, that "there is still no consensus on divisibility of skills", and that in the literature three positions are common: "the first is that reading is a unitary skill; the second is that reading is multidivisible, even though there is no agreement on how many skills might be empirically distinguishable; the third is that there is a two-way split" (Alderson 2000: 95).

Whenever the data deriving from this study could be handled quantitatively, statistical techniques were used. This is true for the test results and the questionnaire data, which are analysed in chapter 5, as well as for the analysis of text complexity presented in chapter 3. When possible and relevant, in chapters 3, 4, and 5 attempts were also made to 'triangulate'[6] the various types of data.

1.3 Outline of Content

In chapter 2 information about the twenty-five reading comprehension tests which form the corpus for this study is given. Information is also given about the testing systems which they belong to, viz. the on-line diagnostic system called DIALANG, the testing systems TOEFL and IELTS, and University of Cambridge ESOL Examinations' Preliminary English Test (PET), First Certificate in English (FCE) and Certificate in Advanced English (CAE).

In chapter 3 a series of quantitative analyses carried out into the texts of the corpus is reported on. They all aim to capture some aspects of the complexity of these texts and to make some inferences

6 In social research the term 'triangulation' is used to indicate the use of data from different sources and obtained by means of different methodologies to increase the quality control and representativeness of a study (see, for example, Bailey 1999: 38, and 5.1 below).

about the difficulty of the tests. The linguistic variables analysed are: the 'type/token ratio', 'lexical density', 'lexical variation', 'vocabulary difficulty', 'grammatical intricacy', and 'text readability'. In order to carry out these analyses, some indices and formulae are made use of, as well as word frequency lists based on large corpora and software designed to process them. Hand tagging of the corpus is performed in order to calculate grammatical intricacy and lexical density following Halliday's method. For each of these procedures and variables the relevant literature is reviewed.

Chapter 4 focuses on some textual aspects of the texts under enquiry which it was thought might be revealing in terms of complexity. In particular, since most of the tasks are 'closed-ended' questions (e.g. multiple-choice, true/false questions), some relations and interactions between the texts and the accompanying tasks are also analysed. The main textual phenomena investigated are 'participant identification' – i.e. the lexico-grammatical realisations of the ways participants are introduced and presumed in the texts and in the tasks – and lexical cohesion. When relevant the contribution of another two aspects of cohesion, thematic structure and lexical cohesion, are taken into account.

Chapter 5 begins with a discussion of the procedures followed and the choices made in the administration of the tests. It then reports on how further information about text and task difficulty was obtained from the administration of a feedback questionnaire about the test takers' perception of the difficulty of each text and of the accompanying tasks. The rest of chapter 5 provides an analysis and discussion of the test results and of the data derived from the feedback questionnaire. In the discussion each piece of data is related to each other and to the data obtained from the analyses reported on in the previous chapters.

The final chapter presents the main conclusions arrived at in this study on the basis of the results obtained by applying the various methods of analysis adopted. The theoretical and practical implications of the results obtained in this study as for text complexity and test difficulty are also discussed. Finally, areas for future research are identified.

2. A Corpus of Reading Comprehension Tests

The corpus the present study is based on is composed of twenty-five reading comprehension input texts and of the accompanying questions/tasks to be answered/performed by test-takers. They were selected from various testing systems that aim to assess the reading skills of students of English as a second or foreign language at different levels of competence and for different purposes. The majority of them were chosen from free sample materials available on-line, i.e. University of Cambridge ESOL Examinations' Preliminary English Test (PET), First Certificate in English (FCE) and Certificate in Advanced English (CAE)[1], and from commercially available preparation courses for the same and other English language examinations, i.e. Test of English as a Foreign Language (TOEFL)[2], the International English Language Testing System (IELTS)[3], and University of Cambridge ESOL Examinations' PET and FCE tests. Some other texts and tasks, on the other hand, were taken from the

---

1 University of Cambridge ESOL Examinations (Cambridge ESOL) is part of Cambridge Assessment <http://www.cambridgeassessment.org.uk>. Cambridge ESOL offers a wide range of exams for learners and teachers of English: General English exams, Professional English exams, Young Learners English, and Teaching Awards. The General English Exams offered are: KET (*Key English Test*), PET (*Preliminary English Test*), FCE (*First Certificate in English*), CAE (*Certificate in Advanced English*), CPE (*Certificate of Proficiency in English*), and *Certificates in ESOL Skills for Life* <http://www.cambridgeesol.org/exams/index.html>. Websites last visited on 25th April 2008.

2 The *Test of English as a Foreign Language* (TOEFL) measures the ability of non-native speakers of English to use and understand English in college and university settings. It is developed and delivered by ETS (Educational Testing Service) <http://www.ets.org/toefl/>, last visited on 25th April 2008.

3 The *International English Language Testing System* (IELTS) is recognised as "secure, valid and reliable indicator of true to life ability to communicate in English in education, immigration and professional accreditation". IELTS is "jointly managed by British Council, IDP: IELTS Australia and the University of Cambridge ESOL Examinations (Cambridge ESOL)" <http://www.ielts.org/general_pages/about_us.aspx>, last visited on 25th April 2008.

reading component of DIALANG, a web-based diagnostic testing system funded by the European Commission[4].

These tests were chosen and focused on for various reasons. Firstly, they are all valid, reliable[5] and constantly monitored tests: evidence of their validity and reliability is provided by studies that are regularly made publicly available (e.g. on the websites devoted to these testing systems). Secondly, although different from each other, their constructs[6] for reading seemed to be relevant for the kind of readers under investigation in this study, i.e. Italian university students of English (including both language and non-language majors) and for the sample population that was actually asked to read the texts, perform the tasks and give feedback on them for this research project[7]. Thirdly, with the exception of TOEFL, the tests range across different levels of language proficiency in English, which helped decide which tests were suited to the current level of English language knowledge of each group of students. In the following paragraphs of this chapter a description of the most relevant characteristics of the testing systems under enquiry and of the texts and tasks selected will be provided. The complete tests are presented in the Appendix.

2.1 The DIALANG Sub-corpus

DIALANG is a web-based language diagnostic and self-assessment system based on the *Common European Framework of Reference for Languages: Learning, teaching, assessment* (CEF) (Council of

---

4 See Alderson (2000: 125-128) and the DIALANG website <http://www.dialang.org/english/index.htm>, last visited on 25th April 2008.

5 For a definition of the concepts of 'validity' and 'reliability' see 5.2 below.

6 "A construct is a psychological concept, which derives from a theory of the ability to be tested" (Alderson 2000: 118). Further information about the concept of construct and construct validity can be found in 5.2 below.

7 The nationality of some of the students that took the tests is actually not Italian. All test-takers, however, were enrolled on undergraduate or post-graduate degree courses at the University of Padua.

Europe, 2001). It aims to assess reading, listening, writing and speaking skills as well as the knowledge of grammatical structures and vocabulary in 14 different European languages by means of separate testing modules. It also has a self-assessment component to be taken before each module, which helps the system 'tune' the test to the test takers' ability[8]. The results test takers are given comprise the level they attained on the Council of Europe scale, an indication of what that level means, i.e. what people can usually do at that level, and a review of the responses to the items listed by sub-skill. DIALANG is therefore principally meant to help language learners diagnose and control their learning progress and increase their language learning awareness. In the reading component "questions and tasks vary from easy to demanding, such that questions asking for specific facts are usually easier than questions requiring synthesis, analysis or inference" (Alderson 2000: 128). For this research project 10 input texts and the corresponding task(s) were chosen from those presented to the present author, who sat the DIALANG reading component himself on 4th May 2005. The 30 items taken were 'screen captured' and so was the feedback received on each one of them and the item review by sub-skill obtained at the end of the module. The selected items were then named and numbered as reported in Figure 2.1. Information about the order in which each one of them was administered, the sub-skill each item is claimed to test and the test method used to elicit test-takers' responses is also provided. As can be seen, four questions are asked about DIALANG text n. 1.

Alderson (2000: 128) specifies that[9]:

> the test is divided into 'overall' and 'analytic' sections. The overall section includes items which 'tap overall reading' – presumably meaning main ideas, gist or the like – but can also include items which tap one or more of the three 'skills' focused on in the analytic section. These are: 'identifying main

---

8 For more information see <http://www.dialang.org/english/index.htm>, last visited on 25th April 2008.

9 Alderson (2000: 128) actually quotes from the DIALANG Assessment Specifications for Reading Comprehension, Version 6, 18th February 1998. He writes that "the Assessment Specifications themselves merely refer to the Common European Framework, the use of which is detailed in a separate document, the DIALANG Assessment Framework (DAF)" (Alderson 2000: 125).

idea(s)/information/purpose; reading for specific detail/information'; and 'inferencing/going beyond the literal meaning (including lexical inferencing).

DIALANG tests	*Order*	*Sub-skills tested*	*Test methods*
1. Eye problems	11-14	identifying main idea	multiple matching
2. Our hands met	5	inferencing	multiple choice
3. It should be fun	10	inferencing	multiple choice
4. Ecotourism	20	identifying main idea	multiple choice
5. Digital cameras	6	inferencing	multiple choice
6. Travel agency	16	inferencing	multiple choice
7. Bangkok	17	inferencing	multiple choice
8. Station	2	inferencing	multiple choice
9. Suzanne Somers	3	inferencing	multiple choice
10. Injuries	25	identifying main idea	multiple choice

Figure 2.1. Information about the DIALANG items chosen for the corpus: their name, order of administration, the sub-skill tested and the test methods used.

2.2 The TOEFL Sub-corpus

The Test of English as a Foreign Language (TOEFL) is "intended to measure English proficiency broadly interpreted, without the engagement of any special background knowledge or specific reference to use […], and would thus be considered a more general purpose language test" (Douglas 2000: 14). It can currently be taken in two different formats: the paper-based TOEFL test and the Internet-based TOEFL test (TOEFL iBT)[10]. The computer-based TOEFL test format is no longer available. The paper-based TOEFL test consists of three separately timed sections, plus a writing test. These sections are called 'Listening Comprehension', 'Structure and Written Expression', Reading Comprehension', and 'Writing (*Test to written*

10 The information about the TOEFL tests given in this section was obtained from <http://www.ets.org/toefl/>, last visited on 25th April 2008, unless otherwise specified.

English)'. The computer-based TOEFL test used to measure English language proficiency in 'Listening', 'Structure', 'Reading' and 'Writing' (essay); the Listening and Structure sections are computer-adaptive and new types of questions were added to the Listening and Reading sections (ETS 2005: 3). The TOEFL Internet-based test (TOEFL iBT) asses all four language skills: speaking, listening, reading, and writing. It is delivered through the Internet and requires one to combine two or more of the above skills to respond to a question.

As can be seen, in all three versions of the test the reading component is present, although the time limit within which the module is to be completed and the number and types of questions vary. In particular, the time limit for the 'Reading Comprehension' module of the paper-based TOEFL test is 55 minutes, the number of texts to be read is 5 and the questions about them are 50 (Mahnke and Duffy 1996: 310, Phillips 2001: 343). By contrast, the maximum time allowed for the completion of the 'Reading' component of the computer-based TOEFL test is 70 minutes, the number of texts to read is 4 and the questions about them are 44 (Phillips 2001: 350). A characteristic that the TOEFL reading tests share with many other reading tests is that they consist of "a number of (usually short) passages, each accompanied by comprehension questions. Since the questions all relate to the same text, they are often regarded as a subtest and are referred to increasingly as 'testlets'" (Alderson 2000: 109).

Enright *et al.* (2000: 18) recommend that the tasks in the TOEFL 2000[11] reading test "be based on a variety of text material. Some could be based on a single text, ranging in length from a short paragraph to a lengthy selection; others could draw on multiple texts, also ranging in length". They also specify that:

---

11 "The TOEFL 2000 project is a broad effort under which language testing and Educational Testing Service (ETS) will evolve into the 21st century" (Enright et al. 2000: i). Enright *et al.* (2000: 49) claim that the most obvious improvement they can make over the reading test is to "articulate the construct [they] want to test and to link the proposed test design to that construct". Furthermore, they say that "one of the goals of the TOEFL 2000 project is to identify variables contributing to task difficulty in order to improve the interpretability of test scores" (Enright *et al.* 2000: 49).

> any subject area that is typical of academic study could provide appropriate material for the reading test. The current TOEFL test covers a range of very general academic topics broadly classified as topics related to Arts, Humanities, Social Sciences, Physical Sciences, or Life Sciences. It seems appropriate to continue to include as much topic variety as possible in the new test. As with the current test, however, care should be taken to ensure that specialized knowledge of a particular field is not necessary to understand the information presented in the passages. (Enright *et al.* 2000: 16)

Enright *et al.* (2000: 20) further recommend that reading passages "be classified according to their dominant pragmatic and rhetorical features", where by dominant pragmatic features they mean the primary intent of the author and by rhetorical features the higher level organization of the text. According to the parameter 'pragmatic features', Enright *et al.* (2000: 20-23) classify the types of text materials as 'exposition', 'argumentation/persuasion/evaluation' and 'historical biographical/autobiographical narrative'. By contrast, different rhetorical patterns help them categorize texts as 'definition', 'illustration', 'classification', 'comparison/contrast', 'cause/effect', 'problem/ solution', and 'analysis' (Enright *et al.* 2000: 23-28). Finally, the types of tasks to be used in the TOEFL 2000 test should test what they call the 'four purposes for reading', i.e. 'reading to find information', 'reading for basic comprehension', 'reading to learn' and 'reading to integrate information across multiple texts' (Enright *et al.* 2000: 31-37).

As TOEFL 2000 is a computer-based test, Enright *et al.* (2000: 37) also list an array of response formats which should be used by item writers for "the TOEFL 2000 reading test and integrated tasks":

- multiple choice;
- open response formats;
- click on a word, phrase, or sentence in the text or graphic;
- click on and drag a word, phrase, or sentence in the text or graphic;
- complete a chart, graph, or table;
- create a chart, graph, or table;
- extended written and/or spoken response.

For this study two texts and the corresponding testlets were selected from those proposed by two preparation courses for the paper-based

TOEFL version and another one from a preparation course for the computer-based TOEFL version. Specifically, text 11 was taken from Phillips (2001: 345) and named 'The next artist', and text 13 was selected from those proposed by Mahnke and Duffy (1996: 311) and named 'One of the most mysterious things'. They are both meant to be used to practise for the paper-based TOEFL version, whereas text 12, taken from Phillips (2001: 351-352) and called 'The final battle', belongs to the reading component of the computer-based TOEFL test. The preparation courses from which the texts and their testlets were taken also provide a classification of the 'sub-skills' tested by each item (Mahnke and Duffy 1996: 585, Phillips 2001: 349). In Figure 2.2 information about the texts, the sub-skills each item is claimed to test and the test method used is given.

TOEFL	*Sub-skills tested*	*Test methods*
11. The next artist (paper-based)		
11.01	answer transition question correctly	multiple choice
11.02	recognize the organisation of ideas	multiple choice
11.03	answer stated detailed questions correctly	multiple choice
11.04	determine meanings from word parts	multiple choice
11.05	answer implied detail questions correctly	multiple choice
11.06	use context to determine meanings of difficult words	multiple choice
11.07	find definitions from structural clues	multiple choice
11.08	determine meanings from word parts	multiple choice
11.09	find 'unstated' details	multiple choice
12. The final battle (computer-based)		
12.01	answer main idea questions correctly	multiple choice
12.02	find pronoun referents	click on a word in the text
12.03	use context to determine meanings of difficult words	multiple choice
12.04	answer stated detailed questions correctly	multiple choice
12.05	answer stated detailed questions correctly	multiple choice
12.06	determine where specific information is found	click on a sentence in the text
12.07	find 'unstated' details	multiple choice
12.08	use context to determine meanings of difficult words	click on a word or phrase in the text

12.09	answer implied detail questions correctly	multiple choice
12.10	determine where to insert a piece of information	click on a 'square' in the text
12.11	recognize the organisation of ideas	click on a paragraph in the text
13. One of the most mysterious things (paper-based)		
13.01	topic and main ideas	multiple choice
13.02	facts and details	multiple choice
13.03	purpose	multiple choice
13.04	referents	multiple choice
13.05	facts and details	multiple choice
13.06	vocabulary in context	multiple choice
13.07	vocabulary in context	multiple choice
13.08	inferences	multiple choice
13.09	organization	multiple choice
13.10	attitude	multiple choice

Figure 2.2. The TOEFL texts chosen for the corpus, the sub-skills tested by each item and the response formats (test methods) used.

2.3. The IELTS Sub-corpus

Texts 14, 15 and 16 and their respective tasks were chosen from the General Training Reading module of IELTS (International English Language Testing System), and text 17 from the Academic Reading module of the same testing system. In particular, the texts and the corresponding tasks were selected from the examination papers found in UCLES (2001): text 14 from UCLES (2001: 98-99), text 15 from UCLES (2001: 102-103), text 16 from UCLES (2001: 116-117), and text 17 from UCLES (2001: 83-86). IELTS is mainly "intended to fulfil English language requirements for entry to English-medium universities, for non-native speakers of English" (Alderson 2000: 130). It consists of four modules: all candidates have to take the same Listening and Speaking modules, but can choose to take either the Academic or the General Training Reading and Writing modules.

The General Training version of the test is for "candidates taking the test for entry to vocational or training programmes not at degree level, for admission to secondary schools and for immigration purposes", while the academic version is "for candidates taking the test for entry to undergraduate or postgraduate studies or for professional reasons" (UCLES 2001: 1). Both the Academic and the General Training Reading modules consist of three sections based on three reading passages and 40 questions. The time limit for both of them is 60 minutes and their sections are claimed to be of increasing difficulty. The General Training Reading module contains texts that are

> based on the type of material candidates would be expected to encounter on a daily basis in an English speaking country. They are taken from sources such as newspapers, advertisements, instruction manuals and books, and test the candidate's ability to understand and use information. The test includes one longer text, which is descriptive rather than argumentative. (Cambridge ESOL *et al.* 2007: 3)

Cambridge ESOL *et al.* (2007: 7) also specifies that the first section of the General Training module is called 'social survival' and contains texts relevant to basic linguistic survival with "tasks mainly about retrieving and providing general factual information." The second section, 'training survival', "involves a text or texts of more complex language with some precise or elaborated expressions". Finally, the third section, i.e. 'general reading', "involves reading more extended prose with a more complex structure but with the emphasis on descriptive and instructive rather than argumentative texts" (Cambridge ESOL *et al.* 2007: 7).

By contrast, in the IELTS Academic Reading module

> texts are taken from magazines, journals and newspapers, all written for a non-specialist audience. At least one of the texts contains a detailed logical argument. One text may contain non-verbal materials such as diagrams, graphs or illustrations. If texts contain technical terms then a simple glossary is provided. (Cambridge ESOL *et al.* 2007: 7)

Alderson (2000: 131) makes reference to de Witt's (1997) account of the IELTS construct and says that "the major skills students need to know, and the things they need to do are: '(Know) how to understand main ideas and how to find specific information; (Do) survey the text;

analyse the questions; go back to the text to find answers; check your answers.'" Figure 2.3 gives information about the four texts chosen: the version of the Reading module and the section they belong to, the number of items accompanying each input text and the test method(s) used to elicit the candidates' responses.

IELTS	*Version; Section*	*N. of items; Test methods*
14. Use of University Ground		
	General Training Reading; Section 1	5; true/false/not given
15. West Thames College		
	General Training Reading; Section 2	7; true/false/not given
16. Central Library		
	General Training Reading; Section 2	7; true/false/not given
17. Green Wave		
	Academic Reading; Section 1	13 questions;
	17.01-17.06	yes/no/not given
	17.07-17.09	multiple choice
	17.10-17.13	summary completion by choosing words from a list

Figure 2.3. The IELTS texts chosen for the corpus, the version of the Reading module and the section they belong to, the number of items accompanying each text, and the test methods used.

2.4 The PET Sub-corpus

Texts 18, 19, 20, and 21 and their tasks are all representative of parts 3 and 4 of the Reading Paper of University of Cambridge ESOL Examinations' Preliminary English Test (PET). PET is a test at level B1 of the CEF (Council of Europe 2001), which is defined as the level at which "learners should be able to cope linguistically in a range of everyday situations which require a largely predictable use of language". Furthermore, language users at level B1 "will be able to use English in their own or a foreign country in contact with native and non-native speakers of English for general purposes" (Cambridge

ESOL 2008: 3). The test consists of Paper 1, i.e. Reading/Writing, Paper 2, i.e. Listening, and Paper 3, i.e. Speaking. The timing for Paper 1 is 1 hour 30 minutes, and its Reading component is made up of 5 parts, "which test a range of reading skills with a variety of texts, ranging from very short notices to longer continuous texts" (Cambridge ESOL 2008: i). In part 1 candidates have to prove they can read and understand various kinds of short texts, and think about the situations in which they would be used in real life. The texts to be read in this part are "authentic notices and signs, packaging information (for example, instructions on a food package or a label on a medicine bottle), and communicative messages (notes, e-mails, cards and postcards)" (Cambridge ESOL 2008: 11). In part 2 test-takers are presented with five short texts which they have to match to five out of eight 'adapted-authentic' factual texts on a particular topic. Part 3 tests the ability to read longer 'adapted-authentic' factual texts and to find specific information in them.

> Frequently these texts take the form of brochure extracts, advertisements in magazines and website information. There are ten questions, which are single-sentence statements about the text. The task is made more authentic by putting these questions before the text, in order to encourage candidates to read them first and then scan the text to find each answer. The information given in the text follows the same order as the content of the questions. (Cambridge ESOL 2008: 11)

In part 4 candidates have to read an 'adapted-authentic' long text,

> which goes beyond the provision of factual information, and expresses an opinion or attitude. There are five multiple-choice questions with four options [...] . In answering these questions, candidates will demonstrate whether they have understood the writer's purpose, the writer's attitude and opinion, or an opinion quoted by the writer, and both the detailed and global meaning of the text. This part requires candidates to read the text very carefully indeed. [...] It may be more practical for candidates to consider the first and last questions together, in that the first focuses on writer purpose and the last on global meaning. The other three questions follow the order of information given in the text and one of the three will focus on attitude or opinion. (Cambridge ESOL 2008: 11)

Finally, in part 5 test participants are presented with an "adaptive-authentic text drawn from a variety of sources [...] of a factual or narrative nature" (Cambridge ESOL 2008: 10). The text in this part contains ten numbered spaces that test-takers have to fill in. In order to do so, they have to answer a four-option multiple choice question for each space. Texts 18, 20 and 21 were chosen from those available in Cambridge ESOL (2008a: 16-17, 25), a collection of Cambridge ESOL's Preliminary English Test (PET) sample papers, while text 19 was chosen from the preparation papers for PET in UCLES (2001b: 10-11). Information about these texts and accompanying tasks is summarised in Figure 2.4.

PET	*Part*	*Sub-skills tested*	*N. of items; Test methods*
18. Exploring the Arctic AND 19. Globewise			
	3	processing a factual text; scanning for specific information while disregarding redundant material	10; true/false
20. Ainsley Harriot AND 21. Doug Allan			
	4	reading for detailed comprehension; understanding attitude, opinion and writer purpose; reading for gist, inference and global meaning	5; four-option multiple choice

Figure 2.4. The PET texts chosen for the corpus, the Part of the Reading paper they belong to, the (sub-)skills tested by each item, the number of items accompanying each text, and the test method(s) used.

2.5 The FCE Sub-corpus

Both tests 22 and 23 were obtained from UCLES (2000: 4-5, 6-7), a collection of past Reading papers from University of Cambridge ESOL Examinations' First Certificate in English (FCE). Recently, both FCE and CAE papers and specifications have undergone an updating process and the changes and improvements made will be introduced in December 2008[12].

12 For more information on this review project see *Research Note 30* at <http:// www.cambridgeesol.org/rs_notes/>, last visited on 25th April 2008.

University of Cambridge ESOL Examinations' First Certificate in English (FCE) is claimed to be a test at level B2 of the CEF (Council of Europe, 2001). At this level "a learner should be able to handle the main structures of the language with some confidence, demonstrate knowledge of a wide range of vocabulary and use appropriate communicative strategies in a variety of social situations" (Cambridge ESOL 2008b: 3). The current test consists of five papers: Reading, Writing, Use of English, Listening and Speaking. The Reading paper takes 1 hour 15 minutes and consists of 4 parts and 35 questions. The input texts are chosen from "newspaper and magazine articles, reports, fiction, advertisements, correspondence, messages, informational material (e.g. brochures, guides, manuals, etc.)" and vary in length from 350 to 700 words (Cambridge ESOL 2008b: 6). In part 1 of the Reading paper candidates have to prove their ability to "identify the main points in a text at paragraph level. One of two different tasks may appear on the paper: headings or summary sentences" (Cambridge ESOL 2008b: 7). The tasks candidates are asked to perform are concerned with reading a list of especially written headings and decide which heading or summary sentence best fits each paragraph. "The headings or summary sentences are printed in a list before the text to encourage candidates to form an impression of the main points they are looking for before they start reading a text" (Cambridge ESOL 2008b: 7). In part 2 what is tested is the candidates' detailed understanding of a text, including the opinions and attitudes expressed in it. The text and each question about them need to be read carefully,

> in order to distinguish between apparently similar viewpoints, outcomes and reasons. […] The multiple-choice questions appear after the text. They are presented in the same order as the information in the text so that candidates can follow the development of the text as they work through the questions. The final question may require candidates to interpret an aspect of the whole text, e.g. the writer's purpose, attitude or opinion. (Cambridge ESOL 2008b: 7)

In part 3 the "candidates' understanding of how texts are structured and their ability to follow the development of the text" are at issue (Cambridge ESOL 2008b: 7). The task requires the selection from a number of options of the correct extract (either a sentence or a paragraph) to fit in each of six or seven gaps in a text. The last part of

FCE, i.e. part 4, aims to test the ability to locate specific information in a text or a group of texts, and "the task consists of a series of prompts, followed by a divided long text or series of texts. Candidates scan the text in order to decide which part of the text each prompt matches" (Cambridge ESOL 2008b: 8). In Figure 2.5 information about the two FCE texts, the questions about them and the sub-skills tested is provided.

FCE	*Part*	*Sub-skills tested*	*N. of items; Test methods*
22. A lot of advice			
	2	detailed understanding of a text, including the opinions and attitudes expressed in it	7; four-option multiple choice
23. In case you lose your luggage			
	1	identification of the main points in a text at paragraph level	7; multiple matching

Figure 2.5. The FCE texts chosen for the corpus, the Part of the Reading paper they belong to, the (sub-)skills tested by each item, the number of items accompanying each text, and the test methods used.

2.6 The CAE Sub-corpus

The last two texts of the corpus, i.e. text 24 and text 25, and the tasks accompanying them are part of University of Cambridge ESOL Examinations' Certificate in Advanced English (CAE) past papers published in Cambridge ESOL (2008c: 10-11). University of Cambridge ESOL Examinations' Certificate in Advanced English (CAE) is a test at level C1 of the CEF (Council of Europe 2001) (Cambridge ESOL 2008d: 3). At this level "learners are expected to be able to use the structures of a language with ease and fluency", and in particular CAE is claimed to offer "a high-level qualification in the language to those wishing to use English for professional or study purposes". CAE consists of five papers: the Reading paper, the Writing paper, the English in Use paper, the Listening paper, and the Speaking test.

The Reading paper contains four parts, approximately 45 questions and takes 1 hour 15 minutes. The texts are taken from newspapers, magazines, journals, non-literary books, leaflets, brochures, etc., and may consist of several short pieces. Their lengths range from 450 to 1,200 words (Cambridge ESOL 2008d: 6). Part 1 of the Reading paper aims to test the candidates' ability to locate specific information in a text. Specifically, "the task requires candidates to scan a single page of text ... and match the [12-18] questions with the relevant information from the text" (Cambridge ESOL 2008d: 7). The questions for the multiple matching task are presented before the text so that the candidates know what information to look for in the text. Part 2, by contrast, "tests candidates' understanding of how texts are structured and their ability to follow text development" (Cambridge ESOL 2008d: 7). Test-takers have to

> select from a number of choices the paragraphs which fit the [6 or 7] gaps in a text. [...] They need to read the gapped text first in order to gain an overall idea of the structure and the meaning of the text, and to notice carefully the information and ideas before and after each gap as well as throughout the whole gapped text. (Cambridge ESOL 2008d: 7)

Part 3 tests candidates' "detailed understanding of a text, including the opinions and attitudes expressed in it" (Cambridge ESOL 2008d: 7). The text needs to be read carefully in order to "distinguish between, for example, apparently similar viewpoints, outcomes, or reasons" (Cambridge ESOL 2008d: 7). The task consists of 6-7 multiple-choice questions that are presented in the same order as the information in the text. Finally, in part 4 the candidate's ability to locate specific information in a text is tested again by means of a multiple-matching task, which requires candidates to scan a text or a group of shorter texts or sections of a text.

CAE	*Part*	*Sub-skills tested*	*N. of items; Test methods*
24. Emotional Intelligence			
	1	locating specific information in a text	12; multiple matching
25. Beginner takes all			
	2	understanding of how texts are structured and ability to follow text development	6 (gaps); gapped text

Figure 2.6. The CAE texts chosen for the corpus, the Part of the Reading paper they belong to, the (sub-)skills tested by each item, the number of items accompanying each text, and the test method(s) used.

University of Cambridge ESOL Examinations' PET, FCE, CAE, along with KET (Key English Test) and CPE (Certificate of Proficiency in English), form a set of general English examinations that UCLES call their 'Main Suite' (Alderson 2000: 287). According to Alderson (2000: 287), a "way of looking at how reading proficiency is thought to develop is to examine sets of language tests, to see what changes as the test levels advance". Unfortunately, though, as can also be seen from the descriptions given above,

> different words are used to describe texts, processes, skills and operations across the suite. It thus becomes somewhat difficult to see what changes as one progresses up through the suite. An inspection of sample papers for FCE and CAE suggests that the language of the texts becomes more difficult, with increasingly arcane vocabulary, the syntax and organisation is unsimplified, and the language of the questions/tasks is less controlled. The CAE tasks are not radically different from FCE, and are not more obviously related to the 'real world', as the specifications claim. Readers at the CAE level may read harder texts and read faster, but only a detailed content analysis, coupled with empirical data on item performance, would throw light on how successful CAE readers have developed beyond their FCE stage of proficiency. (Alderson 2000: 292)

Cambridge ESOL's test developers have provided evidence about how ESOL qualifications, in particular those of the Main Suite, and IELTS scores can be related to each other and to the CEF (Council of Europe, 2001) levels, i.e. A1, A2, B1, B2, C1, C2. This has been done on the basis of "a growing body of internal research combined with long established experience of test use with education and society, as well as feedback from a range of stakeholders regarding the uses of test results for particular purposes"[13]. In Figure 2.7, adapted from the original on the IELTS website, an attempt was made to summarise these correspondences. This figure, however, has to be understood as communicating "relationships between tests and levels in broad terms within a common frame of reference; they should not be interpreted as reflecting strong claims about exact equivalence between assessment products or the scores they generate" .

It should also be noticed that IELTS "is not a level-based test but is designed to stretch across a much broader proficiency continuum" and that it differs from the other tests because of its purpose, measurement scale, test format and test-taker populations. Finally, the "IELTS band scores referred to are the overall band scores, not the individual module band scores" (e.g. the scores of the reading module only).

CEF Level	*Main Suite*	*IELTS*
		9.0
		8.0
C2 (Mastery)	*CPE*	
		7.0
C1 (Effective Operational Proficiency)	*CAE*	
		6.0
B2 (Vantage)	*FCE*	
		5.0
B1 (Threshold)		
	PET	4.0
A2 (Waystage)	*KET*	3.0

Figure 2.7. The relationship of IELTS with the other Cambridge ESOL's tests and with the Common European Framework of Reference (adapted from the figure on the IELTS website, see note 13).

13 The information reported on in the last part of this section was obtained from <http://www.ielts.org/teachers_and_researchers/common_european_framework.asp>, last visited on 25th April 2008.

3. Quantitative Investigation into the Complexity of the Texts of the Corpus

As pointed out in 1.2 above, the nature of reading is affected by two main 'constellations' of variables: the reader and the text. The two of them continuously interact and should not be dealt with separately, since their division "inevitably leads to some degree of distortion" (Alderson 2000: 32-33). Yet, independently of the type of reader, some quantitative analyses aiming at capturing various aspects of the texts' complexity will be undertaken and described in this chapter. These analyses involve the calculation and interpretation of the type/token ratio, of indices of lexical density, lexical variation, vocabulary difficulty, grammatical intricacy, readability formulae, and the use of word frequency lists based on large corpora and of software designed to process them. Also, when relevant, the findings of this investigation will be compared to and integrated with the data derived from the administration of the tests and of the feedback questionnaires.

According to Gervasi and Ambriola (2003: 200-203), two scholars who adopt an engineering approach to the study of text complexity,

> measurement is the process by which numbers or symbols are assigned to attributes of entities in the real world in such a way as to describe them according to clearly defined rules. [Furthermore,] a metric is a direct measure of a criterion or sub-criterion, [...] and must be defined by specifying an attribute of the entity and the exact procedure for measuring it.

The indices and formulae described in this chapter can all be considered as quantitative measures or metrics, in that they are calculated by applying objective, repeatable methods, and in particular by using automatic or semi-automatic tools and procedures.

3.1. The Type/Token Ratio

The type/token ratio (henceforth referred to as TTR) is the ratio of the number of different words (types) to the number of running words (tokens) in a given text or corpus. This index "merely indicates how often, on average, a new 'word-form' appears in the text" (McEnery and Wilson, 1996: 158), and can be used as a way of measuring vocabulary and lexical diversity in a text or corpus (Scott 2006). However, in order to properly interpret the values of the index, there are at least three important factors that need to be taken into account. Firstly, the fact that the index varies widely in accordance with the length of the text (Scott 2006; Biber *et al.* 1999: 53). Secondly, this index does not discriminate between grammatical/function words and lexical items, and therefore all word-forms contribute to lexical diversity in the same way (see 3.2 below). Thirdly, the fact that the index is based on the occurrence of the word-forms in a text, that is, on the occurrence of "any unique string of characters bounded by spaces" (Sinclair 1991: 176), and not on the occurrence of 'lemmas' or 'word families'.

The texts of the corpus under investigation vary considerably in terms of length: as can be seen from Figure 3.1 their lengths range between 52 words (text 5) and 969 words (text 25). As far as the size of texts currently used for testing purposes in concerned, Urquhart and Weir (1998: 145-146) note that they tend to be many and brief. They also claim that text length

> will influence the strategies and skills that the candidate may be asked to deploy. If texts are too short it may not be possible to test expeditious reading strategies (search reading, skimming and scanning), only careful reading.

As seen in chapter 2, all the tests chosen from the DIALANG testing system (tests 1 to 10) actually use short texts to assess either inferencing skills or the ability to identify the main idea in texts, i.e. two skills involving careful reading.

In Figure 3.1 the number of types in each text, the values of the TTR and those of the standardised type/token ratio (henceforth referred to as STTR) (see the discussion below in this paragraph) are also listed. Furthermore, in Figure 3.2 the values of the TTR, those of the number of tokens and those of the types are plotted on the diagram for comparison. It should be clear from the graph in Figure 3.2 that as the number of tokens increases from text 5 to text 25, the TTR tends to decrease, that is, the longer the text, the smaller the index is bound to be. The same tendency can be observed in Figure 3.3. Specifically, it can be seen that as the number of tokens in a text increases, the number of its types is also likely to increase.

Corpus	*Tokens*	*Types*	*TTR*	*STTR50*
Text 01	73	51	69.86	78
Text 02	99	74	74.75	84
Text 03	69	54	78.26	80
Text 04	71	63	88.73	86
Text 05	52	42	80.77	82
Text 06	82	55	67.07	78
Text 07	60	48	80	82
Text 08	55	43	78.18	78
Text 09	73	56	76.71	78
Text 10	76	57	75	76
Text 11	196	108	55.1	73.33
Text 12	324	150	46.3	72.67
Text 13	414	207	50	84
Text 14	154	85	55.19	74
Text 15	379	226	59.63	82.57
Text 16	236	136	57.63	80.5
Text 17	589	280	47.54	86.36
Text 18	445	232	52	79.75
Text 19	395	211	53.42	81.14
Text 20	263	160	60.84	84
Text 21	294	175	59.52	82.8

Text 22	480	244	50.83	80
Text 23	511	262	51.27	85.4
Text 24	666	320	48.05	83.54
Text 25	969	415	42.8	82.32

Figure 3.1. Data set comprising the number of tokens, types and the values of TTR and STTR in the texts.

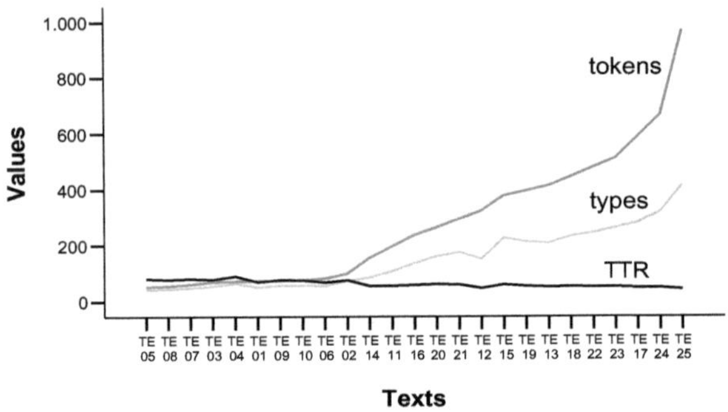

Figure 3.2. The number of tokens in the texts, the types and the values of the TTR.

Biber *et al.* (1999: 53) affirm that:

> in longer texts there is a greater chance that words which have already been used will be repeated. This is true both of the most frequent words which recur in all kinds of texts (*the, of, and,* etc.) and of the words which are connected with the topic of a particular text.

They also report on their analysis of samples of texts belonging to different registers – i.e. conversation, academic prose, news, and fiction – and of different lengths – i.e. 100-word, 1000-word and 10,000-word samples. What they found out is that the TTR scores the lowest values (i.e. there is a tendency to use the same words) in conversation and the highest ones (i.e. there is a tendency to use different words) in news, "which reflects the extremely high density of nominal elements in that register, used to refer to a diverse range of people, places, objects, events, etc." (Biber *et al.* 1999: 54). By contrast, academic prose has the second lowest TTR, and they justify this finding by saying that "a great deal of academic writing has a restricted technical vocabulary and is therefore less variable than fiction and news reportage" (Biber *et al.* 1999: 54).

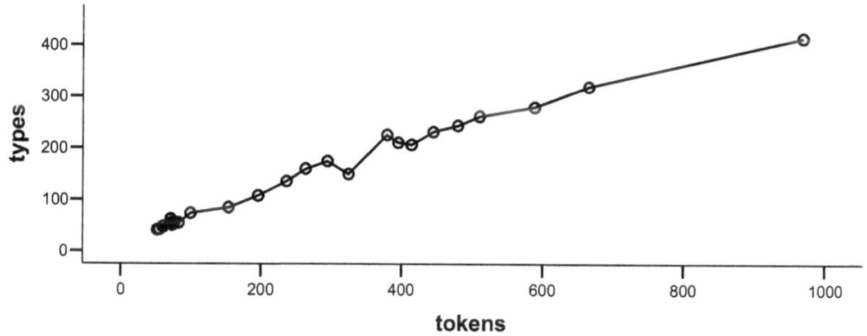

Figure 3.3. The relation between the types and tokens in the texts of the corpus.

As pointed out and exemplified by both McEnery and Wilson (1996: 155-158) and Tuzzi (2001: 266-268), medium-size and large corpora are also characterised by the growth in the number of types as the number of tokens increases. In their studies, though, these scholars consider each single text as contributing to the increase in the total number of types and tokens in a corpus rather than looking at the values of TTR in each single text on its own. In order to do so, they apply the following recursive procedure: they first calculate and plot the values of the types and tokens of the first text on a graph; after this they consider the number of tokens and types of the second text together with those of the first one, and again plot these values on the same graph. This procedure is repeated for all the texts of the corpus, and eventually the curve capturing the increase in the number of types in relation to that of the number of tokens is plotted. By comparing the curves plotted for different corpora, they further manage to demonstrate that corpora representing various sub-languages, e.g. different genres, registers or text types, can be more or less prone to lexical repetition, that is, their lexicons stop growing after variable increases in the number of tokens. McEnery and Wilson (1996: 158), for instance, found that the lexicon of a 1,000,000-word long corpus of IBM manuals "nearly enumerates itself within the first 110,000 words of the corpus", while the number of types in a corpus of transcribed proceedings from the Canadian parliament of the same size continues to grow steadily after the first 200,000 tokens.

According to Scott (2006), when one is dealing with texts of differing lengths and needs to compare them, the TTR is not very informative, while it might be so if one were dealing with a corpus comprising several equal-sized text segments. Consequently, in order to compare texts of different lengths, Scott (2006) proposes making use of the standardized type/token ratio (STTR), which is calculated for the first *n* running words (usually 1,000 words),

> then calculated afresh for the next 1,000, and so on to the end of your text or corpus. A running average is computed, which means that you get an average type/token ratio based on consecutive 1,000-word chunks of text[1]. (Scott 2006)

For the analysis of the present corpus the WordList program in the software *WordSmith Tools* (version 4; Scott 2006), the output of which comprises both the STTR and the TTR, was used. In the settings panel of the software, the *n* number was adjusted to 50, which allowed for the calculation of the STTR for texts at least 50 words long. It was also specified that hyphens do not count as word separators and that numbers are to be included in a word list. In Figure 3.4 the texts have been ordered according to their 'STTR50' values, and bars representing the values of the TTR and those of the STTR50 for each one of them have been plotted. As can be seen, text 17 (589 running words) scores the highest STTR50 value, which is quite noticeable, in that this text is the third longest one. Equally worthy of mention are text 23 (511 words), and of course, the one that scores the lowest value, that is, text 12. It is also interesting to note that the shortest text in the corpus, text 7, scores a much higher STTR50 than longer texts, such as text 17 and 23. In this text, therefore, there is not much repetition. As also reported in Figure 3.1, the STTR50 values of the texts of the corpus range between 72.67 and 86.36, and in most cases they differ remarkably from those of the TTR.

1 Scott (2006) specifies that the *WordSmith* method of computing STTR parallels one of the methods devised by the mathematician David Malvern working with Brian Richards (University of Reading). Information about the research underlying this index can be found at <http://www.personal.rdg.ac.uk/~ehsrichb/ttr.html>, last visited on 25th April 2008.

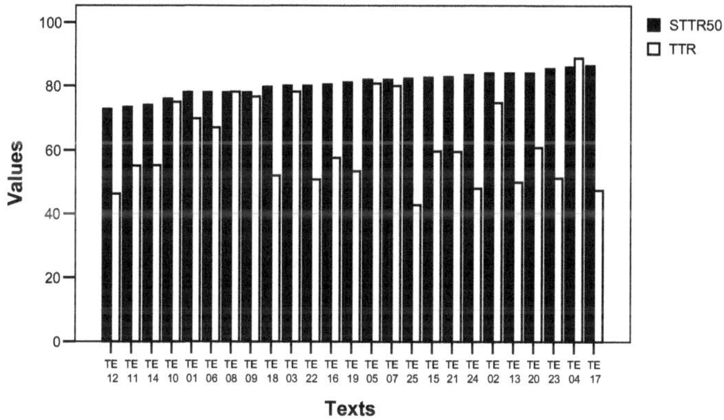

Figure 3.4. The values of the TTR and the STTR in the corpus.

Figure 3.4 also shows that text 4 scores the second highest STTR50 value (viz. 86, which is very close to the STTR50 value of text 17, viz. 86.36) as well as the highest TTR value in the corpus.

As said above, a high STTR50 score indicates that only a few words are repeated in a given text, irrespective of whether they are grammatical or lexical (see 3.2 below for this distinction). In text 4[2], in particular, only the lexical words 'ecotourism' and 'people' and the grammatical words 'in', 'is', 'to, and 'the' are repeated twice each:

(1) People seem to think that ecotourism is about indulging in trips to remote or fragile areas. If you really care about these environments you wouldn't go near them. Real 'ecotourism' is when people spend their free time putting life back into places which have already suffered at the hands of man – in other words not trampling over the rare and beautiful but getting kicks out making good our past mistakes.
This sentence is a summary of the text:
Ecotourism should involve the improvement of damaged environments …
the exploitation of natural areas.
Which is the best group of words for the gap (…) in the sentence and completes the summary of the text?
(a) As well as
(b) In spite of
(c) Instead of
(d) Due to (task 4)

2 Test taken from the DIALANG Reading Module.

As will be shown in 5.4, the test takers perceived both the text and the task as rather difficult, but the high vocabulary diversity that characterises this text does not seem to contribute to its difficulty. Instead, for about 67% of the students it was their lack of knowledge of the large number of infrequent words and multi-word units (see 3.2.1.1 below), e.g. 'indulge', 'to put life back', 'to trample over', and 'to get kicks out of doing something', which were perceived as the main causes of difficulty. As for task 4, some students declared they did not know the meaning of the word 'exploitation', while the majority found it difficult to choose the correct conjunction amongst the four options. 29 out of 60 students (48%) answered the question by choosing 'instead of', 20 students (34%) chose 'due to', 7 chose 'in spite of', and 4 chose 'as well as'.

3.2 Measuring Lexical Density and Lexical Variation

As suggested in 3.1 above, one of the limits of the type/token ratio, or, more precisely, of both the TTR and the STTR, is the fact that they do not distinguish between grammatical/function words and lexical words. This implies, in particular, that they neither provide information about the lexical 'denseness' of a given text nor about the degree of lexical repetition/variation in it. In order to obtain this information, a large number of scholars in a range of disciplines have experimented with and made use of some measures of lexical density and lexical variation for different purposes. In the following sections some methods of calculation of lexical density and some studies in which they were made use of will be reviewed (see 3.2.1), as well as some literature on the distinction between lexical words and function words (see 3.2.1.1). This aims to provide some background information necessary to interpret the results of the calculations of lexical density carried out for the present work (see 3.2.1.2 and also 3.4 below). Finally, in 3.2.2 measures and applications of lexical variation will be reviewed and applied to the corpus under investigation.

3.2.1 Measures of Lexical Density

According to Halliday (1987: 60):

> Lexical density is the proportion of lexical items (content words) to the total discourse. It can be measured in various ways: the ratio of lexical items either to the total running words or to some higher grammatical unit, most obviously the clause; with or without waiting for relative frequency (in the language) of the lexical items themselves.

In text- and corpus-based linguistics both types of measurement of lexical density have been advocated and utilised in studies on register differentiation (e.g. Ure 1971), on the differences between spoken and written modes of meaning (e.g. Halliday 1987, 1989), and to define general quantitative characteristics of spoken and written genres on the basis of large corpora (e.g. Biber 1988, Stubbs 1996: 70-76). Ure (1971), in particular, measured the lexical density of a corpus of 42,000 words in size, and calculated the index as a proportion of running words. This linguist found that the written texts tended to have a lexical density over 40%, while the spoken texts were characterised by a tendency to be under 40% (Ure 1971: 445). Besides textual and register differences, she hypothesised that other situational factors can differentiate between texts. Firstly, she distinguished between impromptu discourse and prepared texts, and according to her data prepared texts have a lexical density of 37% or more. Secondly, she found that the possibility of receiving feedback from the interlocutor seemed to be a powerful discriminating factor, since all her texts with a lexical density above 37% did not allow for a personal response to the speaker (Ure 1971: 448). Thirdly, she claimed that the role played by the social and personal relations between the participants and the kind of subject-matter can all have consequences on the degree of lexical density (Ure 1971: 449). Finally, she pointed out that language-in-action texts, e.g. consulting, describing, discussing, tend to be 'sparser' than, say, narrative texts (Ure 1971: 450).

Stubbs (1996: 70-76) used Ure's method to calculate the lexical density of the LOB and the London-Lund corpora, which total about 1.5 million words together and comprise texts belonging to both written and spoken genres. Stubbs's results confirm that written texts tend to have a lexical density above 40%, whereas spoken texts usually have a lower lexical density. He explains this by saying that spoken texts contain larger numbers of grammatical words, which are necessary to create essentially 'predictable' spoken texts. His original contribution, however, is the definition of the parameter that mainly discriminates between the texts of his large corpus: the possibility of feedback. Specifically, Stubbs (1996: 74) concludes that according to his data

> those uses of spoken language where there is no possibility of feedback from other speakers – radio commentary and telephone answering machines – vary between 46 and 64 per cent, and those where there is feedback [...] vary between 34 and 44 per cent.

Conversely, he found that there are some written texts which are less dense than spoken genres with feedback. He concludes his analysis by saying that "measures of richness of vocabulary, such as lexical density, provide a robust method of distinguishing genres" (Stubbs 1996: 76).

Biber (1988) reports on his investigation into the linguistic variations across the written and spoken genres represented by the texts in the LOB and the London-Lund corpora, with the addition of a collection of professional and personal letters. In his study, however, Biber (1988) does not utilise lexical density as a broad linguistic feature. In contrast, for his 'multi-dimensional' analysis he selected 67 specific linguistic features of the various genres, such as 'nominalisations' (see 3.2.1.2 below), 'modals' and 'passives'. By means of factor analysis he eventually manages to reduce the original variables to six derived variables that he calls dimensions and that are claimed to account for most of the variation among the various genres. Biber *et al.* (1999: 62) discuss the findings of their investigation into the lexical density of the texts in the LSWE Corpus[3]. As said above, they sub-divide the various texts of their corpus into four broad registers: 'news', 'fiction', 'academic prose', and 'conversation'. According to their data, conversation is the register that scores the lowest lexical

---

3 The Longman Spoken and Written English Corpus (the LSWE corpus) contains approximately 40 million words of text, providing a "sound basis for reliable analyses of grammatical patterns" (Biber *et al.* 1999: 24).

density values, while news has the highest lexical density. They expand on these findings, stating that "since a written text is planned and offers the possibility of re-reading, it can tolerate a much higher information load than conversation." They also found that the lowest score among the written registers is found in fiction, which is "presumably partly due to the dialogue passages and partly to the complex purposes of fictional writing (where the informational aspect is combined with aesthetic concerns). The score for academic prose is intermediate, reflecting the purpose of this register."

Some studies in applied linguistics have attempted to investigate the role of lexical density in the prediction and evaluation of the quality and/or difficulty of texts. Gibson (1993: 157-168), for example, investigates the working hypothesis that abstracts characterised by a higher lexical density are generally preferred over abstracts with lower lexical density. In order to test this hypothesis, the author asked eight judges to rank five versions of four abstracts according to how they perceived their quality, and compared these values to the lexical density counts. The correlation analysis falsified the initial hypothesis, and the reverse seemed to hold, that is, "the better abstracts were characterised by lower levels of lexical density" (Gibson 1993: 357). The author, therefore, tentatively suggests that abstracts containing more information are correspondingly harder to process than sparse texts.

On the basis of her investigation into the characteristics of academic writing, Ventola (1995: 187-188) maintains that in order to be successful, academic writers and, in particular, foreign language writers have to "keep a balance in their attempts to create good academic texts." This means, first of all, that academic writers have to be aware of the characteristics of 'dense' and of 'diffused' texts respectively, and that they should use techniques of both 'packing' and 'unpacking' information to write their texts. According to the author's data, good or at least acceptable 'academese' is not necessarily characterised by high values of lexical density, but rather by accurate and conscious linguistic choices that are "essential for balancing the text informationally" (Ventola 1995: 168). Ventola (1995: 188) also makes reference to a study by Linnarud that is relevant to her own work, and reports briefly on this scholar's findings. Linnarud (1986) conducted an investigation into the lexis used by a group of Swedish secondary school pupils and compared it to that used by some British pupils. She noted that the texts produced in English by the two groups did not differ greatly in terms of lexical density, and points out, therefore, that in order to have their work evaluated as native-like, non-native writers were supposed to achieve a native-like threshold of lexical density. However, she also noted that going beyond this threshold might turn against the writers, since in this way their works might sound as 'over-dense' and possibly unnatural. These considerations actually parallel Ventola's (1995) intuition and findings.

In cognitive psychology, lexical density has been studied for its effects on both sentence comprehension and memory retention, "often with seemingly contradictory findings" (Gibson 1993: 156). Gibson (1993: 156), in particular, claims that some scholars found that speakers are generally less able to recall material with a high lexical density than they recall material with a low lexical density, while others concluded that speakers are more able to comprehend material with higher lexical density. Perfetti (1969), for instance, reports on a series of experiments aiming at investigating how the recall probability of 64 individually presented 10-word long sentences was related to their different degrees of lexical density. These sentences were tape-recorded and 40 undergraduates were asked to listen to single sentences followed by a series of numbers. At the signal following the numbers, the students wrote first the sum of the numbers and then the sentence. As hypothesised by Perfetti (1969: 719), the recall probability turned out to be inversely related to lexical density, i.e. "the proportion of lexical, as opposed to grammatical words in the sentence". Perfetti (1969: 723) discusses and interprets these findings by arguing, firstly, that content words have proved to be generally less predictable than function words in cloze procedures, and that

> in general, a greater number of lexical words in the sentence reflects a greater number of underlying 'sentences' or assertions, although not necessarily in a one-to-one relationship. It seems reasonable that the hearer processes these underlying relationships. […] And the greater the density of such relations in the surface structure, the greater the load on memory.

The issue of whether high lexical density facilitates recall or not remains, however, open. Other studies in cognitive psychology such as those conducted by Bradley and Garrett (1983) and Gordon and Caramazza (1985) investigate the specific cognitive mechanisms involved in the processing of closed and open class words in English. According to these and other related studies there seem to be differences in the way open and closed class words are processed, but these differences are likely to depend not only on the class a given word belongs to, but also on factors such as the kind of person under investigation, e.g. on whether normal subjects or aphasics are under investigation, and the frequency of occurrence of the specific words they are asked to process.

In 3.4.2 below, Halliday's method of calculating lexical density will be introduced and applied to the analysis of the texts of the present corpus. In this study the description of this other method was kept separate from that of Ure's for three main reasons. Firstly, as it is based on Halliday's analysis of the clause complex, it involves the distinction between the rather difficult notions of hypotaxis and rankshift (embedding). Secondly, this method is not ready-made, in that its calculation needs the same manual tagging that has to be carried out for measuring grammatical intricacy. Thirdly, since this index is also based on the structure of the clause complex, it is interesting to compare the results not only to those obtained using Ure's method, but also to the index of grammatical intricacy.

3.2.1.1 Lexical Items vs. Grammatical Words

One of the major problems analysts encounter when analysing and tagging a corpus for lexical density concerns the distinction between grammatical and lexical words. In fact, most linguists agree on the fact that there is a continuum from lexis to grammar rather than a clear-cut distinction between the two categories. Matthiessen (1995: 109), in particular, claims that "the move from grammar to lexis is a cline – more specifically the cline of delicacy." He also affirms, following Halliday (1989: 63), that "lexis and grammar form one unified system – lexicogrammar," and that "the most general end of the lexicogrammatical system is grammar," whereas "the more delicate end of the lexicogrammar is lexis" (Matthiessen 1995: 107).

According to Halliday (1989: 63), prepositions and certain classes of adverbs are on the borderline, and for purposes of analysis, "it does not matter exactly where we draw the line provided we do it consistently". Halliday (1989: 61) defines grammatical/function words as

> those that function in closed systems in the language: in English, determiners, pronouns, most prepositions, conjunctions, some classes of adverbs, and finite verbs. (Determiners include the articles).

Lexical words, on the other hand, are defined as content words and are "lexical because they function in lexical sets not grammatical systems: that is to say, they enter into open not closed contrasts" (Halliday 1989: 63). Halliday (1989: 63) also makes it clear that technically the latter are "items (i.e. constituents of variable length) rather than words in the usual sense, because they may consist of more than one word." Phrasal verbs such as 'take over', 'call off' and 'stand up' are given as examples of lexical items. Besides lexical and function words, Biber *et al.* (1999: 55-56) list the class of 'inserts', which are marginal words, simple in form and "marked off by intonation, pauses, or by punctuation marks in writing". Examples of this kind of words are given by 'hm', 'yeah', 'cheers'. Furthermore, they make use of the term 'multi-word lexical unit' to denote any sequence of word forms which functions as a single grammatical unit, and therefore becomes 'lexicalized' (Biber *et al.* 1999: 58). Biber *et al.* (1999: 59) also make use of the term 'idiom' in their work, and specify that it denotes "multi-word units that tend to acquire meanings which are not predictable from the individual parts".

Researchers in the fields of corpus linguistics, lexicography and applied linguistics have also tackled the issue of the kind of structures that should be considered as lexical items. Sinclair (1991: 174-176), in particular, uses the term 'word form' "for any unique string of characters, bounded by spaces", whilst he defines the 'lexis' of a language as the set of its word-forms, and 'lexical structure' as "the organisation of these word-forms into units such as collocations and idioms". Sinclair (1991: 172) also specifies that in his work the term 'idiom' is used to denote "a group of two or more words which are chosen together to produce a specific meaning or effect in speech or writing", and that the notion of 'collocation' partly overlaps with that of idiom. The term collocation, however, should be used if the co-occurrence is interpreted "as the selection of two related words, each of which keeps some meaning of its own" (Sinclair 1991: 172). 'Hold the whip hand' and 'hold sway' are given as examples of idioms, while 'hold a meeting' and 'hold an enquiry' exemplify the phenomenon of collocation (Sinclair 1991: 172). More generally, Sinclair (1991: 109-110) distinguishes between the *open-choice principle* and the *idiom principle* in language. The former accords with the traditional way of viewing language and, in particular, with the fact that language is creative and at "each point where a unit is completed (a word, phrase, or clause), a large range of choice opens up and the only restraint is grammaticalness" (Sinclair 1991: 109). By contrast, by means of the idiom principle he highlights the fact that there are regularities and constraints in language, and that "the choice of one word affects the choice of others in its vicinity" (Sinclair 1991: 173). According to the latter principle, therefore, individual words are not reliably meaningful in themselves and the multi-word groups they form can be considered as single lexical items, in that whole groups of words are required to produce the meaning.

Schmitt (2000: 99) provides a list of some of the most common types of multi-word units in English. The first category in his classification is that of 'compound words', which are "created when two or more words are combined to make a single lexeme". They can be written as multiple orthographic words (e.g. 'freeze dry'), hyphenated words (e.g. 'freeze-dry'), or as a single orthographic word (e.g. 'freezedry)'. The second category is that of phrasal verbs, which are usually made up of a monosyllabic verb and an adverbial or prepositional particle. Sometimes they can be "readily understood from their components (e.g. 'give away'), but quite often their meaning is uninterpretable (e.g. 'give up')" (Schmitt 2000: 99). The third typology concerns binomial and trinomial fixed phrases, which are not "particularly difficult to figure out, but their sequencing is frozen" (Schmitt 2000: 99). Examples are given by 'to and fro', 'back and forth', and 'ready, willing, and able'. The fourth type of multi-word units is that of idioms, which are characterised by 'non-compositionality' (e.g. 'to bite the dust', 'bitten by the love bug', 'to bite off more than you can chew') and by 'frozen collocation'. They do not allow for variation, and "if any variation is inserted into the idiom, it ceases to exist as a unit". For instance, the meaning 'to die' of 'kick the bucket' would be lost if even one element was changed, e.g. 'kick the pail' (Schmitt 2000: 78). Proverbs are the fifth category and "differ from idioms in that they display shared cultural wisdom", and "are usually tied to common situations in a culture" (e.g., 'out of sight, out of mind') (Schmitt 2000: 100). Finally, 'lexical phrases' or 'lexical chunks' is a category listed to comprise "prefabricated language units that can be used as wholes, rather than being composed through vocabulary and syntax" (Schmitt 2000: 101). These multi-word units share the characteristic of being institutionalised and used by speakers and writers to carry out language functions.

By mainly drawing on this literature and on Quirk, Greenbaum, Leech, and Svartvik (1985), Castello (2002: 19-25) attempted to list and summarise the distinctions he drew between grammatical words and lexical items for the manual tagging for lexical density of a 92,500-word corpus of tourist information texts. As reported by Castello (2004) (see also 3.4.2 below), these decisions have also been taken into consideration and only partially modified by other analysts who annotated other specialized corpora manually. In Figure 3.5 these distinctions are listed concisely.

GRAMMATICAL WORDS	*LEXICAL ITEMS*
- *Primary Verbs* (i.e., be, have, do)	- *Nouns* (e.g., land, river, Henry V|||, Dante Gabriel Rossetti – *each nominal group functioning as proper name and made up of 2 or more proper nouns is counted as one lexical item* -)
- *Modal Verbs* (e.g., can, may, seem to, willing to)	- *Adjectives* (e.g., beautiful, cold, nice)
- *Determiners* (e.g., an, which, whatever, many, all, plenty)	- *Full Verbs* (e.g., run, play, look out, look for, look out for – *each group is counted as 1 lexical item* -)
- *Pronouns* (e.g., I, me, both, somebody, other, none)	- *Numerals* (e.g., 2006, second, twice)
- *Prepositions* (e.g., as, at, by, except for, up to)	- *Open-class Adverbs* (e.g., quickly, late, from time to time, every now and then, time and time again – *each group is counted as one lexical item* -)
- *Conjunctions* (e.g., and, or, either … or, and yet, in that, rather than)	- *Three-word sequence Prepositions* (e.g., in view of, in terms of, in accordance with, on behalf of – *each group is counted as one lexical item* -)

- *Some Adverbs* (e.g., yet, so, however, always, very, a bit, too, nearly, such, kind of)	- *Open-class Quantifiers* (e.g., heaps of, a great deal of, a small amount of – *each group is counted as one lexical item* -)
	- *Complex Conjunctions and Subordinators with an open-class word* (e.g., provided that, regardless of, no matter how – *each group is counted as one lexical item* -)

Figure 3.5. Distinctions between lexical and grammatical words (adapted from Castello 2002: 22-25).

As can be seen, there are a number of cases in which groups of subsequent words were counted as single lexical items in the studies: i.e. nominal groups functioning as proper names and made up of two or more proper nouns, phrasal verbs (see also Halliday 2004: 351-354), open-class adverbs, three-word sequence prepositions, open-class quantifiers, and complex conjunctions and subordinators with an open-class noun.

For the present study, however, multi-word lexical items were not tagged as such, in that it was decided to calculate the number of lexical items and grammatical words using an automated procedure, which unfortunately does not allow for the recognition and annotation of multi-word items (or idioms) at present. In practical terms, a 'stop-list' of grammatical words was prepared and used in combination with *WordSmith Tools* (Scott 2006), in order to obtain lists containing only lexical types and tokens. This list is mainly based on the lists of grammatical words presented in Gibson (1993: 152) and in Butler (1985: 219-220), and is reproduced in Figure 3.6. As can be seen, this list also contains strings of letters that are not actual words, but rather just parts of contracted forms. For instance, in the list both the word-form 'I'll' and the strings 'I' and 'll' were included. This was done, for the simple reason that in the *WordSmith Tools* settings panel, the apostrophe was chosen as the 'character within words', that is, the character that occurs between two different words. If this kind of string were not inserted into the stop-list, therefore, the software would count them as separate non-grammatical words. Another disadvantage of this calculation is that words that can be both lexical and grammatical cannot reliably be disambiguated, and in some cases inaccuracies may result. As pointed out by Gibson (1993: 163),

> the problem is most usually handled in terms of statistical likelihood, where each word is assumed to belong to one word class only; so, for example, 'can' is thought to be grammatical, since its lexical use (to denote a receptacle) is altogether more unusual.

List of Grammatical Words
a, about, above, across, after, against, all, along, alongside, although, amid, amidst, among, amongst, an, and, another, any, anybody, anyone, anything, anywhere, apropos, as, at, atop, because, before, behind, below, beneath, beside, besides, between, beyond, both, but, by, can, can't, cos, could, couldn't, dare, daren't, despite, dhi, didn't, doesn't, don't, during, each, ei, either, else, every, everybody, everyone, everything, everywhere, except, few, for, from, he, he'd, he'll, he's, her, here, hers, herself, him, himself, his, how, however, if, I'm, in, inside, into, it, it'd, it'll, it's, its, itself, less, many, may, mayn't, me, mhm, might, mine, minus, more, most, much, must, mustn't, my, myself, needn't, neither, never, nevertheless, no, no-one, nobody, none, nonetheless, noone, nor, not, nothing, notwithstanding, of, off, on, one, or, other, others, ought, oughtn't, our, ours, ourselves, out, outside, over, per, plus, quite, rather, shall, shan't, she, she'd, she'll, she's, should, shouldest, shouldn't, since, so, some, somebody, someone, someplace, something, somewhere, than, that, that'd, that'll, that's, the, thee, their, theirs, themselves, them, then, there, there'd, there'll, there's, there've, therefore, therewith, these, they, they'd, they'll, they're, they've, thine, this, those, thou, though, through, throughout, thus, thy, till, to, too, toward, towards, uhuh, under, underneath, until, up, upon, us, very, via, we, we'd, we'll, we're, we've, what, what'd, what'll, what's, what've, whatever, when, whenever, where, wherever, whether, which, whichever, while, whilst, who, whom, whose, why, will, with, within, without, won't, would, wouldn't, ye, yeah, yes, yet, you, you'd, you'll, you're, you've, your, yours, yourself, yourselves, I, I'd, I'll, I've, ok, be, am, are, is, isn't, aren't, was, wasn't, were, weren't, being, have, haven't, has, hasn't, been, had, hadn't, having, do, did, done, does, don't, doesn't, didn't, doing, frequently, often, usually, always, never, ever, seldom, sometimes, t, s, ll, re, ve, m, d, isn, wasn, weren, hadn, hasn, wouldn, aren, haven, also, such, one's, whereas, couldn, together

Figure 3.6. The list of grammatical words used for the present study.

3.2.1.2 Lexical Density (Ure's method) in the Corpus

As said above, lexical density was calculated automatically by combining the use of the software *WordSmith Tools* (version 4; Scott 2006) with *Microsoft Excel. WordSmith Tools* was used to calculate the total number of lexical words and that of tokens in a text.

Subsequently, the output data was put into an Excel spreadsheet, and the following formula was used to calculate the lexical density (Ure's method, henceforth referred to as LDU) percentage for each text:

$$LDU = \frac{\text{Number of lexical words}}{\text{Total number of words (tokens)}} \times 100$$

In Figure 3.7 the LDU values of each text are represented by means of a histogram with bars of different length (see Figure 3.9 in 3.2.2 below for the exact figures concerning each text).

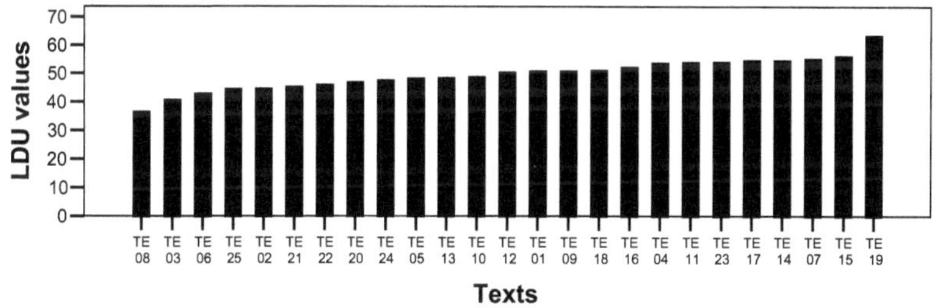

Figure 3.7. Bar graph representing the values of LDU in the texts of the corpus.

As is apparent from Figure 3.7, texts 19, 15 and 7 clearly stand out as the densest ones in the corpus, while texts 8 and 3 as the sparsest ones. It is also interesting to note that both text 19 and text 7 provide tourist information: the former is a leaflet that advertises a guided 22-day tour around Australia, while the latter provides a short description of the city of Bangkok, and is likely to be part of a longer text. Text 19 also contains a picture representing a view of Sydney, and the written information in it is organised into various sections on a single page. As its title clearly states, text 15, the second densest text in the corpus, provides candidates with information about West Thames College. In text 3, on the other hand, the discourse of a person is reported directly by means of some projected clauses, and the clause 'he said' is used to project them all (see, for example, Halliday 2004: 445-448 for a treatment of the projection of direct speech). Finally, text 8 contains the narration of an event, and looks like an extract from a novel. The following are extracts from text 19[4] and text 7[5]:

(2) PERTH Wonderful long, golden beaches, superb restaurants serving fabulous food in delightful surroundings, lush green parks and the beautiful Swan River. City sightseeing tour included. [...]
Sold out for this year – booking now for next year
PHONE US ON 01303 692154
QUOTING REFERENCE GW/398 FOR OUR NEW FULL COLOUR BROCHURE
OUR OFFICES ARE OPEN: Monday to Friday 9am – 8pm Saturday 9am – 4pm
BROCHURE REQUESTS ONLY: Sunday 10am – 2pm. (extracts from text 19)

(3) To the Western visitor, Bangkok is not just another city, it's another way of life. Wooden houses built on stilts over the water stand in the shadows of gleaming tower blocks. The air is filled with different smells. Exotic temples glimmer in the sunlight. Their cool interiors offer a welcome refuge from the heat and noise of the city. (text 7)

It is worth focusing briefly on some lexico-grammatical features of these texts, in that they are likely to be the ones that mainly contribute to their high lexical density. In the extracts from text 19 the second clause occurring under the heading 'Perth', i.e. "city sightseeing tour included", can be classified as an elliptical clause, since it lacks the Finite element 'is' of the Mood (for a definition of Mood in English see, for instance, Halliday 2004: 111-120, and for a definition of elliptical clause see Halliday 2004: 151-153). The clause "brochure requests only: Sunday 10am-2pm" can also be considered as an elliptical clause: in this case, in particular, both the Finite element and the Predicator are lacking (for a definition of Predicator see, for example, Halliday 2004: 121). Finally, in the clause "sold out for this year" both the Finite element and the Subject were omitted, and this clause can also be classified as elliptical. Castello (2002: 62-72) found that elliptical (and minor) clauses are frequent in his corpus of tourist information texts, and in particular in the texts belonging to

4 University of Cambridge Local Examinations Syndicate, Preliminary English Test 2 Student's Book, (2001), © UCLES, reprinted with permission of the publisher (Cambridge University Press) and UCLES.

5 Test taken from the DIALANG Reading Module.

the brochure sub-genre. In this respect, Castello (2002: 83) suggests that the writers of brochures often seem to prefer not to connect the bits of information using the grammar of the clause complex, but rather the grammar of minor and elliptical clauses. The information thus tends to be concentrated and packaged into smaller portions of text, and this is in particular achieved by keeping the lexical words and by omitting most of the function words. It follows from this consideration that this stylistic feature contributes to the increase in lexical density, and, as will be seen in 3.5 below, to the decrease in grammatical intricacy.

Another more general linguistic feature that is bound to contribute to the high lexical density of these texts is the frequent use of complex noun groups containing pre- and post-modifying elements (for more information about the English noun group from a systemic functional perspective, see Halliday 2004: 311-335, and also 4.3 below). According to Halliday (2004: 654), written language is characterised by being lexically dense, since "it packs a large number of lexical items into each clause". In particular, it does so by exploiting the grammar of nominal groups, which allows for the concentration, 'packing', of information inside nominal groups rather than in different clauses, thus making the noun groups longer and more informative (Ventola 1995: 181). Information packaging is often realised through the linguistic phenomenon called nominalisation, i.e. the linguistic resource by which "processes (congruently worded as verbs) and properties (congruently worded as adjectives) are reworded metaphorically as nouns, and function as Thing in the nominal group" (Halliday 2004: 656). The process of nominalisation is illustrated, for example, by Ventola (1995: 182), who gives the example of "this rapid economic growth" and "this rapid growth of economy" as alternative nominalised 'incongruent' versions of the 'congruently coded' clause "the economy grows rapidly". These phenomena are especially typical of academic and scientific English, but text 7 contains two long nominal groups that exemplify the phenomenon of information packing: "wooden houses built on stilts over the water" and "a welcome refuge from the heat and noise of the city". In the former group the Thing 'houses' is pre-modified by the Classifier 'wooden' and post-modified (qualified) by the embedded (rankshifted) clause "built on stilts over the water". The noun 'stilts' is in turn post-modified (qualified) by the prepositional phrase (see Halliday 2004: 359-361) 'over the water', which itself contains the noun 'water'. In the latter group 'refuge' is pre-modified by the Epithet 'welcome' and post-modified (qualified) by the prepositional phrase 'from the heat and noise of the city'. This prepositional phrase contains two noun groups, i.e. 'the heat' and 'the noise of the city'. The latter noun group, in turn, is further qualified by the prepositional phrase 'of the noise', which contains the noun group 'the city'. As illustrated, therefore, noun groups can qualify and be qualified by other nominal groups, rank-shifted clauses and prepositional phrases at the same time, and their Head (Thing) can be the result of nominalisation. According to many studies (see, for example, Ventola 1995 and Halliday 1997), these two linguistic phenomena are especially recurrent in some registers and/or genres, such as those concerned with academic and scientific written production.

The values of LDU found by Castello (2002: 61) in the brochures of his corpus range between 55.06% and 66.12%, and their mean score equals 61.37% (Castello 2002: 61). According to this data, then, both text 19 (LD = 63.29%) and text 7 (LD = 58.33%) score lexical density values typical of the brochure sub-genre. By contrast, text 3 (LD = 43.48%) and text 8 (LD = 38.18%) score low lexical density values, which, as seen above, are typical of spoken texts allowing for feedback (see Ure 1971; Stubbs 1996), and of conversation and some fictional writing (Biber *et al.* 1999: 62, 65). A closer look at text 3 and text 8[6] reported below actually confirms that the two texts are close to the spoken end of the spoken-written continuum.

(4) I do think this course should be fun," he said. "And people should get out of it what they want. Some come here wanting to be serious and even a bit competitive about their work. And that's fine. Others are very hesitant and find that a bit of humour can relax them. *You get all sorts on these courses.* I've taught high court judges and dustmen and the mix is part of the fun. (text 3)

(5) She had gone alone, but the children were to go to the station to meet her. And loving the station as they did, it was only natural that they should be there a good hour before there was any chance of Mother's train arriving, even if the train were punctual, which was most unlikely. (text 8)

6 Texts taken from the DIALANG Reading Module.

As can be seen, not only are these two texts characterised by the occurrence of a larger number of function words per clause than the previous two texts, but also by a large number of ranking clauses per clause-complex (sentence). These texts, therefore, are grammatically more 'intricate' (see also 3.4 below) than the previous ones, in that they contain more clauses per clause-complex, but the density of lexical items per clause is lower. On the basis of his investigation into a representative text sample, Halliday (1987: 66) suggests that "the intricacy of [...] the sentence or clause-complex [...] is more a characteristic of the most unconscious spontaneous uses of language. The more natural, un-self-monitored the discourse, the more intricate the grammatical patterns that can be woven". He admits, however, that "some writers can achieve considerable intricacy in the structure of the clause complex; it can be learnt and consciously developed as a style". Text 8, in particular, seems to belong to this latter type of texts, viz. a piece of fictional writing written in an 'intricate' style.

In comparison to the other tests in the DIALANG sub-corpus (see 2.1 above) neither text 8 nor task 8 were in general perceived as especially difficult (see 5.4 and 5.9 below). However, approximately 35% of the test takers said that they had problems understanding the exact meaning of the text, and added that this was mainly due to the intricacy of the text, to the way punctuation is used in it, and to the length of the second clause complex. They also felt that they had to read between the lines, that is, infer some meanings, and most of them managed to do so only by reading the text twice or even more times. Approximately another 6% of the test takers, on the other hand, found that the main difficulty of this text depends on the fact that it is only a short extract from a longer text and that it is, therefore, out of context. Finally, a couple of students responded by saying that they did not know the exact meaning of the adverb 'unlikely'. As for task 8 (reported below), most of the test takers found that its difficulty depends mainly on the complexity of the text.

(6) What can you say about their mother's train?
(a) It would probably be early.
(b) It would probably be on time.
(c) It would probably be late.
(d) It had been cancelled. (task 8)

35 (58%) students answered the question by choosing option (c), 18 of them (30%) chose (b), 4 chose (d), and 3 chose (a). It is to be noticed that the main distracter, i.e. option (b), states the exact contrary of option (c), that is, the fact that, according to the text, the train is likely to be late.

3.2.2 A Measure of Lexical Variation

Lexical variation measures aim at quantifying the amount of lexical repetition or lexical diversity in a text, and are related to the TTR of the lexical words in a text (Gibson 1993: 157). In lexical variation measures the count is, in fact, limited to lexical content words. The choice of leaving out grammatical items from the count is supported by the fact that "grammatical items tend to be considerably more frequent in occurrence", whereas "lexical items are repeated much less often", and "what is significant is the relative frequency of one lexical item to another" (Halliday 1989: 64). As far as lexical items are concerned, it should also be noticed that their "repetition reduces the effect of density – since even if a word is intrinsically rare, its occurrence sets up the expectation that it will occur again" (Halliday 1989: 65). The formula that was used in this study to quantify lexical variation (henceforth referred to as LV) is as follows:

$$LV = \frac{\text{Number of different lexical words}}{\text{Total number of lexical words}} \times 100$$

If a given text scores low on this index, it means that its author repeated the same lexical words many times. If, on the other hand, it scores high, it means that the same lexical words were not often repeated. As pointed out by Linnarud (1975: 50, quoted by Gibson 1993: 164), however, the length of the text plays an important part, and a high level of LV is more likely to characterise a short rather than a long text. This is confirmed by the data concerning the texts under investigation in the present study and represented in the bar graph in Figure 3.8: it clearly shows that the highest values of the index tend to be associated with the shortest texts, that is, those with a small number of tokens.

The LV index was used by Gibson (1993: 163-164) to calculate the lexical variation in the texts belonging to a corpus of abstracts, and by Linnarud (1975: 46, cited in Gibson 1993: 163) for the evaluation of some essays. The idea behind the use of this formula in these two studies is that lexical variation "may well be a more important factor in influencing evaluators favourably than LD" (Linnarud 1977: 90, cited in Gibson 1993: 157), and it might be supposed to be especially so for abstracts, in which verbal economy is generally assumed to be a highly desirable characteristic. However, Gibson's hypothesis that abstracts scoring high LV, that is, abstracts containing a small amount of lexical repetition, are generally preferred by readers over more 'repetitive' ones was partly falsified by his data (Gibson 1993: 357).

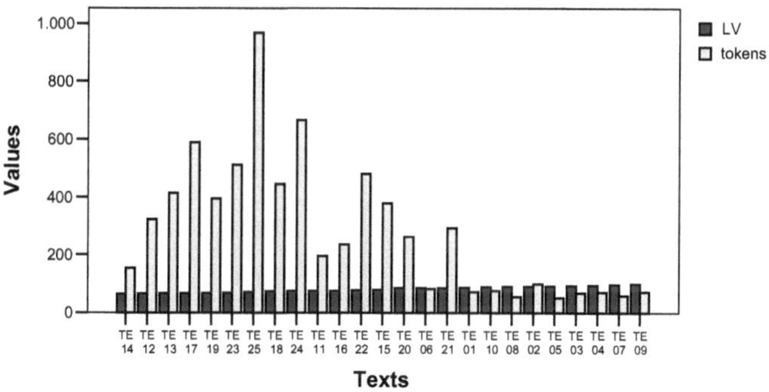

Figure 3.8. Bar graph representing the values of LV and the number of tokens in the texts of the corpus.

For the present study LV was calculated on the basis of the data obtained by using the lists of lexical items generated by the *Word-Smith WordList* tool (Scott 2006), and by excluding the grammatical words contained in the stop-list of grammar words reported in 3.2.1.1 above. Subsequently, the data was put into *Microsoft Excel* and the LV formula was applied to it for the calculation of the values of the index for each text[7]. By applying this formula, however, lexical words belonging to the same lemma, e.g., 'eye' and 'eyes' or 'know', 'knows', 'knew' were counted as different lexical words, and not as word-forms belonging to the same lemma or word-family. On this issue Halliday (1989: 65) remarks that

> normally all the members of a morphological paradigm are the same lexical item: for example, 'differ', 'differed', 'different', 'difference', 'differing', 'differently' are all members of the one lexical item (but not 'differential' in 'differential equation').

An attempt to capture this kind of information as well will be made in 3.3.2 below, in which frequency 'base lists' including both the base forms of words and derived forms will be made use of. In Figure 3.9 the values of LDU and those of LV have been reported, as well as the number of tokens and that of lexical words in each text.

Corpus	*Tokens*	*Lexical words*	*LDU*	*LV*
Text 01	73	37	50.68	86.49
Text 02	99	44	44.44	90.91
Text 03	69	28	40.58	92.86
Text 04	71	38	53.52	94.74
Text 05	52	25	48.08	92
Text 06	82	35	42.68	85.71
Text 07	60	33	55	96.97
Text 08	55	20	36.36	90
Text 09	73	37	50.68	100
Text 10	76	37	48.68	89.19
Text 11	196	106	53.81	74.53
Text 12	324	163	50.31	65.64

7 The procedure followed by Gibson (1993) to quantify lexical variation is similar to the one used for the present study. To this end, however, this scholar used an especially designed computer program (Gibson 1993: 162-163).

Text 13	414	200	48.31	67
Text 14	154	84	54.55	64.29
Text 15	379	212	55.94	78.77
Text 16	236	123	52.12	74.8
Text 17	589	321	54.5	67.29
Text 18	445	227	51.01	74.01
Text 19	395	250	63.29	67.6
Text 20	263	123	46.77	85.37
Text 21	294	133	45.24	85.71
Text 22	480	220	45.83	78.18
Text 23	511	276	54.01	68.84
Text 24	666	316	47.45	74.37
Text 25	969	430	44.38	70.93

Figure 3.9. The number of tokens, lexical words, the values of LDU (Ure's method) and lexical variation (LV) in the texts of the corpus.

According to these data, the least lexically varied text in the corpus is text 14 (LV = 64.29), whilst in text 9 no lexical repetition takes place at all (LV = 100). From the data in Figure 3.9 and from Figure 3.7 it also emerges that text 14 is the fourth densest text in the corpus (LD = 54.55%). As pointed out above, though, the 'effect of density' is likely to be reduced by the frequent use of repetition. Similarly, text 19, i.e. the densest text in the corpus, is also characterised by a relatively low LV, which is bound to partly attenuate its lexical denseness. Task 14, text 14[8], and the list of lexical words generated by *WordSmith Tools* (Scott 2006) are reported below (Figure 3.10):

(7) Look at the information on the following page about the use of vehicles in the University grounds.

8 University of Cambridge Local Examinations Syndicate, Cambridge IELTS 2 Student's Book With Answers. 2001, © UCLES, reprinted with permission of the publisher (Cambridge University Press) and UCLES.

In boxes 1-5 on your answer sheet write
TRUE if the statement is true
FALSE if the statement is false
NOT GIVEN if the information is not given in the passage
14.01 University employees do not need to pay for their parking permits.
14.02 Parking in Halls of Residence is handled by the Wardens of the Halls.
14.03 Having a University permit does not allow staff to park at Halls.
14.04 Parking permits cost £20 a year.
14.05 Students living in Hall do not need permission to park in Hall car parks. (task 14)

USE OF UNIVERSITY GROUNDS BY VEHICULAR TRAFFIC
The University grounds are private.
The University authorities only allow authorised members of the University, visitors and drivers of vehicles servicing the University to enter the grounds.
Members of staff who have paid the requisite fee and display the appropriate permit may bring a vehicle into the grounds. A University permit does not entitle them to park in Hall car parks however, unless authorised by the Warden of the Hall concerned.
Students may not bring vehicles into the grounds during the working day unless they have been given special permission by the Security Officer and have paid for and are displaying an appropriate entry permit. Students living in Halls of Residence must obtain permission from the Warden to keep a motor vehicle at their residence.
Students are reminded that if they park a motor vehicle on University premises without a valid permit, they will be fined £20. (text 14)

As can be seen from the list in Figure 3.10, 18 lexical types out of 54 (33.3%) occur more than once in this text, whilst the remaining 36 occur only once. On the basis of the frequency criterion, in particular, it can also be legitimately presumed that 'university' (7 occurrences), 'grounds' (5 occurrences), and 'permit' (4 occurrences) stand out as the 'key' words, that is, those words the knowledge and understanding of which are likely to be of special importance for the correct and thorough comprehension of the text. This is partly confirmed by the task candidates are asked to perform: in fact, 4 out of 5 of the statements accompanying the text contain the words the 'university permit' and 'permit', but none of them contains the word-form 'grounds'. The comprehension of the word-forms 'hall' and 'halls', on the other hand, turns out to be of special importance for the completion of the task, in that these word-forms occur 5 times in the questions (although only 3 times in the text). In 4.4.3 below text 14 will also be analysed for the reference chains occurring in it. A claim will be made, in particular, about the fact that the identity of the participant 'permit', occurring four times in this rather short text, might not be always clear and unique both for test takers and for the direct receivers of the text (students and members of staff), which could cause processing difficulty. Furthermore, in 5.7 below, the data collected from the administration of test 14 and the feedback questionnaire about it will be analysed and discussed.

WORDS (1)	*Freq.*	*WORDS (2)*	*Freq.*	*WORDS (3)*	*Freq.*
university	7	allow	1	officer	1
grounds	5	authorities	1	only	1
permit	4	car	1	parks	1
students	3	concerned	1	premises	1
vehicle	3	day	1	private	1
appropriate	2	display	1	reminded	1
authorised	2	displaying	1	requisite	1
bring	2	drivers	1	security	1
hall	2	enter	1	servicing	1
members	2	entitle	1	special	1
motor	2	entry	1	staff	1
paid	2	fee	1	traffic	1
park	2	fined	1	use	1
permission	2	given	1	valid	1
residence	2	halls	1	vehicular	1
unless	2	keep	1	visitors	1
vehicles	2	living	1	working	1
warden	2	obtain	1	£20	1

Figure 3.10. The list of lexical words in text 14 generated by *WordSmith Tools*.

A comparison can be made between test (text and task) 14 and test (text and task) 9 (reported below), which, as said above, is the most lexically varied, although only mildly dense (LD=50.68). No keyword can be, therefore, associated with this latter text on the basis of the frequency criterion, and in order to identify the key words, or rather the key concepts in this text, the reader has to rely on other kinds of textual information. If the task about this text is also taken into consideration, it can be reasonably supposed that in order to answer the question correctly, the 'necessary information' (Buck, Tatsuoka, Kostin 1997: 437) which the test taker should focus on is contained in the hypotactically related clauses "Try as she might, she is not particularly convincing here as an ace journalist …".

(8) Suzanne Somers is always going to struggle to lose the image of the superficial character she played in a recent TV drama. Try as she might, she is not particularly convincing here as an ace journalist who sets out to uncover the mystery behind the death of seven people in a night-club. Ed Begley and Michael Nouri offer satisfactory support, but the direction is strictly routine, and the same goes for the script. (text 9)

What is the writer's opinion of the film?
(a) Some of the acting is excellent.
(b) The film is well-directed.
(c) The leading performance is weak.
(d) The script is well-written. (task 9)

Specifically, in the clause 'try as she might' the verb 'try' is used in thematic position, which emphasizes the fact that Suzanne Somers is not convincing as an 'ace journalist', albeit she did her best to become so (for information about Theme in 'dependent bound clauses' see, for example, Halliday 2004: 98-100, 392-395 and section 4.2.2 below). 'Try' also refers back to the information given in the previous clause, that is, to the fact that the actor is always going to struggle to lose the image of the superficial character. According to Eggins (2004: 315), an entire dependent clause such as 'try as she might' can be "seen as acting as Theme to the sentence", since in doing so the writer sets up "thematic expectations for the rest of the sentence." In other words, not only is the verb 'try' in thematic position a textual choice that is highly likely to be key to comprehending the text and correctly answering the question, but the dependent clause 'try as she might' before the main clause also seems to be used to stress the content of the clause that follows.

According to the open feedback obtained from some test takers, however, this clause complex was rather difficult to process. For the majority of the students (about 50%) this is mainly due to the use of the infrequent word 'ace' and the multi-word unit 'set out'. Two other test takers noticed, on the other hand, that the main difficulty of the text is to be related to the fact that the three clause complexes it is made up of are about different people (clause complex 1 and clause complex 3) or 'aspects' (clause complex 2), which suggests that thematic progression (see 4.2.2.2 below) helps the reader infer the quality of the leading performance but not identify the main idea of this text. Finally, some other students (about 7% of the test takers) declared that they had to read the text twice to understand its meaning. As far as task 9 is concerned, about 14% of the students found that both option (a) and option (c) could be correct. In fact, 54 test takers out of 60 (90%) answered task 9 by choosing option (c), while only 5 of them chose (a), and one chose (b). Approximately 7% of the test takers commented on the difficulty of the task by saying that they had to read the text twice to be able to infer how the script was written and how the film was directed (this information is specifically given in the third clause complex). Finally, some students wrote that they had problems understanding the exact meaning of 'weak' and 'leading performance' as used in option (c).

3.3 Measuring Vocabulary Difficulty

In 3.2 above a measure of lexical density, i.e. LDU, and one of lexical variation, i.e. LV, were used to investigate how lexical density and lexical repetition vary among the texts of the corpus under investigation. It was pointed out, in particular, that not only does the 'effect of density' in a text depend on the number of lexical words occurring in it, but also on the amount of lexical repetition. As suggested above, however, at least two other factors are related to the perception of lexical density: the frequency of lexical words in English, and the fact that lexical words can be grouped into lemmas (see, for example, Sinclair 1991: 41-42), and/or word families (see, for example, Nation and Waring 1997), and/or other 'morphological paradigms' (Halliday 1989: 65). In order to account for these aspects with special reference to the needs of the present study, the software *Range 32*[9] (see, for example, Nation 2001 and Heatley, Nation and

---

9 The software is available online at <http://www.lextutor.ca/>, last visited on 25th April 2008.

Coxhead 2002) is used in combination with the three ready-made 'base' word lists that are available along with it, and in combination with fourteen 1000-word family lists created by "using family frequency figures from the 10 million token spoken section of the British National Corpus (BNC)" (Nation 2006).

3.3.1 Range 32 and the Underlying Research

Heatley, Nation and Coxhead (2002: 6) specify that the output of *Range 32* is based on three base word lists: the first one includes the most frequent 1000 word families of English, the second one the second 1000 most frequent word families, and the third one includes words "not in the first 2000 words of English but which are frequent in upper secondary school and university texts from a wide range of subjects." These three lists in fact consist of the base forms of words as well as of derived word forms: the word family 'write' (head word), for example, consists of 'writer', 'writers', 'writes', 'writing', 'writings', 'written', and 'wrote'. The first list thus contains 4119 word forms, the second one 3709, and the third one 3107 (570 head words). The source for the first two lists is the *General Service List of English Words* (GSL)[10] by West (1953), while the AWL (Academic Word List)[11], described in Coxhead (2000), is the source for the word forms contained in the third list. According to Heatley, Nation and Coxhead (2002: 6), the first 1000 head words of the GSL are usually "those in the list with a frequency higher than 332 occurrences per 5 million words, plus months, days of the week, numbers, titles (Mr, Mrs, Miss, Ms, Mister), and frequent greetings (Hello, Hi, etc)". Nation and Waring (1997: 15) give the following reasons for the choice of the GSL:

---

10 The GSL contains 2000 headwords. It was developed in the 1940s, and the frequency figures for most items are based on a 5,000,000 word written corpus (Nation and Waring 1997: 13). The GSL is available on-line at <http://jbauman.com/gsl.html>, last visited on 25th April 2008.

11 The AWL was compiled by Averil Coxhead in 1998 from a corpus of written academic English, which contains approximately 3.5 million words (Schmitt 2000: 182). The list is made up of 570 head words and is available on-line at <http://www.lextutor.ca/lists_download/>, last visited on 25th April 2008.

> although the GSL is in need of replacement because of its age, errors it contains, and its written focus, it is still the best available list, given the range of information it contains about the relative frequency of the meanings of the words. In a variety of studies [...] the GSL has provided coverage of 78 per cent to 92 per cent of various kinds of written texts, averaging around 82 per cent coverage.

Nation and Waring (1997: 9) also report on the work by Francis and Kucera (1982), who similarly maintain that the 2000 most frequent head words in the very diverse 1,000,000-word long Brown corpus allow for 79.7% word coverage of written text. They also say that the knowledge of the first 3000 most frequent head words in English provides 84% word coverage, and that knowing 15,851 word families allows one to understand about 97.8% of written text. The data concerning the percent coverage (up to about 90%) provided by the first 6000 most frequent head words in the Brown Corpus and reported in Brown and Nation (1997: 9) and Cobb (2006) has been represented graphically in Figure 3.11 below. It is important to highlight, however, that according to most researchers, knowing 80% of the words in a text "is not sufficient to allow reasonably successful guessing of the meaning of the unknown words" (Nation and Waring 1997: 11). In fact, 95% of known words in a text

> crops up in the literature frequently, [although] the exact percentage has not been established, and almost certainly depends on factors such as familiarity with topic and whether the unknown words are essential to the gist of the text. (Schmitt 2000: 152)

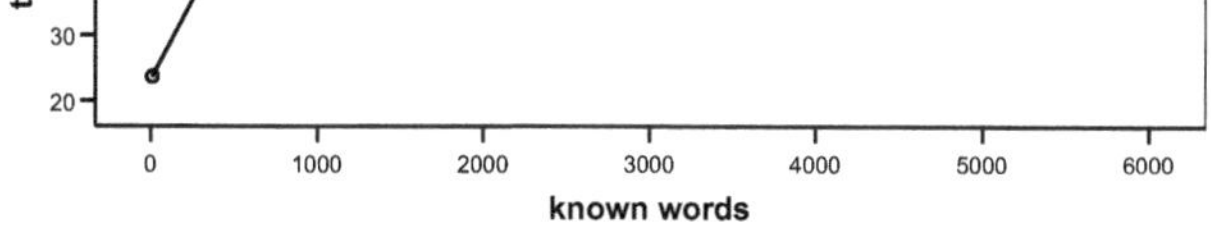

Figure 3.11. Graph representing the percent coverage provided by the first 6000 most frequent word families in the Brown Corpus (adapted from Cobb 2006).

The issue which scholars and educators are especially confronted with, therefore, concerns the choice of the words that should be given priority in the learning, teaching, and testing practice after the first 2000 head words. In order to successfully make this decision, Nation and Waring (1997: 16) propose choosing the words on the basis of what the language learner intends to use English for. In particular, they take the specific needs of university students into account, and make reference to some studies concerning the use of the 836 word list called the 'University Word List' (UWL)[12] (Nation 1990). This list consists of words that are not in the first 2000 words of the GSL, but are "frequent and of a wide range in academic English" (Nation and Waring 1997: 17). In fact, the UWL covered 8.5% of some academic texts (specifically, as the GSL provided 78.1% coverage of these texts, the total coverage equalled 86.6%), and 8.7% of an economics text (specifically, as the GSL provided 82.5% coverage of these texts, the total coverage equalled 91.2%). By contrast, the UWL only covered 1.7% of fictional writing (specifically, as the GSL provided 87.4% coverage of these texts, the total coverage of fictional writing equalled 89.1%). Cobb (2006) reports on the results of similar research carried out by Nation and colleagues, according to which the knowledge of the first 2000 words (word families) contained in the GSL combined with that of the 570 words (word families) in the AWL allows for the coverage of about 90% of the words one meets in most academic texts. He also specifies that the words contained in the AWL, "while not necessarily frequent in the language at large, are very frequent in academic texts. These are normally Greco-Latin terms like 'probability', 'conclusion', and 'hypothesis'".

Nation and Waring (1997: 17-19) also list a set of criteria that should be used for the development of a resource list of high frequency words. These criteria will be briefly reviewed here, in that it is believed that they can help one to appreciate the potentials and the limitations of general service frequency lists, as well as

12 The University Word List (UWL) of high frequency academic vocabulary across disciplines is currently replaced by the shorter yet more powerful Academic Word List (AWL) on the basis of coverage evidence from corpus analysis. (Cobb 2002)

understand how the lists used for this study were conceived of and created. The first factor to be taken into account is the 'representativeness' of the corpora on the basis of which lists are created; this means that the corpora used should contain a wide range of text types, "so that the biases of a particular text type do not unduly influence the resulting list." Two other related factors are 'frequency' and 'range', according to which a word should be included in a general service list because of its frequent occurrence across a wide range of texts in some form or other. Another factor they call 'word families' has to do with the set of criteria concerning what forms and uses (e.g., 'governor' and 'govern') should be "counted as being members of the same family. [...] When making this decision the purpose of the list and the learners for which it is intended need to be considered." The frequency of 'idioms' and that of 'set expressions' (see 3.2.1.1 above) is also an element that needs to be taken into account to decide upon the inclusion of these 'large' items into a list. Other criteria that, for example, West (1953) made use of to create his list concern the evaluation of the 'ease' or 'difficulty' of learning a given word, the 'necessity' of that word to express ideas, the semantic 'cover' of a given word, its 'stylistic level', and the 'emotionality' or 'neutrality' associated with it (Nation and Waring 1997: 18-19).

3.3.2 Using Range 32 and Base Word Lists for the Analysis of the Corpus

For the purposes of the present study, the software Range 32 and the three high frequency word lists, i.e. 'base word lists', available with it were made use of to obtain lists of word tokens, word types and families for each text of the corpus. Besides the first three lists introduced in 3.3.1 above, another list, i.e. list four, was made use of. It contains the names of some countries, the people and the major languages of those countries, and, in particular, the proper nouns of people, cities, mountains, countries, etc. that were found in the texts of the present corpus. The core part of this list was actually 'borrowed' from a list of words found in Nation (2006) and compiled mainly on the basis of the data in the BNC spoken sub-corpus. It was also chosen not to include grammatical words in the lists of words. In order to do so, the list of grammatical words reported in 3.2.1.1 and used for the calculation of LDU and LV was used as a stop list. The software Range 32 classifies all the words that are not contained in any of the abovementioned lists as 'not in the lists', does not include figures in the output, and treats apostrophes as spaces. Part of the output generated by Range 32 for text 13 is reported in Figure 3.12 as an example:

WORD LISTS	*TOKENS/%*	*TYPES/%*	*FAMILIES*
One	143/72.96	84/65.12	72
Two	21/10.71	17/13.18	13
Three	10/ 5.10	8/ 6.20	5
Four	6/ 3.06	5/ 3.88	5
Not in the lists	16/ 8.16	15/ 11.63	?????
Total	196	129	95

Figure 3.12. Part of the output generated by Range 32 (Heatley, Nation and Coxhead 2002) for text 13.

As anticipated in 3.2.2 above, in this study the specific information concerning the base word list and the word family a given word type belongs to was used, first of all, to refine the calculation of lexical variation (LV) in the texts of the corpus. Thus, word types such as 'eye' and 'eyes' in text 1 or 'ability' and 'able' in text 13 were considered as repetitions of the same lexical item, i.e. 'eye' and 'able' respectively, and were not, therefore, counted as different lexical words. In order to include these lexical relations into the measure, the following revised version of the index, i.e. LV(re), was used:

$$LV(re) = \frac{\text{Fam. 1} + \text{Fam. 2} + \text{Fam. 3} + \text{Fam. 4} + \text{Types not in lists}}{\text{Total Number of Lexical Words (Tokens)}} \times 100$$

This formula is based on the sum of the number of families in the first three lists and on the number of types not found in any list. In particular, the number of types not found in any list was used for obvious practical reasons, in that information about the grouping of these words into families was not available. However, by looking at the lists of word types generated by *Range 32*, it was pointed out that in most of the texts of the corpus most types 'not in the lists' could not be associated with any common family, and therefore, the measure seems to be accurate enough. In Figure 3.13 the values of LV(re) are arranged in ascending order for comparison with the values of LV. As predictable, LV(re) consistently scores lower values than LV, and the gap between the two indices depends on the number of inflected and derived forms associated with the same word family present in the text. If a comparison is made with Figure 3.8 above, it can be seen that in most cases the texts are ranked in the same way by both indices. A noticeable difference, though, concerns the fact that the variable LV(re) ranks text 13 as the least lexically varied text in the corpus (as with LV, if a given text scores low on LV(re), it means that a large amount of lexical repetition occurs in it), whilst it is text 14 that scores the lowest LV value. It is also interesting to notice that in text 4, text 7 and text 10 the values of the variables LV and LV(re) coincide, and that, differently from what was stated in 3.2.2 above, in text 9 there is actually some lexical repetition, namely the repetition of the verbs 'going' and 'goes' belonging to the head word 'go' in the first base list.

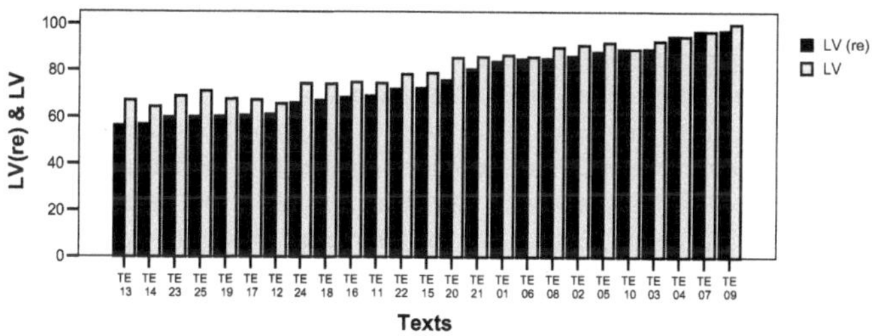

Figure 3.13. Bar graph representing the values of LV(re) and those of LV in the texts of the corpus.

If a closer look is taken at text 13[13] and at (part of) the output generated by Range 32 reported in Figure 3.14 below, it can be seen

13 Heinemann TOEFL Preparation Course. Copyright © M.K. Mahnke, Carolyn Duffy & Jacqeline Flamm, 1996. Reproduced with permission of Macmillan Publishers Ltd.

that the lexical choices made by its author involve the use of a large number of words that belong to the same word families.

(9) One of the most mysterious things in nature is the ability of certain creatures to find their way home, sometimes from great distances. Birds are not the only ones who can do this. Bees, eels, and salmon are able to return to a particular place after long journeys, too.
Most migrations take place between breeding grounds and regions where animals feed. For some animals, such as the lemming, the move is a one-way trip. Some scientists call this movement emigration, because these animals never return to their homes. For other animals, such as birds, the migration includes a return trip home. Birds move in periodic migrations, or at regular times during their lives, and often to the same place year after year.
Many experiments have been made with birds in an attempt to find out what guides them on their way home. In one case, seven swallows were taken 400 miles from home. When they were set free, five of them returned to their nests. In another case, a certain kind of sea bird was taken from its nest off the Welsh coast to Venice by plane. When it was released, it made its way home to its nest, a distance of 930 miles if it flew in a straight line.
Migrating birds offer an even more amazing example of this ability. There are swifts and house martins that migrate from England to South Africa every year. They not only return to England the next spring, but many of them come back to nest in the very same house where they nested the year before. They fly the incredible distance of 6,000 miles, one way.
Certain types of butterflies migrate, too, and find their way home over long distances. In the tropics one can sometimes see great mass flights of butterflies all flying steadily in one direction. They may go a thousand miles and more and then return again in another season.
Despite all the efforts that have been made to explain how these creatures find their way home, we still have no sure explanation. Since many of the birds fly over great bodies of water, we can't explain it by saying they use landmarks to guide them. Just to say they have an 'instinct' doesn't really explain the right conditions. The reason they do it may be to obtain food or to reproduce under the right conditions. But the signals and guideposts they use on their flights are still a mystery to man. (text 13)

The underlined groups of words in text 13 above exemplify some lexical relations occurring between words belonging to the same word families: the words 'fly', 'flew', 'flights', 'flying' all belong to the word family 'fly' in base word list 1; 'explain' and 'explanation' are associated with the word family 'explain' again in word base list 1; the words 'nest', 'nests', and 'nested' belong to the 'nest' word family in base word list 2; and 'migration', 'migrations', 'migrate', ‘migrating’ belong to the word family ‘migrate’ in word list 3. ‘Emigration’ in paragraph 2 is not listed in any one of the three base lists, and, therefore, it is not associated with the family ‘migration’. It is interesting to notice that although some semantic relation between ‘emigration’ and ‘migration’ can be legitimately claimed and supported, question 2 of task 13 explicitly asks test takers about the main difference between animal emigration and animal migration as follows (the correct answer is ‘(d) the final destination’):

(10) 13.02. According to paragraph 2, the main difference between animal emigration and animal migration is
(a) the breeding places
(b) the route taken
(c) the time spent
(d) the final destination

In paragraph 2 of text 13 it is stated that some scientists make a distinction between ‘emigration’ and ‘migration’, whereby ‘migration’ includes a return trip home (that is, the final destination), whilst ‘emigration’ does not. For the correct completion of this task, therefore, it is important to keep the two cognate words separate. As will be shown in 5.7.1, the majority of the students (75%) the test was administered to – 56 out of the 75 students who took the test – answered the question correctly, but 11 went for ‘(b) the route taken’, 6 for ‘(c) the time spent’, and 2 for ‘(a) the breeding places’.

Base 1 fam.	*Type freq.*	*Fam. freq.*
home	7	8
way	7	7
return	5	6
find	4	4
explain	3	4
great	3	3
place	3	3
case	2	2
distance	2	4
fly	2	6
house	2	2
move	2	3
use	2	2
able	1	3
bird	1	7
take	1	3
animal	0	4
mile	0	4

Base 2 fam.	*Type freq.*	*Fam. freq.*
nest	3	5
trip	2	2
guide	1	2
mystery	1	2
plane	1	1
creature	0	2

Base 3 fam.	*Type freq.*	*Fam. freq.*
migrate	1	6
obtain	1	1

Not in lists	*Type freq.*	*Fam. freq.*
butterflies	1	2
amazing	1	1

Figure 3.14. Part of the output generated by Range 32 (Heatley, Nation and Coxhead 2002) for text 13.

As pointed out in 2.2 above, text 13 was selected from a preparation course for the reading component of the TOEFL testing system, and using Enright *et al.*'s (2000: 20-28) terminology, its dominant pragmatic feature can be defined as 'expository' and the main rhetorical pattern it can be associated with as 'illustration'. The text aims to illustrate the phenomenon of migration in nature and exemplifies it by describing the behaviour of some migrating animals. In relation to this, it can be seen that question 13.09 asks test takers to specify the device the author uses to present the information in the passage as follows:

(11) 13.09. Which of the following devices does the author use to present the information in the passage?

(a) Support through exemplification
(b) Narration
(c) Cause and effect
(d) Comparison and contrast

The correct answer is '(a) Support through exemplification', which was chosen by 66 participants (88%). Nine participants, however, chose '(d) Comparison and contrast', two '(b) narration' and two '(c) Cause and effect'. '(d) Comparison and contrast', therefore, turned out to be quite good a distracter, which can be explained by the fact that the behaviours of the animals reported on for exemplification are, to a certain extent, also compared and contrasted. A consideration can also be made here about question 13.01 and, in particular, about the correct option (d):

(12) 13.01. What does this passage mainly discuss?
(a) How migratory creatures find their way home
(b) Popular places for migration
(c) Why animals migrate
(d) The emigration patterns of butterflies

As can be seen, all the content words in the first option belong to frequently occurring word families, that is, 'migrate', 'creature', 'way', 'find', and 'home'. The same holds true for option (c), in that this option contains the words 'animal' and 'migrate'. Option (b) and option (d), by contrast, contain lexical words that are not even used in the input text, that is, 'popular' and 'patterns'. It might be, therefore, hypothesised that the use of frequently occurring words in option (a) can be regarded as one of the clues candidates might make use of to answer the question correctly (see also 1.1 above).

Another application of the information obtained from word frequency lists is suggested by Halliday (1989: 65), who says that

> for a systematic, formal investigation of lexical density in texts we should have to adopt some weighting whereby lexical items of lower frequency 'scored' more highly than common ones. Word-frequency lists have been available for some time, and there are now large bodies of written and spoken text in machine-readable form in various places from which such information can readily be obtained.

In order to capture and 'weigh' the contribution of lexical items to the lexical density of the texts under examination (see Figure 3.7 in 3.2.1.2 above and reproduced below as Figure 3.18), two pieces of information have been represented by means of bar graphs. Firstly, the percentages of lexical words belonging to the word families in frequency lists 1, 2, 3 and 4, calculated as the ratio of the number of occurrences of each family to the total number of word families in each text (see Figure 3.15 and Figure 3.16). Secondly, the percentages of lexical word types that do not belong to any family, calculated as the ratio of the occurrences of the lexical word types not found in any list to the total number of lexical word types in each text (see Figure 3.17).

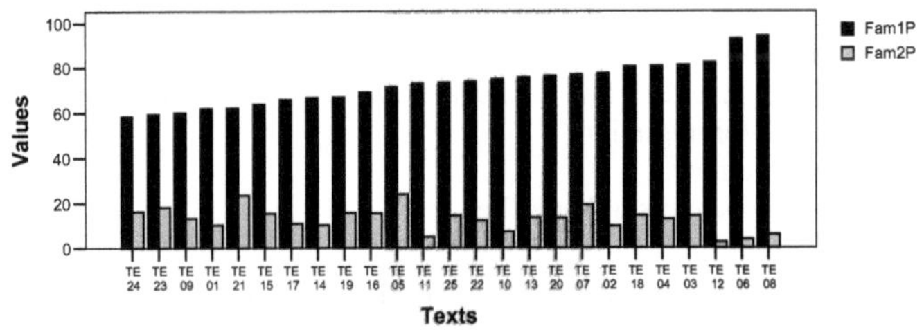

Figure 3.15. Bar graph showing the percentages of lexical words belonging to the first and second base word lists (Fam1P and Fam2P) in the texts of the corpus. The values of Fam1P are in ascending order.

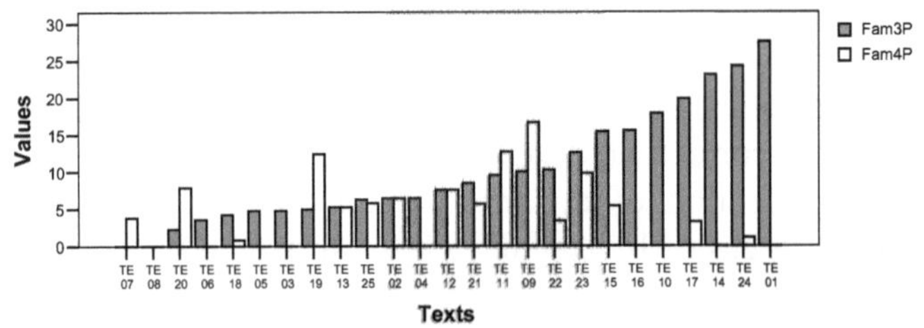

Figure 3.16. Bar graph showing the percentages of lexical words belonging to the third and fourth base word lists (Fam3P and Fam4P) in the texts of the corpus. The values of Fam3P are in ascending order.

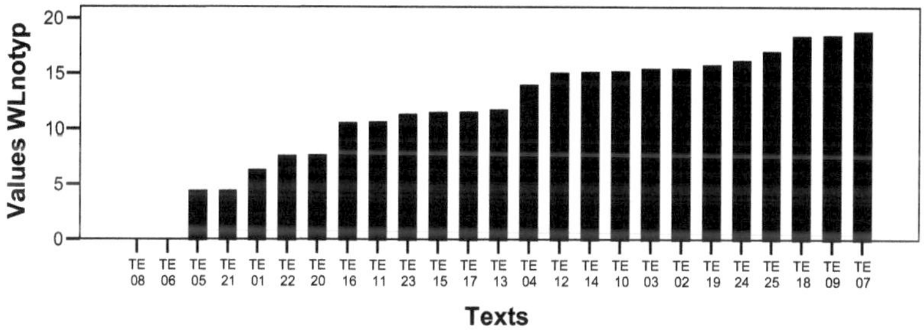

Figure 3.17. Bar graph showing the percentages of types that do not belong to any word base list in the texts of the corpus. The values of WLnotyp are in ascending order.

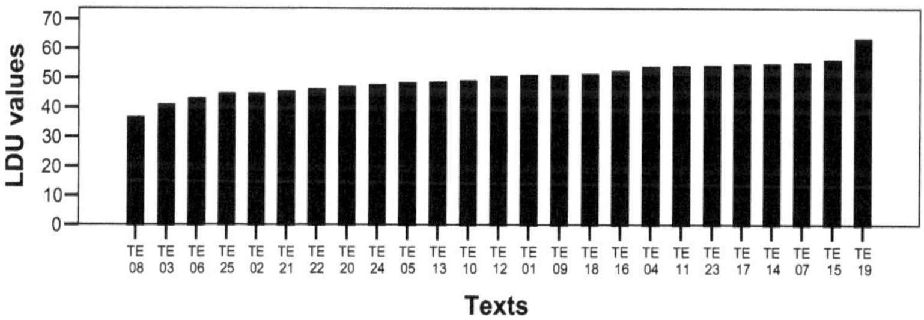

Figure 3.18. Bar graph representing the values of LDU (Ure's method) in the texts of the corpus in ascending order.

Some considerations can be made on the basis of the data plotted on these graphs. First of all, the analysis of the variation of lexical density in the texts of the corpus and reported on in 3.2.1.2 above can be refined thanks to the information gained from this data. Specifically, it can be seen that the least dense text in the corpus, i.e. text 8, is also the one that contains the highest percentage of high frequency words, that is, words belonging to the first base word list (Fam1P). In contrast, the densest one, i.e. text 19, is characterised by a high percentage of proper names of countries and places, e.g. 'Australia', 'Perth', 'London', 'Manchester' (Fam4P), and by quite a large number of words that do not belong to any word list, e.g. 'sightseeing', 'barrier', 'lush', 'reef', 'fabulous' (WLnotyp). Text 7, the third densest text in the corpus, contains the highest percentage of lexical words not occurring in any one of the lists, such as, 'gleaming', 'glimmer', 'interiors', and 'stilts', but does not contain any word belonging to the third word list. Finally, text 3, the second most lexically sparse text in the corpus, contains quite a high percentage of very frequent lexical words, but also some infrequent types not belonging to any list, e.g. 'dustman', 'hesitant'.

By examining the graphs above, it can also be seen that text 24 and text 1 stand out as those texts that contain the highest percentages of academic words, that is, words belonging to word families to be found in the third word list (Fam3P). According to the word frequency criterion, therefore, these texts can be defined as the most 'academic' of the corpus. Text 24 also contains a high percentage of types not occurring in any list, and is also characterised by the lowest percentage of words belonging to list 1 (Fam1P). The following extract from text 24[14] and the reproduction of text 1[15] actually confirm that the language used in these texts has some typical characteristics of academic English.

(13) 'Emotional intelligence' is generally defined as the ability to monitor and regulate one's own and others' feelings, and to use feelings to guide thought and action. In our work-life it comprises five basic elements: self-awareness, self-regulation, motivation, empathy and adeptness in social relationships. There is a common assumption that it may demand not 'being nice', but rather, for example, bluntly confronting someone with the uncomfortable truth. Nor does emotional intelligence mean giving free rein to feelings – 'letting it all hang out'. Rather, it means managing feelings so that they are expressed appropriately and effectively, enabling people to work together smoothly towards their common goal. (extract from text 24)

(14) 1. Since eye problems are related to physical health and hereditary factors, problems currently or previously experienced by you or your family will be identified and taken into account. *What is the best title for text 1?*

2. This checks your all round vision and can indicate neurological problems. *What is the best title for text 1?*

3. This determines if your eyes are coordinated and working together properly. *What is the best title for text 1?*

---

14 University of Cambridge ESOL Examinations. Excerpts adapted from 'Sample Paper R CAE.pdf' downloaded from <http://www.candidates.cambridgeesol.org/>, © University of Cambridge ESOL Examinations.

15 Test taken from DIALANG Reading Module.

4. Examination of the various parts of the eye can give an overall picture of the condition of your eyes and your general health. *What is the best title for text 1?* (text 1)

Besides the occurrence of academic words such as 'regulate', 'monitor', 'assume', and 'coordinate', the most salient linguistic features on the basis of which these texts can be defined as academic are: the use of nominalisation, that of pre- and post-modification of noun groups, and the frequent use of the passive voice. In these two texts there are numerous examples of the phenomena of nominalisation and of post-modification of nominal groups, dealt with in 3.2.1.2 above. In text 24, for instance, the word 'intelligence' can be considered as a nominalised or incongruent version of 'intelligent', 'ability' as a nominalised version of 'able', 'awareness' as a nominalisation of 'aware', and 'regulation' as one of 'regulate'. In text 1 the word 'examination' can be analysed as a possible nominalised version of 'examine', and 'an overall picture of the condition of your eyes and your general health' as a complex nominal group, in which 'picture' is the Thing (Head), 'an' is a non-specific Deictic, 'overall' functions as Epithet, and the prepositional phrase 'of the condition of your eyes and your general health' as Qualifier. In both text 1 and in text 24 the passive voice is frequently used (see, for example, the underlined verb groups in the extract above), namely 25 per cent of finite verbs in text 1 are passive and 22 per cent in text 24. According to Biber, Conrad and Leech (2002: 167-168) and to their investigation into the 40,000,000 word LSWE corpus, the passive voice is actually most common in academic prose, in which it accounts for about 25 per cent of all finite verbs, it is also common in news, in which it accounts for about 15 per cent of all finite verbs, but is rare in conversation.

With Figure 3.19 an attempt is made to convey the information given through the various bar graphs above in a single graph. To this end, the values of the lexical density index (LDU), the percentages of words belonging to the word families contained in the four base word lists, and the percentages of lexical word types that do not belong to any list are plotted on a line graph.

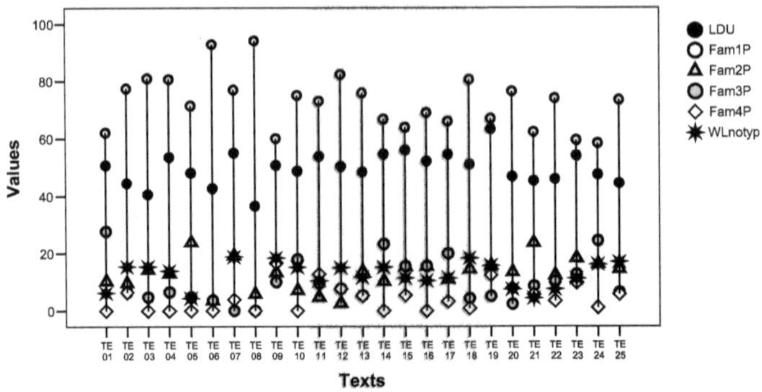

Figure 3.19. Line graph showing the values of LDU, the percentages of lexical words belonging to Fam1P, Fam2P, Fam3P, and Fam4P and that of lexical types that do not belong to any list (WLnotyp).

In order to complete the present investigation into the vocabulary difficulty of the 25 tests of the corpus, it is interesting to compare the lexical characteristics of the texts with those of the corresponding tasks. To this end, the Pearson product-moment correlation coefficients were calculated for the pairs of variables[16] reported in Figure 3.20. The assumptions underlying the Pearson Correlation and listed by Hatch and Lazaroton (1991: 436-437) are met, in that all the data is measured as scores and are continuous, the scores on the two variables are independent, the data is (rather) normally distributed through their range, and the relationship between the variables is linear. According to the data listed above, there are only two statistically significant correlations (see also Pallant 2001: 120-121): the medium positive correlation between the values of lexical density in the texts and those in the corresponding tasks (LDU (texts) / LDU (tasks)), and the medium negative relationship between the percentage of words in the first base list in the texts and that in the tasks (Fam1P (texts) / Fam1P (tasks)). All the other relationships are not statistically significant. On the basis of these observations, therefore, it can be affirmed that in the tests under examination the lexical density of a given input text tends to be similar to the lexical

16 The coefficients have been calculated by means of SPSS (version 13.0 for Windows).

density of the associated task, while the percentage of frequent lexical words (i.e. those belonging to the first word base list) in a given input text is likely not to correspond to the percentage of those used in the task.

Pairs of variables	*Correlations*
LDU (texts) / LDU (tasks)	.425 (N=25, p < .05)
Fam1P (texts) / Fam1P (tasks)	-.421 (N=25, p < .05)
Fam2P (texts) / Fam2P (tasks)	.186 (N=25, p > .05)
Fam3P (texts) / Fam3P (tasks)	-.261 (N=25, p > .05)
Fam4P (texts) / Fam4P (tasks)	.067 (N=25, p > .05)
WLnotyp (texts) / WLnotyp (tasks)	-.016 (N=25, p > .05)

Figure 3.20. Pearson Product-Moment correlations between LDU, Fam1P, Fam2P, Fam3P, and WLnotyp.

3.4 Measuring Grammatical Intricacy and Halliday's method of Measuring Lexical Density

In 3.2.1 above, it has been suggested that written texts tend to be lexically dense, while more unconscious, spontaneous and usually spoken texts share a tendency to be grammatically 'intricate'. Starting from this consideration, in 3.4.1 below the concept of grammatical intricacy and the grammar of the English clause complex as conceived of in systemic functional linguistics will be illustrated in some detail. In 3.4.2 below, Halliday's proposal for calculating lexical density (henceforth referred to as LDH) and a measure of 'grammatical intricacy' (henceforth referred to as GI) will be applied to the analysis of the present corpus and compared with the findings of other studies.

3.4.1 Halliday's Method of Calculating Lexical Density (LDH), Grammatical Intricacy, the Grammar of the Clause Complex, and Embedding

In 3.2.1.2 above, not only was it established that text 8 is the sparsest text in the corpus, but also that this text can be considered as a grammatically intricate piece of fictional writing. In general, Halliday (2004: 654) associates these two textual characteristics with two different kinds of *complexity*:

> typically, written language becomes complex by being lexically dense: it packs a large number of lexical items into each clause; whereas spoken language becomes complex by being grammatically intricate: it builds up elaborate clause complexes out of parataxis and hypotaxis.

Halliday (1987: 66) also points out that:

> the more natural, un-self-monitored the discourse, the more intricate the grammatical patterns that can be woven. Usually, this kind of discourse will be spoken, because writing is in essence a more conscious process than speaking. But there are self-conscious modes of speech, whose output resembles what we think of as written language, and there are relatively spontaneous kinds of writing; spoken and written discourse are the outward forms that are typically associated with the critical variable, which is that of consciousness.

In line with Halliday (1987: 66), in this study grammatical intricacy and lexical density are conceived of as complementary 'paired properties', and so are their measures. In particular, it is believed that both of them ought to be considered for the accurate 'placement' of a given text on the spoken-written continuum. In fact, Halliday (1989: 66) reconsiders the notion of 'density', which has to do with "how closely packed the information is", from this broader perspective, and concludes by saying that:

> words are not packed inside other words; they are packaged in larger grammatical units – sentences, and their component parts. It is this packaging into larger grammatical structures that really determines the informational density of a passage of text.

To this he adds that the most relevant grammatical structure "where meanings are organised and wrapped up together" is the clause, and that the clause provides "a more relevant organising concept [...] for enabling us to capture the special properties of both spoken and written language" (Halliday 1989: 66-67). Halliday (1989: 66) also clarifies that he prefers using the term 'clause complex' instead of 'sentence', in that the former "will allow us to refer both to written and to spoken language in a way that makes the two comparable". In order to calculate lexical density, Halliday (1989: 67), therefore, suggests that:

> instead of counting the number of lexical items as a ratio of the total number of running words, we will count the number of lexical items as a ratio of the total number of clauses. Lexical density will be measured as the number of lexical items per clause.

By following this methodology, the 'discrepancy' between spoken-like and written-like texts "stands out more sharply" (Halliday 1987: 61), although this method is more difficult to apply, since "it is not always easy to recognise what a clause is" (Halliday 1989: 67). In particular, Halliday (1989: 80) makes it clear that "embedded clauses are not counted separately, since they function inside another clause," which implies that for the count an analyst is supposed to distinguish reliably and consistently between embedded (or rankshifted, or down-ranked) clauses and 'ranking' clauses, i.e. 'independent', 'paratactic' and 'hypotactic' clauses. In order to make these distinctions clearer, paragraph 3.4.1.1 will be devoted to a brief description of the resources of clause complexing, and paragraph 3.4.1.2 to embedding.

3.4.1.1 The Clause Complex

'Clause complex' is the term used in systemic functional linguistics to refer to the "grammatical and semantic unit formed when two or more clauses are linked together in certain systematic and meaningful ways" (Eggins 2004: 255). Two basic systems determine how one clause is related to another: the tactic system and the logico-semantic system. The two of them, and the related more delicate systems, are represented in Figure 3.21. Eggins (2004: 258-269) defines the tactic system as the system that "describes the type of interdependency relationship between clauses linked into a clause complex", and 'taxis' as "a basic language resource for expanding units at any rank to make more meaning at that same rank". As can be seen from Figure 3.21, there are two different degrees of interdependency: parataxis (equal status) and hypotaxis (unequal status). "Parataxis is the relation between two like elements of equal status, one initiating and the other continuing", whilst "hypotaxis is the relation between a dependent element and its dominant, the element on which it is dependent" (Halliday 2004: 374-375). Paratactic relations are usually represented by numbers, and hypotactic structures by Greek letters, with the symbol α reserved for the dominant clause. Whilst making a 'tactic' choice, a language user has to make another simultaneous choice concerning the logico-semantic relation which holds "between a primary and a secondary member of a clause nexus" (Halliday 2004: 376-377). Specifically, one has to choose between 'projection' and 'expansion'. Projection is the relation by means of which the secondary clause is projected through the primary clause, and that allows for the quotation or report of speech ('locution') or thought ('idea'). 'Expansion', on the other hand, is the "logico-semantics of developing on previous meanings" (Eggins 2004: 271), by means of which "the secondary clause expands the primary clause, by (a) elaborating it, (b) extending it or (c) enhancing it" (Halliday 2004: 377). Clauses can be either 'finite' or 'non-finite'. "A finite clause is in principle independent; it becomes dependent only if introduced by a binding (hypotactic) conjunction. If it is joined in a clause complex, its natural status is paratactic" (Halliday 2004: 423). On the other hand,

> a non-finite clause is by its nature dependent [...]. Hence when a non-finite clause occurs without a conjunction, there is no doubt about its hypotactic relation in a clause complex; but there may be no indication of its logical-semantic function. (Halliday 2004: 424)

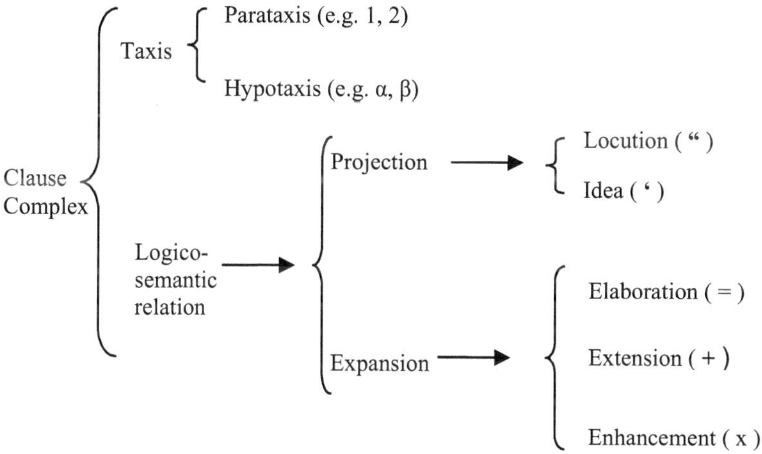

Figure 3.21. The basic systems of clause complexing (adapted from Halliday 2004: 373 and Eggins 2004: 259). Next to the most delicate choices of this system network, the notations used in systemic functional linguistics to mark them have been indicated.

The following are some examples of clause complexes taken from the corpus of input texts under investigation and analysed using the notation suggested by Halliday (see, for example, Halliday 2004: 377-378) and also sketched in Figure 3.21 above[17]:

- Paratactic expansion (elaboration):

(15) ||| 1 Leave the ship in Spitsbergen || +2 and fly to London from Tromsø. ||| (from text 18);

(16) ||| Whereas one's IQ undergoes few changes, 1 emotional intelligence continues to develop as we go through life and learn from our experiences; || =2 our competence in it can keep growing. |||[18] (from text 24);

---

17 Following Halliday (2004: 401), the sign || will be used at the beginning of a clause, and the sign ||| at the beginning and at the end of a clause complex. The angle brackets << … >> will be used to denote enclosure. They are doubled because they denote a clause.

18 The hypotactic relations between some of the clauses in examples 1, 3, and 10 have not been indicated since the purpose here is to point out parataxis.

- Paratactic expansion (extension):

(17) ||| 1 They not only return to England the next spring, || +2 but many of them come back to nest in the very same house where they nested the year before. ||| (from text 13);

- Paratactic expansion (enhancement):

(18) ||| 1 I was a part-time comedian for years, || x2 so I learned how to stand in front of audiences. ||| (text 20);

- Paratactic projection (locution):

(19) ||| "2 "I do think this course should be fun," || 1 he said. ||| (from text 3);

- Paratactic projection (locution) of proposal, finite (free indirect speech):

(20) ||| "β He would be involved <<,α he said, >> but at arm's length. ||| (from text 25);

- Hypotactic expansion (elaboration), finite:

(21) ||| α Since 1 April 1993, <<β = when it became independent of the local authority and adopted its new title,>> West Thames College has continued to build on that first class reputation. ||| (from text 15);

- Hypotactic expansion (elaboration), non-finite:

(22) ||| α Four in ten people are 'ethical spenders', || =β buying goods which do not, for example, involve dealings with oppressive regimes. ||| (from text 17);

- Hypotactic expansion (extension), finite:

(23) ||| +β Whereas one's IQ undergoes few changes, || α emotional intelligence continues to develop [...].||| (from text 24);

- Hypotactic expansion (extension), non-finite:

(24) ||| α Rather, it means managing feelings so that they are expressed appropriately and effectively, || β enabling people to work together smoothly towards their common goal. ||| (from text 24);

- Hypotactic expansion (enhancement), finite:

(25) ||| xβ When they were set free, || α five of them returned to their nests. ||| (from text 13);

- Hypotactic expansion (enhancement), non-finite :

(26) ||| α The new rules can be used || xβ to indicate who is likely to become a star performer and who is most prone to mediocrity. ||| (from text 24);

(27) ||| α A University permit does not entitle them to park in Hall car parks however, || xβ unless authorised by the Warden of the Hall concerned. ||| (from text 14);

- Hypotactic projection (idea) of proposition[19], finite:

(28) ||| α I think || 'β that I would be foolish not to write another novel. ||| (from text 25);

- Hypotactic projection (locution) of proposal, finite:

(29) ||| α Students are reminded that || 'ββ if they park a motor vehicle on University premises without a valid permit, || 'βα they will be fined £20. ||| (from text 14);

- Hypotactic projection (locution) of proposition, finite:

(30) ||| α He said || "β that he had since turned down an offer to write the screenplay of The Horse Whisperer. ||| (from text 25);

- Hypotactic projection (locution) of proposal, non-finite:

(31) ||| α Airlines insist on exaggerated check-in times || (βα which require passengers || ββ to report to the airport at a given time before departure) ||| (from text 23);

---

19 "When language is used to exchange information, the clause takes on the form of a *proposition*. It becomes something that can be argued about [...], affirmed, denied, and also doubted, contradicted, insisted on [...]". When language is used to exchange goods-and-services, the clause takes the form of a *proposal* (Halliday 2004: 110).

- Hypotactic projection (idea) of proposal, non-finite:

(32) ||| α Don't forget || 'β to include the flight numbers. ||| (from text 23).

3.4.1.2 Embedding of Clauses

Halliday (2004: 426) defines embedding as:

> a semogenic mechanism whereby a clause or phrase comes to function as a constituent within the structure of a group, which itself is the constituent of a clause. [...] Embedding is thus the 'rank shift' by which a clause or phrase comes to function within the structure of a group. The characteristic function of an embedded element is as Post-modifier in a nominal group [...]. Other functions are: as Head of a nominal group (i.e. as a nominalization); and as a Postmodifier in an adverbial group.

Like clauses in paratactic and hypotactic relations, an embedded clause can be either an expansion or a projection (Halliday 2004: 427), and can be either finite or non-finite. As Taylor Torsello (1992: 189) observes, "a clause-level event that is rankshifted down into an embedded clause in the nominal group seems to have been 'predigested', processed and packaged, together with the rest of the nominal group, by the speaker for the hearer", and is therefore "out of the range of critical argument or disagreement on the part of the hearer." As far as the role of nominal constructions and in particular of nominalization is concerned, Halliday (1989: 73-75) points out that:

> there are a lot of things that can only be said in nominal constructions: especially in registers that have to do with the world of science and technology, where things, and the ideas behind them, are multiplying and proliferating all the time. [...] The structure of the modern world and the structure of the language combine together to make the written language what it is: a language with a high lexical density, measured in the number (and informational load) of lexical items per clause, and a strong tendency to encode this lexical content in a nominal form: in head nouns, other items (nouns and adjectives) in the nominal group, and nominalised clauses. It is these nominalised structures that give the clause its enormous elasticity.

In the remaining part of this section some embedded clauses taken from the corpus of texts under investigation will be given as examples of the various types of embedding. The examples will be organised according to the classification proposed by Taylor Torsello

(1992: 188-206), and following Halliday (2004: 426-441, 467-480), embedded clauses will be represented with the symbols [[ ]] and embedded phrases with [ ].

- Nominalisation, finite, with 'that':

(33) || It is a fact [of life] [[that sooner or later regular travellers will be parted from their luggage]] || (from text 23);

- Nominalisation, finite, 'wh-clauses':

(34) || We are being judged by a new yardstick – not just by [[how clever we are]] || (from text 24);

(35) ||| α This yardstick is increasingly used || xβ in choosing [[who will be hired and who will not]] …||| (from text 24);

(36) || These rules have little to do with [[what we were told at school was important]] || (from text 24);

- Nominalisation, non-finite (imperfective):

(37) || Nor does emotional intelligence mean [[giving free rein to feelings]] || (from text 24);

- Defining relative clauses, finite:

(38) || Daniel Goleman examines the 'people skills' [[that are essential for a place at the top of your profession]] || (from text 24);

(39) || In a time [[when few guarantees of job security have led to the very concept of a 'job' being rapidly replaced by 'portable skills']], personal qualities begin to play an important role in the workplace. || (from text 24);

(40) ||| α He waved his left hand impatiently || xβ to silence the loquacious Proprietor [[who had begun the parrot formula [[he had repeated at least fifteen times so far]]]]. ||| (from text 2);

(41) || They let you review your pictures the moment [[you take them]] || (from text 5);

- Defining relative clauses, non-finite:

(42) || The ability [[to do well in examination]] is largely irrelevant to this standard || (from text 24);

(43) || Problems [[currently or previously experienced by you or your family]] will be identified and taken into account. || (from text 1);

- Restrictive appositive clauses[20]:

(44) || The risk [[of baggage being lost when changing planes]] is higher than average at certain airports || (from text 23);

- Embedded comparative clauses:

(45) || They can choose from a far wider range of subjects [[than it would be practical for a sixth form to offer]] || (from text 15);

(46) || My job title is the same [[as it was when I started]] || (from text 22);

- Embedded result and purpose clauses:

(47) || Please note that staff are not available [[to train people or give a lot of detailed instruction on [[how to use the programs]]]] || (from text 16);

(48) || The person at the other end has to be ready [[to pull you out fast]] || (from text 21).

---

20 "The restrictive appositive is most often a projection, since its major function is as a modification of a noun which expresses a verbal or mental metaphenomenon. These nouns can be names for locutions, such as *news*, *proposal*, *report*, or *question*. They can be names for ideas or feelings such as *principle*, *belief*, [...]. They can also be facts, such as *fact*, *proof*, *case*, *possibility*, *certainty* or *need*. [...] There are also appositive post-modifying embedded clauses which are expansions. These are the ones which modify nouns which are names for phenomena, such as *act*, *action* or *process*" (Taylor Torsello 1992: 200-203).

3.4.2 Further investigation into the Lexical Density of the Texts and into their Grammatical Intricacy

Halliday's method of calculating lexical density introduced in 3.4.1 above can be expressed by the following formula (henceforth referred to as LDH):

$$LDH = \frac{\text{Total number of lexical words}}{\text{Total number of ranking clauses}}$$

As can be seen, the information needed to count this LDH index concerns the number of lexical words and the total number of ranking clauses in each text.

As far as the quantification of grammatical intricacy is concerned, the index chosen for this study will be calculated "in terms of the average number of ranking clauses per clause complex" (Taylor Torsello 1996: 120):

$$GI = \frac{\text{Total number of ranking clauses}}{\text{Total number of clause complexes}}$$

By applying this formula (henceforth referred to as GI), the intricacy of grammar is captured in terms of how many clauses join together to form a clause complex, and the higher the index, the more intricate the text. Obviously, for counting this index, information about the number of clause complexes as well as about that of ranking clauses in a given text is necessary. In order to gain further information about the grammatical complexity of the texts under investigation, the H/P index will also be used:

$$H/P = \frac{\text{Total number of hypotactic clauses}}{\text{Total number of clause complexes}}$$

This formula aims to quantify grammatical complexity as the ratio of the total number of secondary clauses in hypotactic structures to that of independent, paratactically related clauses, and primary clauses in hypotactic structures. The more secondary clauses in hypotactic structures in a text, the higher the value of this index.

In order to distinguish between these types of clauses, all the input texts in the corpus were annotated (tagged) manually using the following tags:

- '*$ccc*' standing for clause complexes; number the pages progressively
- '*$mc*' standing for independent clauses, paratactically related clauses, and primary clauses in hypotactic structures;
- '*$h*' standing for secondary clauses in hypotactic structures.

No tagging was done for embedded clauses, in that these clauses do not count in any of the above counts. However, as said above, the distinction between embedded and ranking clauses was of the utmost importance whilst tagging. The following tagged version of text 6 exemplifies how tagging was carried out:

(49) $ccc $mc With your travel agency, the holiday you book is the holiday you get. $ccc $h If you arrive $h and find we've failed to live up to our promises, $mc let us know what the problem is within one day of your arrival. $ccc $mc We'll spend 24 hours $h doing everything possible $h to sort the problem out. $ccc $mc In the unlikely event that we can't resolve your problem and make you happy within 24 hours, we'll fly you home $mc and give you your money back.

Once the tagging procedure was completed, the annotated texts were processed by means of *WordSmith Tools* (Scott 2006), thus obtaining lists containing the occurrences of these tags. This data was then put into *Microsoft Excel*, and the above formulae were applied. Figure 3.22 represents the distributions of the LDH index across the texts of the corpus. These values can also be compared to the LDU values in Figure 3.23 (already given in Figure 3.7 above). Figure 3.24 shows the values of Grammatical Intricacy (GI), and Figure 3.25 those of the H/P index.

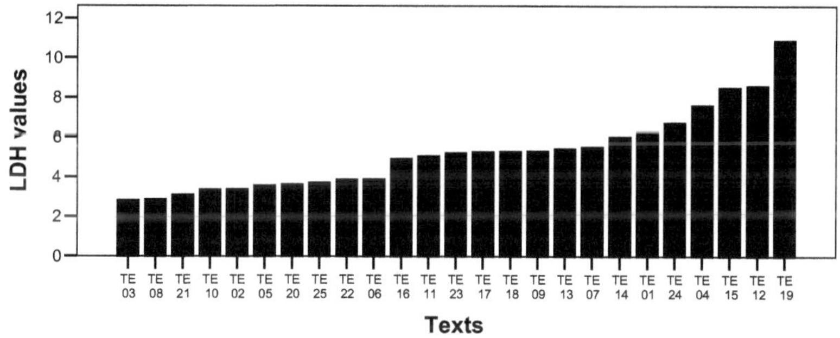

Figure 3.22. Bar graph showing the values of the LDH index in the texts of the corpus (see also Figure 3.7 above).

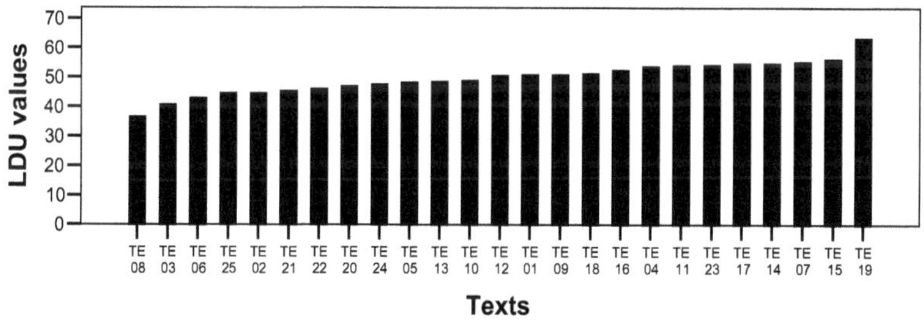

Figure 3.23. Bar graph showing the values of the LDU index in the texts of the corpus.

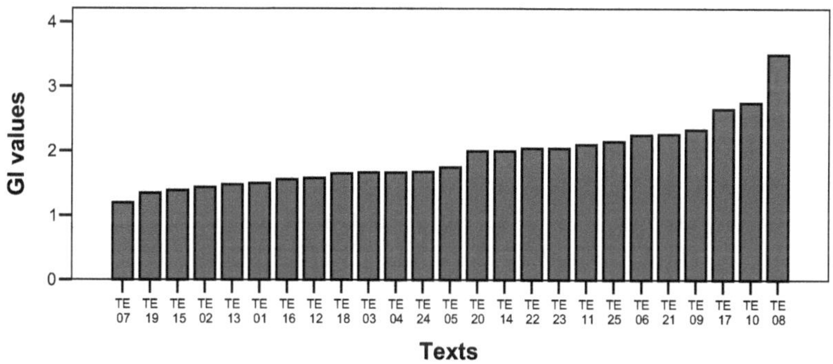

Figure 3.24. Bar graph showing the values of the grammatical intricacy index (GI) in the texts of the corpus.

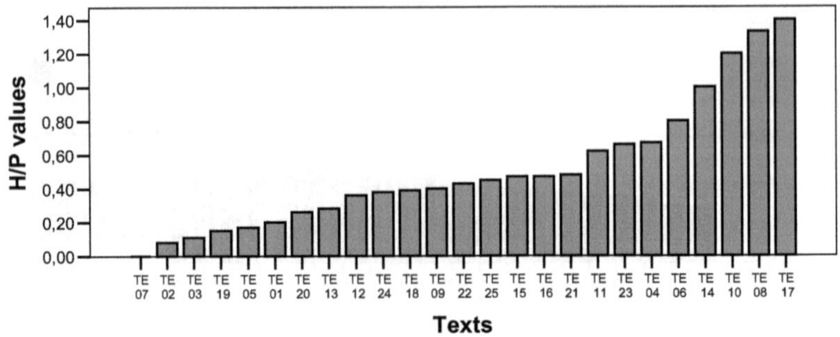

Figure 3.25. Bar graph showing the values of the H/P index in the texts of the corpus.

If Figure 3.22 is compared to Figure 3.23, it can be seen that text 19 scores the highest lexical density values, irrespective of the index used (LDH = 10.87; LDU = 63.29%). It can also be seen that text 3 and text 8 score very similar LDH low values (text 3 scores 2.8 and text 8 scores 2.86), thus confirming the results discussed in 3.2.1.2 above. What is remarkable, however, is the sharper discrepancy between text 19 and text 3 as shown by the distribution of the LDH index, which confirms Halliday's (1989: 61) findings. According to Figure 3.24, text 8 also scores the highest grammatical intricacy value (GI = 3.50), and can, therefore, be classified as the most spoken-like text in the corpus. The fact that it scores the second highest H/P value in the corpus (H/P = 1.33) means that it also contains a relative large number of secondary clauses in hypotactic structures.

By looking at the values of these indices, it also emerges that two other texts are worth comparing: text 7 and text 12. In particular, text 7 stands out as the least grammatically intricate text in the corpus (GI = 1.20), in that it contains 6 ranking clauses and only 5 clause complexes (i.e., there is nearly one ranking clause per clause complex only). Moreover, it is the only text that does not contain any hypotactic clause. Finally, as seen in 3.2.1.2 above, it scores rather high on the LDU index (LDU = 55%), but not remarkably so on the LDH index (LDH = 5.50). By contrast, text 12 scores the second highest LDH value (LDH = 8.58), although it scores an intermediate LDU value (LDU = 50.31%). It also scores an intermediate GI index (GI = 1.58), since it contains 19 ranking clauses (5 secondary clauses in hypotactic structures) and 12 clause complexes. This situation can be summarised by means of Figure 3.26, which should help bring out the relationships between the three variables:

Variables	*Text 7*	*Text 12*
GI	low: 1.20	intermediate: 1.58
LDU	high: 55%	intermediate: 50.31%
LDH	intermediate: 5.5	high: 8.58

Figure 3.26. Values of GI, LDU, and LDH in text 7 and text 12.

It can be claimed that in text 7 the intermediate LDH score is due to the presence of a proportionally high number of ranking clauses, as compared to the total number of lexical words. This fact is not captured by the GI index, in that the text also contains a large number of clause complexes that contribute to the decrease in the values of this index. In contrast, in text 12 the high value of the LDH index is due to the proportionally low number of ranking clauses, which is again not captured by the GI index. This shows that, as stated in 3.4.1 above, grammatical intricacy and lexical density are paired properties, and, in particular, that the values of the three indices proposed above to quantify them (the H/P index does not measure grammatical intricacy, but rather the textual complexity caused by the use of hypotaxis) should be all taken into account and interpreted critically.

The LDH and the GI indices have been used by various scholars and researchers to analyse different types of texts and corpora: to conclude this section devoted to the two indices, results that have emerged from other studies will be reviewed. First of all, Halliday (1989: 80) says, that on the basis of various samples, he found that a typical average lexical density (the LDH index) for spoken English "is between 1.5 and 2, whereas the figure for written English settles down somewhere between 3 and 6, depending on the level of formality in the writing." He also says that

> the figures themselves will vary considerably according to the theoretical basis of the analysis [...]. But provided whatever criteria are adopted are applied consistently, the lexical density of written language is likely to be of the order of twice as high as that of speech. (Halliday 1989: 80)

Halliday and Martin (1993: 76) report on their analysis of some clause complexes taken from *Scientific American*, and show that their lexical density (LDH index) varies between 10 and 13, which are much higher values than those found for less specific texts. Castello (2002: 57-59) found that the LDH values in a 92,500-word long corpus of tourist publications range between 5.11 and 11.75 (average score = 7.40), and those of the GI index between 1.29 and 2.92 (average score = 1.79). He (Castello 2002: 68-73) also found out that the sub-corpus of brochures is characterised by the highest LDH mean score (LDH = 9.16), whereas the sub-corpus of tourist guides by the lowest mean score (LDH = 6.69). Finally, he found that the sub-corpus of tourist guides scores the highest average GI value (GI = 1.86), while brochures score a lower value (GI = 1.65). Montin (2002, quoted in Castello 2004: 145-146; see also Taylor Torsello and Baldry 2005: 320-321) counted the LDH values of a 23,000-word corpus of scientific texts (eight scientific articles in English), and found that the average LDH score is 9.67 and that the average GI score is 1.79. However, her study also revealed that the values of these indices vary considerably according to the section of the article they belong to: the sections 'methods and materials', for instance, score the highest LDH score (LDH mean value = 11.09), whilst the sections 'results and discussion' are characterised by the lowest values (LDH mean value = 8.65). By contrast, 'introduction' and 'discussion' are characterised by the highest GI values (GI mean value = 1.71), whilst 'conclusions' by the lowest ones (GI mean value = 1.4). Brombin (2002, cited in Castello 2004: 146-147; see also Brombin 2004: 401-427, and Taylor Torsello and Baldry 2005: 320-321) reports on the results of her investigation into the lexical density and grammatical intricacy of a 16,000-word corpus of popular science texts. As predictable, the average LDH value of this corpus turned out to be remarkably lower than that of the scientific corpus reported on above (i.e. 5.28 vs. 9.67), while the average GI value found is higher (i.e. 1.91 vs. 1.57).

3.5 Measuring Readability

This section will be devoted to the discussion and application of some measures belonging to the so called 'readability formulae'. They were used with the aim of gaining some further insights into the characteristics of the tests under examination in this study, that is, information about the readability of the input texts.

3.5.1 Readability Formulae

Bruce and Rubin (1988: 5) define 'readability' as 'ease of reading', 'interest' or 'ease of understanding', and a readability formula as a method of assigning a numerical estimate of readability to a text. From this definition it can be deduced that the concept of readability concerns many different factors including "reader-specific factors, such as motivation, interest, values, or purpose" (Bruce and Rubin 1988: 8), but also text-related factors, such as sentence structure and the legibility of print, e.g. the layout, typographical features, and the reading conditions (Johnson 1998: 1-3). Readability is, therefore, a multifaceted concept, and is mainly concerned with the problem, already raised in chapter 1 above and central in reading comprehension research, "of matching between reader and text" (Johnson 1998: 1). Research into readability began in the 1920s, and an array of metrics have since been designed and made use of in different fields. In general, a readability formula is intended as a quick and conventional measurement, which usually

> takes into account only easily measurable aspects of a text such as word difficulty and average sentence length. A weighted combination of these measurements yields a number for each text. Some readability formulas produce estimates that represent grade levels; others range over a 100 point scale where higher numbers indicate greater readability. (Bruce and Rubin 1988: 5-6)

Anderson and Davison (1988: 25) point out that "scholars of readability are aware of the impossibility of reducing all text properties to formula variables", and that the formula values should not be taken as "anything but rough predictions of text ease or difficulty." In particular, they affirm that:

> the presence of long sentences and complex words in a text in some way reflects or is correlated with complexities of subject matter, but need not directly cause a text to be difficult. While these factors may impede comprehension for some readers who have difficulty segmenting words and parsing sentences or who have limited working memory capacity, these very same factors also provide the reader with explicit information about the composition of a word or the relations between sentences. (Anderson and Davison 1988: 48)

They, therefore, welcome alternative and/or complementary approaches to text readability, such as "field-testing on a sample of readers and the judgement of experienced readers" (Anderson and Davison 1988: 26). In fact, in the 'new Dale-Chall' readability formula revised by Chall and Dale (1995), approaches of this latter kind actually coexist with traditional readability factors. Specifically, this formula

> also takes into account reader-specific variables (especially motivation and the background knowledge required), and factors such as reading purposes, availability of assistance from educators, and so on. The new formula also provides guidelines for making qualitative judgements about content aspects (e.g. conceptual difficulty and density), and about organisational features contributing to difficulty. (Masi 2003: 125)

In practical terms, the Dale-Chall formula rates text on a U.S. grade-school level based on the average sentence length and the number of *unfamiliar* words, using an updated list of words commonly known by 4th grade students[21].

Other scholars in various fields also warn against uncritical reliance on readability formulae. Gibson (1993: 204), for example, maintains that all such measures can be easily discredited, and that "many formulae which purportedly measure syntactic complexity are in fact doing no such thing: this is because the sentence is defined

---

21 <http://www.rfp-templates.com/search/for/Dall-Chall-Readability.html>, last visited on 25th April 2008.

orthographically, rather than syntactically". Urquhart and Weir (1998: 146) claim that

> the literature abounds with warning against reliance on readability formulae for estimating text difficulty [and that] the difficulty of the text will be largely determined by its linguistic, organisational, propositional and discoursal attributes […] and is obviously relative to individual variation across a number of parameters (e.g. background knowledge, purpose(s) for reading, etc.).

Alderson (2000: 73-74) observes that "readability formulae give only crude measures of text difficulty, and are rarely suitable for second- or foreign-language readers, even of English texts". Alderson (2000: 72) also mentions studies, e.g. Taylor (1953), that report on the use of cloze techniques, i.e. the random or pseudo-random deletion of words from text, as possible measures of text readability. He says that "many studies have established high correlations between readability as measured by formulae, and readability measured by cloze" (Alderson 2000: 72); in language testing, however, "cloze techniques are themselves a testing procedure, and so will give a biased estimate of text difficulty" (Alderson 2000: 74).

Bruce and Rubin (1988: 18-19) list a series of criteria that should be met, in order to properly use readability formulae. Firstly, the time spent on reading and the manner of reading should be determined by the reader, and not by external factors. Secondly, the texts to be read should be 'honestly' written, that is, they should be written to satisfy a communicative goal, not formula constraints. Thirdly, "features such as higher-level organisation, conceptual load, […], must be generally predictable from the word and sentence characteristics of the text". Fourthly, the target reader's purposes and strategies should not be altered from those of the readers in the validation studies. Fifthly, readability scores are statistical averages applicable to large numbers of texts and readers and not to an individual text for an individual person. Finally, "any attempt to expand the use of the formula to evaluate materials for readers whose background, dialect, purpose in reading, etc. differs from that of the readers used in validation is likely to lead to difficulties".

3.5.2 Investigating the Readability of the Corpus

For the present study the following three readability metrics were made use of:

1. Flesch Reading Ease (*FRE*)	= 206.835 – (1.015 x ASL*) – (84.6 x ASW**)
2. Flesch-Kinkaid Grade Level (*FKGL*)	= (0.39 x ASL) + (11.8 x ASW) – 15.59
3. Gunning-Fog Index (*GFI*)	= (ASL x percentage of words of three or more syllables) x 0.4

** ASL = average sentence length in words or average number of words in sentence (number of words divided by the number of sentences);*
*** ASW = average syllables per word (the number of syllables divided by the number of words).*

As Gervasi and Ambriola (2003: 208) specify, these as any other readability metric are "*density-like* measures; they do not depend on the size of the text". Furthermore, "since they are based exclusively on a count of the number of words, sentences, syllables, etc., all the metrics are clearly unstructural and order-independent" (Gervasi and Ambriola 2003: 210).

The 'Flesch Reading Ease' (FRE) index was devised by Rudolf Flesch (see, for example, Flesch 1948). The output of the 'Flesch Reading Ease' formula ranges between 0 and 100. "Zero means practically unreadable and 100 means extremely easy. The minimum score for Plain English is 60, or about 20 words per sentence and 1 ½ syllables per word. Conversational English for consumers [advertisements] should score at least 80, or about 15 words per sentence and 1 ½ syllables per word" (Flesch 2002). Flesch (2002) also reports on some of his tests of other materials: *Reader's Digest* magazine, for instance, scored 65, *Time* magazine about 52, and the *Harvard Law Review* about 32. For the present study, the 'Flesch Reading Ease' readability scores concerning the texts of the corpus were obtained automatically from *Microsoft Word* and they gave an average score of 55.76 (minimum value = 34.4; maximum value = 84.3).

The 'Flesch-Kinkaid Grade Level' (FKGL) formula 'translates' the 0-100 score into a U.S. grade level. The following is the conversion of the scores into school grades as specified by Flesch (2002):

FKGL Scores	*School levels*
90 to 100	5th grade
80 to 90	6th grade
70 to 80	7th grade
60 to 70	8th and 9th grade
50 to 60	10th and 12th (high school)
30 to 50	college
0 to 30	college graduate

Figure 3.27. The correspondence between FKGL scores and U.S. school or grade levels.

For the present study, the 'Flesch-Kinkaid Grade Level' for each text of the corpus was obtained automatically from Microsoft Word revealing an average score of 9.6 (minimum value = 4.2; maximum value = 12).

The 'Gunning-Fog Index' (GFI), devised by Robert Gunning (see, for example, Gunning 1952), is a rough measure of the age someone ought to be at to understand the content of a text and takes the form of a number. The lower the number, the more understandable the content. By making specific reference to the readability of academic texts and quoting Ellins and Hopkins (1985: 36), Ventola (1996: 157-158) claims that if the index is

> over 13, then the writer is approaching 'a danger level of difficulty'. [...] The basic principle seems to be that writers ought to avoid sentences which are too long (otherwise the average length percentage goes up) [and they should also avoid] words that have more than two syllables (otherwise the respective percentage goes up).

For the present study, the Gunning-Fog Index was obtained automatically from a website[22] which provides this index as a possible

22 <http://www.ilovejackdaniels.com/resources/readability-score/>, last visited on 25th April 2008.

tool for checking the readability of texts written to be published on the web. In particular, people writing texts for web sites are advised to write content that falls roughly in the 11-15 range for this test, and anything over 22 is roughly equivalent to post-graduate level[23]. The average of the texts of the present corpus for the Gunning Fog Index was 18.48 (minimum value = 9; maximum value = 24).

The values of the three indices have been plotted on the three graphs in Figures 3.28, 3.29 and 3.30. In order to make the comparison of the three outputs easier, the values of FRE are shown in ascending order, whilst those of FKGL and FGI in descending order. As can be seen, the three indices grade the readability of the texts in similar ways, which is also proved by the high correlations that occur between the three metrics, and especially between the FRE and the FKGL indices. Figure 3.31 specifically shows that the negative correlation between the FRE and the FKGL is high (-.900), but also that between the FKGL and the GFI (.872), and the one between the FRE and the GFI (-.866)[24].

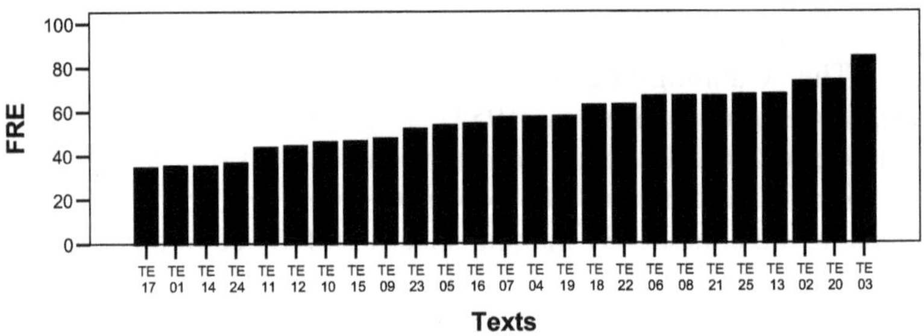

Figure 3.28. Bar graph showing the values of the Flesch Reading Ease index in the texts of the corpus. The values are in ascending order.

23 <http://www.ilovejackdaniels.com/php/gunning-fog-function/>, last visited on 25th April 2008.

24 The Pearson product-moment correlation coefficients were calculated by means of *SPSS* 13.0 for Windows.

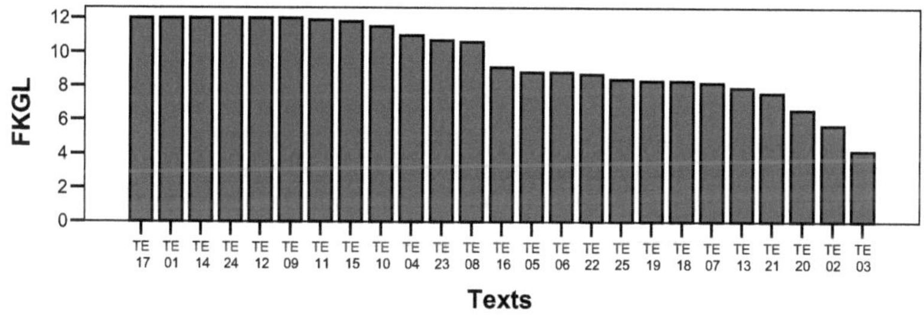

Figure 3.29. Bar graph showing the values of the Flesch-Kinkaid Grade Level in the texts of the corpus. The values are in descending order.

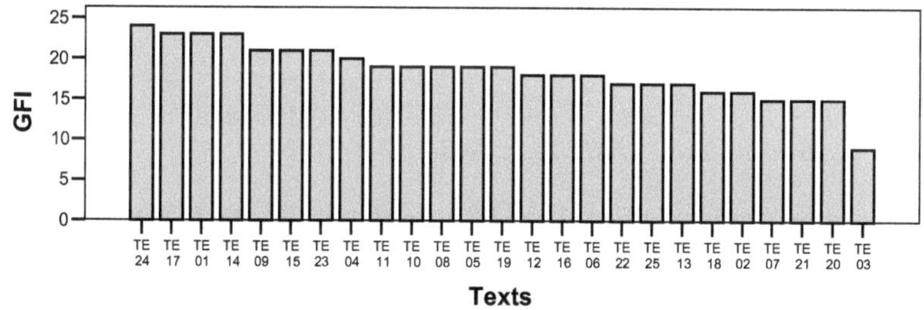

Figure 3.30. Bar graph showing the values of the Gunning-Fog Index in the texts of the corpus. The values of this index are in descending order.

Pairs of variables	*Pearson Product-Moment Correlations*
FRE / FKGL	-.900 (N=25, p < .05)
FRE / GFI	-.866 (N=25, p < .05)
FKGL / GFI	.872 (N=25, p < .05)

Figure 3.31. Correlations between FRE, FKGL, and GFI.

According to the FRE and FKGL figures, the least readable texts in the corpus are text 17, text 1, text 14, and text 24, whilst the most readable ones are text 3, text 20 and text 2. Similarly, according to the GFI, the least readable texts are text 24, text 17 and text 1, and the most readable ones are text 3, text 20, and text 21. In order to investigate the differences, if any, between one of the 'least readable' and one of the 'most readable' texts in the corpus, the following two almost equally long sample passages were chosen from text 17[25] and text 20[26] respectively:

(50) (1i) A report from Mintel, the market research organisation, says (1ii) that despite recession and financial pressures, (1iii) more people than ever want to buy environmentally friendly products (1vi) and a 'green wave' has swept through consumerism, (1v) taking in people previously untouched by environmental concerns. (2i) The recently published report also predicts (2ii) that the process will repeat itself with 'ethical' concerns, (2iii) involving issues such as fair trade with the Third World and the social record of businesses. (3i) Companies will have to be more honest and open in response to this mood. (4i) Mintel's survey, (4ii) based on nearly 1000 consumers, found (4iii) that the proportion who look for green products and are prepared to pay more for them has climbed from 53 per cent in 1990 to around 60 per cent in 1994. (extract from text 17)

Words: 125; syllables = 196; 3+syllables = 13; clauses = 12 (4 clause complexes) ; FRE: 28.2; FKGL: 12; GFI: 24

(51) (1i) I've lived in London all my life (1ii) and have just moved to a larger house with my wife Clare and our two children, Jimmy and Madeline. (2i) We spend a lot of time just singing and dancing around the house. (3i) I grew up with music (3ii) because my dad is the pianist, Chester Harriott – (3iii) who's still playing, by the way. (4i) My working day is divided between television and writing cook books, (4ii) though TV takes most of my time. (5i) I spend about five days a fortnight (5ii) working on the cooking programmes I appear in. (6i) I eat all sorts of things at home (6i) but I only buy quality food. (7i) When I'm cooking, (7ii) I experiment with whatever is in the fridge – (7iii) it's good practice for my TV series. (extract from text 20)

Words: 123; syllables = 166; 3+syllables = 4; clauses = 15 (7 clause complexes) ; FRE: 71.3 ; FKGL: 7.6 ; GFI: 16

The clause complexes in the extracts were analysed and segmented following Halliday's framework (see, for example, Halliday 2004: 363-484, and 3.4 above), and the indices were obtained as explained above. As can be seen, the extract from text 17 contains fewer clauses

25 University of Cambridge Local Examinations Syndicate, Cambridge IELTS 2, Student's Book With Answers (2001). © UCLES, reprinted with permission of the publisher (Cambridge University Press) and UCLES.

26 University of Cambridge ESOL Examinations. Excerpts adapted from 'Sample paper R&W PET.pdf' downloaded from <http://www.candidates.cambridgeesol.org/>, © University of Cambridge ESOL Examinations.

and clause complexes than the extract from text 20: it is made up of 12 clauses and 4 clause complexes, whilst the extract from text 20 contains 15 clauses and 7 clause complexes. This means that each clause in text 17 is bound to be longer and to contain more words than a given one in text 20. Furthermore, the extract from text 17 is made up of 196 syllables and of 13 three-or-more syllable words (e.g. 'environmentally', 'consumerism', 'companies'), whilst the one from text 20 contains fewer syllables (166) and only 4 words with three or more syllables (e.g. 'television', 'experiment'). This means that the words used in text 17 are generally longer and contain more syllables than those in text 20. These facts are all captured by the three formulae, in that text 17 scores remarkably low on the FRE index (28.2), and remarkably high on the FKGL (12) and the GFI (24) indices. By contrast, text 20 scores quite high on the FRE index (71.3) and quite low on the FKGL (7.6) and the GFI (viz. 16) indices. As pointed out in 3.4.1 above, however, one has to proceed with caution when interpreting the values of these indices. As said above, although the presence of long clause complexes and complex words in text 17 might be correlated with the complexity of this text, they need not be considered as the direct cause of the possible lack of readability perceived by some types of readers. As will be seen in 5.8 below, text 17 was in fact deemed difficult. Some of the reasons the candidates added to account for their judgements had to do with the large number of figures and percentages in the text, their lack of knowledge of some infrequent multi-word units, and the particular way the contents are developed in the text. Text 20, by contrast, was perceived as rather easy, and, in particular, as easier than text 21 (see 5.5 below). However, some test takers had the impression that text 20 contains too much 'unconnected' information, which caused them processing difficulty.

According to the three readability formulae, one of the least readable texts in the corpus is text 1[27]. Text 1, in fact, is not a single text, in that it is made up of four different sub-texts, each consisting of one clause and/or clause complex only. Furthermore, while the four options in task 1 are titles (see the instructions), the sub-texts can

27 Test taken from the DIALANG Reading Module.

be thought of as statements, each dealing with different aspects of the same topic, that is, sight and eye problems.

(52) Here are four texts (1, 2, 3, 4) which talk about an eye examination. Choose the best title ('a', 'b', 'c', 'd') for each text.

1. Since eye problems are related to physical health and hereditary factors, problems currently or previously experienced by you or your family will be identified and taken into account. *What is the best title for text 1?*
2. This checks your all round vision and can indicate neurological problems. *What is the best title for text 2?*
3. This determines if your eyes are coordinated and working together properly. *What is the best title for text 3?*
4. Examination of the various parts of the eye can give an overall picture of the condition of your eyes and your general health. *What is the best title for text 4?* (text/s 1)

(a) Interior and exterior of eyes.
(b) Muscle balance analysis.
(c) Visual field screening.
(d) Full history and symptoms. (text/s 1)

The low degree of readability of this peculiar input text is explained mainly by the large number of polysyllabic words it contains, e.g. 'hereditary', 'previously', 'neurological', 'coordinated', 'examination', most of which belong to the Academic Word List (see 3.3.1 above). This consideration, in particular, accords with the findings reported on in 3.3.2 above, where it was stated that text 1 contains the highest proportion of academic words in the corpus. According to approximately 40% of the test takers, the presence of these field-specific words caused them processing difficulty, in that they had to read the texts carefully to understand what they exactly mean. Approximately 25% of the test takers, on the other hand, found that the main problems were due to the fact that the four sub-texts are in fact extracts from longer texts, and this is clearly confirmed by the anaphoric use of the demonstrative pronoun 'this' (see 4.1.3 below) in sub-texts 2 and 3. Specifically, they found it difficult to decide about the appropriate title for texts of which they can only read an extract. In relation to this consideration, some test takers explicitly complained about the fact that the sub-texts deal with different but related topics, which caused them problems while they were trying to assign each sub-text the appropriate title. In order to account for task difficulty, most test takers provided these same arguments. As for the actual performance of the test takers on task 1, it can be claimed that this task turned out to be rather easy (see also 5.4 below). Specifically, 98% (59) of them matched sub-text 1 with title (d), 78% of them (47) managed to match sub-text 2 with title (c), 82% (49) correctly matched sub-text 3 with title (b), and, finally, 83% of the test takers (50) matched sub-text 4 with title (a).

3.6 On the Relationship between the Variables

In order to pull the results of these quantitative analyses together, in this section statistical techniques will be used to explore the relationships between them. Firstly, to compare the results of the calculations with each other, *z-scores* will be made use of. In particular, the z-scores calculated for each linguistic variable and text will be plotted on some graphs for comparison. z-scores will also be plotted for the percentages of academic word types, i.e. WL3typ, and for the percentages of word types which do not belong to any list, i.e. WLnotyp. Secondly, 'factor analysis' will be used to reduce these variables to 'factors' (see, for example, Hatch and Lazaroton 1991: 492-498; Pallant: 152-167). Thirdly, 'multidimensional scaling' will be used to show the differences and/or similarities between the values of the variables as distances (see, for example, Hatch and Lazaroton 1991: 498-502). Since some variables are 'redundant', that is, they involve the calculation of ratios between the same quantities (e.g. the FRE, the FKGL, and the GFI), only one among those in each group of similar variables will be taken into account for the last two analyses. Therefore, only STTR50, LDU, LDH, H/P, GI, WLnotyp, WL3typ, LV(re), and FKGL will be under scrutiny in this last section.

According to Hatch and Lazaroton (1991: 196-201), z-scores are useful in locating values in a distribution, in that they give information about how many standard deviations above or below the mean score any given score or observation might be[28]. In Figures 3.32-3.36 histograms representing the z-scores for each variable and each text are plotted. The straight vertical lines indicate the standard deviation of the normal distribution (-1 and +1), and the mean score equals 0. The values of a given variable can be compared with each other and with those of the other variables by making reference to their distance from the mean score and to the standard deviation.

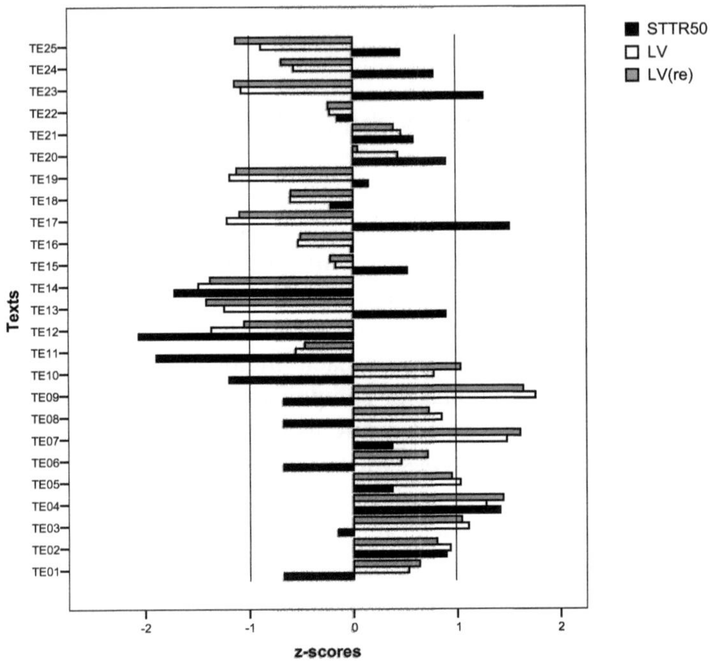

Figure 3.32. z-scores for the variable 'STTR50' (Standardised Type/Token ratio – see 3.1 -), 'LV' (Lexical Variation index – see 3.2.2 -), and 'LV(re)' (Lexical Variation, Revised version – see 3.3.2 -).

28 z-scores are calculated using the following formula: z = (X – M)/SD, where 'X' is the single value, 'M' is the mean score, and 'SD' is the standard deviation.

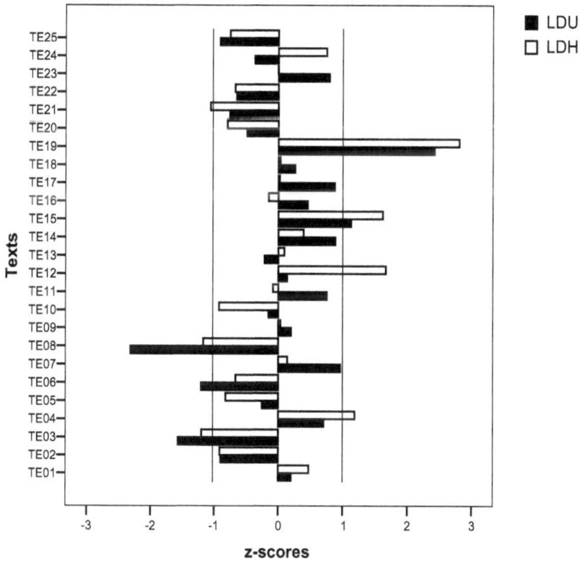

Figure 3.33. z-scores for the variable 'LDU' (Lexical Density, Ure's method – see 3.2.1.2 -) and 'LDH' (Lexical Density, Halliday's method – see 3.4.2 above -).

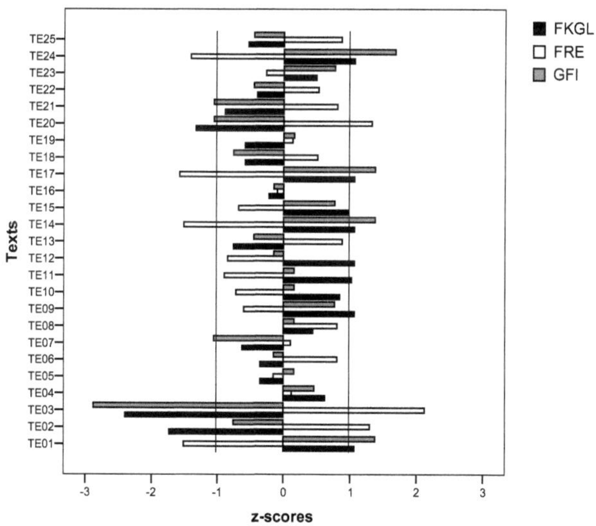

Figure 3.34. z-scores for the variable 'FKGL' (Flesch-Kinkaid Grade Level), 'FRE' (Flesch Reading Ease), and 'GFI' (Gunning-Fog Index). See 3.5.2 above.

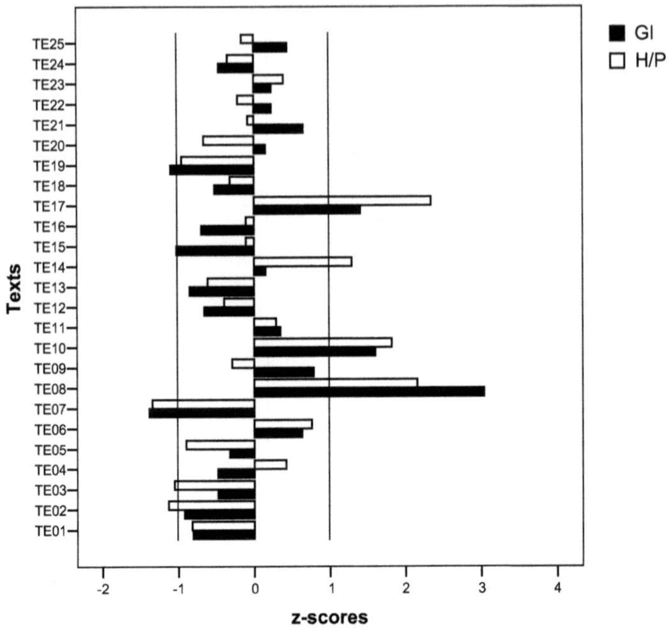

Figure 3.35. z-scores for the variable 'GI' (Grammatical Intricacy) and 'H/P' (the ratio of the total number of secondary clauses in hypotactic structures to that of independent, paratactically related clauses, and primary clauses in hypotactic structures). See 3.4.2 above.

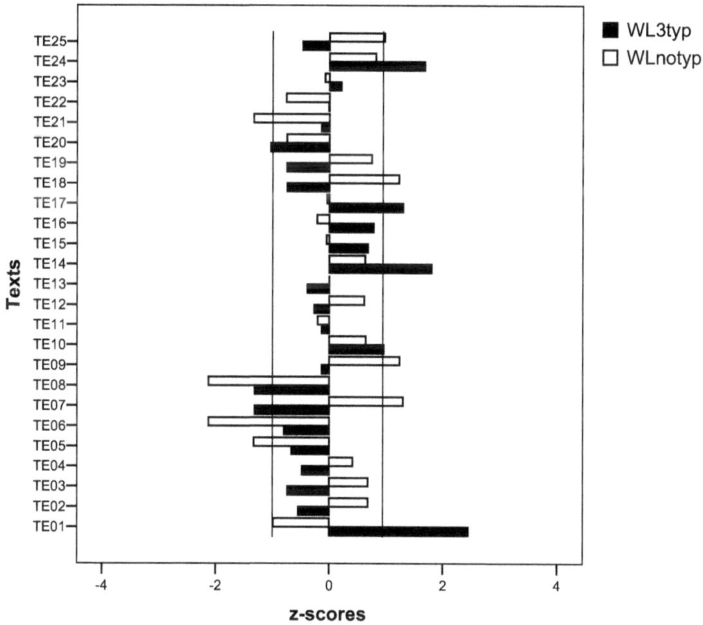

Figure 3.36. z-scores for the variable 'WL3typ' (the percentages of lexical words belonging to the third and fourth base word lists (Fam3P and Fam4P), 'WLnotyp' (the percentages of lexical types that do not belong to any word base list in the texts of the corpus). See 3.3.2 above.

The nine variables STTR50, LDU, LDH, H/P, GI, WLnotyp, WL3typ, LV(re), and FKGL were subjected to principal components analysis (PCA) using the software SPSS, in order to check whether they can be grouped together into factors on the basis of the frequency counts. Prior to performing factor analysis the suitability of the data for this type of analysis was assessed[29]. The analysis revealed the presence of two main factors or components, which are defined

29 An inspection of the correlation matrix revealed the presence of many coefficients of .3 and above, the Kaiser-Meyer-Oklin was .619, and the Barlett's Test of Sphericity reached statistical significance (see Pallant 2001: 165). Principal components analysis revealed the presence of two components with *eigenvalues* exceeding 1, explaining 32.2 and 29 per cent of the variance respectively.

and represented by the 'component matrix' in Figure 3.37 below. It will be argued that the first factor tends to capture those aspects of text complexity that have to do with the amount of content words in the texts and will be called 'Density'. By contrast, the second factor seems to account for the complexity due to the number of words, clauses, and clause complexes in the texts, that is, their length. It will, therefore, be called 'Length'.

Variables	*Factors*	
	(1) Density	*(2) Length*
LDH	.883	
LDU	.880	
GI	-.692	.615
Wlnotyp	.621	
LV(re)	-.430	
FKGL	.340	.848
H/P	-.325	.844
WL3typ		.709
STTR50		-.406
% of variance explained	32.2%	29%

Figure 3.37. Varimax rotation matrix of the two factors for the 9 quantitative variables calculated in this chapter (only loadings above .3 are displayed).

Both components show strong 'loadings', and most variables load substantially on only one component. Specifically, LDU, LDH, and WLnotyp load strongly and positively on component 1, while LV(re) loads negatively but less strongly on it. FKGL, H/P, and WL3typ, on the other hand, load strongly and positively on component 2, while STTR50 loads negatively on component 2. GI is the only exception, in that it loads strongly on both components, but contributes negatively to the first one and positively to the second one. This latter result, however, is consistent with the discussion in 3.4 above about the characteristics of the indices used to measure lexical density and grammatical intricacy and, in particular, about the fact that they usually capture complementary aspects of text complexity.

The two components that emerged from factor analysis are thought of here as two possible groupings of these nine variables. A linguistic interpretation of the reasons why the variables turned out to be statistically related in this way will be attempted on the basis of the underlying aspects of text complexity measured by each set. As for component 1, since both LDU and LDH are respectively the ratio of the number of lexical words to the total number of words and the ratio of the number of lexical words to the total number of ranking clauses, their underlying common feature is clearly the number of lexical words in a text. The LV(re) index, on the other hand, is a measure of the amount of different lexical words in a text, and is negatively correlated with both LDU and LDH. In fact, if a large number of lexical words belonging to the same word families is used in a text, its LV(re) value gets lower, whilst its lexical density increases. Finally, since the WLnotyp variable accounts for the amount of infrequent words in a text, and since these are always lexical words, this variable is positively correlated with both LDH and LDU, and contributes positively to component 1. As anticipated above, on the basis of this argumentation factor 1 was labelled 'Density'.

As far as component 2 is concerned, the variable that stands out as loading strongly on it is FKGL (Flesch-Kinkaid Grade Reading Ease), that is, a measure of readability. As seen in 3.5.2 above, the readability metrics discussed in this study are characterised by the presence of two 'density factors' (see Gervasi and Ambriola 2003: 208), namely 'ASW' and 'ASL'. The former is the ratio of words to sentences in a text, while the latter is the ratio of syllables to words. The underlying 'trait' measured by these formulae, therefore, concerns textual denseness in terms of the number of words per sentence and of syllables per word. Also the variable WL3typ – i.e. the variable that accounts for the number of academic words in a text – loads positively on this component, which can be explained by its high and positive correlation with FKGL. This correlation can be explained by the fact that, as seen in 3.3.1 above, academic words are usually Greco-Latin terms that tend to be long, that is, to contain a large number of syllables (e.g. 'appropriate', 'facilitate', 'investigate'). In 3.5.2 above, in particular, it was shown that text 1 contains a large number of academic words and that it is also one of the least readable texts in the corpus. H/P – i.e. the ratio of the total number of secondary clauses in hypotactic structures to independent, paratactically related clauses, and primary clauses in hypotactic structures – is positively correlated with both GI and FKGL. These correlations account for its strong loading on component 2, and can be explained by the fact that the more secondary clauses in hypotactic structures in a text, the larger the number of words and, therefore, the number of syllables per sentence, irrespectively of whether the words are lexical or grammatical. Finally, the negative loading of STTR50 on component 2 can be explained by the negative correlations of this variable with all the other variables which load on component 2. These observations support the choice of labelling component 2 as 'Length'.

In order to investigate how each text in the corpus differs from the others in terms of the combined effect of these nine variables, 'multidimensional scaling' was made use of. According to Hatch and Lazaroton (1991: 498) "multidimensional scaling is a data reduction/ clustering procedure similar to factor analysis." In particular, with this technique the correlations between the variables are viewed as similarities,

> where high correlations represent greater similarity, and the lower correlations show 'dissimilarity'. The procedure will convert these similarities and dissimilarities to distances in space. (Hatch and Lazaroton 1991: 498)

In Figure 3.38 the 'distances' between the texts are represented in two dimensions. The information about the 'position' of a given text can be gained by reference to two 'macro-variables', namely one represented in terms of distances in dimension 1 and the other in terms of distances in dimension 2. So text 7, for example, is noticeably distanced from text 1 by reference to dimension 2, but it is also noticeably distanced from text 19 by reference to dimension 1. In order to find out which variables mainly contribute to the difference between these texts, in Figure 3.39 the z-scores for each one of the nine variables for texts 7, 1, and 19 are listed. As can be seen, text 7 and text 19 differ mainly in terms of their values of LDU, LV(re) and WLnotyp. Text 7 and text 1, on the other hand, differ mainly in terms of the values of WL3typ, H/P, and FKGL. If this data is compared to the findings of factor analysis reported on above – and in particular to the figures in Figure 3.37 –, it can be seen that the difference between text 7 and text 19 is mainly explained by the variables which load on factor 1, while the difference between text 7 and text 1 is mainly explained by factor 2. Like the two factors identified by factor analysis, dimension 1 could, therefore, also be labelled 'Density', and dimension 2, 'Length'.

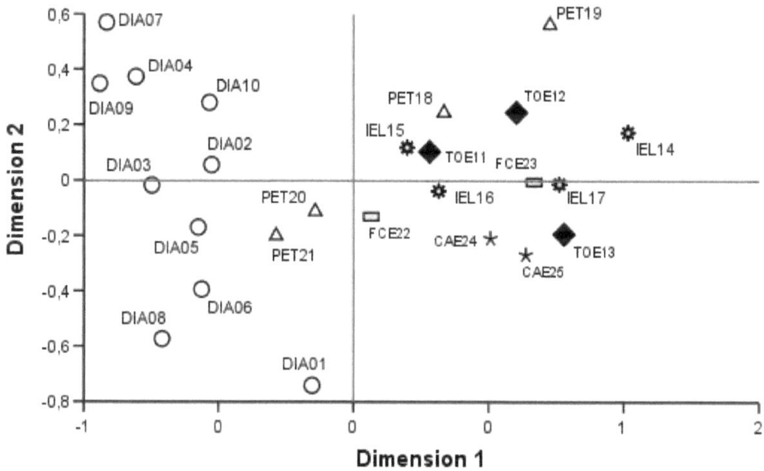

Figure 3.38. Similarities and dissimilarities between the texts of the corpus expressed as distances in two dimensions.

VARIABLES	*text 7 (z-scores)*	*text 1 (z-scores)*	*text 19 (z-scores)*
LDU	.96	.20	2.417
LDH	.14	.47	2.8
GI	-1.38	-.81	-1.09
LV(re)	1.60	.63	-1.11
WLnotyp	1.30	-.98	.747
WL3typ	-1.31	2.44	-.746
FKGL	-.62	1.06	-.578
H/P	-1.34	-.82	-.948
STTR50	.37	-.67	.147

Figure 3.39. z-scores of the nine variables for text 7, text 1, and text 14.

As a way of concluding this section, it can be stressed that each measure used in this chapter captures a specific aspect of the complexity of the texts under enquiry. As shown in 3.4 above, in particular, grammatical intricacy and lexical density capture paired properties of the texts, and can, therefore, be considered as complementary variables. In a sense, however, all the aspects quantified are complementary, and each calculation measure contributes to the definition of the general profile of the text-inherent complexity of the texts. As shown in this paragraph, the two factors and the two dimensions identified by performing factor analysis and multidimensional scaling can be thought of as two macro-variables that define two 'macro-aspects'. It can also be claimed that, on a statistical basis, the measures chosen tend to capture one of two distinct 'macro-aspects' of complexity: namely, either a macro-aspect that has to do with the number of lexical words in the texts, or a macro-aspect which concerns the 'length' of the words or of a higher unit, in terms of either the number of syllables, the number of clauses or the number of clause complexes (sentences) in the text.

4. Text Complexity, Participant Identification and Cohesion in the Corpus

The investigation conducted in this chapter aims to explain how the choice and use of certain cohesive items and structures contribute to the complexity of some tests of the corpus. It also focuses on the bearing that these choices are likely to have on the successful completion of the tasks, that is, on test difficulty. The analyses carried out will mainly concern reference items, but in some cases other relevant cohesive structures and patterns, including thematic structure and lexical cohesion, will be taken into account. The framework for the analysis of reference is mainly based on Martin's (1992) *Participant Identification* system, and the main working hypothesis underlying this part of the study is that, generally speaking, correctly retrieving the identity of the 'participants' in a given input text is fundamental to the successful completion of the accompanying reading comprehension tasks. The term 'participant' will be used here to indicate the 'people', 'entities', 'places', and 'things' that are introduced in a text and are made reference to in the text itself and in the related tasks.

The present chapter is organised as follows: in section 4.1 the 'tracking system' network and the central Participant Identification system network will be introduced and the relevant literature on them reviewed; in order to clarify some relevant features of the linguistic model within which this part of the research project was conducted, in section 4.2 the systemic functional perspective on semantics, lexicogrammar, and the textual metafunction (including cohesion and thematic structure) will be briefly illustrated; in section 4.3 the most salient characteristics of the English nominal group, as well as of the elements functioning within it, will be discussed from a systemic functional perspective; finally, in 4.4 the role played by participant identification and cohesion in some tests of the corpus will be taken into account, and specific arguments will be brought forward to support hypotheses about the difficulty due to specific textual choices. When relevant, the indications that will emerge will be confronted with the data derived from the administration of the tests and of the feedback questionnaires.

4.1 Participant Identification in English: from Discourse Semantics to Lexicogrammar

In paragraph 4.1.1 some contributions to the phenomenon of participant identification and, therefore, relevant to the present study will be reviewed; in paragraph 4.1.2 the concept of 'phoricity' will be dealt with and the tracking system network will be presented; finally, paragraph 4.1.3 will be devoted to the central Participant Identification system network.

4.1.1 Participant Identification in English: a Review of some Relevant Contributions

Martin (1992: 95) defines Participant Identification as "the strategies languages use to get people, places and things into a text and refer to them once there". Martin and Rose (2003: 16) enumerate 'identification' as one of the "five key sets of resources for making meaning as text" along with 'appraisal', 'ideation', 'conjunction', and 'periodicity' and give the following definition:

> identification is concerned with tracking participants: with introducing people and things into a discourse and keeping track of them once there. These are textual resources, concerned with how discourse makes sense to the reader, by keeping track of identities. (Martin and Rose 2003: 17)

As Martin (1992: 95) specifies, his own work has been largely influenced by the work of the Hartford stratificationalists (see, for example, Gutwinski 1976 and Gleason 1973, cited in Martin 1992: 95), who gave a discourse orientation to the study of semantics. In a nutshell, the stratificationalists' proposal

> involved setting up semantic structures for texts referred to as *reticula*. These consisted of events and the connections between them (an *event-line*) and participants and the roles they play in events. [...] Reticula enabled the Hartford stratificationalists to focus on the way in which discourse structure conditioned the realisation of participants, sequences of actions (or *event-clusters* [...]) and role relations in grammar. (Martin 1992: 95)

Martin (1992: 95), however, enumerates some important differences between the framework elaborated by these researchers and his own work on participant identification. Firstly, the fact that reticula concerned three linguistic phenomena that systemic functional linguistics[1] usually describes separately under the labels 'Identification', 'Conjunction' and 'Transitivity'; secondly, the fact that while systemic functional linguistics treats grammar as a meaning-making resource and does not "dualise meaning and form", stratificationalists typically did so (for more information about the semantic and the lexicogrammatical strata from a systemic functional perspective, see 4.2.1 below); thirdly, the fact that "the work of the Hartford school remained at a very programmatic stage as far as semotactics was concerned", that is, they did not place much emphasis on the definition of the Identification system network, and worked mainly on discourse structures; fourthly, stratificationalists were especially concerned with examining the ways in which discourse structures are realised variously across languages, and not just in English (Martin 1992: 96).

Martin (1992: 129) defines a participant as "a person, place or thing, abstract or concrete, capable of functioning as Agent or Medium in transitivity" (see Halliday 2004: 280-305 for a recent introduction to the transitive and ergative models, with special emphasis on their similarities, differences and complementarity). He (Martin 1992: 93-157) then furnishes a description of the English Identification system network as well as of the "structures deriving from phoric options in the Identification system", i.e. 'reference chains' (Martin 1992: 140-153). Martin and Rose (2003: xi) take Martin's 1992 book *English Text* as a point of departure, and extend

1 Both Martin (1992) and Martin and Rose (2003) are mainly concerned with discourse analysis, but their works are carried out within the framework of and informed by systemic functional linguistics.

this work taking into account "developments throughout the 1990s, presenting it for a general audience interested in discourse analysis". The standpoint from which they account for these systems is therefore discourse semantic, and their works are principally concerned with the recoverability of participant identity from the context and, in particular, with the opposition between introducing participants into discourse and presuming their identity. Martin (1992: 97) points out that "the work of identifying participants in English is done mainly through the nominal group", and that "all participants are realised through nominal groups; but not all nominal groups realise participants and some of them realise more than one" (Martin 1992: 129; see also 4.3 below). Semantically, Martin (1992: 98) distinguishes between 'phoric' and 'non-phoric' nominal groups, according to "whether their grammar signals the identity of the participant they realise as recoverable or not". Clearly, this distinction also holds true for all the other reference items, e.g. personal pronouns and possessive determiners (see also Halliday 2004: 554), and "several locative expressions which presume information about time and space: now, then, here, there, hereby, thereby, herewith, therewith" (Martin and Rose 2003: 172-173). Phoric items are, therefore, those items "presuming information from their context", those associated with presuming reference, whereas non-phoric items are associated with presenting reference and signal that the identity of the participant in question cannot be recovered from the context (Martin 1992: 101).

In line with the stratificationalists, Grimes (1975: 35) maintains that different parts of discourse communicate different types of information, which he distinguishes between "event information and identificational and other kinds of information". In accordance with Gleason, in his work he sub-categorises event information into 'events' and 'non-events', since

> sometimes entire paragraphs are devoted to non-events, as in the description of a scene or a person. At other times, […] long stretches of speech may be devoted to nothing but event information, the rules of the reference system being such that the hearer always knows by deduction who is doing what. (Grimes 1975: 36)

The other kind of information to be found in discourse is information "that identifies the participants in an event." This information "not only links participants to events, but also links one mention of a participant with other mentions of the same participant. It obeys rules of its own in addition to combining with event information" (Grimes 1975: 43). Grimes also proposes a distinction between 'participants' and 'props', or rather, he conceives of the two as forming

> a scale of relative involvement in an action, from deliberate involvement expressed by the agent, to being acted upon in the patient and instrument, and from there on down to zero involvement. (Grimes 1975: 44)

Finally, Grimes (1975: 45) points out that in his work he keeps the concept of 'reference' distinct from 'identification', in that "reference has to do with who or what is being talked about, and goes back to the speaker's assumption that the hearer knows who or what is involved." Identification, on the other hand, concerns "the linguistic means that the speaker uses to communicate to the hearer who or what is involved": e.g. naming; anaphoric elements; nouns used as umbrella terms to cover a wide area of more specific nouns, such as 'thing', 'person', 'idea'; pronouns; inflectional reference; zero or implicit identification "to know who the participants are by deducing it from the context" (Grimes 1975: 50). In Grimes' model reference can be made to either individuals or to groups of individuals. As reference sometimes shifts during the course of a text (Grimes 1975: 46), what at first is treated "as reference to individuals may change to reference to groups, or one pattern of grouping may be replaced by another through merging and splitting" (Grimes 1975: 47). Grimes (1975: 46) also lists the three kinds of shifts that, in his view, can be found in discourse: "introduction and deletion, recombination, and scope change." Introduction and deletion involve "expanding and contracting reference by adding or subtracting individuals from a group": Recombination occurs when the original group is split, that is, when "all its members remain in view", although the groups they are assigned to are not the same as the one in the earlier identification. Finally, scope change is like "the effect of a zoom lens on a camera. It changes the area that is under attention. It may combine individuals who were formerly seen as individuals and treat them as a group" or vice versa. In a narration, for example, this can happen for the duration of 'a scene' only, and later the scope can zoom back out to the entire group or back in to the individual (Grimes 1975: 46).

From their psycholinguistic standpoint, Garrod and Sanford (1990: 465) argue that "a richer account of what needs to be represented about text characters and a richer account of how different referential devices interact with these different forms of representation" are needed. In their view, these accounts should rely on two different pieces of information that the reader "may draw on to individuate the characters portrayed in a story: entity-individuating information and role-individuating information." Readers, therefore, need to be able to

> recognize not just the characters in a narrative as individuals but also the roles that these characters are playing in the situations portrayed. Hence, in reading any narrative we seem to assign the characters to roles automatically. (Garrod and Sanford 1990: 465)

According to their investigation into the matter, at least three types of clues can help readers identify roles in texts: "stereotypic situations of which the events described form a part", text titles, which can be more or less explicit in suggesting the roles enacted, and verbs. An example of a verb acting as a clue to interpretation is given by the clause "Keith drove to London". In this clause the verb 'to drive' "should identify a situation establishing a role for the vehicle used", and "assign an immediate referential interpretation for a subsequent definite noun phrase identifying the vehicle, e.g. 'the car' (Garrod and Sanford 1990: 475). They also suggest that "pronouns have a privileged status as anaphoric devices for maintaining reference to discourse individuals, whereas other definite descriptions serve the more general function of identifying discourse roles" (Garrod and Sanford 1990: 478). Moreover, as exemplified by the clause quoted above, definite descriptions may not be associated with a given entity at a given stage in discourse. Due to the different mechanisms supposedly involved in the distinction between entities and roles, Garrod and Sanford (1990: 478) propose a specific "memory focus framework", which includes "two dynamic systems operating in parallel, implicit and explicit focus." Specifically, they propose to associate entities with explicit focus, in that "for an entity to be represented in focus and subsequently be directly accessible by a pronoun, it must at least have been explicitly referred to in the context" (Garrod and Sanford 1990: 479). Implicit focus, on the other hand, is associated with roles, because "role representations […] have their origins in the reader's knowledge of the situations referred to" (Garrod and Sanford 1990: 482).

4.1.2 Phoricity and the Tracking System Network

The fundamental discourse-semantic concept underpinning participant identification is that of 'phoricity', i.e. the semantic relationship occurring between a given presuming item and another element "already present in the verbal or non-verbal context" (Halliday 2004: 89). According to the kind of information that needs to be retrieved, Martin (1992: 99-101) distinguishes between 'reminding', 'relevance' and 'redundancy' phoricity, and points out that "nominal groups may depend on their context with respect to any one, or any combination of the three". Reminding phoricity "signals that the identity of the participant being realised is recoverable" and is glossed by the author himself as 'you know my identity'. Reminding phoricity is the main way to present participants and to refer back to them once they are introduced. Relevance phoricity, on the other hand, "signals that the identity of one or more participants related to the participant being realised is recoverable", or in common parlance, 'you know the identity of related participants'. In Martin's (1992) framework, relevance phoricity partly corresponds to what Halliday (2004: 560-561) and Eggins (2004: 35) call 'comparative reference'. Under this label Martin (1992: 101, 230-234, 159-270) also includes 'conjunction' (the system realised by, for example, 'and', 'because', 'so') and 'continuity' (the system realised by, for example, 'too', 'still', 'either', 'neither', 'as well', 'so'). In Halliday and Hasan's (1976: 320-322) and in Halliday's (2004: 538-549) works, these are resources that belong to the category of 'conjunction', i.e. one of the four ways in which grammatical cohesion is created in English (see also 4.2.2 below). Finally, redundancy phoricity is concerned "not with tracking the identity of participants but with signalling (in the context of nominal groups) that experiential meaning needs to be recovered from the context", that is to say, "you know my experiential content". In English, redundancy phoricity is mainly realised through 'substitution and ellipsis', which concern relationships "in the wording rather than directly in the meaning" (Halliday 2004: 562). Consequently, "there is always some significant difference between the second instance and the first (between presuming item and presumed)" (Halliday 2004: 570), as exemplified by the following fabricated short dialogue: "A: Where is your rain coat? B: I can't find it. A: Take this *one*". The substitution item 'one', in fact, calls up a member of the class 'rain coat', but not the same rain coat, or participant: there is co-classification, but not co-reference (see Hasan 1989: 84-85).

The 'area'[2] within the context that relevant information is to be retrieved from is another parameter according to which phoricity can be sub-categorised. Drawing mainly on Malinowski's work (see, for example, Malinowski 1923), Martin (1992: 121-127) distinguishes between context of situation and context of culture, and defines the former as "relevant information that can be perceived (seen, heard, felt, tasted, smelled), including text", and the latter as "relevant information which cannot be perceived, but which can be assumed because of shared knowledge among interlocutors deriving from their membership in some definable community". Martin's context of culture broadly corresponds to Halliday and Hasan's (1976: 305) 'social context', which they define as:

> a kind of semiotic structure within which meaning takes place; the 'realities' of which it is made up may be of an entirely intangible kind. But equally they may reside in the persons and the objects that figure in the immediate vicinity; and, if so, reference will have to be made to them.

When an "interlocutor's membership in a particular community means that certain participants can be treated as inherently 'given'", the type of reference can be called 'homophora' (Martin 1992: 122). In the extract from text 14 reported below[3], for example, the

2 This expression is used by Hasan (1996: 210).
3 See also 3.2.2 above.

participants 'the University grounds' and 'the University authorities' can be considered as homophoric, in that their intended readers are themselves members of the University community and this is clearly indicated linguistically by the use of the definite article 'the'. In fact, both 'grounds' and 'authorities' are pre-modified by the Classifier 'University', which not only indicates a sub-class of grounds and authorities respectively, but also makes reference to the same individual, that is, the same institution (see also 4.3 below). In this text, therefore, 'University' is not only common knowledge shared by all the intended readers, but is also text cohesive.

(53) USE OF UNIVERSITY GROUNDS BY VEHICULAR TRAFFIC
The University grounds are private.
The University authorities only allow authorised members of the University, visitors and drivers of vehicles servicing the University to enter the grounds. (extract from text 14)

Indeed, in text 14 as we find it in the test no information is provided about which university is at issue. This may have been a choice of the author of task 14, but it is also quite likely that it was not necessary at all, since all the intended readers would have understood the reference homophorically.

(54) University employees do not need to pay for their parking permits.
Parking in Halls of Residence is handled by the Wardens of the Halls.
Having a University permit does not allow staff to park at Halls.
Parking permits cost £20 a year.
Students living in Hall do not need permission to park in Hall car parks. (task 14)

At any rate, this textual choice contributes to the 'homophoricity' in the text and conveys a feeling of membership to the readers.

When participants are not identifiable homophorically, reference to the context of situation – either verbal (co-text) or non-verbal – is to be made. Technically, reference to the co-text is referred to as 'endophora', and reference to the non-verbal context as 'exophora' (Martin 1992: 122). Martin, however, points out that in some respects the realisations of homophora, exophora and endophora overlap, and he distinguishes his view from that of Halliday and Hasan (1976: 71-74), who treat homophoric reference as a type of exophora and generic reference realised by 'the' as a kind of homophora. In his work he prefers to distinguish homophora from exophora/endophora, since in his conception homophora includes the class of proper names. He also prefers not to consider generic reference as a type of homophora, since "generic reference depends on the knowledge of the language and culture as a whole", e.g. 'The/a cat is a feline', and not on the "knowledge of the relevant cultural context", e.g. 'Have you fed the cat?'. In the former case, in particular, definite and indefinite articles function as 'alternative realisations' (Martin 1992: 122-123). An example of exophoric expression is given by:

(55) This checks your all round vision ... (extract from text 1)

where the possessive deictic 'your' is referred to the reader and makes reference to the non-verbal context. By contrast, 'a vehicle' in the following extract from text 14 is an example of generic reference, since this noun group makes reference to a "whole class or category" (see also, for example, Taylor Torsello 2003: 246):

(56) Members of staff who have paid the requisite fee and display the appropriate permit may bring a vehicle into the grounds. (extract from text 14)

It is important to point out that "exophoric reference links the text to its environment, but does not contribute to the cohesion of the text, except indirectly when references to one and the same referent are repeated, forming a chain" (Halliday 2004: 552). According to Hasan (1996: 209), exophoric 'encoding devices' may involve various degrees of implicitness, and when a highly implicit device is chosen, "its precise intended meanings become fully available when the listener has access to the material situational setting."

Exophora, since it involves retrieval of the referent in the context, is context-dependent and, in this sense, involves implicitness and sharing of a context. But the context may not always be completely shared, or there might be more than one possible referent, or it might not always be clear where the referent is to be found in the context or in the co-text. Merlini Barbaresi (2003: 26-27), in her study on language and text, focuses, in particular, on the complexity involved in the use of referential (and other types of) devices used in different contexts, and uses the notion of "functional complexity" to account for the complexity due to a "departure from unique interpretability". In order to illustrate this point this linguist gives the following examples:

> In conversation, the use of expressions such as *look* (imperative) or *I mean* may be either referential (*Look there, isn't it nice?; I mean something different*) or have an extra function as discourse markers (*Look, I don't intend to accept all this; Well, I mean, I'm not sure*). Similarly, *yes*-answers may be referential (A: *Are you sure?* B: *Yes*) or phatic, i.e. functioning as mere *back-channel* signals (A: *Well, I went there* B: *yes* A: and *I met her*). [...] In these cases the addition of the non-referential function increases its complexity in terms of departure from unique interpretability, whereas reduction is in terms of greater economy of language elements in the system.
> Merlini Barbaresi (2003: 26-27)

These examples not only illustrate functional complexity, but also how the other side of the same coin is a decrease in material complexity – in this case in terms of the use of fewer expressions in the system.

Endophoric reference is concerned with presuming information from the preceding co-text, i.e. 'anaphora', or from the following co-text, i.e. 'cataphora'. The former points 'backwards' to "the history of the unfolding text, that is, to a referent that has already been introduced and is thus part of the text's system of meanings". The latter points 'forwards' "to the future of the unfolding text, that is, to a referent that is yet to be introduced" (Halliday 2004: 552), and is less frequently used than anaphora. The pronoun 'this' in the first clause complex of the following paragraph from text 23 exemplifies cataphora, in that it is used to anticipate the content of the subsequent passage:

(57) The system works like <u>this</u>. Airlines insist on exaggerated check-in times (which require passengers to report to the airport at a given time before departure) designed to allow sufficient time for baggage to pass through the airport and be loaded on to the plane. Minimum connecting times (MCTs) are the shortest time it takes to transfer between two flights. These, too, are exaggerated to allow for baggage transfers (extract from text 23)

The expression 'the following' in task 24 is also cataphoric, and is used to introduce a list of statements about text 24:

(58) In which section is the following mentioned? (extract from task 24)

Forward reference within the same nominal group is called 'esphora' (Martin 1992: 123) or 'structural cataphora' (cf. Halliday 2004: 552 and Halliday and Hasan 1976: 72). Esphora is commonly used to introduce participants into a text, and occurs when one of the Deictics 'the', 'that', 'those' are used in a nominal group to indicate that the Qualifier within it is to be considered as defining. An example of esphoric reference is given by the noun group 'the Warden of the Hall concerned' (extracted from text 14), in which the identity of 'the Warden', a new participant, is defined by 'of the Hall concerned'. Another example is given by the following extract from text 2:

(59) He waved his left hand impatiently to silence the loquacious Proprietor who had begun the parrot formula he had repeated at least fifteen times so far. 'I have the honour, sir, to introduce ...

In actual fact, in this extract an interplay between esphora and cataphora seems to be at issue: not only is the noun phrase 'the parrot formula' qualified by the embedded clause 'he had repeated at least fifteen times so far', but it also points forward to the actual quotation – direct speech – in the following clause complex.

According to Botley and McEnery (1999: 3), the phenomenon of anaphora

> has received a great deal of attention from linguists, cognitive psychologists, philosophers and computer scientists. Firstly, in linguistics, anaphoric phenomena are of interest because they tell us about how discourse is constructed and maintained – how linguistic patterning above and beyond the sentence is arranged. Secondly, at the sentence level, anaphoric features function to bind structural elements together, and therefore can play an important role in the syntactic description of languages. [...] Thirdly, as well as being of interest to linguists, anaphora has been of great interest to computational linguists, because of the immense challenges that anaphora presents to natural language processing. [...] However, for now, a computer system that can resolve all anaphoras is effectively some way off. Finally, cognitive psychology, and those branches of linguistics that share many of the aims and assumptions of cognitive psychology, are also interested in anaphora because it tells us some things about how language is understood, and processed. In particular, such workers are interested in how anaphoric

> phenomena function to recall concepts in discourse to the consciousness of the hearer or reader, to facilitate understanding.

Givón (1995: 64-75), for example, accounts for both the psycho-cognitive and the linguistic aspects of anaphora and cataphora. In the model he proposes, text comprehension is equivalent to the construction of a structured mental representation of the text, and text is represented, at least partly, as a network of connected 'nodes' (chunks of information) in episodic memory (vs. working memory buffer). The more connections a node has with other adjacent nodes, i.e. the more grounded it is, the more mentally accessible it is. In the mentally represented text, the grounding of information nodes involves two main directions: anaphoric (backward) and cataphoric (anticipatory) grounding. The former involves "connecting incoming new information chunks to some existing mental representation – either of the text or of other mental entities". The latter, by contrast, "involves the opening of pending connections in yet-to-be completed structure, in anticipation of text that is in the process of being constructed" (Givón 1995: 64-65).

As pointed out by Martin (1992: 124), phoric nominal groups may "presume information that is implied rather than directly retrievable". Making use of the term introduced by Clark and Haviland (1977: 6), Martin refers to indirect reference of this kind as 'bridging'. Similarly, Eggins (2004: 36) uses the term 'bridging reference', "when a presuming reference item refers back to an early item from which it can be inferentially derived," and Martin and Rose (2003: 160) use the expression 'inferred anaphoric expressions' to denote the same phenomenon. Clark and Haviland (1977: 5-6), in particular, define 'bridging' as one of the procedures that readers (or listeners) can deploy to make up for their lack of knowledge of the "only direct antecedent for any given information" in a given text. Specifically, they suppose that

> when the listener cannot find a direct antecedent, most commonly he will be able to form an indirect antecedent by building an inferential bridge from something he already knows. (Clark and Haviland 1977: 6)

Following Martin (1992: 124), in this study bridging will be conceived of as dependent on "experiential connections between presuming and presumed which facilitate the recovery of an implied identity". In practical terms, since the lexical relations of hyponymy, meronymy, and synonymy[4] (with identity of reference), in particular, involve implicational relations between phoric items, in this study they will be analysed as sources of 'bridged' reference or bridging. Besides these ones, there are other experiential connections on which bridging might depend, such as those occurring between the items 'the seeds' and 'sowing the fields', 'the weight' and 'carrying a heavy load', and 'the bus' and 'set off to work' (Martin 1992: 125). The following extract from text 11 contains an example of bridging:

(60) The next artist in this survey of American artists is James Whistler; he is included in this survey of American artists because he was born in the United States, although the majority of his artwork was completed in Europe. Whistler was born in Massachusetts in 1834, but nine years later his father moved the family to St. Petersburg, Russia, to work on the construction of a railroad.

In this passage there is no previous mention of 'the family', yet the existence of the family can be presumed via bridging from the occurrence of 'his father' and by the fact that everybody is likely to have a family.

When the information presumed by a given nominal group is not directly retrievable from either the verbal or the non-verbal context, interlocutors/readers can process the phoric item as a 'first mention', that is, as having been mentioned for the first time. In this respect, Clark and Haviland (1977: 7) maintain that

> sometimes it is impossible to find any way of bridging the gap between known information and the appropriate antecedent. Then the listener must add to memory, perhaps hypothetically, a new node (a nominal associated with one or more propositions) to serve as the antecedent to the given information. This often occurs, for example, at the beginning of stories.

Again following Clark and Haviland (1977), Martin (1992: 125) calls this strategy 'addition'. An example of addition is contained in the second clause of text 2[5]:

---

4 For a definition of these relations see 4.2.2.1 below.

5 Test taken from the DIALANG Reading Module.

(61) Our hands met. I looked him straight in the face. The smile slowly creased up into lines of thought. He waved his left hand impatiently to silence the loquacious Proprietor who had begun the parrot formula he had repeated at least fifteen times so far. 'I have the honour, sir, to introduce ...'
'That's right,' said the Minister not to anyone in particular, but to some mechanism of memory inside his head. 'You are ... no, not Nwege, ... it's Odili, isn't it?'
'Yes, sir'
'Odili, the great,' said the Minister boyishly. 'Where have you been all this time?' (text 2)

As will be seen in 4.4.2.1, in text 2 the participants realised by 'I' and 'him' are not presented, and need, therefore, to be processed as a first mention. According to Clark and Haviland (1977: 8), both bridging and addition involve 'detours' that could cause processing difficulty. As they put it, "for each detour the listener must do something extra, and this ought to lead to extra processing time or to a feeling that the sentence was difficult to understand" (Clark and Haviland 1977: 8). As will be discussed in 4.4.2.1 below, the feedback collected from the students on the difficulty of text 2 bears out Clark and Haviland's claim, since they admitted the difficulty they experienced whilst trying to retrieve the identity of the participants and, in particular, the identity of the writer.

Another possible situation envisaged by Martin (1992: 125) which might be a source of text difficulty is likely to occur when more than one referent is available in the context for a given item. In this case 'ambiguity' may arise and the reader/speaker has to make decisions in order to successfully retrieve the referent. While there seems to be no case of this type of ambiguity in the corpus, test 17[6] contains two nominal groups the exact identities of which are not clear until the end of the text, namely 'green consumers' and 'green products'. It is suggested here, in particular, that in order to successfully process both items, the readers should apply both the strategy of bridging and that of addition throughout the unfolding text, which might cause processing difficulty. As for 'green consumers', this

6 University of Cambridge Local Examinations Syndicate, Cambridge IELTS 2 Student's Book With Answers. 2001, © UCLES, reprinted with permission of the publisher (Cambridge University Press) and UCLES.

‘generic’[7] participant is introduced in the first paragraph of text 17 as follows:

(62) Research in Britain has shown that 'green consumers' continue to flourish as a significant group amongst shoppers.

As can be seen, the Classifier ‘green’ indicates the sub-class of consumers characterised by the fact of being environmentally friendly. At this stage, the reader is likely to interpret this type of consumer as people who buy ‘environmentally friendly products’ by applying the strategies of addition and bridging, that is, by building an inferential bridge from what they know is usually meant by ‘green’. Actually, this interpretation is confirmed in the second paragraph:

(63) A report from Mintel, the market research organisation, says that despite recession and financial pressures, more people than ever want to buy environmentally friendly products and a 'green wave' has swept through consumerism, taking in people previously untouched by environmental concerns.

In the fourth paragraph, the generic participant ‘green products’ is introduced, but it is not stated exactly what these products are, or at least, what distinguishes them from other products, and again the strategies of bridging and/or addition ought to be applied to process this nominal group:

(64) Mintel’s survey, based on nearly 1,000 consumers, found that the proportion of those who look for green products and are prepared to pay more for them has climbed from 53 per cent in 1990 to around 60 per cent in 1994. On average, they will pay 13 per cent more for such products, although this percentage is higher among women, managerial and professional groups and those aged 35 to 44.

After a paragraph containing information about the research carried out by Mintel and some statistics, the participant ‘green products’ occurs again in the fifth paragraph as part of the definition of a particular sub-type of consumer, i.e. ‘pale green’ as follows:

---

7 For the definition of ‘generic participant’ see 4.1.3 and 4.3 below.

(65) Mintel's 1994 survey found that 13 per cent of consumers are 'very dark green', nearly always buying environmentally friendly products, 28 per cent are 'dark green', trying 'as far as possible' to buy such products, and 21 per cent are 'pale green' – tending to buy green products if they see them.

In the sixth paragraph the generic participant 'ethical spenders' is introduced, but this time an exact explanation of what is meant by ethical spenders in the text is given in the subsequent clause:

(66) Four in ten people are 'ethical spenders', buying goods which do not, for example, involve dealings with oppressive regimes.

In the remaining part of the sixth paragraph, the participant 'armchair ethicals' is introduced and some statistics are given. Finally, in the last paragraph an exact definition of what, according to the market research organisation, 'green consumers' do not conceive as 'green products' is given, which also clarifies who 'green consumers' are:

(67) Among green consumers, animal testing is the top issue – 48 per cent said they would be deterred from buying a product if it had been tested on animals followed by concerns regarding irresponsible selling, the ozone layer, river and sea pollution, forest destruction, recycling and factory farming.

On the basis of the results of the feedback questionnaire, it can be said that some students perceived this textual feature as contributing to their feeling that some information was missing in the text (see 5.8 below).

The resources available for tracking participants described in this paragraph are organised systemically in Figure 4.1:

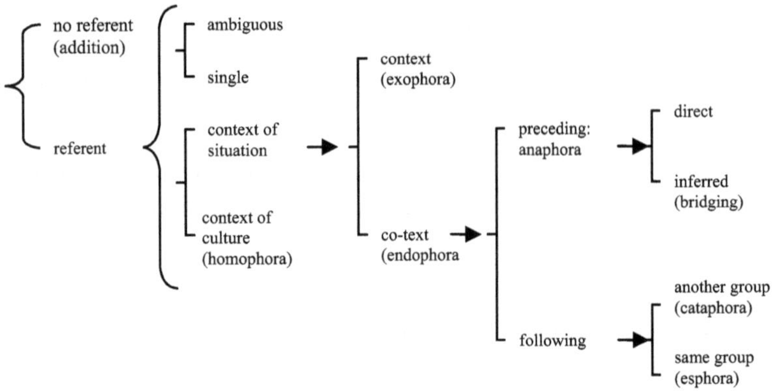

Figure 4.1. The tracking system network (adapted from Martin 1992: 126 and Martin and Rose 2003: 172).

4.1.3 The Participant Identification System Network

Before discussing the ways the semantics of participant identification is realised lexicogrammatically in the nominal group (see 4.3 below), a quick overview of the central Participant Identification system network, adapted from Martin (1992: 102) and Martin and Rose (2003: 171), will be given here. As can be seen from Figure 4.2, the central Participant Identification system network can be thought of as consisting of the following three least delicate simultaneous systems:

- presenting/presuming;
- comparison/-;
- generic/specific.

The [presenting/presuming] system involves a choice between presenting reference and presuming reference. The former occurs when the identity of a given participant is presented, whilst the latter when it is recoverable from the context. An example of a non-phoric item associated with presenting reference is the noun group 'a new yardstick' occurring in the following extract from text 24:

(68) We are being judged by a new yardstick – not just by how clever we are, or by our training or expertise, but also by how well we handle ourselves and each other.

The same participant is presumed in the subsequent clause by the phoric item 'this yardstick', which exemplifies presuming reference:

(69) This yardstick is increasingly used in choosing who will be hired and who will not …

In particular, the combination of demonstrative reference and lexical repetition should make the reference chain particularly easy to follow. In this text, however, the lexical item 'yardstick' created problems, as many students did not understand the meaning of the word (see 5.8 below).

The [comparison/-] system captures the paradigmatic relation between reference to the identity of related participants [comparison] or reference to the same participant [-]. When comparison occurs, comparative expressions presume "some standard of reference in the preceding text" (Halliday 2004: 560), and this may be on the basis of similarity (also called semblance) or difference (Martin 1992: 116-117). In the following extract from text 17, the phoric item 'such products' is an example of similarity of reference:

(70) Mintel's survey, based on nearly 1,000 consumers, found that the proportion of those who look for green products and are prepared to pay more for them has climbed from 53 per cent in 1990 to around 60 per cent in 1994. On average, they will pay 13 per cent more for such products, although this percentage is higher among women, managerial and professional groups and those aged 35 to 44.

Comparative difference, on the other hand, is exemplified by 'other animals' in this passage extracted from text 13:

(71) Some scientists call this movement emigration, because these animals never return to their homes. For other animals, such as birds, the migration includes a return trip home.

Finally, the [generic/specific] system indicates that a choice needs to be made between generic reference and specific reference (Martin

1992: 103). The former is selected when “the whole of some experiential class of participants is at stake”, e.g. ‘birds’, whereas the latter refers to “a specific manifestation of that class”, e.g. ‘seven swallows’ in the following extract from text 13:

(72) Birds move in periodic migrations, or at regular times during their lives, and often to the same place year after year. Many experiments have been made with birds in an attempt to find out what guides them on their way home. In one case, seven swallows were taken 400 miles from home.

Further and more delicate systems within this system network account for more delicate semantic options that a speaker/writer can choose from once her/his primary choices are made. In particular, the system network provides various resources that make it possible to presume the identity of participants, and these can be divided into ‘pronominal’ and ‘nominal’. Pronominal reference is in turn divided into first and second person (speaker and addressee roles) and third person (other persons). Presuming nominal reference, on the other hand, involves a choice between proper names (e.g. ‘Suzanne Somers’ in text 9) and determined nouns, which in turn can be either preceded by the definite article (e.g. ‘the mix’ in text 3) or by demonstratives (e.g. ‘this course’ in text 3).

(73) “I do think this course should be fun,” he said. “And people should get out of it what they want. Some come here wanting to be serious and even a bit competitive about their work. And that’s fine. Others are very hesitant and find that a bit of humour can relax them. ... I’ve taught high court judges and dustmen and the mix is part of the fun.” (text 3[8])

8 Text taken from the DIALANG Reading Module.

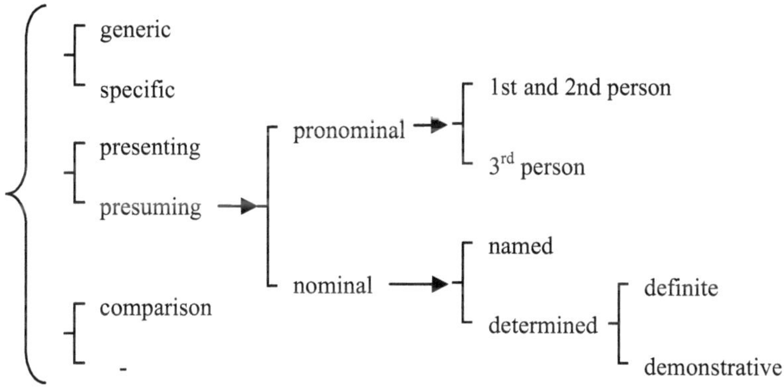

Figure 4.2. The central Participant Identification system network (adapted from Martin 1992:105 and Martin and Rose 2003: 171).

As maintained by Martin and Rose (2003: 151-152), in English there are indefinite expressions that seem to be realising presenting reference, even though they are not, in that it is already clear who or what they refer to. The following extracts from three texts of the corpus exemplify these 'apparent anomalies':

(74) 'Try as she might, she is not particularly convincing here as an ace journalist who sets out to uncover the mystery behind the death of seven people in a night-club.' (from text 9)

(75) 'The *Northern Star* is a very special ship and our past voyages have been very popular.' (from text 18)

(76) 'Juliette now works as a Personal Assistant to Brenda Cleverdon, the Chief Executive of Business in the Community.' (from text 22)

As can be seen, the expressions 'an ace journalist', 'a very special ship' and 'a personal assistant to Brenda Cleverdon' do not have the function of introducing new participants into the texts or of identifying them; they describe them rather, as in the case of 'a very special ship', or assign roles to them, as in the case of 'an ace journalist' and 'a personal assistant'.

An important resource for identifying participants which is not captured by the above system network has been dealt with by Martin

(1992: 131-132), Martin and Rose (2003: 149-151, 165), and Eggins (2004: 36), and will be referred to in this study as the category of 'possessive reference'. Possessive reference consists in the use of possessive adjectives such as 'my', 'your', 'his', 'her', 'its', 'our', 'their', or the respective possessive pronouns 'mine', 'yours', 'his', 'her', 'its', 'ours', 'theirs', or nouns used in Deictic position with the possessive marker *'s* or *'* to identify participants. Possessive reference is exemplified by the following extract from text 11:

(77) Whistler worked in various art forms, including etchings and lithographs. However, he is most famous for <u>his</u> <u>paintings</u>, particularly Arrangement in Gray and Black No. 1: …

As can be seen, the expression 'his paintings' realises two identities. The first one is lexicogrammatically realised by the possessive adjective 'his', which presumes and can be tracked back to 'Whistler', and the other one by the possessed Thing, that is, 'paintings', which was not previously mentioned, despite the fact that it is semantically related to 'various art forms'.

Two other relevant types of phoric reference that will be distinguished in this study are called 'text reference' and 'locational reference'. With the category of text reference the referent "may be a sequence of actions or events mentioned previously, […] or 'the previous text up to this point'" (Eggins 2004: 36). Martin and Rose (2003: 169) point out that "with text reference, tracking tends to be fairly local, with just one or two references back to what's been said," as in the following extract from text 17:

(78) Research in Britain has shown that 'green consumers' continue to flourish as a significant group amongst shoppers. <u>This</u> suggests that politicians who claim environmentalism is yesterday's issue may be seriously misjudging the public mood.

In this particular text the demonstrative pronoun 'this' refers to what the writer has argued in the previous clause complex, and, as Martin and Rose (2003: 154) maintain, "the advantage of this kind of tracking is that stretches of meaning can be packaged up to play a new role as the argument unfolds." Besides demonstrative pronouns, another type of referential devices that are made use of to refer back to a previous portion of discourse are nouns preceded by a determiner. In particular, according to Halliday and Hasan (1976: 274-281), 'general nouns' such as 'thing', 'object', 'place', or more specific nouns such as 'business', 'affair', 'move', 'process', 'initiative', 'measure', 'change' can be used to this end. Francis (1986) uses the label 'anaphoric noun' (A-noun) to refer to the specific category of nouns that not only summarise previously presented information, but also present "the given information for the next argument" (Morley 1998: 142; see also Partington 1998: 96-101). Specifically, in Francis' model anaphoric nouns can be either 'metastatements', that is, "statements used to comment on the ongoing discourse", such as 'framework', 'argument', 'point' and 'statement', or, according to Francis (1994), nouns that are not metastatements, e.g. 'problem' or 'issue'. An example of a general noun preceded by a determiner and used to make reference to a previous stretch of text is given by the noun 'position' preceded by the noun 'Evans' used in Deictic position in text 25 (see also 4.4.2.2 below):

(79) The first months of the year were not kind to Nicholas Evans, screenwriter, producer and aspiring director. The year began badly when Life and Limb, a film project he had been working on for months, fell through 'almost overnight'. His disappointment mingled with stomach-churning worry: it had been two years since he had earned any money and the promise of that film had been the only buffer between him and an increasingly irate bank manager. A wise man, finding himself in Evans' position, would have got a job.

Another example of an anaphoric and text cohesive noun phrase is given by the general noun 'problem' preceded by the demonstrative 'this' used in the following extract from the BNC[9]:

(80) There is much concern in the US about the addiction to 'crack' (purified cocaine) within inner-city communities. Doctors are worried that women turning to prostitution to finance the habit are at risk of contracting Aids. This problem has spread to the UK. (BNC)

As for the degree of processing difficulty (or rather processing ease) involved in the use of general nouns used as text cohesive resources, Bruti (2003: 293) writes that:

---

9 British National Corpus.

> if reference is made to a previous portion of discourse by using such lexical items a fact is clearly identified for the benefit of the reader, because despite the fact that they are widely applicable, these terms are semantically descriptive within the textual worlds in which they occur. Consequently, they can often unambiguously identify their discourse antecedents, thanks to the interplay of all the elements in the nominal group.

Similarly, Merlini Barbaresi (2003: 27) claims that:

> When the anaphoric function is added to the referential function of a nominal, complexity arises, but generally this is accompanied by a quantitative simplification in the coding material and a semantic restriction within the range of possible meanings. In the same text, for example, an entity may be first referred to as a *£9bn plan to build nuclear stations* and later co-referred to as *the plan for new stations* or simply *the plan* (The Guardian 27-2-02). Material simplification is jointly triggered by a strong tendency towards economy of coding efforts and the necessity of leaving as much room as possible for new information. This may also be in terms of new semantic elements added locally to the nominal phrase.

Finally, locational reference involves "the identification of a location in time or space", and in written texts "locational referents such as 'here', 'there', 'then', 'above', 'below' are usually retrieved endophorically from surrounding text" (Eggins 2004: 36). Martin (1992: 153-155) and Martin and Rose (2003: 173) specify that the demonstrative adverbs 'here', 'now', 'there', 'then', 'hereby', 'thereby', 'herewith', 'therewith' functioning as circumstances of location in clause structure or as Qualifiers in nominal groups (e.g. 'that person *there*') involve reminding phoricity, in that they contribute to the recovery of the identity of a given participant by presuming location in space or time (Martin 1992: 153). An example of a demonstrative adverb used as a locational referent for reminding phoricity is given by the following extract from text 7:

(81) I do think <u>this course</u> should be fun. [...] Some come <u>here</u> wanting to be serious ...

In this extract 'here' is endophoric and refers back to 'this course'.

Martin (1992: 19) pursues Halliday and Hasan's (1976) definition of text as a semantic unit, and adds that "the discourse semantics both generalises across grammatical resources and accounts for relations between as well as within clause complexes. The discourse semantics is thus more abstract than, and deals with larger units than, lexicogrammar" (Martin 1992: 19). In particular, Martin (1992: 23) affirms that the structures that result from the choices made within the Identification system network, i.e. 'reference chains'[10], are of a non-grammatical kind, and are discourse semantic (and textual) in nature. In his view, reference chains, as any other discourse structure, are closely related to 'univariate' grammatical structures, and share, therefore, some characteristics with, for instance, hypotactic structures within the clause complex (see 3.4.1.1 above). The main distinguishing feature of Identification structures is that a presuming item, which by definition depends on a presenting item, is "typically itself depended on as the text unfolds, dependents becoming Heads in their turn" (Martin 1992: 24). Martin (1992: 25), in particular, chooses to refer to these structures as 'covariate', and calls 'phoric' those covariate structures in which one item presumes another, and 'expectant' those that are not so presumed.

4.2 Semantics, Lexicogrammar and the Textual Metafunction: a Systemic Functional Perspective

This section will be devoted to a brief description of some aspects of the theoretical framework, i.e. systemic functional linguistics, on which the present study and, in particular this chapter, are based. Paragraph 4.2.1 will focus, in particular, on the most salient characteristics of the 'semantic' and the 'lexicogrammatical' strata and on the relationships occurring between the two. It will also provide definitions of the concepts of 'system', 'system network', and 'choice'. Paragraph 4.2.2, on the other hand, will be devoted to the structural and non-structural textual resources that combine to make up the 'textual metafunction'.

---

10 Hasan (1989: 84) calls these chains 'identity chains'. See also 4.2.2.1 below.

4.2.1 *The Semantic and the Lexicogrammatical Strata in Systemic Functional Linguistics*

As said above, in his work Martin (1992) opts primarily for a perspective on discourse-semantics. This means that he focuses on the meanings made by language users and, at the same time, on how they are realised lexicogrammatically. In Systemic Functional Linguistics, 'semantics' (the meaning) is thought of as one of the three strata that constitute the linguistic system, the other two being 'lexicogrammar' (the 'wording', i.e. syntax, morphology and lexis) and 'phonology' (the 'sound') (Halliday 1978: 128-133)[11]. As far as the semantic system is concerned, Halliday (1978: 128) assumes that it consists of four functional components or 'metafunctions', i.e. the 'experiential', the 'logical', the 'interpersonal' and the 'textual' component (see 4.2.2 below), and that each stratum and component can be described as a network of options, that is, as sets of interrelated choices. Halliday (1978: 128) also advocates that "it is the function of the lexicogrammatical stratum to map the structures one onto another so as to form a single integrated structure that represents all components simultaneously". Finally, he (Halliday 1978: 129) assumes that the lexicogrammatical system is organised by rank, that each rank is "the locus of structural configurations", and that each structural composite or combination of structures, e.g. clause, verbal group, nominal group, "derives from one or other component of the semantics". Halliday (2004: 531) reiterates this concept and hypothesises that "lexicogrammar stands in a natural relation to semantics, and not in an arbitrary or conventional one" and that "logogenetic patterns that emerge at the level of lexicogrammar are also at the same time semantically significant." In fact, although Halliday conceives of grammar as "not extending its compositional organisation beyond the rank of clause, the resources of lexico-grammar make two fundamental contributions beyond this upper

---

11 Besides these strata, Halliday (2004: 20-31) includes the stratum of 'phonetics', and distinguishes between 'phonology' and 'phonetics' by saying that the former concerns "the organisation of speech sound into formal structures and systems" (internal organisation), whilst the latter accounts for "the interfacing with the body's resources for speech and for hearing" (Halliday 2004: 25).

limit of grammatical units: (i) the creation of 'logogenetic' patterns and (ii) the marking of cohesion" (Halliday 2004: 586). The patterns developed in this way are, therefore, "patterns of meaning, not patterns of wording", and are to be considered as "patterns at the level of semantics rather than at the level of lexicogrammar". Furthermore, a text does not consist of clauses (clause complexes), but rather "a text is realised by clauses (clause complexes), the two being located on different strata – semantics (the stratum of meaning) and lexicogrammar (the stratum of wording) respectively" (Halliday 2004: 587).

As far as the lexicogrammatical stratum is concerned, it is worth expanding on the fact that systemic functional linguistics views grammar and lexis not as two different strata, but rather as "the two poles of a single continuum, properly called lexicogrammar" (Halliday 2004: 24). Halliday (1976: 55) specifies that 'grammar' is "that level of linguistic form at which operate closed systems", and 'system' is by definition closed. Specifically, in his work 'system' is a set of terms with the following characteristics: the number of terms captured by a given system is finite and the items can be listed; each item is exclusive of the others, that is, a given term cannot be identical to others; and if a new term is added to the system, this changes the meaning of all the others. Any linguistic form that is not captured by closed systems belongs to the level of 'lexis'. As said above, the 'closed systems' of a language, i.e. its grammar, and the 'open set' patterns in language, i.e. its lexis, form the 'lexico-grammar cline' (Halliday 2004: 43). On this continuum there are also patterns "which lie somewhere between structures and collocations, having some properties of both", such as '*take + pride/pleasure/ delight + in + ... -ing*'. Patterns of this kind can be found by the use of standard concordancing techniques and have to be dealt with as "moderately delicate choices in the grammar" (Halliday 2004: 45; see also 3.2.1.1 above). In relation to this, Halliday (1976: 69) makes the following incisive remark:

> The grammarian's dream is (and must be, such is the nature of grammar) of constant territorial expansion. He would like to turn the whole of linguistic form into grammar, hoping to show that lexis can be defined as the 'most delicate grammar'. (see also Hasan 1987)

Central to the systemic functional linguistic model are also the concepts of 'choice' and that of 'system network'. Halliday (2004: 23) stresses their importance in the theory as follows:

> a text is the product of the ongoing selection in a very large network of systems – a system network. Systemic theory gets its name from the fact that the grammar of a language is represented in the form of system networks, not as an inventory of structures. Of course, structure is an essential part of the description; but it is interpreted as the outward form taken by systemic choices, not as the defining characteristic of language. A language is a resource for making meaning, and meaning resides in systemic patterns of choice.

4.2.2 The Textual Metafunction: Cohesion and Thematic Structure in English

In systemic functional linguistics the context of situation, that is, the immediate environment in which a text operates, is schematically described in terms of 'field', 'tenor', and 'mode'. The field refers to what is happening, to the nature of the social action that is taking place, and what the participants are engaged in; the tenor refers to who is taking part in the event, to the nature of the participants, their statuses and roles; the mode refers to what part the language is playing, and to the specific organisation of the text in terms of its function, channel, rhetorical mode, cohesion, coherence, etc. (Halliday 1985: 12). These contextual variables are recoverable from the text itself thanks to language, since language is functionally organised in such a way as to simultaneously construct reality (i.e. the context of situation) along three basic metafunctions: the 'ideational' (subdivided into 'experiential' and 'logical'), the 'interpersonal', and the 'textual'. In systemic functional linguistics, therefore, every text and every sentence in a text is regarded as multifunctional (Halliday 1985: 23). Specifically, the experiential metafunction has to do with the fact that any text involves some doings or happenings and some participants; the ideational metafunction with the fact that doings and happenings appear in a logical sequence; the interpersonal metafunction concerns the text producer's beliefs as to the role of the text in the interaction and his/her relation with the interlocutor; finally, the textual metafunction concerns the organisation of the text and "can be regarded as an enabling or facilitating function, since both the others – construing experience and enacting interpersonal relations – depend on being able to build up sequences of discourse, organising the discoursive flow and creating cohesion and continuity as it moves along" (Halliday 2004: 30).

The textual metafunction, in particular, concerns the text as a message and its internal organisation, both at the clause level and beyond. The textual features that combine to make up the textual resources of the lexicogrammar of English can be subdivided into cohesive and structural ones (Halliday 2004: 579). Cohesive resources are non-structural in nature (Hasan 1989: 81-82; Halliday 2004: 550) and are usually dealt with within the framework of 'cohesion'. They are used "for managing the flow of discourse: for creating semantic links across sentences – or rather, semantic links which work equally well either within or across sentences" (Halliday 2004: 87). Structural resources have to do with units up to the rank of the clause complex, but they nevertheless contribute to the creation of discourse: the two most important such structural resources are 'thematic structure' and 'information structure' (Halliday and Hasan 1976: 325-326; Halliday 2004: 579). Cohesive resources will be illustrated briefly in 4.2.2.1 and thematic structure in 4.2.2.2 below.

4.2.2.1 Cohesion in English

Halliday and Hasan (1976: 4) define the concept of 'cohesion' as:

> a semantic one; it refers to relations of meaning that exist within the text and that define it as a text. Cohesion occurs where the interpretation of some element in the discourse is dependent on that of another. The one presupposes the other, in the sense that it cannot be effectively decoded except by recourse to it.

In their model cohesion is realised lexicogrammatically by the resources they discuss under the headings 'reference', 'substitution and ellipsis', 'conjunction', and 'lexical cohesion'. "Reference, substitution and ellipsis are clearly grammatical, in that they involve closed systems", whereas "conjunction is on the borderline of the grammatical and the lexical" (Halliday and Hasan 1976: 303).

Lexical cohesion, by contrast, is lexical and involves open-ended choices. As already discussed in 4.1.2 and 4.1.3 above, in Martin's (1992: 101) discourse semantic model the lexicogrammatical resources that realise reference are among those that also realise reminding and relevance phoricity, those realising conjunction contribute to relevance phoricity, and, finally, those realising substitution and ellipsis also realise redundancy phoricity. What remains to be dealt with and discussed here, therefore, is lexical cohesion.

Halliday and Hasan (1976: 318) define lexical cohesion as "'phoric' cohesion that is established through the structure of lexis, or vocabulary, and hence […] at the lexicogrammatical level." According to Halliday's (2004: 570-571) categorization, "the primary types of lexical relations derive from either the paradigmatic or the syntagmatic organisation of lexis." Specifically, if the nature of the relation is paradigmatic (lexical set), the lexical relation between two items can be one of 'repetition', 'synonymy', 'hyponymy', or 'meronymy'. If, on the other hand, the relation is syntagmatic, the lexical relation is one of 'collocation'. Repetition occurs when a given lexical item is repeated, albeit "it need not be in the same morphological shape", in that inflectional variants such as 'migrating', 'migrate, 'migration', "always belong together as one item" (Halliday 2004: 571). Synonymy involves the choice of a lexical item that is in some sense synonymous with a preceding one. If synonymy occurs with identity of reference, then "the range of potentially cohesive items includes synonyms of the same or higher level of generality: synonyms in the narrower sense and superordinates" (e.g. 'swallows' and 'birds') (Halliday 2004: 573). Hyponymy is a kind of lexical relationship based on classification (specific to general), in which one lexical item represents a class of things (e.g. 'tree') and the other either a superclass or a subclass, or another class at the same level of classification, i.e. the items are co-hyponyms (e.g. 'oak', 'pine', and 'elm'). Thus, its general sense is 'be a kind of' (Halliday 2004: 575). By contrast, the relation of meronymy is glossed as 'be a part of', in that the relationship at issue is one among parts (e.g. 'trunk', 'branch', 'leaf') of a more general class (e.g. 'tree'). Between the two concepts, however, there is no clear line, "especially with abstract terms; and a given set of items may be co-hyponyms of one term but co-meronyms of another" (Halliday 2004: 576). Finally, the syntagmatic relation of collocation is defined as one occurring between two items that tend to co-occur (e.g. 'cold' and 'ice'), as already discussed in 3.2.1.1 above.

Grammatical and lexical cohesive resources do not occur in isolation in texts. On the contrary, in order to be effective, they need to support one another, thus forming 'cohesive chains' (Hasan 1989: 82-83), that is, chains

> formed by a set of items each of which is related to the others by the semantic relation of co-reference, co-classification, and/or co-extension. Taking the type of relation into account, we can sub-categorise chains into two types: identity chains and similarity chains. (Hasan 1989: 84)

In particular, the relation between the members of an identity chain is that of "co-reference: every member of the chain refers to the same thing, event, or whatever" (Hasan 1989: 84). An example of a reference chain can be found in 4.4.2.3 below (Figure 4.14), where the reference chain concerning the participant 'James Whistler' is represented. As will be seen, some of its members are 'James Whistler', 'he', 'his', and 'Whistler'. The members of similarity chains, by contrast, are "related by co-classification and/or co-extension", and the relationship between them is "not identity of reference but similarity of reference, so that the referents lie within the same general field of meaning" (Hasan 1989: 85). Therefore, if one knows the 'field of discourse', i.e. one of the parts of the context of situation relevant to an interaction, "it will be possible to predict that some selection from this or that semantic grouping will appear in the shape of similarity chains in the text generated; equally, selections from given semantic groupings are constitutive of the field of discourse". As said in 4.1.3 above, Martin (1992: 333) calls identity chains 'reference chains' and similarity chains taxonomically based lexical strings.

4.2.2.2 Thematic Structure

Thematic structure – Theme/Rheme – is one of the structural and lexicogrammatical resources at the clause level, but it is equally important at the textual level, in that it helps create semantic links across clause complexes, thus contributing to the cohesion and coherence of a text. The Theme

> is the element which serves as the point of departure of the message; it is that which locates and orients the clause within its context. The remainder of the message, the part in which the Theme is developed, is called in Prague school terminology the Rheme. As a message structure, therefore, a clause consists of a Theme accompanied by a Rheme, and the structure is expressed by the order – whatever is chosen as Theme is put first. (Halliday 2004: 65)

Theme is, therefore, a semantic element that is encoded in the first position of the clause structure in order to provide the reader with a point of departure for the message. The choice of Theme depends primarily on the Mood of the clause: the 'unmarked' Theme in a main clause is Subject in a declarative, Finite in a yes/no interrogative; WH- element in a WH- interrogative; Predicator in an imperative. The writer/speaker can, however, select a 'marked' Theme, that is, an element that does not typically occur at the beginning of the respective clause type. The most usual elements chosen as marked Themes in declarative clauses are adverbial groups (e.g. 'today', 'luckily') and prepositional phrases (e.g. 'in normal circumstances') functioning as Adjunct in the clause (Halliday 2004: 72-73). The Theme of a clause always ends with the end of "the first constituent that is either participant, circumstance, or process [in the Transitivity system]. We refer to this constituent, in its textual function, as the 'topical' Theme" (Halliday 2004: 79). There may be, however, other elements in the clause preceding the topical Theme, namely 'interpersonal' and 'textual' themes, as exemplified by the following clauses represented in Figures 4.3 and 4.4:

Unfortunately,	the news […]	was not	communicated	in a timely manner …
Adj.: comment	*Subject*	*Finite*	*Predicator*	*Adj.*
	Mood		*Residue*	
interpersonal	*topical*			
Theme		*Rheme*		

Figure 4.3. An example of multiple Theme: interpersonal Theme followed by topical Theme (taken from text 11).

However,	he	is	most famous	for his paintings
Adj.: conjunction	*Subject*	*Finite*	*Predicator*	*Adj.*
	Mood		*Residue*	
textual	*topical*			
Theme		*Rheme*		

Figure 4.4. An example of multiple Theme: textual Theme followed by topical Theme (taken from text 11).

The choice of clause Themes constitutes the 'method of development' of a text, and plays a fundamental part in the way discourse is organised (Ghadessy 1995: 132). A number of 'arguments' (research hypotheses) about the nature and the characteristics of Theme development have been put forth by Fries (1983: 119) and dealt with by Fries himself and other linguists. Ghadessy (1995: 132), for example, elaborates on the fact that thematic progression[12] correlates with the structure of a text and that this correlation concerns similarity and identity chains in the Themes. He also affirms that "thematic content correlates with the method of development of a text (and the nature of the text)" and that this relation concerns the lexico-semantic content of Themes. Two further hypotheses posit that "grammatical and lexico-semantic properties of Themes are different for different registers/genres", and that "thematic content correlates with different generic structures within a text" (Ghadessy 1995: 132). Fries (1983: 135) also focuses on the notion of method of development in a paragraph, that is, a short text segment, and writes that

> the information contained within the Themes of all the sentences of a paragraph creates the method of development of that paragraph. Thus if the

12 Fries (1983: 121-124) reproduces the three major patterns of thematic progression as found by Daneš (1974: 118-119) within successive clauses of passages. Eggins (2004: 324) describes the 'zig-zag pattern' as the one in which an element which is introduced in the Rheme of one clause gets promoted to become the Theme of the next clause. She calls 'Theme reiteration' the pattern involving the reiteration of the same Theme for several consecutive clauses. Finally, she defines 'multiple-Rheme pattern' as the one in which "the Theme of one clause introduces a number of different pieces of information, each of which is then picked up and made Theme is subsequent clauses". (Eggins 2004: 325)

> Themes of most of the sentences refer to one semantic field [...] then that semantic field will be perceived as the method of development of the paragraph.

'Information structure' is another resource that contributes to textual coherence. However, "the domain of the set of options related to information structure as defined by Halliday and Hasan's model is not a unit of syntax at all, but a phonological unit, with one information unit being one tone group in the intonation pattern in the discourse" (Taylor Torsello 1984: 156). As the present study concerns written texts, information structure has not been taken into consideration.

These structural resources, along with the non-structural ones described in 4.2.2.1 above, make up the semantic system of English (cf. Halliday 1978: 132), and combine to create 'texture' (Halliday and Hasan 1976: 26). Halliday and Hasan (1976: 324), however, point out that texture involves much more than merely cohesion, and that it concerns the relationship the clause complex and its parts have to its 'environment' and the "macrostructure of the text, that establishes it as a text of a particular kind"[13]. As far as the differences between structural and non-structural textual resources are concerned, Halliday (2004: 579-580) specifies that:

> the systems of *theme, information, reference* and *ellipsis* are all concerned with textual statuses. Speakers assign these statuses to components of messages to help themselves produce texts and to help their listeners interpret them.

However,

> whereas Theme and New are parts of textual structures – Theme ^ Rheme in the clause and Given ^ New in the information unit respectively, … *reference* and *ellipsis*, are not. That is, while an element is marked cohesively as identifiable by means of a grammatical item such as the personal pronoun *they* or as continuous by means of a grammatical item such as the nominal substitute *one*, the textual statuses of identifiability and continuity are not structural functions of the clause or of any other grammatical unit. They can occur freely within Theme or Rheme, and within Given and New (although there are certain unmarked associations). (Halliday 2004: 550)

13 For further information on 'texture' as conceived of within the systemic functional model see, for example, Hasan (1989) and Martin (1992: 381-491).

4.3 Participant Identification and the English Nominal Group

As already pointed out above, most of the resources for identifying participants in English are realised by one or more functional elements of the nominal group (Martin and Rose 2003: 172). Knowing how the English nominal group is structured and the functions carried out by its parts is, therefore, of utmost importance in understanding how the process of participant identification works. In systemic functional terms the nominal group can be defined as a 'class', that is, as

> a set of items that are in some respect alike [...]. The most familiar, in our traditional grammar, are classes of words: verb, noun, adjective, adverb, pronoun, preposition, conjunction (and sometimes also interjection) in the usual list. But every unit can be classified: there are classes of group and phrase, classes of clause and, at the other end of the rank scale, classes of morpheme. (Halliday 2004: 50)

Halliday (2004: 52) also makes it clear that

> the class of an item indicates in a general way its *potential range of grammatical functions*. Hence, words can be assigned to classes in a dictionary, as part of their decontextualized definition. But the class label does not show what part the item is playing in any actual structure. For that we have to indicate its function.

Halliday (2004: 312) lists the following elements that carry out a specific function in the nominal group: 'Deictic', 'Numerative', 'Epithet', 'Classifier', 'Thing', and 'Qualifier'. He also affirms that "they serve to realise terms within different systems of the system network of the nominal group" (Halliday 2004: 312), which means that if the entry condition 'nominal group' applies, then the simultaneous 'sub-systems' of Deictic, Numerative, Epithet and Classifier describe the nominal group in more detail. In order to illustrate this point, the sub-systems associated with the element 'Deictic', that is, the ones that form the so called 'Determination' system network, will be focussed on briefly. First of all, according to Halliday (1976: 86)

> the class of words defined as operating with this value [Deictic] is the class sometimes known as 'determiner'. This includes some forty items that are fully grammatical (i.e. reducible to one-member classes by successive steps in delicacy; *the this that which whose my its a any some either each other same certain* etc.), together with other items forming an open set (i.e. that cannot be so reduced; *John's*, etc., including compound ones as in '*the railway company's* property).

This class of determiners can be subdivided in various ways (see, for example, Halliday 1976: 87 and Halliday 2004: 314-316), but there can also be secondary groupings which occur in different parts of the nominal group and give rise to secondary chain classes. For instance, the nominal group 'all my other friends' contains three secondary chain classes, that is, the 'pre-determiner' 'all', the 'determiner' 'my' and the 'post-determiner' 'other'. Within each one of these three classes further choices can be made. For instance, a determiner may be either 'specific' or 'non-specific' (e.g. 'my' vs. 'every'), 'selective' or 'non-selective' (e.g. 'my' vs. 'a'), and/or 'possessive' or 'demonstrative' (e.g. 'my' vs. 'this', as a further subdivision of the class formed by the intersection of 'specific' and 'selective'). Finally,

> these and various other systems eventually yield, by their subdivisions and intersections, one-member classes: thus, *my* can be uniquely classified as 'determiner: specific, possessive: personal: first person'. (Halliday 1976: 87)

If the nominal group is analysed as an experiential structure, that is, as representing 'experience', it can be described as made up of one or more of the functional elements listed in the first row of Figure 4.5. As can be seen from the information in the second row of Figure 4.5, the classes of words that most typically realise these functions are those of determiner (associated with Deictic), adjective (associated with 'post-Deictic' – Deictic_2 –, Epithet, and Classifier), numeral and adverb (associated with the Numerative element), and noun (associated with Classifier and Thing). Besides the possibility of being preceded by some functional elements, the Thing in a nominal group can be followed by an element called 'Qualifier', which is nearly always a rank-shifted element, i.e. an element embedded in the Thing (e.g. a clause or a prepositional phrase). If approached experientially, therefore, the nominal group is to be considered as a 'multivariate' structure, in that it is conceived of as "a configuration of elements each having a distinct function with respect to the whole" (Halliday 2004: 331).

Element	*Deictic*	*Deictic$_2$*	*Numerative*	*Epithet*	*Classifier*	*Thing*	*Qualifier*
Class	determiner	adjective	numeral, adverb	adverb, adjective	noun, adjective	noun	clause, groups
Function	identify	compare	compare	compare, evaluate	-	identify	compare, locate

Figure 4.5. Functional elements in the English nominal group, the classes that realise them, and the identification function they carry out (adapted from Halliday 2004: 320/561 and Martin and Rose 2003: 173).

As far as the role of the nominal group in identifying participants is concerned (see the third row of Figure 4.5), the function of both Deictic and Thing is that of identifying participants by means of a determiner and of a noun respectively (e.g. in 'a/the road'). Both Numerative and Epithet, on the other hand, can be used to compare participants: the former compares them by their order and does so by means of a numeral (e.g. in 'the first road'), whilst the latter compares them according to their qualities with either an adverb or an adjective (e.g. in 'the broadest road'). Another function of Epithets concerns evaluation (Martin and Rose 2003: 164), such as in 'that busy uneven road'. The Qualifier compares or locates the participant it qualifies by means of embedded clauses or groups (e.g. 'the road that you can see from here'), and the Classifier is not involved in identifying participants (see, for example, Halliday and Hasan 1976: 147-154 and Martin 1992: 134-135), but rather in indicating "a particular subclass of the thing in question" (e.g. 'a country road') [14]

14 Taylor Torsello (2003: 243), however, suggests that there may be exceptions to this rule and that this is a matter worthy of further research. As seen in 4.1.2 above, for example, not only does the Classifier 'University' in 'the University grounds' and in 'the University authorities' indicate a sub-class of

(Halliday 2004: 319). Finally, the post-Deictic (Deictic$_2$) usually has a comparative function and is realised by an adjective that comes after the Deictic but before a Numerative (Martin and Rose 2003: 172), such as in 'the other two roads'.

If the focus of the analysis is on the logical relationships between the constituent parts of the nominal group, then the structure of the nominal group is to be considered as 'univariate' (see also 4.1.3 above), in that it is generated as an iteration of the same functional relationship. As with any other univariate structure, the 'Head' of the noun group is labelled α, whereas the modifying elements, that is, 'Premodifiers' and 'Postmodifiers', are labelled with β, γ or with any other Greek letter, as in Figure 4.6 (see also Halliday 2004: 329-335).

Premodifier (β)	Head (α)	Postmodifier (γ)

Figure 4.6. The logical structure of the English nominal group.

The analysis of the nominal group as a logical structure is of special relevance when the Head of a given nominal group is dissociated from the Thing. In English this can happen when, for instance,

> one of the pre-modifying functions is taken on by something that is itself a nominal group, in such a way that the Thing gets embedded in a prepositional phrase with *of,* which then functions as post-Head Qualifier, as in *a cup of tea*. (Halliday 2004: 332)

What, in fact, happens in these cases is that the Head of the nominal group is conflated with any of the premodifying functions, i.e. non-specific Deictics, Numeratives, Epithets, or Classifiers (Halliday 2004: 332-333). The nominal group 'most of the students' taken from text 15 exemplifies how a Numerative can become the Head of a nominal group. Specifically, in this group the Head is 'most', that is, the Numerative, whereas the Thing is 'students', as represented in Figure 4.7:

'authorities' and 'grounds' respectively, but also a specific 'individual' (viz. a particular institution).

most	of	the	students
Numerative	*preposition*	*Deictic*	*Thing*
Head	*Postmodifier*		
α	γ		

Figure 4.7. The experiential and logical structures of a nominal group with a Numerative as Head.

In other cases the Head of the nominal group can 'delimit' the Thing "in terms of generality, some species of it, some aspect of it, or its composition" (Halliday 2004: 334). In these cases a few very general terms such as 'kind', 'sort', 'type', 'example', 'instance', 'specimen' are used. In Figure 4.8 the structure of the nominal group 'certain types of butterflies' taken from text 13 exemplifies the use of a 'type item' as Head:

Certain	types	of	butterflies
Deictic	*'type item'*	*preposition*	*Thing*
Premodifier	*Head*	*Postmodifier*	
β	α	γ	

Figure 4.8. The experiential and logical structures of a nominal group with a 'type item' as Head.

According to Martin (1992: 107-108), the way a participant is introduced into a text is to be related to the importance that this participant is going to have in it. When the nominal group contains a Deictic, this encodes important information about the role the text participant is going to play, as schematically and tentatively represented in Figure 4.9:

	Marked	
	Singular	*Plural/Mass*
Unrestricted	any	any
Non-particular	/some/	∅
Particular	one	/some/
Major role	this (*spoken*), a certain	these (*spoken*), certain
	Unmarked	
	Singular	*Plural/Mass*
-	a	/^sm/

Figure 4.9. English Deictics graded according to the importance of the participant introduced (adapted from Martin 1992: 107).

According to Martin's model, therefore, if a participant is introduced by a noun group containing 'any' (used for both singular and plural countable nouns and mass/uncountable nouns), "it does not matter which member of the experiential class in question is being referred to" (Martin 1992: 108), which is the case of the 'carrier' in the following extract from text 23:

(82) Even the best airlines slip up from time to time, and it is impossible for any carrier to guarantee that a passenger's checked luggage will go on the same flight, particularly when a journey calls for one or more changes of aircraft.

Furthermore, the use of 'a/an' (with singular nouns) or 'some' (/^sm/) (with plural or mass nouns) signals that there is no intention to comment on the centrality of the participant being introduced. They should be, therefore, considered as the unmarked realisations of participants. In the following extract from text 17, for instance, the participant 'report' is introduced using the partial, non-selective and singular determiner 'a', and is presumed later on in the text with the specific determiner 'the'.

(83) A report from Mintel, the market research organisation, says that […] . The recently published report also predicts that […] Mintel's survey, based on nearly 1000 consumers, found that […]

In fact, in the remaining part of text 17 Mintel's report is no longer presumed; in contrast, 'Mintel's survey', on which the report is based, is at issue and is repeatedly presumed. In the extract from text

13 reported below 'some scientists' presents a group of scientists who share the same opinion and beliefs about the meaning of the word 'emigration'. These scientists, however, are not presumed later on in the text, which confirms their non-centrality.

(84) Some scientists call this movement emigration, because these animals never return to their homes. For other animals, such as birds, the migration includes a return trip home. (extract from text 13)

As seen in 4.1.3 above, one of the least delicate sub-systems concerning participant identification involves the choice between either a generic or a specific participant. According to Martin (1992: 103), generic participants refer to "the whole of some experiential class of participants" and their context is simply that of "knowledge of the language being used". One of the distinguishing linguistic features of generic reference is the 'neutralisation' of definiteness and number[15] (Martin 1992: 103-104). Furthermore, generic participants can be introduced either with definite or indefinite deixis, as exemplified by the occurrences of both the non-phoric noun 'a vehicle' and of 'vehicles' in the following extract from text 14:

(85) Members of staff who have paid the requisite fee and display the appropriate permit may bring a vehicle into the grounds. [...] Students may not bring vehicles into the grounds during the working day unless they have been given special permission.

As can be seen, both expressions make reference to the same generic category, viz. 'vehicles'.

In relation to what was said above about the function of the Classifier, it is important to note here that generic participants presented as 'Classifier + Thing' introduce a sub-class of a broader class of participants, as can be seen in the following extract from text 1:

(86) Since eye problems are related to physical health …

15 'Number' is the sub-system of the noun group system network signalling whether the noun group is count or mass, and if count, if it is singular or plural (see, for instance, Halliday 1976: 131).

In this clause, in particular, the Classifier 'eye' limits the experiential extension of 'problems', and 'physical' that of 'health'.

4.4 The Role of Participant Identification and of Textual Cohesion in the Corpus

This section will focus on the linguistic resources used in some tests of the corpus to identify the participants relevant to the completion of the tasks, as well as on the contribution of other cohesive elements to text/test cohesion. That is, even though this section will be mainly concerned with the resources used to track the identity of the participants introduced in the texts and presumed afterwards also in the accompanying tasks, other phenomena such as lexical cohesion and thematic structure will be taken into account.

The rationale behind this investigation is suggested by and based on research conducted in the field of language testing. One area of research which is deemed especially important here involves the study of the relationship between input texts and test takers' 'responses'. In Bachman and Palmer's (1996: 43-59) framework for test construction and analysis, for instance, the 'relationship between input and response' is one of the variables used to describe tasks, along with the characteristics of the setting, those of the 'test rubric', the characteristics of the input, and those of the 'expected response'. Another area of research on language testing relevant to the present study has to do with the characteristics of the test methods used to elicit the candidates' responses. As described in chapters 1 and 2, the testing techinques or test methods mainly used in the tests under examination are: multiple-choice, multiple-matching, true/false, and true/false/not given. According to Buck, Tatsuoka and Kostin (1997: 450), these techniques, and in particular multiple-choice, require specific processes and 'sub-skills' that are "perhaps uncommon in other comprehension situations." One of them concerns the way the 'necessary information', that is, "the information in the text which the reader must understand to be certain of the correct answer" is retrieved (Buck, Tatsuoka and Kostin 1997: 437; see also the discussion in Chapter 1). Furthermore, Buck, Tatsuoka and Kostin (1997: 450) claim that in order to answer this type of question, test takers need to consider the "lexical overlap between the necessary information and the correct option". In the present section, the phenomenon of 'lexical overlap' is interpreted and analysed in terms of the textual resources, including those for participant identification, which are used to link the various parts of a given test to each other. On the basis of these analyses, hypotheses will be formulated about the difficulty that the retrieval of the necessary information is likely to cause.

4.4.1 Test 12

In this paragraph an analysis of test 12 will be carried out. As specified in 2.2 above, test 12 belongs to the reading component of the computer-based TOEFL test, and adopting Enright *et al.*'s (2000: 20-23) classification of pragmatic features, the input text can be defined as a 'historical narrative'. The following is the first clause of the text:

(87) The final battle of the War of 1812 was the Battle of New Orleans

<table>
<tr><td>The final battle</td><td>of the War</td><td>of 1812</td><td>the Battle</td><td>of New Orleans</td></tr>
<tr><td>D + N + T</td><td>Qualifier (pr.+D+T)</td><td>Qualifier (pr.+T)</td><td>D + T</td><td>Qualifier (pr.+T)</td></tr>
<tr><td></td><td>Participant (1ba)</td><td>Participant (1bb)</td><td></td><td></td></tr>
<tr><td>Participant (1a)</td><td colspan="2">Participant (1b)</td><td>Participant (1a)</td><td>Participant (1c)</td></tr>
<tr><td colspan="3">Participant (1)</td><td colspan="2">Participant (1)</td></tr>
</table>

Figure 4.10. The participants realised by the 2 nominal groups in the first clause of text 12 (D=Determiner; N=Numerative; T=Thing; pr.=preposition).

As can be seen from Figure 4.10, in this clause the participant 'the final battle of the War of 1812' is presented and equated with 'the Battle of New Orleans'. The specific reference 'the final battle of the

War of 1812' is realised by means of the Deictic 'the', the Numerative 'final', the Thing 'battle', i.e. participant (1a), followed by the Qualifier 'of the War of 1812'. It can also be seen that participant (1) contains forward reference within itself, i.e. esphora, in that the Qualifier provides enough information to identify this participant. The Qualifier 'of the War of 1812' consists of the preposition 'of', the Deictic 'the' and the Thing 'war', and is in turn qualified by 'of 1812', that is, by the preposition 'of' and the Thing '1812'. The first Qualifier realises participant (1ba), i.e. 'the War', and the Qualifier of this Qualifier realises participant (1bb), i.e. '1812'. The Value of this identifying clause is the proper name of the battle, i.e. 'the Battle of New Orleans'. It consists of the participant 'the Battle', i.e. again participant (1a), but written with capital 'B', and the Qualifier 'of New Orleans', which contains the preposition 'of' and the homophorically presumed participant 'New Orleans', i.e. participant (1c).

Most of the participants introduced in this first clause are realised as rank-shifted structures, viz. 'the War of 1812', 'the War', '1812', 'New Orleans', and might, therefore, be thought of as being less relevant to the test than those introduced by means of the two primary/ranking nominal groups, viz. 'the final battle …' and 'the Battle …' (capitalized). This hypothesis might also be formulated in the following semiotic terms: whilst participants realised as primary Thing could be perceived as 'foregrounded' information, i.e. "salient information that is fundamental to the interpretative process" (Baicchi 2003: 333), those realised as rank-shifted structures might seem to be part of the 'more pallid and less differentiated ground' (Merlini Barbaresi 1988: 90), and to provide backgrounded and less salient information. However, if the tasks are also taken into consideration, it can be seen that in question 12.01 (test method: multiple-choice) both 'The War of 1812' and 'the Battle of New Orleans' are presumed along with other 'similar' participants, viz. 'The War of Independence' and 'The Treaty of Ghent':

(88) 12.01 The main idea of this passage is that …
the War of Independence was unnecessary
the War of 1812 was unnecessary
the Treaty of Ghent was unnecessary
the Battle of New Orleans was unnecessary

‘The War of 1812’ is also at issue in questions 12.04, 12.05 and 12.06; in question 12.04, in particular, it occurs along with ‘independence from Britain’:

(89) 12.04 According to the passage, when did the United States win its independence from Britain?
(a) Shortly before the War of 1812
(b) During the War of 1812
(c) Just after the War of 1812
(d) Long after the War of 1812

(90) 12.05 According to the passage, some U.S. sailors were
(a) taken forcibly to Shanghai
(b) made to go to Ghent
(c) forced to work on British ships
(d) responsible for causing the War of 1812

(91) 12.06 Click on the sentence in paragraph 2 that indicates when the War of 1812 officially ended.

On the basis of these observations, two more specific working hypotheses can be formulated rather than the initial and more tentative one above. Firstly, that in test 12 both the participants introduced as ranking nominal groups and those introduced by means of rank-shifted structures (e.g. qualifying noun groups, embedded clauses) are equally at issue and therefore important for the comprehension of the text and the successful completion of the tasks. Secondly, that having to deal with and to distinguish between participants identified with rank-shifted structures and others realised by means of ranking structures might cause cognitive difficulty and possibly confusion to inexperienced or inaccurate readers.

Further insights into the ways these participants are identified in test 12 can be gained by examining the ways they are presumed, that is, referred to, as the text unfolds. To this end, the two reference chains formed by the occurrences of the participants ‘the battle’ and ‘the War of 1812’ in the text are represented in Figure 4.11[16].

---

16 The symbols [[ ]] indicate that the elements inside them are embedded in a higher ranked structure; ‘Th’ stands for Theme; the dependency arrow ↑ indicates the dependency of a presuming element on a preceding one.

The battle	
The final battle of … (Th)/ the Battle of New Orleans	*TEXT*
↑	
This battle (Th)	
↑	
it (Th)	
↑	
[[the ensuing battle]] (Th)	
↑	
The Battle of New Orleans	
↑	
[[a battle]]	
↑	
Q1: The Battle of New Orleans	*TASKS*
↑	
Q7: [[the final battle]]	
↑	
Q9: The Battle of New Orleans	
↑	
Q11: the battle	

The war of 1812	
[[the War of 1812]]	*TEXT*
↑	
The War of 1812 (Th)	
↑	
The war (Th)	
↑	
[[… the war]]	
↑	
[[… the war]]	
↑	
[[… the war]]	
↑	
Q1: the War of 1812	*TASKS*
↑	
Q4: [[the War of 1812]] (*4 times*)	
↑	
Q5: the War of 1812	
↑	
Q6: the War of 1812	

Figure 4.11. The reference chains for the participants 'the battle' and 'the war of 1812'.

The reference chains concerning the two other participants presumed in question 1, i.e. 'the Treaty of Ghent' and 'The War of Independence', are represented in Figure 4.12.

The Treaty of Ghent	
a peace treaty (Th)	*TEXTS*
↑	
[[the Treaty of Ghent]] (Th)	
↑	
[[the treaty]] (Th)	
↑	
[[the treaty]]	
↑	
[[the treaty]] (Th)	
↑	
the treaty	
↑	

The War of Independence	
(its) independence	*TEXT*
↑ *(BR)*	
Q1: the War of Independence	*TASKS*
↑	
Q4: (its) independence	

Ghent	
[[Ghent]]	*TEXT*
↑*(BR)*	
[[Ghent]]	

[[the peace treaty]]
↑
[[the peace treaty]]
↑
Q1: the Treaty of Ghent — *TASKS*
↑
Q7: the treaty
↑
Q11: [[the treaty]]

↑
[[Ghent]]
↑
Q1: [[Ghent]] — *TASKS*
↑
Q5: [[Ghent]]
↑
Q7: Ghent

Figure 4.12. The reference chains for the participants 'the Treaty of Ghent', 'the war of Independence' and 'Ghent'.

According to the counts presented in Figure 4.13 below, the most recurrent participant in the text is 'The War of 1812' (13 occurrences), followed by 'the Battle' and 'The Treaty of Ghent', both of them occurring 11 times. As can be seen, 'the War of 1812' is mainly realised by rank-shifted nominal groups (8 out of 13 times), while 'the Battle' is mainly realised by means of ranking structures (8 out of 11 times). Furthermore, in the text 'Battle' is presumed in thematic position 4 times, while the 'War' only twice. In the tasks, on the other hand, both participants are presumed by means of ranking nominal groups 3 times and by means of rank-shifted ones 1 and 4 times respectively.

	Ranking	*Rank-shifted*	*Total*	*Thematic*
The Battle			11	
Text	5	2		4
Tasks	3	1		-
The War of 1812			13	
Text	2	4		2
Tasks	3	4		-
The Treaty of Ghent			11	
Text	2	6		4
Tasks	2	1		-
(The War of) Independence			3	
Text	1			-
Tasks	2			-
'Ghent'			6	
Text		3		-
Tasks	1	2		-

Figure 4.13. Table representing the total number of occurrences of the main participants in test 12 and their status.

From Figure 4.12 it emerges that 'the Treaty of Ghent' is presented by means of the generic nominal group 'a peace treaty' (with 'peace' functioning as Classifier), and that it is mainly presumed by means of rank-shifted structures. 'The War of Independence', on the other hand, is presented in the first question, and not presumed at all afterwards; there is, however, at least a lexical cohesive relation between 'the War of Independence' and 'independence' occurring in the text, due to the repetition of the word 'independence'. Furthermore, in the text the word 'independence' is preceded by the possessive Deictic 'its', which presumes 'the United States' through possessive reference (see 4.1.3). In this way, the actual occurrence of the 'War of Independence' (and even the identity of the winner) can be implied and retrieved indirectly by 'bridging' (BR) (see 4.1.2 above). Similarly, because of the repetition of the word 'Ghent', the participant 'the Treaty of Ghent' can be bridged back to 'Ghent'.

On the basis of the foregoing considerations, it can be affirmed that even though 'the War of 1812' is presented as a rank-shifted structure, it is presumed more often than 'the Battle', especially in the tasks. Also, 'the War of 1812' is mainly tracked with rank-shifted nominal groups, while 'the battle' occurs more often in thematic position. 'The Treaty of Ghent' is also recurrent in the test, although mainly realised with rank-shifted structures. As will be seen in 5.7.2 below, the feedback obtained from a large percentage of test takers adds reinforcement to the suggestions made in this paragraph. Specifically, they mentioned their difficulty in distinguishing between the many and similar events and dates that are presented and presumed in the text and in the tasks. They also found it difficult to relate events and dates, that is, to locate the events in time. Further reinforcement to these suggestions is given by the fact that both text 12 and its accompanying tasks were evaluated as the most difficult ones in battery 4 (see 5.7.2 below).

4.4.2 Tests 2, 25 and 11

In this paragraph three tests will be taken into consideration: test 2 ('Our hands met') taken from DIALANG, test 25 ('Beginner takes all') taken from University of Cambridge ESOL Examinations'

Certificate in Advanced English (CAE), and test 11 ('The next artist'), chosen from the reading component of the paper-based TOEFL version. Text 2 seems to be a short extract from a longer narrative text, and consists of narration and quoted speech. Text 25 is an apparently complete magazine article, which also contains both narrative sections and reported speech. Finally, text 11 can be defined as a historical narrative, and consists in short the narration of the life and artwork of the American painter James Whistler.

4.4.2.1 Test 2

In test 2 the only task (test method: multiple-choice) that test takers are explicitly asked to perform is retrieving the identity of the participant that wrote the input text:

(92) 2.01 Who wrote this?
The Proprietor.
The Minister.
Nwege.
Odili.

As text 2 is likely to be part of a longer text, the parts preceding this extract probably contain information about the identity of these participants that a test taker does not know when approaching the passage. In fact, the extract begins with first and third person pronouns, that is, presuming devices, as follows:

(61) Our hands met. I looked him straight in the face. The smile slowly creased up into lines of thought. He waved his left hand impatiently to silence the loquacious Proprietor who had begun the parrot formula he had repeated at least fifteen times so far. 'I have the honour, sir, to introduce …'
'That's right,' said the Minister not to anyone in particular, but to some mechanism of memory inside his head. 'You are … no, not Nwege, … it's Odili, isn't it?'
'Yes, sir'
'Odili, the great,' said the Minister boyishly. 'Where have you been all this time?'

The pronouns and the nominal groups used, however, provide important linguistic clues that the reader must properly interpret to make decisions about the protagonists' identities. In particular, the successful reader should adopt and apply the strategy of addition (see

4.1.2 above) and interpret the phoric items in the first two clause complexes as a first mention. Accordingly, the possessive and determinative specific deictic ‘our’ and the personal pronouns ‘I’ and ‘him’ should be dealt with as if the participants were introduced at the beginning of the extract. The word ‘hand’ in the first clause complex, on the other hand, should be processed not only by using the strategy of addition, but also that of bridging, since ‘a hand’ is part of the human body (meronymy). The subsequent realisations of all these participants should then be interpreted as anaphoric.

As said above, a specific characteristic of this text is that it contains quoted speech as well as narration. This implies that the personal pronouns used within the projected clauses, e.g. ‘I’, ‘you’, and those used in other clauses, e.g. ‘he’, ‘Odili’, are different, even though they refer to the same participants.

Odili	
↑ *(AD)*	*TEXT*
I	
↑	
Odili	
↑	
you	
↑	
Odili	
↑	
it	
↑	
Odili the Great	
↑	
you	
↑	
Odili	*TASK*

the Minister	
↑ *(AD)*	*TEXT*
him	
↑	
He	
↑	
his left hand	
↑	
sir	
↑	
The Minister	
↑	
his head	
↑	
sir	
↑	
The Minister	
↑	
The Minister	*TASK*

the Proprietor	
the loquacious Proprietor	*TEXT*
↑	
he	
↑	
I	
↑	
The Proprietor	*TASK*

Figure 4.14. The reference chains concerning the participants ‘Odili’ and ‘the Minister’ and ‘the Proprietor’ in test 2.

The pronoun 'I' used in the second clause complex, for instance, refers to Odili, as does 'you' in the last clause in which speech is quoted. On the basis of these considerations, it can be hypothesised that at least part of the difficulty involved in test 2 is due to the complexity of tracing all the mentions of each participant back to the correct reference chain. In order to visualise these textual relations, the reference chains concerning three of the participants mentioned in the options of the question are reported in figure 4.10. As 'Nwege' is only introduced but not presumed in the extract and does not, therefore, form a chain, it was not reported in the figure. In Figure 4.14[17] 'AD' is the notation used to indicate addition. This is an example of what Masi (2003) calls 'polyphonic' discourse. This type of discourse can be ambiguous, in that "the text producer's voice meets up with other voices, which are inevitably changed through such an encounter" (Masi 2003: 402).

Although most test takers performed well on test 2, they evaluated text 2 as the most difficult one to read among the ten DIALANG tests and task 2 as the second most difficult one (see 5.4 below). Furthermore, about 46% of them admitted that their main difficulty was concerned with understanding the exact identity of the participants, the relationships between them, and, in particular, the identity of the story teller. To this, some of them added that they needed to read the text twice to retrieve this information, but they eventually managed to understand the text and, consequently, to answer the question. The feedback from about 40% of the test takers on the difficulty of task 2 similarly suggests that the main problems they encountered were concerned with the identification of the writer and that they needed to read the text twice. They also found that the very fact that the input text is a short extract from a longer text and that it contains reported speech as well as narration contributed to increasing the difficulty of the task. This feedback added, therefore, further reinforcement to and confirmed the hypotheses posited above.

Martin and Rose (2003: 165-167) claim and show that there can always be indeterminacy whilst tracking the participants in any kind of text, since

17 The dependency arrow ↑ indicates the dependency of a presuming element on a preceding one, and 'AD' stands for 'addition'.

> participant tracking is a dynamic device that is very sensitive to just where we are as a discourse unfolds. In analysis it is important not to lose sight of indeterminacy that eventually gets resolved because we can look backwards, carefully, taking as much time as we like, and with hindsight weigh up the evidence and make sense of the meaning overall. The way in which meaning accumulates in discourse is just as significant as what we finally decide a discourse meant. Reading opens; readings close. (Martin and Rose 2003: 167)

Unfortunately, though, testing situations are usually characterised by time constraints that do not allow for multiple readings. Furthermore, as exemplified by test 2, in order to test reading comprehension, incomplete texts are often used as input materials, which makes it necessary for readers to use 'special' tracking strategies in order to successfully comprehend a passage. Finally, the tasks test takers are asked to perform and the test methods used to elicit a given answer can increase the difficulty involved in the process of participant identification. This last consideration, in particular, leads us on to the discussion of test 25.

4.4.2.2 Test 25

Test 25 is based on a text about "Nicholas Evans, screenwriter, producer and aspiring director", and the author of 'The Horse Whisperer' (test 25). As pointed out in 3.1 above, it is the longest input text in the corpus, and seems to have been used in its entirety. As explicitly stated in the instructions, the text is a gapped magazine article and contains two characteristic elements of articles, namely the headline 'Beginner takes all', and the following lead paragraph:

(93) Even before it was published, The Horse Whisperer was the hottest book of the year. A first novel by <u>British screenwriter Nicholas Evans</u>, it has earned its author record-breaking sums. He talks here about his inspiration and his triumph.

It can be seen that the determiner in the headline was deleted, which is a typical characteristic of nominal groups in headlines (see, for example, Morley 1998: 137 and Martin and Rose 2003: 173). The lead paragraph, on the other hand, contains an instance of the particular type of nominal group that results from the mechanism that Morley, following Bell (1991: 195-196), calls 'titleness'. In particular, the nominal group 'British screenwriter Nicholas Evans' is characterised by "the deletion of the determiner from a first descriptive noun phrase in apposition with a second noun phrase containing a proper name", thus acquiring "a status akin to titles" (Morley 1998: 137). Titleness is also a summary device (Morley 1998: 138; Bell 1991: 196-197), and the whole lead paragraph can be considered as a short summary of the contents of the article, as well as an indicator of its main participants, namely Nicholas Evans and his novel 'The Horse Whisperer'. These two participants are actually presumed by means of different types of devices throughout the whole article. As in test 2, in test 25 the narration of events is intertwined with reported speech, a strategy that, as seen above, compels the author to use both first and third personal pronouns to refer to the same participant. For example, in order to refer to Nicholas Evans, the following tracking resources are used in the text:

Nicholas Evans
British screenwriter Nicholas Evans, 1 occurrence
Nicholas Evans, 1 occurrence
he, 22 occurrences
his, 8 occurrences
him, 4 occurrences
my, 2 occurrences
Evans, 7 occurrences
Nick, 1 occurrence
I, 18 occurrences
we (Evans and his agent), 5 occurrences
us (Evans and his family), 2 occurrences
us (Evans and his agent), 1 occurrence
they (Evans and his agent), 1 occurrence
they (Evans and his 2 agents), 2 occurrences

Figure 4.15. Linguistic resources used in text 25 to keep track of the participant 'Nicholas Evans'.

As can be seen from Figure 4.15, the third person singular pronoun 'he' (22 occurrences) and the first person singular pronoun 'I' (18 occurrences) are both used to identify Nicholas Evans in the narrative and in the reported speech sections respectively. The same is true for the possessive determiners 'my' (e.g. 'my life') and 'his' (e.g. 'his agent'). It also emerges that the devices 'we', 'they', and 'us' include reference to Evans, as exemplified by the following extracts:

(94) When pushed, he ventured that Evans might get $30000 as an advance on the book. 'I had in mind how much I needed to pay off a bit of the overdraft and keep us going, […]

(95) We couldn't believe it; we sat there with our jaws gaping. We'd never sent the manuscript to New York, we still don't know how it got there,' Evans says. Nor did they send it to Hollywood, but within that same week the major studios were fighting over it. […]

(96) Evans told me all this as we sat drinking coffee on a wooden verandah perched above the leafy garden of his home. He said that he had since turned down an offer to write the screenplay of The Horse Whisperer.

As far as 'the novel' is concerned, it can be claimed that the nominal groups used to identify this participant contribute to "framing the phases" of the 'story' told in this article, whereas the pronouns "sustain reference within the phases" (Martin and Rose 2003: 164). An attempt to organise the framing and the tracking devices used in this text has been made in Figure 4.16.

THE HORSE WHISPERER
- PHASE 1 (lead paragraph)
The Horse Whisperer – *A first novel by British …* ← it ← its author
(end of 2nd paragraph)
His first novel
- PHASE 2 (3rd paragraph)
The story (BR)
- PHASE 2/3 TRANSITION (4th paragraph)
The story as a book
- PHASE 3 (6th and 7th paragraph)
The book – The Horse Whisperer – The book
- PHASE 4 (8th and 9th paragraph)
The novel – the novel
- PHASE 5 (10th and 11th paragraph)
The manuscript ← it ← it ← it – An unfinished novel – The Horse Whisperer-
PHASE 6 (13th paragraph)
his novel – another novel

Figure 4.16. Linguistic resources used in text 25 to keep track of the participant 'The Horse Whisperer' in the various paragraphs.

From Figure 4.16 it emerges that the noun groups used in the lead paragraph and in the first three paragraphs to identify the novel are 'The Horse Whisperer' and 'his first novel'. 'A first novel by British screenwriter Nicholas Evans' does not identify the participant, but rather classifies and evaluates it (see also 4.3 above). The second phase in the text is not exactly about the novel, but about 'the story' the novel is based on. The relation between the two is in fact indirect, in that the latter participant can be bridged back to the previous one on the basis of the lexical relation of meronymy occurring between the novel and the story it is about. In the transition between the second and the third phase, it is said that Evans is prompted to write the story as 'a book': this statement classifies 'the story' and at the same time links it to 'the book'. 'The book' is repeated in paragraph six and seven (phase three), whilst in paragraphs eight and nine the term 'the novel' is chosen. In phase five (i.e. paragraphs ten and eleven) the identification is realised by means of 'the manuscript', sustained by the use of the pronoun 'it', and classified as 'an unfinished novel'. The use of 'manuscript' and that of the evaluative Epithet 'unfinished', in particular, convey a sense of incompleteness, and stress the fact that in this phase the piece of writing at issue is not yet a full-fledged novel. In phase six (13th paragraph) 'his novel' is used as the Qualifier of 'the success', and, finally, comparative reference is made to 'another' novel Evans might write, but has not written yet.

Cambridge ESOL's CAE test takers are asked to read a version of text 25 from which six paragraphs have been removed. The missing paragraphs are arranged in a jumbled order after the text, and the candidates' task is to reconstruct the text by choosing the right paragraph, in order to fill each single gap in the text. As said in 2.6 above, in UCLES (2006f: 7) test takers are advised to read the text quickly (skimming) first, so that they can gain an overall idea not only of its meaning, but also of its structure. After this, they are instructed to read the text carefully, and to focus, in particular, on the information given before and after each gap. Similarly, Alderson and Cseresznyés (2003: 73), who call this test method 'matching sentences to gaps in text', give the following suggestions to prospective test takers:

You need now to read the text very carefully, because there are clues in the passage both before and after each gap. The development of ideas can be seen in different lexical and grammatical relations between sentences and paragraphs of the text and, therefore, you should pay careful attention to both lexical words and phrases, and grammatical structures that may indicate such relations in the text. Thus, for example, the repetition of certain nouns or adjectives, the use of synonym words and phrases, the use of one verb tense rather than another, the use of reference words, like personal pronouns (referring to either people or things), demonstratives *(this, those)*, possessives *(its, their)*, relative pronouns *(who, that, which)*, temporal markers (*when* or *while*), adverbs of frequency *(often, never)*, of place *(here, there)*, logical connectors *(besides, however)*, or other linking words indicating contrasts *(but, while)*, or reasons *(because, since)* – all might help you to see how information in the text is organized, and decide which sentence fits where in the text.

In other words, Alderson and Cseresznyés recommend paying careful attention to all cohesive markers in a text, and especially to those occurring before and after each gap. Consequently, reference devices are among the clues that have to be taken into account to carry out this type of task, although they are not the only ones. This holds true for test 25 too, as will be shown with the following discussion of the first five gaps in the text.

As far as the first gap is concerned, it can be seen that the link between the first and the second paragraph is not only provided by the phoric reference occurring between 'him' and 'Evans', but also, and more importantly, by the word 'position' used in paragraph 2:

(97) (1) […] it had been two years since <u>he</u> had earned any money and the promise of that film had been the only buffer between <u>him</u> and an increasingly irate bank manager.

(2) A wise man, finding himself in <u>Evans'</u> <u>position</u>, would have got a job. […]. <u>He</u> bought a ticket to America and set off <u>for three months</u> to research <u>his first novel</u>. (1st gap)

In fact, as said above, the name 'Evans' and the pronoun 'him' occur frequently in text 25, which prevents them from being reliable clues as to where in the gapped text a given paragraph is to be inserted. The word 'position' in 'Evan's position', on the other hand, clearly summarises the content of the last clause complex of the previous paragraph, that is, the one describing Evans' difficult situation. In particular, it can be claimed that 'position' functions as an 'anaphoric noun' (see 4.1.3 above), since it can be thought of as 'encapsulating a foregoing piece of text' (Partington 1998: 96). The same function is carried out by 'the subject' used in the second clause complex of the fourth paragraph to sum up and refer backwards to the previously mentioned 'story of a docile horse that had turned, no one knew why, into a fiend':

(98) (3) Although he was acting very much on impulse, the seeds for the story had been with him for some time, sown by a farrier he met on Dartmoor while staying with a friend. [...] Its owners were desperate until they heard of a gypsy who, simply by talking to the animal, transformed its temperament in a matter of hours. Such men, the farrier said, were known as 'horse whisperers'.

(4) Evans' imagination was captured. He began researching the subject with a view to writing a screenplay – he was, after all, a film-maker. [...] And so throughout the spring he drove across the US, stopping at ranches and learning about horses and the men who work with them. (2nd gap)

'Evans' position' and 'the subject' (defined by their respective Deictics), therefore, can be regarded as two of the most important devices that link the first with the second paragraph and the third with the fourth one respectively. This latter paragraph and the fifth one, on the other hand, cohere mainly thanks to the use of the evaluative nominal group 'a funny time', which sums up the description of Evans' drive across the US, whilst the main link between the fifth and the sixth paragraph is provided by the text reference marker 'that' pointing back to Evan's thought in the fifth paragraph.

(99) (5) 'It was a funny time,' he says now. [...] here I was, hugely in debt and wondering how I was going to feed the children, and thinking maybe it was all just folly.'

(6) He thought that again towards the end of August, by which time he [...] I plucked up the courage to show it to a friend who was a literary agent; he read it and said it was "fine".' (3rd gap)

Both in the seventh and eighth paragraphs the 'local' participant 'synopsis' is identified as follows:

(100) (7) I was also advised to write a 12-page synopsis of the remainder of the book.' (8) In October, together with the first two hundred pages of the novel, this was sent to seven UK publishers [...]"It's OK. I just sense something is happening".' (4th gap)

As can be seen, this participant is introduced with the nominal group 'a 12-page synopsis' in the former paragraph, and referred back to as 'this' in the latter, thus creating an important cohesive relation. A clue that confirms this relation is provided by the occurrence of the following 'semantically complementary' expressions at the end of paragraph 7 and at the beginning of paragraph 8 respectively: the Qualifier 'the remainder of the book' and the prepositional phrase 'with the first two hundred pages of the novel.'

The correct choice of the paragraph that fills in the fifth gap is made possible by the following underlined devices occurring between the ninth and the eleventh paragraphs:

(101) (9) The events that followed have become publishing history. Within a week [...] the novel had been sold to Transworld Publications in the UK for $550000 and to Delacorte in the US for $3.15 million, both record-breaking advances for a first novel.

(10) 'We couldn't believe it; [...] 'My agent in the UK wisely involved an agent over there and when he phoned us to say, "I think we can get $3 million outright," we laughed in disbelief.' (5th gap)

(11) As they all agreed to this sum, it was decided that they should each 'pitch' to Evans. [...] He said that he had since turned down an offer to write the screenplay of The Horse Whisperer.

Three of these devices, in particular, are worthy of special attention, namely 'the events that followed' in the ninth paragraph, 'it' in the tenth paragraph, and 'this sum' in the eleventh one. The first nominal group is cataphorically related to the remaining part of the ninth paragraph, in that it presumes all the events reported on in it. The pronoun 'it' in the tenth paragraph refers back to these events, thus providing an important link between the two paragraphs, and, finally, the definite nominal group 'this sum' clearly presumes the figure '$3 million'. It should also be seen that the circumstantial Adjuncts referred to time (see, for instance, Halliday 2004: 125 and Eggins 2004: 158-160) underlined in the various paragraphs of the text, viz.

‘for three months’, ‘throughout the spring’, ‘towards the end of August’, ‘in October’, ‘within a week’, are valuable resources that provide temporal connections between the various parts of the text.

4.4.2.3 Test 11

An input text that is interesting to compare to text 2 and text 25[18] is text 11, reported below:

(102) The next artist in this survey of American artists is James Whistler; he is included in this survey of American artists because he was born in the United States, although the majority of his artwork was completed in Europe. Whistler was born in Massachusetts in 1834, but nine years later his father moved the family to St. Petersburg, Russia, to work on the construction of a railroad. The family returned to the United States in 1849. Two years later Whistler entered the U.S. military academy at West Point, but he was unable to graduate. At the age of twenty-one, Whistler went to Europe to study art despite familial objections, and he remained in Europe until his death.
Whistler worked in various art forms, including etchings and lithographs. However, he is most famous for his paintings, particularly Arrangement in Gray and Black No. 1: Portrait of the Artist’s Mother or Whistler’s Mother, as it is more commonly known. This painting shows a side view of Whistler’s mother, dressed in black and posing against a gray wall. The asymmetrical nature of the portrait, with his mother seated off-center, is highly characteristic of Whistler’s work.

First of all, it can be seen that text 11 also begins with an identifying clause, a representation of which is given in Figure 4.17.

<table>
<tr><td>The next artist</td><td>in this survey</td><td>of American artists</td><td>James Whistler</td></tr>
<tr><td>D + N + T</td><td>Qualifier
(pr.+D+T)</td><td>Qualifier
(pr.+C+T)</td><td>T</td></tr>
<tr><td></td><td>Participant (1ba)</td><td>Participant (1bb)</td><td></td></tr>
<tr><td>Participant (1a)</td><td colspan="2">Participant (1b)</td><td></td></tr>
<tr><td colspan="3">Participant (1)</td><td>Participant (1)</td></tr>
</table>

Figure 4.17. The participants realised by the 2 nominal groups in the first clause of text 11 (D=Determiner; N= Numerative; C=Classifier; T=Thing; pr.= preposition).

18 From the ‘Longman Complete Course for the TOEFL Test, Preparation for the Computer and Paper Tests’, by Deborah Phillips. Copyright © 2001 by Addison-Wesley Longman, Inc. Reprinted by permission of Pearson Education, Inc., pp. 345-346.

As pointed out in 4.4.1 above, identifying clauses are a way of introducing participants into texts. In the text under investigation 'James Whistler', i.e. participant (1), is equated with 'the next artist in this survey of American artists', that is, a nominal group that contains three participants: the specific participants 'the next artist' (participant 1a), 'this survey' (participant 1ba), and the generic participant 'American artists' (participant 1bb). Moreover, 'this survey of American artists' can be considered as a single nominal group (and, therefore, participant 1b), in that 'this survey' is qualified by 'of American artists'. The use of the Numerative 'next' in the first nominal group and of the Deictic 'this' in the second nominal group are worthy of special mention, since they are cohesive and make reference to other related participants occurring in some preceding parts of the text, from which text 11 is likely to have been extracted. Specifically, the Numerative 'next' carries out a comparative function, and postulates that a different artist from the one currently under consideration, i.e. 'this artist', has been dealt with previously. These textual choices are at issue in the first question accompanying this text, i.e. question 11.1 (test method: multiple-choice), which asks the reader about the 'paragraph preceding this passage' as follows:

(103) 11.01 The paragraph preceding this passage most likely discusses
(a) a survey of eighteenth-century art
(b) a different American artist
(c) Whistler's other famous paintings
(d) European artists

As seen in 2.2 above, the authors of the TOEFL preparation manual define this question as a 'transition question', and in order to answer it correctly (correct answer: "b) a different American artist"), the linguistic clues that test takers should draw on are precisely the numeral 'next' in the first nominal group, and 'this' in the second one. It can be claimed that starting off a text in this way is a marked textual choice, in that what is presumed is the identity of participants (and/or classes of participants) that have not yet been introduced, or that appear in previous parts of a longer text. Obviously, though, testing purposes and needs can justify this choice.

James Whistler	
The next artist .../James Whistler	*TEXT*
↑	
he (Th)	
↑	
He (Th)	
↑	
... [[his (artwork)]]	
↑	
Whistler (Th)	
↑	
his (father)	
↑	
Whistler	
↑	
he (Th)	
↑	
Whistler	
↑	
he (Th)	
	(continued)

... *James Whistler*	
↑	*TEXT*
his (death)	
↑	
Whistler (Th)	
↑	
he (Th)	
↑	
... [[his (paintings)]]	
↑	
[[... Whistler's (mother]]	
↑	
[[... his (mother)]]	
↑	
[[... Whistler's (work]]	
↑	
Q11.1: Whistler's (paintings)	*TASKS*
↑	
Q11.2: Whistler's (family)	
↑	
Q11.3: Whistler (Th)	
...	

Figure 4.18. The reference chain concerning 'James Whistler' in test 11.

Text 11 is characterised by an expository, descriptive and possibly also didactic rhetorical mode (see, for example, Halliday 1989: 12), in that it is mainly concerned with the description of Whistler's life and artwork. As can be seen from Whistler's reference chain represented in Figure 4.18, the artist is presented by means of the identifying clause analysed above and presumed afterwards in various ways. Firstly, it is presumened by the use of the personal pronoun 'he', mainly occurring in thematic position; secondly, by the possessive pronoun 'his' used in Deictic position and preceding other nouns; thirdly, by the repetition of the proper name 'Whistler', also often occurring in thematic position; and, finally, by 'Whistler' used in Deictic position. In the questions the same presuming devices are used, and often in thematic position (however, only the first three of them are shown in Figure 4.18). In contrast to what emerged from the analyses of text 2 and text 25, therefore, the identity of this participant is easily retrievable throughout the text, and this is so thanks to the transparency of the presuming devices used. As expected, no explicit question is asked about Whistler's identity.

4.4.3 Test 14

As specified in 2.3 above, test 14 was chosen from the first section of the General Training Reading module of IELTS. Text 14 can be defined as an informative text, in that it provides factual information about the regulation on the use of university grounds in a university, the name of which is not specified. The participant primarily at issue in test 14 is 'permit'. This claim can be made, firstly, because in the short input text (154 words, title included) the word is repeated four times (see also 3.1 and 3.2.2 above). Secondly, because three of the following five questions (test method: true/false/not stated, with the questions preceding the input text) are about this participant:

(7) 14.01 University employees do not need to pay for their parking permits.
14.02 Parking in Halls of Residence is handled by the Wardens of the Halls.
14.03 Having a University permit does not allow staff to park at Halls.
14.04 Parking permits cost £20 a year.
14.05 Students living in Hall do not need permission to park in Hall car parks. (Task 14)

USE OF UNIVERSITY GROUNDS BY VEHICULAR TRAFFIC
The University grounds are private.
The University authorities only allow authorised members of the University, visitors and drivers of vehicles servicing the University to enter the grounds.
Members of staff who have paid the requisite fee and display the appropriate permit may bring a vehicle into the grounds. A University permit does not entitle them to park in Hall car parks however, unless authorised by the Warden of the Hall concerned.
Students may not bring vehicles into the grounds during the working day unless they have been given special permission by the Security Officer and have paid for and are displaying an appropriate entry permit. Students living in Halls of Residence must obtain permission from the Warden to keep a motor vehicle at their residence.
Students are reminded that if they park a motor vehicle on University premises without a valid permit, they will be fined £20. (Text 14)

Despite the repetition of the word 'permit', which creates textual cohesion, it is hypothesised here that this text might be perceived as opaque, in that it is not always completely clear whether these occurrences realise exactly the same (class of) permit(s) or not. The first time this participant is mentioned in this text is in the third clause complex, where it is presumed homophorically by means of the nominal group 'the appropriate permit'. As can be seen from Figure 4.19, this nominal group occurs within the embedded and defining relative clause 'who have paid the requisite fee and display the appropriate permit', which qualifies the Thing 'members'.

Members	of staff	who have paid the requisite fee	and display the appropriate permit
T	*Qualifier (1) (pr.+ T)*	*embedded clause (1), containing «the requisite fee»*	*embedded clause (2), containing «the appropriate permit»*
	Qualifier (2)		
Head	*Postmodifier*		

Figure 4.19. The structure of a nominal group in text 14 (D=Determiner; E=Epithet; T=Thing; pr.=preposition).

As explained in 3.4.1.2 above, in systemic functional linguistics embedding is defined as "the 'rank-shift' by which a clause or phrase comes to function within the structure of a group" (Halliday 2004: 426), and its "function is to specify which member or members of the class designated by the Head noun are being referred to" (Halliday 2004: 428). Furthermore,

> when embedding occurs, sharedness may be implied. The clause is the most congruent grammatical realization of an event, so when an event is not given full clausal status, then there can easily be an implication that the event is shared. (Martin/Matthiessen 1991: 366, quoted in Taylor Torsello 2002: 30)

In this text it is the identity of the 'members of staff' that is further specified, and up to the third clause complex the participant 'permit' seems to be completely embedded in the context of situation. In the following clause complex, however, this participant is not presumed, but rather presented through the generic nominal group 'a University permit', that is, an item indicating generic reference and occurring in thematic position. By means of this nominal group, in particular, general information is provided about what a permit does not entitle members of staff to do. Subsequently, in the first clause complex of the fourth paragraph, the participant is introduced for the second time as 'an appropriate entry permit'. On this occasion the participant's realisation occurs in the clause complex represented in Figure 4.20[19] (the notation used is from Halliday 2004: 363ff.):

α		*Students* may not bring vehicles into the grounds during the working day
β	1	unless *they* have been given special permission by the Security Officer
	2	and *[ ]* have paid for
	3	and *[ ]* are displaying *an appropriate entry permit.*

Figure 4.20. The structure of the first clause complex in paragraph 4.

As can be seen, four distinct clauses can be identified in this clause complex: a primary clause in hypotactic relation (α), and three secondary clauses in hypotactic relation in respect of (α), but at the same time paratactically related with respect to each other, i.e. clauses (β 1), (β 2), (β 3). In all these clauses 'students', variously realised as 'students', 'they' and 'omitted Subject', perform the functions of Subject and Theme, which signals that the people at issue in this segment of text are no longer members of staff but the students (see also 4.2.2.2 above). At this point, therefore, readers (including the intended direct receivers of this text, that is, the students and employees in the campus) might wonder whether the appropriate permits available to the students are the same entry permits that members of staff are supposed to display. This is, in fact, what could cause confusion and possibly processing difficulty for the readers. The last realisation of this participant occurs in the last clause complex and is represented in Figure 4.21. As can be seen, for the fourth time the generic participant 'permit' is introduced in this clause complex. Specifically, in this clause complex 'a valid permit'

19 Notation used: 'α' stands for primary clause in hypotactic relation; 'β' stands for secondary clause in hypotactic relation; '1, 2, 3' stand for paratactically related clauses.

is part of a condition expressed by means of the secondary clause in hypotactic relation (β β). In this case, too, though, confusion is likely to arise, since the identity of the permit is not clearly stated.

α		*Students* are reminded that
β	β	if *they* park a motor vehicle on University premises without *a valid permit*,
	α	*they* will be fined £20.

Figure 4.21. The structure of the first clause complex in paragraph 4.

In particular, the valid permit at issue might be interpreted as a different permit, namely one enabling students (not members of staff) to park on university premises, rather than just to bring their vehicles into the grounds.

As said above, the word 'permit' also occurs in questions 14.01, 14.03, and 14.04. In question 14.01, in particular, the noun group 'their parking permits', with 'parking' functioning as Classifier, presumes anaphorically the sub-class of permits that allow their owners (the employees in this question) to park within the university. The participant 'parking permits', however, is not presented in the text, although, as pointed out above, in the final clause complex the verb 'to park' is used within the conditional clause introduced by 'if' (β β). The verb 'to park' also occurs in the fourth clause complex, whilst in the sixth clause complex the verb 'to keep (a motor vehicle)' is used. In other words, the nominal group 'their parking permits' presumes information that is not directly retrievable, but rather to be implied from the context, and bridging its identity from the co-text is necessary to fulfil this part of the task. In question 14.03 'a university permit' occurs within the imperfective embedded clause 'having a University permit'. In this way, the congruent fact of possessing a university permit (a non-participant meaning) is nominalised and turned into a participant-like meaning (Martin 1992: 138). This 'incongruent' realisation is interpreted here as one that anaphorically presumes the participant 'a university permit' via bridging. Finally, in question 14.04 'parking permits' presents the sub-class of permits already presumed in the first question. On theoretical bases, therefore, it can be hypothesised that the uncertainty about the identity of the participant 'permit' and of 'parking permit' might cause difficulty. In fact, as will be seen in 5.7.1 below, this is partly confirmed by the feedback obtained from the students, in that a high percentage of them responded by saying that they found it difficult to understand whether the text is about one or more permits. Furthermore, some of the feedback about the difficulty of the tasks in test 14 suggested that one of the main difficulties encountered concerned the retrieval of the necessary information in the text, since the questions are worded differently from the text. As for the performance on the items, question 14.04 is in fact the only question that turned out to be difficult, whilst the other questions proved to be relatively easy. As will be argued in 5.7.1 below, the main difficulty of this question is, however, likely to be caused by the specific problems concerned with the 'true/false/not stated' technique, rather than by the problematic issues raised above.

In Figure 4.22[20] the reference chains formed by 'permit' are aligned under one another, following the strategy that Martin (1992: 146) uses to represent 'generic chains' (i.e. chains generated by generic participants). As can be seen from Figure 4.22, non-phoric nominal groups are mainly used to realise generic participants, while the phoric groups used need bridging to successfully presume implied information.

For these reasons, a remark can be made about the fact that these reference chains share many characteristics with 'similarity chains' or 'taxonomically based lexical strings'. As seen in 4.2.2.1 above, Hasan (1989: 85) describes this latter kind of chain as "related by co-classification and/or co-extension" and those in which "the relationship between items is not identity of reference, but similarity of reference, so that the referents lie within the same field of meaning". 'A University permit' can thus be interpreted as the general class of permits provided by the university, i.e. the superset, whilst 'an (appropriate) entry permit' and 'parking permits' as two subclasses of permits, taxonomically related to it. In this respect, Martin (1992: 419) further points out that there is often "considerable overlap between some strings and chains". In fact, in the test under

20 In Figure 4.22 'Th' stands for Theme, the dependency arrow ↑ indicates the dependency of a presuming element on a preceding one, 'BR' stands for 'bridging', 'HO' stands for 'homophoric'.

enquiry understanding and deciding whether identity or similarity of reference is at issue seems to have important implications for the successful completion of the tasks. As already anticipated, even though some students perceived these textual characteristics as causing confusion and, therefore, processing difficulty, the majority of them eventually managed to answer the questions correctly.

Permit	
↑ (HO)	*TEXT*
the appropriate permit	
A University permit (Th)	
An appropriate entry permit	
a valid permit	
↑ (BR)	*TASKS*
their parking permits	
↑ (BR)	
(Having) a University permit	
Parking permits (Th)	

Figure 4.22. Reference chain(s) concerning the participant 'permit' in test 1.

4.4.4 Test 22

Test 22 was chosen from the second part of University of Cambridge ESOL Examinations' First Certificate in English. Text 22, in particular, is an article that aims to provide advice to 'college leavers heading for their second job'. The following are the first two clauses in text 22:

(104) A lot of advice is available for college leavers heading for their first job. In this article we consider the move to <u>a second job</u>.

An element which deserves special attention in this first part of the text is the Premodifier of the word 'job', 'second', occurring in the second clause. According to Halliday (2004: 318) and Matthiessen (1995: 703-704), the ordinal numeral 'second' could be interpreted as realising the function of ordinative Numerative, and could be used as the Premodifier of the Head 'job'. In this particular case, however, the nominal group 'a second job' presents a generic participant that seems to be independent of the specific participant presumed (through bridging) by 'their first job' in the preceding clause. According to this latter interpretation, therefore, 'second' should be considered as a Classifier, in that 'second' does not presume a group of jobs of which the participant in question is the second (Martin 1992: 115). In the nominal group 'a second temporary position', on the other hand, 'second' is actually used as an ordinative Numerative, in that it pre-modifies the sub-class 'temporary position', which is, of course, a different participant from 'a second job'. Another consideration supporting this latter interpretation is the fact that three out of seven questions (test method: multiple-choice) contain nominal groups presuming this participant and asking the reader to disambiguate between 'first', 'second', 'future', 'permanent', 'temporary' job/work, viz. questions 22.02, 22.03 and 22.04.

(105) 22.02 According to the writer, why is the choice of your second job important?
(a) It will affect your future job prospects.
(b) It will last longer than your first job.
(c) It will be difficult to change if you don't like it.
(d) It should give you the opportunity to study.

(106) 22.03 'it' in line 26 refers to your
(a) first job.
(b) second job.
(c) application.
(d) career.

(107) 22.04 If you have a desire to travel, when does the writer suggest that you do it?
(a) straight after you have left college.
(b) when you are unable to find a permanent job.
(c) after you have done some temporary work.
(d) between the first and second job.

On the basis of these considerations, a series of hypotheses can be formulated. Firstly, the fact that 'second job' is so often at issue in the tasks might be an indicator of its centrality in text 22. Secondly, and more generally, that deciding whether a given participant is generic, i.e. referring to a generic sub-class, or whether it is a particular instance of a class, might cause difficulty. Thirdly, that the similarity of some of the participants in this text might cause confusion. Finally, but related to this last point, the fact that in some parts of the text presuming the identity of 'second job' might be difficult, in that, as can be seen from Figure 4.23[21], one of the most often recurrent presuming devices used is the personal pronoun 'it', which might presume any of the types of job mentioned. Furthermore, the nominal group 'what sort of job' in 'What sort of job should you look for?' was analyzed as ambiguous ('AM'), since the kind of job it presumes is unclear.

Second job	
(to) a second job	*TEXT*
↑ (AM)	
what sort of job	
↑	
your second job	
↑	
It	
↑	
This job (Th)	
↑	
It (Th)	
↑	
(most important), it (Th)	
↑	
the second job	
↑	
it	
↑	
my job (Th)	

Figure 4.23. Reference chain concerning the participant 'second job' in test 22.

As will be seen in 5.6 below, the feedback obtained from a large number of test takers supports these suggestions. In fact, some of them said that they found it rather difficult to distinguish between the

21 In Figure 4.23 'Th' stands for Theme; the dependency arrow ↑ indicates the dependency of one presuming element on a preceding one; 'AM' stands for 'ambiguous'.

notions of 'first' and 'second' job, and, more generally, to understand the first two paragraphs of text 22. Some test takers also found it difficult to retrieve the identity of the participant presumed by 'it' in line 26, and did not manage, therefore, to answer question 22.03 correctly.

In test 22 at least one similarity chain is likely to provide test takers with an important clue, namely the one formed by the nominal groups 'career ladder' (embedded in 'a real step in the career ladder'), 'serious career planning' (embedded in 'evidence of serious career planning'), and 'a career'. The first two nominal groups occur in text 22 as follows:

(108) Instead, we will be dealing with those of you taking a real step on the career ladder, choosing a job to fit in with your ambitions [...] This job will be studied very carefully when you send your letter of application for your next job. It should show evidence of serious career planning.

The general participant 'a career', on the other hand, is introduced in question 22.01, option C as follows:

(109) 22.01 Who is intended to benefit from the advice given in the article?
(a) students who have just finished their studies.
(b) people who are unhappy with their current job.
(c) those who are interested in establishing a career.
(d) people who change jobs regularly.

In particular, the noun 'career' functions as Classifier in the first two nominal groups, in which it neither introduces nor presumes participants (see 4.3 above), and as Thing in the last one. It can be claimed, therefore, that the information provided by the repetition of the noun 'career' is important to answer question 22.01 correctly, irrespective of its function in the nominal group. In fact, as will be seen in 5.6 below, this item turned out to be rather easy, in that 84% of candidates answered it correctly.

4.4.5 Tests 15, 3 and 13

In some tests of the corpus what candidates are implicitly asked to do is to decide upon the identity of related participants. In these cases a given reference item in the tasks "is interpreted not by being identified with what it presupposes, but by being compared with it" (Halliday and Hasan 1976: 313). As said in 4.1.2 above, Martin (1992: 100) calls this kind of phoric relation 'relevance phoricity', and captures it systemically by means of the system [comparison/-] (see 4.1.3 above). Halliday (2004: 560) claims that:

> in comparative reference, the reference item still signals 'you know which'; not because the same entity is being referred to over again but rather because there is a frame of reference – something by reference to which what I am now talking about is the same or different, like or unlike, equal or unequal, more or less. Comparative reference items function in nominal and adverbial groups; and the comparison is made with reference either to general features of identity, similarity and difference or to particular features of quality or quantity.

The principal comparative reference item types in English are organised in Figure 4.2. As can be seen, comparative reference items can function either in nominal groups or in adverbial groups, and if the former is the case, they can occur in post-Deictic ($Deictic_2$), Numerative, or Epithet position. Moreover, the nature of the comparison can be of identity, similarity, difference, quality or quantity.

There are reasons to hypothesise that decoding the semantic information conveyed by lexicogrammatical items realising relevance phoricity might cause difficulty to test takers. In fact, in these cases not only reminding phoricity is involved, but also relevance and even redundancy phoricity can be at issue. If this last type applies, the three forms of phoricity have to be processed simultaneously. In test 15, test 3 and test 13, in particular, the comparative phoric relations occurring between some relevant participants identified in the text and presumed later on in some of the accompanying questions stand out as especially relevant for the successful completion of the tasks. They therefore deserve the special treatment that will be devoted to them in the following paragraphs.

	Nominal group			Adverbial group
	Deictic$_2$ (adj.)	*Numerative (adv.)*	*Epithet (adj., adv.)*	
Identity	e.g. same, equal, identical		e.g. identical	e.g. identically, (just) as
Similarity	e.g. similar, additional		*Comparative adj*.: e.g. such	e.g. so, likewise, similarly
Difference	e.g. other, different		e.g. different	e.g. otherwise, else, differently
Quality or Quantity		*Submodifier*: e.g. more, fewer, less, further; so, as … + *Subhead*: *Numeral quantifier*, e.g. much, many	*Comparative adj*.; bigger OR *Submodifier*: e.g. more, less, so, as … + *Subhead*, adj.	*Comparative adv*.: e.g. better OR *Submodifier*: e.g. more, less, so, as … + *Subhead*: adv.

Figure 4.24. Comparative reference items in English (adapted from Halliday 2004: 561 and Martin 1992: 114-119).

4.4.5.1 Test 15

Test 15 was chosen from the second section of the general training Reading Module of IELTS (see 2.3 above), a testing system managed by British Council, IDP: IELTS Australia and University of Cambridge ESOL Examinations. It can be claimed that in this test relevance phoricity is especially at issue, since in 6 out of 7 questions (test method: true/false/not stated, with the task preceding the text) there are phoric resources that refer backwards to participants introduced in the input text and that code either relevance phoricity (comparative reference, in particular) or both reminding and relevance phoricity. All the following underlined nominal groups contain this type of elements:

(110) 15.01 Chiswick Polytechnic was closed at <u>the same time West Thames College was opened</u>.
15.02 <u>Most of the students at the college</u> come from outside the local area.
15.03 The college changed its name to West Thames College in 1993.

15.04 There are currently 6000 students over the age of 19 attending the college.
15.05 Students under the age of 16 cannot attend any of the courses offered by the college.
15.06 The college offers a more mature environment in which to learn than a school.
15.07 There are fewer subjects to study in the sixth form of a school than at the college. (Tasks 15)

The nominal group 'the same time West Thames College was opened' in question 15.01 contains the post-Deictic 'same', the Thing 'time', and the Qualifier (embedded clause) 'West Thames College was opened'. Specifically, the Thing and the Qualifier presume the moment when West Thames College was opened (reminding phoricity), and the post-Deictic 'same' indicates identity of reference (relevance phoricity). In order to verify the truth of the statement made in question 15.01, test takers need to take the beginning of the text into account:

(111) West Thames College (initially known as Hounslow Borough College) came into existence in 1976 following the merger of Isleworth Polytechnic with part of Chiswick Polytechnic.

As can be seen, the information necessary to check that what is stated in 15.01 is 'not stated', is that the text mentions the merger of 'Isleworth Polytechnic with part of Chiswick Polytechnic', but not the closure of Chiswick Polytechnic. What is, in fact, at issue in the text is just a 'portion' of the college, which is indicated linguistically by the use of a nominal group with the Numerative 'part' as Head. No information is, therefore, 'stated' about the whole college.

In question 15.02 it can be seen that two different functional elements in the nominal group 'most of the students at the college' realise two types of phoricity. Specifically, the Qualifier (prepositional phrase) 'at the college' embedded in this nominal group limits ('locates') the identity of the students to those attending West Thames College, thus facilitating reminding phoricity. The Numerative 'most', on the other hand, signals that it is the majority of the students at the college that is at issue in this question (relevance phoricity). The necessary information that has to be retrieved from text 15 is contained in the following extract:

(112) In fact, about eight per cent of its students come from continental Europe and further afield, whilst a further 52 per cent are from outside the immediate area.

It can be seen that in this clause complex the sum of the percentage of students from continental Europe, i.e. 'about eight per cent of its students', and that of the students from outside the immediate area, i.e. 'a further 52 per cent', corresponds to the majority of college students, which provides the required information. As already discussed in 4.3 above, the Head of the nominal group 'most of the students' is the Numerative 'most' and the Thing is 'students'. The same is true for 'about eight per cent of its students', in which 'about eight per cent' is the Numerative and 'students' is the Thing. In 'a further 52 per cent', on the other hand, the Head is '52 per cent' and the Thing, i.e. 'students', is ellipted.
In order to answer question 15.04, the following clause in text 15 is to be focused on:

(113) More than half of West Thames College' s 6000 students are over 19 years old.

According to Halliday (2004: 215-226), this clause can be described as an 'intensive attributive relational' clause, in that the entity 'students' is ascribed or attributed to the class made up of over 19-year-old students. The nominal group 'over 19 years old', in particular, occurs as the 'Attribute' in the clause and is characterised by the use of the Epithet 'old' as Head (see also Halliday 2004: 331). The nominal group 'more than half of West Thames College's 6000 students' can be represented as follows:

more than	half	of	West Thames College's 6000 students
Deictic$_2$	*Numerative*	*prep.*	*Deictic + Numerative + Thing*
Premodifier	*Head*	*Postmodifier*	
β	α	γ	

Figure 4.25. The structure of 'more than half of West Thames College's 6000 students'.

As can be seen, in this nominal group 'half' and 'students' function as the Head and the Thing respectively. The Numerative 'half' is pre-modified by the pre-Deictic 'more than', which carries out a comparative function. On the other hand, the statement contained in question 15.4, i.e. 'there are currently 6000 students over the age of 19 attending the college', is an existential clause, that is, a clause representing the fact that "something exists or happens" (Halliday 2004: 256-259). In the nominal group '6000 students over the age of 19 attending the college' can be represented in Figure 4.26:

6000	students	over the age of 19 attending the college
Numerative	*Thing*	*Embedded prepositional phrase (Qualifier)*
Premodifier	*Head*	*Postmodifier*
β	α	γ

Figure 4.26. The structure of '6000 students over the age of 19 attending the college'.

In this nominal group the noun 'students' functions as both Head and Thing, and the prepositional phrase embedded in it conveys the same information contained in the Attribute 'over 19 years old' in text 15 but further qualified by the embedded clause 'attending the college'. It is interesting to notice that the item writer 'manipulated' the structure of the original nominal group, packaged most of the information contained in the above clause in a single nominal group, and used the noun 'students' as both Thing and Head. The statement in 15.4 above is, of course, 'false', in that the actual number of West Thames College students specified by the former nominal group from, i.e. the necessary information, are more than 3000 (half of 6000), but definitely not 6000.

In question 15.05 it is stated that 'students under the age of 16' cannot attend any course at the college. In order to verify the correctness of this claim, the following extract from text 15 is to be processed:

(114) The college is also very popular with <u>16-18 year olds</u>, who see it as a practical alternative to a further two years at school.

As can be seen, in the text no information about 'students under the age of 16' is given, but only about '16-18 year olds': the correct answer is, therefore, 'not stated'. The structures of these two nominal groups are represented in Figures 4.27 and 4.28 below:

Students	under the age of 16
Thing	*Embedded prepositional phrase (Qualifier)*
Head	*Postmodifier*
α	β

Figure 4.27. The structure of 'students under the age of 16'.

16-18	year	olds
Numerative	*Classifier*	*Thing*
Premodifier		*Head*
β		α

Figure 4.28. The structure of '16-18 year olds'.

It can be claimed that both the Qualifier 'under the age of 16' in the first nominal group and the Numerative '16-18' in the second one perform a comparative function, in that they both refer backwards to the class of 'students' and compare them with reference to the feature of quantity.

Question 15.06 reads: 'The college offers a more mature environment in which to learn than a school'. In order to decide, the structure of the nominal group contained in question 15.06 and represented in Figure 4.29 is to be processed. As can be seen, the structure of this nominal group is complex, in that it contains a 'cascade' of pre- and post-modifying relations. First of all, the Epithet 'mature' is pre-modified by the adverb 'more', and this complex Epithet in turn pre-modifies 'environment'. 'Environment' is post-modified by both the embedded relative clause 'in which to learn' and by the embedded (elliptical) comparative clause 'than at school'. The semantic domain of the embedded relative clause is the Thing 'environment', whilst the 'strict semantic domain' of the embedded comparative clause is the pre-modifying Epithet 'more mature' (see, for instance, Halliday 2004: 434). Another aspect of this nominal group to be considered is that the reference item 'more' is used cataphorically within the nominal group, since 'the reference point' for 'more' lies in the two embedded clauses that follow (see Halliday 2004: 561). This type of comparative reference can therefore be defined as 'structural'.

a	more	mature	environment	in which to learn	than a school
Deictic	β	α	*Thing*	*embedded relative clause (1)*	*embedded comparative clause (2)*
	Epithet			*Qualifier*	
Premodifier			*Head*	*Postmodifier*	
β			α	γ	

Figure 4.29. The structure of 'a more mature environment in which to learn than a school'.

The necessary information to verify the truth of this statement, on the other hand, is to be retrieved from the following portion of text:

(115) These days there is no such thing as a typical student. More than half of West Thames College's 6000 students are over 19 years old. Some of these will be attending college part-time under their employers' training schemes. Others will want to learn new skills purely out of interest, or out of a desire to improve their promotion chances, or they may want a change in career.

(116) The college is also very popular with 16-18 year olds, who see it as a practical alternative to a further two years at school. They want to study in the more adult atmosphere the college provides.

Special attention should, in particular, be paid to the second paragraph and to the nominal group represented in Figure 4.30 below:

the	more	adult	atmosphere	the college provides
Deictic	β	α	*Thing*	*embedded relative clause (Qualifier)*
	Epithet			
Premodifier			*Head*	*Postmodifier*
β			α	γ

Figure 4.30. The structure of 'the more adult atmosphere the college provides'.

As can be seen, not only are the two nominal groups under investigation characterised by a similar structure, but also by the use of synonyms carrying out the same function, viz. 'mature' and 'more adult' functioning as Epithet, and 'environment' and 'atmosphere' functioning as Thing (and Head). Furthermore, in both groups the Numerative 'more' is used, and both Things (Heads) are qualified by an embedded clause. What differs in the two structures is the

Determiner: in the former the use of the indefinite article 'a' identifies a participant, i.e. 'environment', which is compared to other similar 'school' environments by indicating forward reference within the nominal group, i.e. esphora (see 4.1.2 above). By contrast, in the latter nominal group the definite article 'the' is text cohesive, in that it refers back to the previous paragraph, where the characteristics of the students of the college are described (see 4.1.2 above).

Question 15.07 reads: 'There are fewer subjects to study in the sixth form of a school than at the college.' The nominal group 'fewer subjects to study in the sixth form of a school than at the college', in particular, can be analysed and represented as in Figure 4.31. As can be seen, the structure of this nominal group is similar to the one of 'a more mature environment in which to learn than a school' represented in Figure 4.29 above. In particular, the semantic domain of the elliptical embedded comparative clause 'than [there are] at the college' is the Numerative 'fewer' in the nominal group: the comparative reference is therefore structural. The information that is necessary to answer question 15.07 is to be found in the following clause in text 15:

fewer	subjects	to study in the sixth form of a school	than at the college
Numerative	*Thing*	*emb. relative clause (1)*	*emb. comparative clause (2)*
		Qualifier	
Premodifier	*Head*	*Postmodifier*	
β	α	γ	

Figure 4.31. The structure of 'fewer subjects in the sixth form of a school than at the college'.

(117) They can choose from <u>a far wider range of subjects than it would be practical for a sixth form to offer.</u>

As can be seen, the word 'subjects' is used in both question 15.07 and in the text, which should make it easier to retrieve the passage and should create lexical cohesion in the test as a whole. The relevant nominal group in the text is represented in Figure 4.32.

A	far	wider	Range	of	subjects [[than it would be practical [[for a sixth form to offer]]]]
Deictic	β	α	*'measure item'*	*pre.*	*Thing + Qualifier (the embedded comparative clause contains another embedded clause)*
	Epithet				
Premodifier			*Head*	*Postmodifier*	
β			α	γ	

Figure 4.32. The structure of 'a far wider range of subjects than it would be practical for a sixth form to offer'.

In the former nominal group (see Figure 4.31) 'subject' is both the Thing and the Head of the nominal group and is preceded by the Numerative 'fewer'. In contrast, in the latter group (see Figure 4.32) the Head of the group is the 'measure item' 'range', pre-modified by 'a far wider', and the Thing is 'subjects'. In both cases 'subjects' is followed by a rank-shifted structure. Specifically, in the former nominal group 'subjects' is followed by an embedded relative clause followed by an elliptical embedded comparative clause, whilst in the latter it is followed by an embedded comparative clause containing another embedded clause. Both Qualifiers also contain the participant 'sixth form'.

As far as the correct completion of the task is concerned, it can be claimed that test takers need to pay careful attention to the structures of these two groups and verify that the statement made in question 15.07 is 'true'. Specifically, they need to focus on the relevance phoricity (comparative reference) realised in the first nominal group by the Numerative 'fewer' and in the second one by the intensified Epithet 'far wider'. They also need to consider the information contained in the two embedded structures. As will be seen in 5.7.1 below, the feedback obtained from a high percentage of students suggests that the considerations made in this paragraph about the difficulty caused by the structural complexity of some nominal groups in the questions and in the text are correct. The tasks turned out to be rather easy, apart from question 15.01, which proved to be very difficult, and question 15.05. As suggested in 4.4.3 above for question 14.04, however, the main problem of question 15.01 has to do with the fact that the correct answer to this question is 'not stated'.

4.4.5.2 Test 3

Test 3 is part of the DIALANG sub-corpus (see 2.1 above). As explained in 3.2.1.2 and in 3.5.2 above, the findings of the investigation carried out into the lexical density and the readability of the texts in the corpus revealed that text 3 is the second least dense text and the most readable one. In test 3 (test method: multiple-choice gap-filling) test takers have to decide which sentence out of four best fills the gap in the following text:

(118) "I do think this course should be fun," he said. "And people should get out of it what they want. Some come here wanting to be serious and even a bit competitive about their work. And that's fine. Others are very hesitant and find that a bit of humour can relax them. [...] I've taught high court judges and dustmen and the mix is part of the fun." (text 3)

(119) They tend to learn more.
(b) You get all sorts on these courses.
(c) I prefer the serious student.
(d) We only accept professional people. (task 3)

In the short input text reported above the general participant 'people' is introduced and presumed later on in the two subsequent clause complexes by means of two elliptical nominal groups, viz. 'some' and 'others'. The former nominal group is characterised by the use of the Deictic 'some' as Head, whilst the latter by the post-Deictic 'other(s)' used as Head. In the second option the nominal group 'all sorts' can be described as made up of the non-specific Deictic 'all' and the 'type noun' 'sorts' used as Head. This last group is also an elliptical nominal group, in that it does not contain the Thing 'people', even though the reference made to it is inferable. As pointed out in 4.1.2 above, in general, ellipsis signals redundancy phoricity (see also Martin 1992: 100), but this nominal group also signals reminding and relevance phoricity. The three nominal groups are shown in Figure 4.33:

some
Deictic
Head
α

others
Deictic$_2$
Head
α

all	sorts
Deictic	*'Type noun'*
Premodifier	*Head*
β	α

Figure 4.33. The structures of 'some', 'others', and 'all sorts'.

On the basis of these considerations, it can be claimed that the information coded by the first two comparative items covers most of the semantic reference entailed by 'all sorts', which is an important clue as to the choice of the correct option. In fact, approximately 52% of the test takers explicitly said that they found the four options ambiguous, and that they therefore had problems in choosing the correct one. While the majority of them (70%) chose option (b) as the answer to the question, 28.3% answered by choosing option (a), and 1.7% went for option (c) (see also 5.4 below). Option (a) proved then to be the main distracter. Option (c) basically contradicts what is stated in the first clause complex and is actually not correct, while option (a) could be acceptable. In this case, however, the pronoun 'they' would refer back to 'others' and not to 'some', which would (slightly) affect the total coherence of the text. Specifically, it can be argued that the choice of the participant 'they' would be less cohesive with the preceding and following co-text. In this respect, approximately 20% of the candidates complained about the fact that the text is only an extract from a longer text and that it is to be read 'out of context'. They found, in particular, that their lack of knowledge of the context of situation prevented them from clearly understanding the communicative intentions of the speaker (some of them even wondered whether s/he is ironic), and made it difficult for them to decide upon the correct answer. Approximately another 27% of the candidates admitted their lack of knowledge of the precise meaning of the following words and multi-word units: 'high court judges', 'dustmen', 'to get something out of something'.

4.4.5.3 Test 13

As seen in 2.2 above, test 13 was chosen from the reading module of the paper-based TOEFL test, and the test method used in all 10 questions is multiple-choice. Furthermore, as reported on in 3.3.2

above, the findings of the investigation into the lexical variation of the texts of the corpus revealed that text 13 is the least lexically varied, since it contains a large number of words that belong to the same word families. Question 13.03 is worthy of special attention in this paragraph, in that relevance phoricity is involved:

(120) 13.03. In paragraph 5 the author mentions butterflies in order to
(a) include the tropics in the discussion of migration
(b) give another example of migratory creatures
(c) show that these small flying creatures don't get lost
(d) show that butterflies are sturdy creatures

As explicitly stated in the question, test takers have to focus on the following paragraph:

(121) Certain types of butterflies migrate, too, and find their way home over long distances. In the tropics one can sometimes see great mass flights of butterflies all flying steadily in one direction. They may go a thousand miles and more and then return again in another season.

It can be claimed that the most important clue candidates ought to rely on to answer this question is the conjunctive Adjunct 'too', which in Martin's (1992: 101) model realises 'continuity' and, consequently, relevance phoricity (see also 4.1.2 above). In practical terms, not only does 'too' signal that migrating birds can 'find their way home over long distances', but also that butterflies have this ability. The rhetorical aim of paragraph 5 is, therefore, that of further exemplifying the phenomenon of animals' migration, viz. option (b). As will be seen in 5.7 below, item 13.03 turned out to be rather easy, since 84% of the candidates answered it correctly, whilst 16% chose option (c). The high percentage of correct responses to this item, in particular, suggests that the conjunctive Adjunct 'too' was actually perceived as a resource that relates the participant 'birds' to the participant 'butterflies', that is, as a resource that realises relevance phoricity.

Question 13.04 is also interesting to be focussed on, even though it does not involve relevance phoricity:

(122) 13.04. In line 17 the phrase 'this ability' refers to
(a) the ability to fly without stopping

(b) the ability to build nests year after year
(c) the ability to find their way home
(d) the ability to fly in a straight line

The information necessary to answer this question can be retrieved, first of all, from the very first paragraph of text 13:

(123) One of the most mysterious things in nature is the ability of certain creatures to find their way home, sometimes from great distances.

The subsequent narration of what some birds managed to do during two experiments provides test takers with further textual information that helps them choose the correct option, i.e. option (c):

(124) [...] Many experiments have been made with birds in an attempt to find out what guides them on their way home. In one case, seven swallows were taken 400 miles from home. When they were set free, five of them returned to their nests. In another case, a certain kind of sea bird was taken from its nest off the Welsh coast to Venice by plane. When it was released, it made its way home to its nest, a distance of 930 miles if it flew in a straight line.
Migrating birds offer an even more amazing example of this ability. [...]

Specifically, in this extract the congruent processes 'returned (to their nests)' and 'made (its way home to its nest)' are coded as verbs and describe the particular abilities of the birds under investigation. In line 17 of the fourth paragraph these processes are reworded and nominalized as 'this ability', thereby creating textual development and cohesion. It is also to be seen that the clause complex immediately preceding the fourth paragraph contains an appositive construction, viz. 'a distance of 930 miles if it flew in a straight line', on the basis of which option (d), viz. 'the ability to fly in a straight line', was probably devised. It can be claimed that the position in which this last construction occurs enhances the potential for this option to 'distract' test takers. In fact, as will be seen in 5.7 below, only two test takers out of 73 were distracted by this option, which confirms the easiness of this item.

4.4.6 Tests 18, 19, 23, and 21

The present section will concern the role played by participant identification in those tests of the corpus characterised by tasks in which the true/false and the multiple-matching techniques are used, viz. test 18, test 19, test 23, and test 24.

4.4.6.1 Tests 18 and 19

Text 18 and text 19 are tourist leaflets that provide information about a cruise around the Arctic and about an organised trip to Australia respectively. As pointed out in 3.2.1.2, text 19 is the densest text in the corpus, mainly because it contains a large number of lexical words occurring in elliptical and minor clauses. Both tests are part of the PET sub-corpus (see 2.4 above), and their tasks consist of 10 true/false questions presented before the respective input texts. As far as this testing technique is concerned, Alderson (2000: 222) claims that with this type of items

> students are presented with a statement which is related to a target text and have to indicate whether this is True or False, or whether the text agrees or disagrees with the statement. The problem is, of course, that students have a 50% chance of getting the answer right by guessing alone. To counteract this, it is necessary to have a large number of such items.

In order to locate the specific information needed and verify the truth of the statements, test takers ought to read the texts selectively and subsequently read the identified text portions carefully. Most of these statements contain resources that encode reference to participants whose identity is to be recovered from specific parts of the accompanying texts. In question 18.01, for instance, it is stated that:

(125) FOR ONLY £1985 OUR PRICE INCLUDES: Scheduled <u>flights</u> by Australia's national airline, Qantas, from London or Manchester to Perth, <u>returning from Melbourne</u>.

As can be seen, in the nominal group 'scheduled flights by Australia's national airline, Qantas, from London or Manchester to Perth, returning from Melbourne' the Thing 'flights' is pre-modified by the Epithet 'scheduled' and qualified (post-modified) by a series of qualifying elements, among which the embedded clause 'returning from Melbourne'. It can be claimed that the nominal groups 'return flights from Melbourne' and 'flights returning from Melbourne' presume the same sub-class of flights, which confirms the truth of the statement. As will be seen in 5.5 below, item 19.02 also proved to be very easy, in that 98% of the test takers managed to answer it correctly.

4.4.6.2 Tests 23 and 24

Test 23 and test 24 belong to the FCE and CAE sub-corpora respectively (see 2.5 and 2.6 above), and are characterised by the use of the same test method, that is, multiple-matching. Both text 23 and text 24 are newspaper articles that contain a headline, a lead paragraph, and the names of their respective authors. As shown in 3.3.2, text 24 is the second most 'academic' text in the corpus and is about the notion of 'emotional intelligence', whereas text 23 is about the problem of lost luggage at airports. Test 23 consists of 7 headings that have to be matched with the seven paragraphs the text is made up of. An extra summary sentence is also listed with the other headings as a distracter. Test 24, by contrast, contains 12 statements that have to be matched with one of the four paragraphs of the text. In both tests the lists of options are presented before the text to be read. According to Alderson and Cseresznyés (2003: 69) this type of task

> tests the ability to find specific information in a text by scanning, and to follow and understand the main points in a text by skimming. [...] The best approach to these task types is to scan the text for *words in the text* which match *words in the question* – looking for specific information. [...] You should read selectively, ignoring information in the text that is not relevant to your purpose in reading (to answer the questions!).

In order to successfully perform this type of task, therefore, scanning for specific words in the text that match words in the tasks is highly recommended. In many cases this also means looking for participants or classes of participants in the tasks matching participants in the input text. This can be exemplified by taking question 23.04 into consideration:

(126) 23.04 H Busy airports are likely to have more problems.

In this heading reference is made, first of all, to the generic sub-class of participants identified as 'busy airports'. Moreover, the nominal group 'more problems' encodes relevance phoricity, involving as it does contrast in terms of quantity (see 4.4.5 above). In the fourth paragraph of the text these two participants are also mentioned, which provides the clue necessary to successfully perform the task:

(127) these problems can become severe at large transfer airports, known as 'hubs', because of the large number of bags that are processed.

In fact, in this extract 'these problems' is used anaphorically, since it refers back to all the problems listed in the previous paragraph, viz. 'extra security checks', 'problems with air traffic', 'the possibility of human error', and 'an accident in which the destination label is torn off'. Furthermore, the class of 'large transfer airports' in which 'a large number of bags are processed' can be bridged back to that of 'busy airports'. As will be shown in 5.6 below, item 23.04 turned out to be rather easy, since 85% of the test takers managed to match heading H with paragraph 23.04. However, 5.2% answered by choosing heading G, 4.1% chose heading C, 2.1% chose heading E, another 2.1% heading B, 1% heading F, and another 1% heading D.

In eight of the twelve questions (statements) in test 24 the participant 'emotional intelligence' is mentioned. Its occurrences in the tasks, therefore, do not provide a reliable clue as to the text section in which each statement is made. In question 24.01, for instance, the words that are likely to somehow match other words in the text are: the noun 'work', the adjective 'challenging' functioning as Attribute, and the nominalization 'significance':

(128) In which section is the following mentioned?
24.01 The significance of emotional intelligence in work that is challenging

In paragraph C the two following clauses contain evaluative Epithets and Attributes that can actually be related to those used in question 24.01:

(129) And the less straightforward the job, the more emotional intelligence matters – if only because a deficiency in these abilities can hinder the use of whatever technical expertise or intellect a person may have. There are many examples of people who have risen to the top notwithstanding flaws in emotional intelligence, but as work becomes more complex and collaborative, companies where people work together best have a competitive edge.

Specifically, the Epithet 'straightforward' and the adverb 'the less' pre-modifying the adjective 'straightforward' in the nominal group 'the less straightforward the job' allow for bridging the Thing 'job' back to the nominal group 'work that is challenging'. The same identifying function is carried out by the adjectives 'complex and collaborative', pre-modified by the Numerative 'more' and used as Attribute in the clause 'as work becomes more complex and collaborative'. Despite the presence of these lexical overlaps which seem to account clearly for the correct answer, as will be seen in 5.8 below, item 24.01 turned out to be the second most difficult item of test 24. Specifically, 21 students (35%) managed to answer it correctly by choosing paragraph C, 16 students (26.7%) chose Paragraph A, 18 of them (30%) chose Paragraph B, and 5 of them went for Paragraph D (8.3%). According to the feedback obtained from some candidates, paragraphs B and C are the most difficult ones to read in the text, and it can be hypothesised that the particular structure of the first clause complex in the above extract is especially difficult to process:

(130) And the less straightforward the job, the more emotional intelligence matters

Following Biber, Conrad and Leech (2002: 374), the first clause in this clause complex can be described as an adverbial clause used to express proportions; it begins with '*the* + comparative' and requires '*the* + comparative in the main clause'. The use of 'the' here is very strongly marked, as opposed to the unmarked use as a Determiner, which is the one that students expect to find. This very marked use is therefore likely to create interpreting difficulty.

5. Assessing Test Difficulty: Data Analysis of Tests and Feedback Questionnaires

In this chapter the data derived from the administration of the tests and of the feedback questionnaires will be analysed. The chapter will start with a section (section 5.1) devoted to a general overview of the methodologies adopted for the collection and analysis of the data. In section 5.2 the concepts of test reliability and validity as conceived of in language testing will be illustrated briefly. Section 5.3 will be devoted to the presentation of detailed information about the test takers under investigation, the way the tests and the questionnaires were administered, and the way the gathered data was analysed. In sections 5.4 to 5.8, the data concerning each one of the five test batteries into which the twenty-five tests were subdivided will be presented and discussed. Finally, in section 5.9 the discussion of the data for each battery will include comparison with the data for the other batteries, as well as with the indications of text complexity presented in chapters 3 and 4.

5.1 Setting the Scene for this Part of the Study

It can be claimed that the methodologies followed and the techniques used in this part of the study share some characteristics with the pre-testing procedures in Bachman and Palmer's (1996: 234-237) framework, which aim at "collecting feedback for assessing usefulness[1]". In particular, Bachman and Palmer (1996: 237-238) affirm that:

---

1 Bachman and Palmer (1996: 17-18) claim that 'usefulness' is the most important quality of a test, in that it accounts for the use it is intended for. In their model, 'usefulness' includes six test qualities – 'reliability', 'construct validity', 'authenticity', 'interactiveness', 'impact', and 'practicality'.

a fourth[2] purpose of pre-testing is to obtain preliminary information on how test takers respond to the test tasks in three areas: test-taking processes, their perceptions of test tasks, and test performance. To obtain information on the first two, we will typically rely upon qualitative assessment procedures. Feedback on test-taking processes typically includes observations of test takers as they take the test and various kinds of self-reports. Feedback on how test takers perceive a test and react to it typically comes from self-reports of perceptions of the relevance of the test, its difficulty, the appropriateness of the time allocation and administration procedures, and so forth. To obtain feedback on test performance test takers' scores are collected for individual tasks, for individual parts, and for the test as a whole. These scores can be analysed statistically to discover how difficult the individual tasks, the parts, and the whole test are, the degree to which individual tasks are functioning like other tasks, the reliability of the test as a whole, and so forth. These scores can also be used to help judge whether the test is providing the information it was designed to provide.

Clearly, the present study neither aims to provide evidence for the validity and reliability of the testing systems under investigation (see 5.2 below), nor to make inferences about the test takers' reading ability and/or linguistic proficiency. In fact, as explained in chapter 2, the 25 tests chosen belong to testing systems which aim to measure different constructs for reading at different levels of linguistic competence. The main field of enquiry, as already pointed out, is concerned with an attempt to gain insights into the test takers' performance on the various tests, into their perceptions of text and task difficulty, and, finally, into the relationship that this data might have with the data presented in chapter 3 and the analyses carried out in chapter 4. That is, the scope of this study covers text and task, or rather, test difficulty. Research into test difficulty is advocated by Alderson and Banerjee (2002: 84) as follows:

one of the goals of language testing research is to explore the nature of difficulty – of tests, of tasks and items – and the causes of such difficulty. […] Difficulty is relative to readers, and thus the reader's ability and other characteristics have to be taken into account.

---

2 The other purposes of pre-testing in their model are "determining the adequacy/appropriateness/efficiency of administrative procedures", "determining appropriate time allocations", "identifying problems in task specification and clarity of instructions."

In order to investigate the difficulty of the tests under enquiry, both quantitative and qualitative data was collected and different methodologies were followed to obtain it: the quantitative data obtained from the calculations in the third chapter; the qualitative evaluations which emerged from the analysis of the cohesive patterns in the fourth chapter; and the quantitative data, the quantitative feedback, as well as the qualitative feedback obtained from the administration of the tests and of the questionnaires introduced in 5.3 below. Since data from more than one source and various eliciting techniques are used to study the complexity and difficulty of the tests under investigation, it can be claimed that this study is characterised by the "triangulation of research methods across the so-called qualitative/quantitative divide" (Alderson and Banerjee 2002: 83). As Bailey (1999: 38), partly citing from Hammersley and Atkinson (1983: 198), explains:

> the term *triangulation* derives from a loose analogy with navigation and surveying. For someone wanting to locate their position on a map, a single landmark can only provide the information that they are situated somewhere along a line in a particular direction from the landmark. With two landmarks, however, their exact position can be pinpointed by taking bearings on both landmarks; they are at the point where the two lines cross.

In fact, as Bailey (1999: 38) points out

> in social research, the error inherent in one type or source of data may go undetected, but if different types of data lead to the same conclusion, one can be a little more confident in that conclusion. Various approaches to triangulation may be employed to increase the quality control and representativeness of a study.

5.2 Reliability and Validity in Language Testing

As pointed out in the foregoing discussion (see chapter 2, and 5.1 above), reliability and validity are two of the fundamental characteristics of a test and need to be proved by language testers. Reliability can be defined as a "measure of accuracy, consistency, dependability, or fairness of scores resulting from the administration of a particular examination" (Henning 1987: 74). According to Alderson, Clapham and Wall (1995: 87-89), in order to make their test reliable, test constructors have to "reduce the causes of unsystematic variation to a minimum". They should ensure, for example, that the test is administered and marked consistently by applying reliable inter- and intra-rater procedures, that the instructions are clear, and that there are no ambiguous items.

Reliability can be estimated in various ways, among which the so called 'test-retest' measure of reliability, the 'parallel-form' reliability measure, and the 'inter-item consistency' measure (see also Henning 1987: 80-87). 'Test-retest' is the easiest measure to conceptualise, although there are problems with the concept (Alderson and Banerjee 2002: 101). It involves two administrations of the same test to the same candidates on two different occasions, and

> in theory if a person takes the same test on a second occasion, and the test is reliable, the score should remain constant. But the score might have changed because candidates have learned from the first administration or because their ability has changed in some way. In which case, a somewhat lower test-retest correlation might be expected, and this would be a valid indication of the change in ability. (Alderson and Banerjee 2002: 101)

The 'parallel-form' measure, on the other hand, involves the use of parallel (very similar) forms of the same test. However, the "parallel forms of a test are often *validated* by correlations (concurrent validity), and so high correlations between parallel forms would be a measure of validity, not reliability" (Alderson and Banerjee 2002: 101). Finally, 'inter-item consistency' involves a single administration of a given test, the division of the test into two parts, and the measurement of the correlation between the two, as if they were parallel versions. The more strongly the two halves correlate, the higher the reliability (Alderson, Clapham and Wall 1995: 89).

The concept of validity is central to the whole process of language testing and concerns the extent to which a test measures what it is intended to measure. Validity is related to the purpose, the uses and the interpretations of test scores. According to Henning (1987: 89), "validity in general refers to the appropriateness of a given test or any of its component parts as a measure of what it is purported to measure. A test is said to be valid to the extent that it measures what it is supposed to measure. [...] Any test may be valid for some purposes, but not for others". Validity can be proved in a number of different ways by using different methods, and "there are a variety of different perspectives from which evidence for validity can be accumulated, and thus in a sense, validation is never complete: more evidence can always be gathered for or against a particular interpretation of test scores" (Alderson and Banerjee 2002: 102). Alderson, Clapham and Wall (1995: 171-172) distinguish between 'internal', 'external', and 'construct' validity. Internal validity relates to studies of the perceived content of the test and its perceived effect, whilst external validity relates to studies comparing students' test scores with measures of their ability gleaned from outside the test. Construct validity has many of the attributes of both internal and external validity.

Internal validity can be assessed in three main ways: 'face' validation, 'content' validation, and 'response' validation. Face validity refers to the extent to which the test is perceived to reflect the stated purpose, the test's surface credibility, and public acceptability. Data concerning face validity may be gathered by interviewing students or asking them to complete questionnaires about their attitudes or reactions to a test. Content validity concerns the extent to which the content of a test can be said to be sufficiently representative and comprehensive. This type of validation involves the use of expert judgements gathered by following reliable procedures for collecting, recording, and analysing the data obtained. Response validity is the extent to which test takers respond in the way expected by the test developers. Response validation involves gathering introspective data concerning the processes test takers go through and the reasoning they engage in when responding to test items (Alderson, Clapham and Wall 1995: 176). External validity can be of two main types: 'concurrent' and 'predictive'. Concurrent validity is the extent to which test takers' scores on one test relate to those on another externally recognised test, also called 'criterion measure' (Council of Europe 2003: 109). Predictive validity, on the other hand, is the extent to which scores on test Y predict test takers' ability to do test X (Alderson, Clapham and Wall 1995: 181-182), where Y could be, for instance, the score obtained on a placement test for a language course and X the score obtained on the achievement test taken at the end of the course. Construct validity concerns the extent to which evidence can be provided to support the underlying theoretical construct on which the test is based. It basically involves an assessment of the extent to which a test is successfully based on its underlying theory and whether it is a successful operationalisation of the theory. One way of assessing construct validity is to correlate the different test components with each other by calculating their 'internal correlation'. These correlations should be expected to be fairly low (possibly between +3/+5), in that the different test components are designed to measure different abilities. The correlation between each subtest and the whole test, on the other hand, might be expected to be higher (possibly around +.7 or more). It is common practice to correlate a given test component with the test total minus the component in question, since this would artificially inflate the correlation (Alderson, Clapham and Wall 1995: 183-184).

Validity and reliability are often claimed to be complementary concepts, since "a test needs to be reliable to be valid, although the reverse is not necessarily true" (Alderson and Banerjee 2002: 101). Alderson and Banerjee (2002: 101), citing Davies (1978), also argue that "if we maximise reliability, it may be at the expense of validity, and if we maximise validity, it is likely to be at the expense of reliability". This, in other words, means that a balance ought to be kept between the two variables.

5.3 Test and Questionnaire Administration and Data Analysis

The tests described in chapter 2 were organised into five test batteries and administered to groups of students at different levels of language proficiency during the spring semester of the 2005/2006 academic year. A short questionnaire was also attached to the end of each test to obtain feedback from test takers. Most of the participants were language majors enrolled on either a three-year undergraduate course or on a two-year postgraduate course at the University of Padua. They were all attending one of the spring semester English courses targeted at the specific curricular needs of their year of study. Their language teachers were able to provide information about the language proficiency of their group on the basis of the ongoing coursework. Three other small groups of students were recruited from English courses for Engineering, Education Studies and Communication Science students. The data analysis concerning each test battery will be reported on in the following paragraphs along with an interpretation of each data set. A relevant aspect of the test administration, however, is that all the tests were inserted in the software *QuestionMark Perception*, version 3.3.1[3], and that all the language majors and the Communication Science students sat their respective tests in a language laboratory and used the computer to respond to the questions. The students who took battery 1 read the input texts on screen and answered the questions on the computer, whilst the groups of students taking the other test batteries read printed versions of the input texts and gave their responses using the computer. The Engineering and the Education Studies students who took test batteries 1 and 2 read the texts on paper and gave their answers to the questions on an answer sheet.

As suggested above, in order to gather information about how test takers evaluated the difficulty of the texts and tasks they were asked to read and perform, short feedback questionnaires were administered after each single test. The questionnaire consists of the following four questions:

1) Come valuti la difficoltà di questo testo?[4]
(*1 = non difficile*; *2 = non molto difficile; 3 = abbastanza difficile; 4 = molto difficile*)

O 1 O 2 O 3 O 4

2) Come valuti la difficoltà della/e domanda/e[5]?

3 Information about the software is available online at <http://www.question mark. com/uk/perception/index.htm>.

4 English translation: How difficult did you find this text? (1 = not difficult; 2 = not very difficult; 3 = rather difficult; 4 = very difficult)

(1 = non difficile/i; 2 = non molto difficile/i; 3 = abbastanza difficile/i; 4 = molto difficile/i)

O 1 O 2 O 3 O 4

3) Cosa hai trovato di difficile nel testo[6]?

4) Cosa hai trovato di difficile nella/e domanda/e[7]?

As can be seen, the first two questions in the questionnaire are 'closed-ended', whereas the last two are 'open-ended'. The questionnaire is in Italian, so as to allow the students to focus on the content of their responses rather than on the form. According to Dörnyei's (2003: 36) terminology, the closed-ended questions used for this study are 'rating scales', in that they

> require the respondent to make an evaluative judgement of the target by marking one of a series of categories organised into a scale. [...] The various points on the continuum of the scale indicate different degrees of a certain category.

Specifically, they are 'Likert scales', i.e. "series of statements all of which are related to a particular target (which can be, among others, an individual person, a group of people, an institution, or a concept); respondents are asked to indicate the extent to which they agree or disagree with these items by marking one of the responses [...]" (Dörnyei 2003: 37). For this study, four-point Likert scales were used, although the optimal number of response options is debatable. In fact, originally Likert scales contained five response options, but

> some researchers prefer using an even number of response options because of the concern that certain respondents might use the middle category [...] to avoid making a real choice, that is, to take the easy way out" (Dörnyei 2003: 37)

Henning (1987: 23) also affirms that for scoring purposes it is not important whether the scale runs from 1 to 4 or from 4 to 1, "so long

5 English translation: How difficult did you find this/these question/s? (1 = not difficult; 2 = not very difficult; 3 = rather difficult; 4 = very difficult)

6 English translation: What did you find most difficult about the text?

7 English translation: What did you find most difficult about the question/s?

as the direction is borne in mind and the same procedure is consistently applied."[8] Bachman and Palmer (1996: 243) claim that:

> a strength of rating scales is that they elicit responses to specific questions in the form of scaled, quantifiable data which can then be subjected to powerful statistical analyses. A weakness is that they restrict the range and content of stakeholders' responses.

By contrast, in their view

> a strength of the open-ended questionnaire format is that it elicits responses that might not be anticipated. A weakness is that it does not insure that the respondent has considered specific responses which are of interest" (Bachman and Palmer 1996: 243)

Open-ended questions were used for this study, in that pre-prepared response categories could not be provided that would cover these points; moreover, it was thought that they might have led "us to identify issues not previously anticipated" (Dörnyei 2003: 47) related to text and task difficulty.

Questionnaires are one of the most important methods of data collection in second language research, and their "popularity is due to the fact that they are easy to construct, extremely versatile, and uniquely capable of gathering a large amount of information quickly in a form that is readily processable" (Dörnyei 2003: 1). This holds true for closed-ended questions, while the processing of open-ended questions is less straightforward and requires some coding. In order to overcome this problem, Dörnyei (2003: 116-117) suggests that with

> short-answer questions [...] the categorization process involves more potentially subjective elements on the part of the coder. In order to avoid the harmful effects of such rater subjectivity, these items are to be processed by means of some systematic *content analysis*, whereby the pool of diverse responses is reduced to a handful of key issues in a reliable manner. This is usually achieved through a stepwise process that involves two broad phases:

---

8 Weir, Huizohong and Yan (2000: 260-264) also used a Likert scale running from 1 to 4 to gather data about the test takers' evaluation of the texts and tasks in their test. In their questionnaire survey the value label 1 also stood for 'not difficult' and the label 4 for 'very difficult'.

> […] 1. Taking each person's response in turn and marking in them any distinct content elements, substantive statements, or key points. 2. Based on the ideas and concepts highlighted in the texts […], forming broader categories to describe the content of the response in a way that allows for comparisons with other responses. The categories obtained in phase 2 can be numerically coded and then entered into the data file to be treated as quantitative data.

For practical reasons, the responses to the open-ended questions in the questionnaire were not assigned a number. Rather, the detailed information contained in the responses was condensed "into a limited number of categories; thus, the assigned codes can be seen as shorthand symbols standing for the longer replies" (Dörnyei 2003: 99). Ongoing decisions were made, in particular, about how to label similar but not identical responses, which was sometimes difficult. As for the distinction between quantitative data and qualitative data, Bayley (1999: 39) affirms that in the field of language learning and teaching this distinction is somewhat simplistic. She, instead, makes the case for considering two different dimensions: "whether the data are quantitatively or qualitatively collected, and whether they are quantitatively or qualitatively analysed".

The data collected for the present part of the study through the administration of the tests and of the questionnaires can be organised into the following typologies:

- Responses to items;
- Responses to closed-ended questionnaire questions;
- Responses to open-ended questionnaire questions;
- The university course attended and an evaluation of the proficiency level of each group of test takers.

This data was analysed by applying Classical Test Theory methods, and the software *SPSS*, version 13.0.1, was used to carry out some classical test theory analyses. In order to refine them, Item Response Theory was used, with the aid of the software *Winsteps*[9], which is

9 More information about this software can be found at: <http://www.rasch.org>. To be precise, the corresponding free version, *Ministep*, was used. This version of the programme, however, can only analyse a maximum of 30 items and 200 persons (cases).

based on the 'Rash' model. As far as the distinguishing characteristics of the two theories are concerned, it can be said that:

> Classical Test Theory investigates how items perform based on a given test population, and how test takers perform based on a given set of test items. In other words, an item's difficulty is directly related to the abilities of the persons who responded to the item; and a person's performance is directly related to the level of difficulty of the items used. Classical Test Theory usually involves determining such aspects as facility values, levels of discrimination, the contribution each item/part makes to the test's internal reliability, the degree of overlap between the various parts of a test, *inter alia*. It relies heavily on the correlation coefficient as a statistical procedure. (Green 2004: 1)

Item Response Theory, on the other hand,

> is based on probability theory, i.e. the chances of a person answering an item correctly is a function of his/her ability and the item's difficulty. […] The more able the person, the better his chances for success with any item. The easier the item, the more likely any person is to solve it. (Green 2004: 1)

By making reference to Thorndike (1982: 10), Baker (1997: 33) explains that:

> item response models share the view of a test item as providing an indication of the individual's standing on the ability/difficulty scale: a person who succeeds on an item is likely to be positioned above it on the scale, while a person who answers the same item incorrectly is likely to fall below it. Estimating a person's ability which is the IRT equivalent of scoring a test, is therefore a matter of finding the point on the scale that best summarises his/her performance on a set of items. Similarly, estimating the difficulty of an item involves using the information contained in responses to it by a group of persons in order to find its likely location on the scale.

In order to obtain reliable and detailed information by using one of the models belonging to the IRT family, however, large numbers of candidates and items are needed. Specifically, the One-parameter logistic model (Rash Model), which is a "scale of person ability and item difficulty", is fully operative with a sample of 100 to 200 persons. The Two-parameter logistic model requires 200-400 subjects, but besides including parameters for difficulty, it also includes a continuous estimate of discriminability. The Three-parameter logistic model includes parameters for difficulty, discrimination and guessing, and "depends on the availability of 1,000 to 2,000 persons for parameter estimation to proceed meaningfully" (Henning 1987: 107-116). Furthermore, one of the major assumptions made in these models is that "the item set is unidimensional, i.e. that the items measure a single ability or trait", which requires that the item set be relatively homogeneous (Baker 1997: 30). In the following paragraphs of this chapter each test battery will be dealt with on its own, and the quantitative data – i.e. the test results and the answers to the closed-ended questions – as well as the test takers' responses to the open-ended questions, will be analysed. When relevant, reference will also be made to the data and indications of complexity emerging from the analyses discussed in chapters 3 and 4.

5.4 Test Battery 1 (from Test 01 to Test 10)

The test battery under investigation in this section is made up of the ten tests chosen from the DIALANG testing system. With the exception of item 01, each test in test battery 10 consists of one item only. They were all administered in the order in which they are listed in Figure 5.1 (see also 2.1 above) to 60 language majors at various levels of language proficiency and attending different language courses. 24 test takers were attending an intermediate first-year undergraduate English course ('MLC1int'), 7 students were attending a first-year undergraduate advanced English course ('MLC1adv'), 22 a third year undergraduate English course ('MLC3'), and 7 a second year postgraduate English course ('MLCpost2'). MLC1int was a class of students at level B2 of the CEF, 'MLC1adv' was made up of 5 students at about level B2 and of other 2 students at 'B2+' level, that is, 'strong' B2 students working towards C1 level. 'MLC3' was also a class of students at B2+ level, and 'MLCpost2' one made up of students at C1 level. In Figure 5.1 the population is sub-categorised pictorially according to the language course attended by each test-taker and their estimated CEF level. The time limit set for the whole battery was 50 minutes.

TEST BATTERY 1

Test 01. Eye problems
Test 02. Our hands met
Test 03. It should be fun
Test 04. Ecotourism
Test 05. Digital cameras
Test 06. Travel agency
Test 07. Bangkok
Test 08. Station
Test 09. Suzanne Somers
Test 10. Injuries

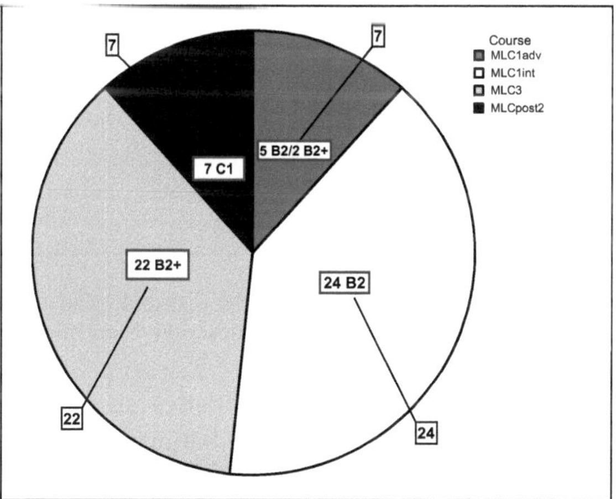

Figure 5.1. Test battery 1 and the characteristics of the population to which test battery1 was administered (the course attended and the estimated CEF level of the students).

Figure 5.2 displays the distribution of the total score obtained on the test as a whole (i.e. 'totbattery1') by the 60 test takers under investigation, along with some descriptive statistics of 'central tendency' and of 'variability'. The former, in particular, are used to "talk about the central point in the distribution of values in the data" (Hatch and Lazaroton 1991: 159), and are the Mean, the Median, and the Mode. The latter, on the other hand, are used to give information about the "variability of scores within the data set" and are the Standard Deviation, the Minimum and the Maximum (Hatch and Lazaroton 1991: 168-176).

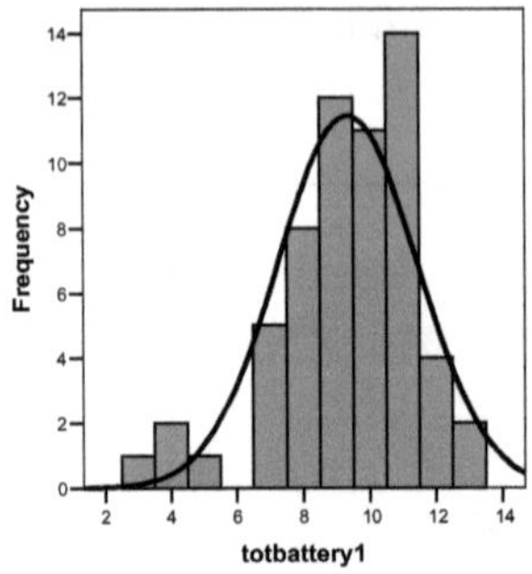

DESCRIPTIVE STATISTICS
N = 60
Mean = 9.35
Mode = 11
Median = 10
Standard Deviation = 2.090
Minimum = 3
Maximum = 13

Figure 5.2. Histogram displaying the distribution of the total score on 'totbattery1' for the whole population and some descriptive statistics.

As the histogram and the statistics suggest, most of the test takers performed well on this test battery, and the Median indicates in particular that half of them obtained a score of 10 or higher. In order to investigate which of the test items were more problematic, in Figure 5.3 below the 'facility values'[10] (henceforth referred to as 'F.V.') of each one of the ten DIALANG tests/items have been listed, along with information about the number of candidates, the lowest and highest values (i.e. Minimum and Maximum), and the Standard Deviation of each item. Since the first DIALANG test involves the matching of 4 different concurrent titles with 4 short texts, it consists of 4 sub-items, i.e. 'test01a', 'test01b', 'test01c', 'test01d' (see also 3.5.2 above). As can be seen from Figure 5.3, the item in the battery that stands out as the most difficult one is test 06 (its F.V. equals .40), since only 40% of the candidates answered it correctly; the next most difficult ones are test 04 (F.V.=.48), test 10 (F.V.=.57), and test 08 (F.V.=.58). 'Test01a' (F.V.=.98) and test 09 (F.V.=.90), by contrast, turned out to be the easiest items in the battery.

As for the most difficult test in the battery – test 06 – 75% of the test takers found that its difficulty was due mainly to their lack of knowledge of the semantic denotations of the lexical words used in the optional choices, namely 'vow', 'commitment', 'appeal', and

10 The 'facility value' of a given item is calculated by dividing the number of candidates who answered that item correctly by the total number of candidates.

‘warning’. In fact, 24 test takers (40%) chose option ‘(a) a commitment’, 14 of them (23.3%) chose ‘(d) a vow’, 11 of them (18.3%) chose ‘(c) a warning’, and another 11 went for ‘(b) an appeal’. The data about the second most difficult test, test 04, was discussed in 3.1 above and the data about test 08, the fourth most difficult one, in 3.2.1.2 above. Test 10, the third most difficult test in the battery, is to be discussed in detail in this section. According to the majority of the test takers’, the main difficulty of text 10 was due to their lack of knowledge of the multi-word unit ‘to be stacked against somebody’ and the word ‘burdensome’, while some of them found it ‘grammatically intricate’. As for task 10, the majority of the candidates said that they found it difficult to decide which option best summarises the main idea of the text. In particular, they found that all the optional choices are somehow related to the contents of the text and that the question is therefore tricky. The question in task 10 was answered correctly by 34 test takers (56.7%) who chose option (d), while 14 test takers (23.3%) chose option (b), other 9 test takers (15%) chose option (a), and finally 3 of them (5%) chose option (c).

In order to check whether there are statistically significant differences in the responses given by the candidates to different pairs of items, a series of analyses were conducted: the paired-samples t-test; the Wilcoxon Signed Rank Test, a non-parametric alternative to the t-test; and the Z-test of proportions. The first test compares the mean scores for the same group on two different items (see, for example, Hatch and Lazaroton 1991: 249-270 and Pallant 2003: 92; 181-183), whereas the second one compares the performance on two items “on the basis of their ranks above and below the median” (Hatch and Lazaroton 1991: 274), which allows for conducting the test on data which are not normally distributed data. Finally, the Z-test of proportions compares two facility values by dealing with them as proportions or percentages and not as mean scores. The analyses were run on the six pairs of items listed in Figure 5.4.

Battery 1	*N*	*Min*	*Max*	*F.V.*	*S.D.*
test01a	60	0	1	.98	.129
test01b	60	0	1	.78	.415
test01c	60	0	1	.82	.390
test01d[11]	60	0	1	.83	.376
test02	60	0	1	.77	.427
test03	60	0	1	.70	.462
test04	60	0	1	.48	.504
test05	60	0	1	.78	.415
test06	60	0	1	.40	.494
test07	60	0	1	.75	.437
test08	60	0	1	.58	.497
test09	60	0	1	.90	.303
test10	60	0	1	.57	.500
Totbatt1	60	3	13	9.35 (.71)	2.090

Figure 5.3. Basic descriptive statistics of test battery 1.

As said in 2.1 above, tests 04 and test 10 aim to test the ability of identifying the main idea in the respective input texts, whereas test 05 and test 06 are claimed to be testing inferencing skills. One would expect, in particular, that test takers perform similarly on items assessing the same skill and differently on items assessing a different one. According to the results of the three statistical tests, however, the differences between test 04 and test 05 (pair 2; p=.001) and between test 05 and test 06 (pair 3; p=.000) are statistically significant, while the ones between test 04 and test 06 (pair 1) and test 04 and test 10 (pair 4) are not. On the basis of these observations, therefore, it can be concluded that the test takers' performance on a given item in battery 1 does not always depend on the ability tested, although this might be the case.

11 The statistics for the sum of the four items in test 01 (i.e. dia01) are the following ones: N=60; Min=1; Max=4; Mean=3.42; St. Dev.=1.046. If 3.42 is divided by 4, the figure obtained (i.e., .85) expresses the mean score as a percentage, thus allowing for the comparison with the facility values.

PAIRS	*Paired t-test*			*Wilcoxon S.R.*		*Z-test of prop.*
	T	*df*	*Sig. (2-t.)*	*Z*	*Asymp. Sig. (2-t.)*	
test04 – test06	.869	59	.389	-.870	.384	1.81
test04 – test05	-3.461	59	.001	-3.182	.001	4.83
test05 – test06	4,457	59	.000	-.388	.000	6.03
test04 – test10	-.869	59	.389	-.870	.384	1.4
test07 – test03	-.725	59	.471	-.728	.467	0.88
test07 – test08	1.932	59	.058	-1.890	.059	2.83 (p=.0023)
test1a – test1b	3.494	59	.001	-3.207	.001	3.52
test1a – test1c	3.085	59	.003	-2.887	.004	4.10

Figure 5.4. Paired t-test, Wilcoxon Signed Rank Test, and Z-test of proportion for some pairs of tests in battery 1.

Pairs 5 and 6 were also deemed worthy of special investigation for various reasons: firstly, the fact that, as seen in 3.2.1.2 and 3.4.2 above, text 7 is the most lexically dense text in the DIALANG sub-corpus, while text 8 and text 3 are respectively the first and second most sparse texts in the corpus, and text 8 is also the most grammatically intricate one; secondly, the consideration that even though the three tests aim to test test takers' inferencing skills, task 3 and task 7 involve the choice of the sentence that best fills the gap in the respective input texts, while task 8 consists in a multiple-choice question, the answer to which is to be inferred by reading the input text. Unlike test 08, therefore, tests 03 and 07 especially focus on the recognition of the textual cohesive elements in the input texts. As can be seen from Figure 5.4, the difference between the performance on test 03 and that on test 07 (pair 5) is not statistically significant, whereas the difference between test 07 and test 08 (pair 6) is significant according to the Z-test of proportions (Z=2.83; p=.0023). According to the t-test and the Wilcoxon S.R. test their difference is not significant, in that the p-values (i.e. Sig.) are .058 and .059 respectively, and they should be equal or less than .05 to be significant. These values, however, are very close to the level of significance, and the performance on the two items can be regarded as dissimilar. Finally, it is interesting to note that the difference between the performance on 'test01a' and 'test01b' (pair 7) and the difference between the performance on 'test01a' and 'test01c' (pair

8) are statistically significant, even though they belong to the same test (i.e. test 01) and involve the choice of one out of the four titles contained in the same list.

In order to investigate into whether there are statistically significant differences between the performance of the four groups on the whole test battery as well as on the single items, a series of one-way between-groups analyses of variance (ANOVA) or Kruskal-Wallis Tests were conducted (see, for example, Pallant 2001: 186-195, and Hatch and Lazaroton 1991: 312-318 for information about this type analysis and the software used to run it)[12]. From the analyses carried out it emerged that there are statistically significant differences at the $p < .05$ level in the performance of the four groups on the whole test (totbattery1) and on test 04, test 05, test 06, and test 07. As for 'totbattery1', it is the mean score for the 'MLC3' students (M[13]=8.68, SD=2.147) and the one for the 'MLCpost2' students (M=11, SD=1.291) that are significantly different from one another at the $p<.05$ level, while those for the other groups are not. An exploration of the performance on test 04 reveals that the difference between the mean score for the 'MLC3' students (M=.23, SD=.429) and the one for the 'MLC1int' students (M=.67, SD=.482) are statistically significant. It also reveals that the 'MLC3' candidates performed the worst on this item. On the basis of the responses given to test 05, the performance of the 'MLC1int' students (M=.96, SD=.204) differs again significantly from the one of the 'MLC3' students (M=.55, SD=.510), in that this latter group's performance was the worst on this item as well. On test 06 the 'MLCpost2' students (M=.86, SD=.378) performed the best, whilst the 'MLC1int' students (M=.25, SD=.442) performed the worst, and the difference is

12 The difference between variances was considered significant only when the Sig. value was less or equal to .05 and the Levene's test for homogeneity of variance was greater than .05. If the Levene's test was less than .05, then the Kruskal-Wallis Test, i.e. the non-parametric alternative to One-Way between-groups ANOVA, was used (see, for example, Pallant 2001: 190; 263-265, and Hatch and Lazaroton 1991: 328-337). The detailed SPSS outputs from the One-way between groups ANOVA analyses and the Kruskal-Wallis Tests for this and the four other batteries are available, but only the statistically significant results will be reported on and commented in this study.

13 'M' stands for Mean Score and 'SD' for Standard Deviation.

statistically significant. Finally, the performance on 'test07' is characterised by statistical differences between all four groups: the 'MLC1int' group (M=.83, SD=.381), the 'MLC1adv' group (M=1, SD=.0), the 'MLC3' group (M=.50, SD=.512), and the 'MLCpost2' group (M=1, SD=0). All that can be learnt from these figures can be summarised by saying that the 'MLCpost2' students obtained the best scores on the whole test and on most of the items, and that the contrary is true for the 'MLC3' students. The 'MLC1int' students, however, obtained the lowest average score on test 06. These differences are likely to be mainly due to the higher proficiency level of the 'MLCpost2' students and to the fact that the 'MLC1int' and the 'MLC1adv' students appeared to tackle this test particularly seriously.

It is interesting to examine how the students evaluated the difficulty of the texts and of the tasks in test battery 1 and to verify whether the difficulty as they perceived it is related to their actual performance on the tests. To this end, first of all, the frequency percentages of the students' perceptions of the difficulty of each DIALANG text have been organised in Figure 5.5 and those of each DIALANG task in Figure 5.6. As suggested by these figures, the majority of students endorsed a score of 2 on the difficulty of both the texts and the tasks of most of the tests. The exceptions to this tendency are 'test05q1', 'test06q1', 'test08q1', 'test08q2', and 'test09q2', in that the majority of candidates assigned the lowest value, 1, in their evaluation of difficulty. On the other hand, 8.3% of them endorsed a score of 4 on 'test06q2', which suggests that 'task06' was perceived as particularly difficult by some test takers. The percentages of the candidates' responses to 'test02q2' are also peculiar, in that 3.3% of the candidates endorsed 4 on it and the remaining endorsements are equally distributed among the values 1, 2, and 3 (31.7% each). In order to gain insights into the central tendency and the variability of these data, some descriptive statistics have been reported on in Figures 5.7 and Figure 5.8. If the Mean column (M) in Figure 5.7 is taken into account, it will be seen that the text which was evaluated as the most difficult one is text 2, while the one that was judged as the easiest one is text 5.

Test/Values[14]	*1*	*2*	*3*	*4*	*miss*
te01q1	21.7	43.3	31.7	1.7	1.7
te02q1	18.3	41.7	35.0	3.3	1.7
te03q1	36.7	45.0	16.7	1.7	
te04q1	21.7	43.3	31.7	1.7	1.7
te05q1	73.3	20.0	5.0		1.7
te06q1	65.0	23.3	11.7		
te07q1	28.3	43.3	26.7	1.7	
te08q1	53.3	30.0	15.0		1.7
te09q1	38.3	38.3	21.7		1.7
te10q1	35.0	41.7	21.7		1.7

Figure 5.5. Frequency percentages of the students' perception of the difficulty of each DIALANG text.

Test/Values	*1*	*2*	*3*	*4*	*miss*
te01q2	25.0	50.0	25.0		
te02q2	31.7	31.7	31.7	3.3	1.7
te03q2	31.7	38.3	25.0	1.7	3.3
te04q2	25.0	48.3	23.3	1.7	1.7
te05q2	36.7	43.3	18.3		1.7
te06q2	18.3	36.7	36.7	8.3	
te07q2	16.7	53.3	25.0	3.3	1.7
te08q2	50.0	35.0	13.3		1.7
te09q2	55.0	25.0	16.7		3.3
te10q2	23.3	48.30	26.7		1.7

Figure 5.6. Frequency percentages of the students' perception of the difficulty of each DIALANG task.

On the other hand, the task that was rated as the most difficult is task 6, whereas the one perceived as the easiest is task 9 (see Figure 5.8).

14 Value labels: 1= not difficult; 2 = not very difficult; 3 = rather difficult; 4 = very difficult.

	N	*Min.*	*Max*	*M*	*Mode*	*Median*	*S.D.*
test01q1	59[15]	1	4	2.14	2	2	.776
test02q1	59	1	4	2.24	2	2	.795
test03q1	60	1	4	1.83	2	2	.763
test04q1	59	1	4	2.14	2	2	.776
test05q1	59	1	3	1.31	1	1	.565
test06q1	60	1	3	1.47	1	1	.700
test07q1	60	1	4	2.02	2	2	.792
test08q1	59	1	3	1.61	1	1	.743
test09q1	59	1	3	1.83	1, 2	2	.769
test10q1	59	1	3	1.86	2	2	.753

Figure 5.7. Descriptive statistics of question 1 on the difficulty of the texts in test battery 1.

	N	*Min.*	*Max*	*M*	*Mode*	*Median*	*S.D.*
test01q2	60	1	3	2.00	2	2	.713
test02q2	59	1	4	2.07	1, 2, 3	2	.888
test03q2	58	1	4	1.97	2	2	.816
test04q2	59	1	4	2.02	2	2	.754
test05q2	59	1	3	1.81	2	2	.730
test06q2	60	1	4	2.35	2, 3	2	.880
test07q2	59	1	4	2.15	2	2	.738
test08q2	59	1	3	1.63	1	1	.717
test09q2	58	1	3	1.60	1	1	.771
test10q2	59	1	3	2.03	2	2	.718

Figure 5.8. Descriptive statistics of question 1 on the difficulty of the tasks in test battery 1.

It is important to point out that the answers to Likert scales can be treated either as continuous variables or as ordinal or 'rank-order'

15 Some students did not answer all the questions in the questionnaire. Consequently, the actual number of respondents varies from item to item and has been specified in column 'N'.

variables[16]. If they are dealt with as ordinal variables, then the "the points on the scales, and the numbers used to represent those points, are not equal intervals" (Hatch and Lazaraton 1991: 57), which implies that the Mean and the Standard Deviation might not be appropriate to describe the data. For this reason the Mode, i.e. the most frequent score obtained in the data, and the Median were also used to describe this type of data. In fact, it can be seen that there are two Modes for 'test09q1' (namely, 23 students chose '1', 23 chose '2', and 13 '3'), and three for 'test02q2' (namely, 19 students went for '1', 19 for '2', another 19 for '3', and 2 for '4').

In order to investigate the relationships between the test takers' evaluations of the difficulty of the texts, those of the corresponding tasks and their actual performance on the tests, a series of correlation analyses were carried out. Correlation analyses were, first of all, conducted between the responses given to survey question 1 and survey question 2 for each test. They were also conducted between the performance on each single test and the answers given to the corresponding survey questions 1 and/or 2. Two important aspects need to be pointed out, however: firstly, the responses given to the survey questions are ordinal variables (either 1, 2, 3, or 4), and they are appropriate data for the Product Moment Correlation Coefficient only if they "approach equal-interval measurement and a normal distribution" (see Hatch and Lazaroton 1991: 436); secondly, the answers given to a single test item are either right or wrong (1 or 0), and can be, therefore, best described as dichotomous nominal variables (see Hatch and Lazaroton 1991: 448). Unfortunately, the

16 Pallant (2001: 95) distinguishes between 'continuous' (also referred to as 'interval level data'), 'ordinal' (also referred to as 'rankings' or 'rank-order'), and 'categorical' variables (also referred to as 'nominal level data'). An example of a continuous variable is given by 'age' expressed in years, an example of an ordinal variable are the answers to a Likert-type scale, and an example of categorical variable is 'sex: male/female'. Hatch and Lazaraton (1991: 180) maintain that "statisticians most often argue in favour of treating rank-order data as though they were interval data. They see ordinal scales as being continuous (with the mean score as the appropriate measure of central tendency). Many argue that statistically it makes no real difference and that it is best to treat ordinal and interval as the same. In many cases we believe it is better to consider them as ordinal rather than as interval data, and especially so if the data are not normally distributed".

Pearson Product Moment Correlation Coefficient is not appropriate to find out if there are statistically significant relations between this latter type of variable. To this end, two non-parametric statistical techniques – which are "ideal when you have data that is measured on nominal (categorical) and ordinal (ranked) scales" (Pallant 2001: 255) – were used along with the Pearson Correlation Coefficient, namely the 'Spearman Rank-Order Correlation (Rho)' index and the 'Kendall's tau' index (see also Hatch and Lazaroton 1991: 451-454). Figure 5.9 shows the statistically significant correlations between the performances on a given item and the responses to the corresponding survey questions. They are all negative, although the significance of the direction of the evaluation and that of the actual performance are the same. In fact, as seen in 5.3 above, a high rank score, e.g. 4, signals high difficulty, whilst a high facility value, e.g. .85 or .90, indicates low difficulty. Therefore, if the correlation between the two is negative, then the higher the evaluation of difficulty, the lower the score obtained, and the lower the evaluation, the higher the score. Figure 5.10, on the other hand, lists the statistically significant correlations between the responses given to survey question 1 and those given to survey question 2. For each pair of variables the three correlation coefficients have been reported on in the tables, as well as the corresponding levels of significance.

Pairs of variables	*Significant Correlations*
test02/test02q2	Pearson: -.364 (N=59, p=.005) Kendall's tau: -.362 (N=59, p=.003) Spearman's rho: -.387 (N=59, p=.002)
test07/test07q1	Pearson: -.331 (N=60, p=.010) Kendall's tau: -.326 (N=60, p=.008) Spearman's rho: -.346 (N=60, p=.007)
test07/test07q2	Pearson: -.357 (N=60, p=.006) Kendall's tau: -.350 (N=59, p=.005) Spearman's rho: -.370 (N=59, p=.004)
test08/test08q2	Pearson: -.402 (N=59, p=.002) Kendall's tau: -.391 (N=59, p=.002) Spearman's rho: -.408 (N=59, p=.001)

Figure 5.9. List of statistically significant correlations between the answer given to a given item and those given to questions 1 or 2.

Pairs of variables	*Significant Correlations*
test01q1/test01q2	Pearson: .433 (N=59, p=.001) Kendall's tau: .377 (N=59, p=.001) Spearman's rho: .417 (N=59, p=.001)
test03q1/test03q2	Pearson: .364 (N=58, p=.005) Kendall's tau: .326 (N=58, p=.006) Spearman's rho: .362 (N=58, p=.005)
test04q1/test04q2	Pearson: .363 (N=58, p=.017) Kendall's tau: .259 (N=58, p=.036) Spearman's rho: .272 (N=58, p=.039)
test05q1/test05q2	Pearson: .311 (N=58, p=.005) Kendall's tau: .259 (N=58, p=.005) Spearman's rho: .272 (N=58, p=.005)
test07q1/test07q2	Pearson: .563 (N=59, p=.000) Kendall's tau: .496 (N=59, p=.000) Spearman's rho: .537 (N=59, p=.000)
test08q1/test08q2	Pearson: .499 (N=59, p=.000) Kendall's tau: .514 (N=59, p=.000) Spearman's rho: .550 (N=59, p=.000)
test09q1/test09q2	Pearson: .559 (N=58, p=.000) Kendall's tau: .529 (N=58, p=.000) Spearman's rho: .562 (N=58, p=.000)
test10q1/test10q2	Pearson: .550 (N=59, p=.000) Kendall's tau: .497 (N=59, p=.000) Spearman's rho: .546 (N=59, p=.000)

Figure 5.10. List of statistically significant correlations between the answers given to questions 1 and 2 on the various tests.

The statistics in Figure 5.9 suggest that the students' performances on tests 02, 07 and 08 are significantly correlated with the respective perceived degrees of task difficulty, and that only the evaluation of the difficulty of text 7 is significantly correlated with the performance on the corresponding test. Specifically, the easier they found text 7 and the tasks in tests 02, 07 and 08, the higher the scores they obtained. As can be seen by inspecting Figure 5.10, the evaluation of the difficulty of text 7 and that of task 7 are also significantly correlated. It can also be seen that the students' evaluations of the difficulty of the other texts are significantly correlated with those of the accompanying tasks, with the exception of those concerning tests 02 and 06. These latter findings can be explained by the fact that, as seen above, text 2 was on average perceived as the most difficult text in the battery, while task 2 was not perceived in the same way. By contrast, task 6 was on average evaluated as the most difficult task, which was not the case for text 6. As far as test 6 is concerned, it is important to notice that it also proved to be the most difficult item in the battery. In Figure 5.11 the statistically significant correlations between test performance, and/or the perceived text difficulty, and/or the perceived task difficulty have been indicated by putting a tick under the column of the correlated elements.

	(1) Performance	*(2) Perceived text diff.*	*(3) Perceived task diff.*
test01		√	√
test02	√		√
test03		√	√
Test04		√	√
test05		√	√
test06			
test07	√ (neg. with 2 and 3)	√	√
test08	√ (neg. with 2)	√	√
test09		√	√
test10		√	√

Figure 5.11. Statistically significant correlations between the performance on the DIALANG tests, the perception of text difficulty, and the perception of task difficulty ('neg.' stands for negative correlation).

In order to find out whether the evaluations of the texts and the tasks of the DIALANG sub-corpus varied according to the group that the respondents belonged to, one-way between-groups ANOVA were carried out on the evaluations of the various texts and tasks. The analyses revealed that the four groups of students differed significantly at the $p<.05$ level in their responses to the following survey questions: 'test03q1', 'test04q1', and 'test06q2'. These findings are also confirmed by the results of the Kruskal-Wallis Tests run on the three variables. In particular, the 'MLCspec2' students were the ones that endorsed the lowest average score (i.e. they rated it as rather easy) on 'test03q1' (M=1.14, SD=.378), while the 'MLC1adv' students endorsed the highest average score on this item (M=2.43,

SD=.535). In contrast, the 'MLC1adv' students average evaluation of 'test04q1' was the lowest among the four groups (M=1.33, SD=.516), while the 'MLC1int' students' evaluation was the highest (M=2.46, SD=.658). Finally, the groups that differed significantly in their responses to 'test06q2' are the 'MLC1int' and 'MLCpost2' groups. In particular, the 'MLCpost2' students were the ones that assigned the lowest average rate score to task 6 (M=1.57, SD=.787), while the 'MLC1int' students endorsed the highest (M=2.71, SD=.908). In Figure 5.12 the significant different average performances and text and/or task evaluations between the four groups are displayed.

Test and F.V.	*Groups*	*Performance*	*textev*	*taskev*
test03	MLC1int		1.96	
.70	MLC1adv		2.43	
	MLC3		1.73	
	MLCpost2		1.14	
test04	MLC1int	.67	2.46	
.48	MLC1adv	.57	1.33	
	MLC3	.23	2.05	
	MLCpost2	.57	2.00	
test05	MLC1int	.96		
.78	MLC1adv	.86		
	MLC3	.55		
	MLCpost2	.86		
test06	MLC1int	.25		2.71
.40	MLC1adv	.43		1.86
	MLC3	.41		2.36
	MLCpost2	.86		1.57
test07	MLC1int	.83		
.75	MLC1adv	1.00		
	MLC3	.50		
	MLCpost2	1.00		

Figure 5.12. Statistically significant different performances of the four groups and the corresponding different evaluations of the texts and the tasks.

As can be learnt from these figures, the main discrepancies between the groups concern the two items which turned out to be the most difficult ones on the basis of the test results, namely 'test04' and 'test06'. Specifically, the 'MLC1int' students found text 4 difficult but performed well on the task, while the 'MLC3' students found the text of mid-difficulty but performed poorly on it. On the other hand, the 'MLC1int' students evaluated task 6 as difficult and performed badly on it, whereas 'MLCspec2' found the task rather easy and performed well on it. From these considerations we can conclude that there is not always a correspondence between the average (statistically different) performances of the various groups and their average evaluations of the corresponding texts and tasks.

In order to complete the present investigation into the difficulty of test battery 1, some further analyses were carried out by applying Item Response Theory. It must be pointed out, however, that the low number of test takers (60) and items (13) in this study militates against the robustness of these methods, and the results of these analyses must be, therefore, treated with caution. In fact, a first warning bell is the low 'level of person reliability' index, i.e. the accuracy of the picture of the test takers, which equals .33 on a scale of -1 to +1. The 'item reliability' index, on the other hand, is quite high, .87, which suggests that the information on item difficulty is more reliable. Information about the test takers' ability and the difficulty of the items is displayed pictorially in the 'variable map'[17] shown in Figure 5.13. As can be seen, in this type of map test takers are placed on the left-hand side of the dividing line, and the numbers of persons are indicated by #. The more able persons are placed towards the top of the map and the less able ones towards the bottom. Items, by contrast, are placed on the right-hand side of the dividing line: the more difficult items appear towards the top of the map, whereas the easier ones towards the bottom. The map confirms that test 6 turned out to be the most difficult test/item in the battery, while 'test01a' and test 9 the least difficult ones. Test 2 and test 7 and 'test01b' and test 5 appear next to each other on the same line, which means that they have the same or very similar facility values. It is also to be noticed that the same 'logit'[18] scale measures both item difficulty and person ability, which makes it possible to predict the probability of a test taker's success on a given item.

---

17 A 'variable map' is one of the many output tables available in Ministep.

18 "A logit is a unit of measurement on an interval latent trait scale. It is derived by natural logarithmic transformation of proportion-correct/incorrect ratios, and scalar adjustment for test length and sample size" (Henning 1987: 194).

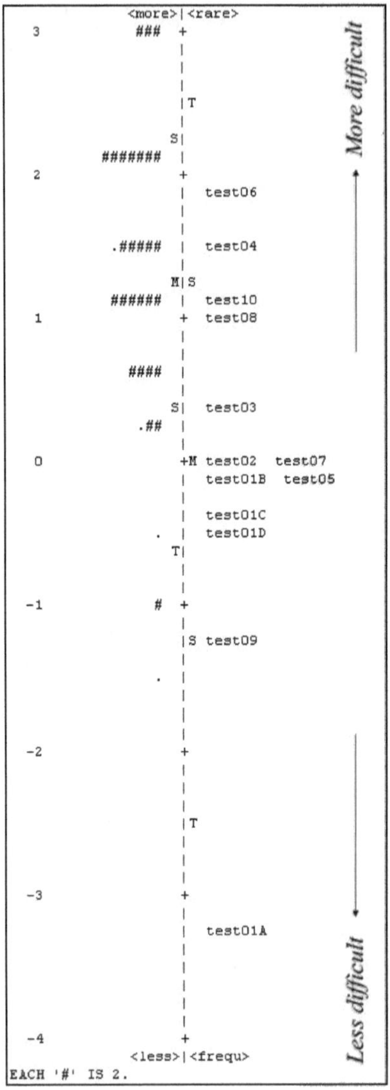

Figure 5.13. Variable map of the items in test battery 1.

Unfortunately, however, as said above, the population under consideration in this study is too small to obtain reliable information on the ability of each single test taker. The letter '+M' opposite to the zero logit value is the mean score for item difficulty, whereas the mean score for person ability is higher up on the left-side of the dividing line. Finally, the letters ‘T’ and ‘S’ indicate the spread of the items/persons from their respective mean scores and represent standard deviations.

A variable map has also been created for the 20 closed-ended questions in the questionnaire and is represented in Figure 5.14.

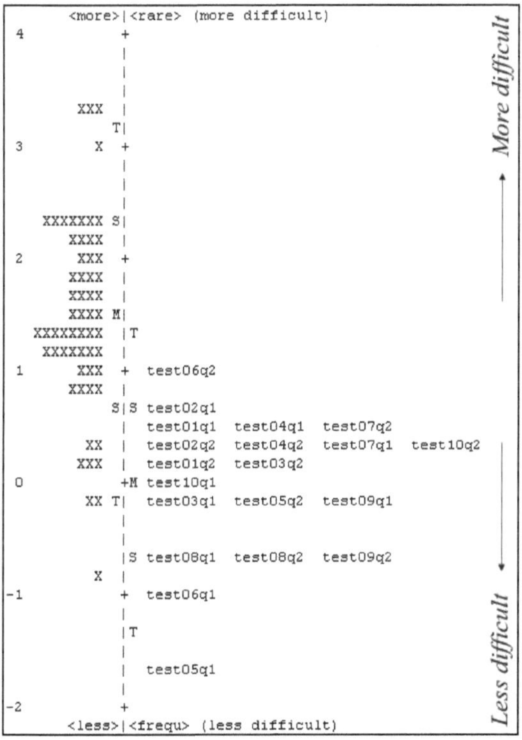

Figure 5.14. Variable map of the closed-ended survey questions about test battery 1.

It must be said, first of all, that both the person and the item reliability indices of the questionnaire are higher than those of the item battery, and score .80 and .89 respectively. This suggests, in particular, that the probability of a given person endorsing a particular score on a given question is rather high. On the basis of the position of 'test06q2' on the logit scale, for example, it can be claimed that task 6 was perceived as the most difficult one, while text 6 was perceived as the second least difficult text in the battery. Text 2, on the other hand, was perceived as the most difficult text, and the perceived difficulty of task 2 equals that of task 4, task 10 and text 7. Information about the evaluation of the other items can be gained from their positions on the map[19].

5.5 Test Battery 2 (from Test 18 to Test 21)

Test battery 2 is made up of the four tests belonging to the PET sub-corpus described in 2.4 above. Using the terminology introduced in 5.2 above, test 18 and test 19 can each be defined as a parallel or equivalent form of the same (sub-)test, as too can test 20 and test 21 (Alderson, Clapham and Wall 1995: 96-97). All these tests were administered to 87 students whose general language proficiency was estimated at about level B1 of the CEF as shown in Figure 5.15. As can be seen from Figure 5.15, the population of test takers consists of four groups: the largest one is the 'MLC1preint' group and is made up of 46 language majors who were attending a first-year pre-intermediate English course; the 'ENG' group is made up of 21 students of Engineering; the 'EDSC' group consists of 15 Education Studies students; and the 'COSC' group of 5 Communication science students. The students belonging to the last three groups were all attending an English course at B1 level to fulfil a language requirement. The time limit for the completion of the four tests in battery 2 was 50 minutes.

---

19 Detailed information about each test taker could also be obtained from other variable maps available in *Ministep*.

TEST BATTERY 2

Test 19. Globewise
Test 18. Exploring the Arctic
Test 20. Ainsley Harriot
Test 21. Doug Allan

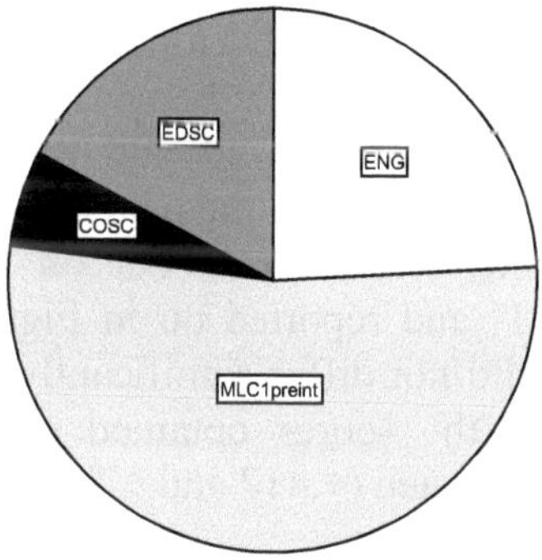

Figure 5.15. Test battery 2 and the characteristics of the population which it was administered to.

In Figure 5.16 the distribution of the total score for the whole population (i.e. 'totbattery2') is displayed along with some descriptive statistics.

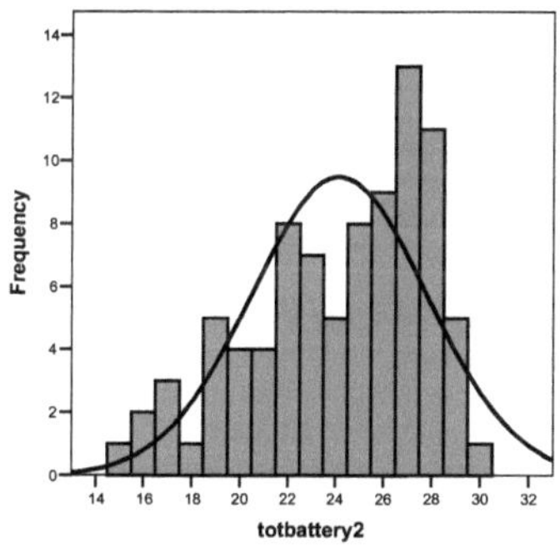

DESCRIPTIVE STATISTICS
N = 87
Mean = 24.03
Mode = 27
Median = 25
Standard Deviation = 3.665
Minimum = 15
Maximum = 30

Figure 5.16. Histogram displaying the distribution of the total score on 'totbattery2' for the whole population and some descriptive statistics.

As suggested by the information in Figure 5.16, most of the test takers performed well on test battery 2. Further details about the average performance on the four tests is shown in the form of a table in Figure 5.17: by inspecting the column displaying the mean scores expressed as percentages (M%), it can be seen that the most difficult test turned out to be test 21, followed by test 20, test 18, and test 19. Furthermore, the difference between test 18 and test 19 is not large, while that between test 20 and test 21 is noteworthy. In fact, according to the results of the paired t-tests and of the non-parametric Wilcoxon S.R. tests carried out on the pairs 'test18-test19' and 'test20-test21' and reported on in Figure 5.18, the members of the former pair did not differ significantly, while those of the latter did. Nevertheless, the scores obtained on the tests of both pairs are positively correlated (+.417 and +.555 respectively).

	N	*Min*	*Max*	*M*	*M%*[20]	*S.D.*
test18	87	5	10	8.34	83.4	1.328
test19	87	3	10	8.43	84.3	1.369
test20	87	2	5	4.08	81.6	1.003
test21	87	0	5	3.18	63.6	1.290

Figure 5.17. Descriptive statistics of the 4 tests in test battery 2.

	Paired t-test						*Wilcoxon S.R.*	
	M	S.D.	Correl.	t	df	Sig. (2-t.)	Z	Asy. S. (2-t)
t18–t19	0.80	1.456	.417	.515	86	.608	-.323	.747
t20–t21	.897	1.111	.555	7.530	86	.000	-3.838	.000

Figure 5.18. Paired t-test and Wilcoxon Signed Rank Test for two parallel forms of tests in battery 2, tests 18 and 19 and tests 20 and 21.

On the basis of this data, it can be speculated that tasks (and tests) constructed by following the same specifications and guidelines can nonetheless turn out to be of different difficulty. If the same population of test takers is taken into account, this discrepancy might depend, among other things: on the input text chosen; on the way tasks are constructed and the questions formulated; on the familiarity that test takers have with the test method(s) used; and on the

---

20 The symbol 'M%' stands for the values of a given mean score expressed as a percentage. These values are calculated by using the following formula: (M/Max)x100.

possibility of some candidates not attempting some of the questions in a test due to time limitations. In the case of the present study, since test 21 was administered after test 20 but nonetheless turned out to be the most difficult, the 'familiarity' factor can be excluded. The two main factors that remain are, therefore, the texts and the tasks used in the two tests. In this respect, Cambridge ESOL[21] stresses that for the creation of their tests input texts are chosen and tasks are constructed according to strict guidelines and that all tasks are pre-tested to ascertain the suitability of their level of difficulty.

In order to account for task difficulty, in Figure 5.19 some descriptive statistics for each single item/question in the tests are displayed along with a variable map that illustrates their degree of difficulty in terms of their position on the map (see 5.4 above). However, while the information on item difficulty is reliable (item reliability equals .89), person reliability does not score very high (.59). It can be seen, in particular, that the easiest item in the battery is task 18.01, that is, the first item in test 18, while the most difficult one is task 21.03, that is, the third item in test 21. Furthermore, all the items in test 21 appear above the zero logit value, whereas those of the other tests appear below it, which illustrates the fact that the tasks in test 21 proved to be the most difficult ones. Another important point that emerges is that there is not always a correspondence between the difficulty of two 'parallel' items, although there are cases in which such pairs are characterised by similar facility values. Tasks 20.03 and 21.03, for example, are the two most difficult items in the tests they belong to, as well as the most difficult ones in the battery as a whole. Using the PET authors' terminology reported on in 2.4 above, these two items can be described as aiming to assess the candidates' ability to "understand attitude, opinion and writer purpose", and it can be hypothesised that the cause of the test takers' poor performance was their lack of this underlying ability. However, a closer look at the questions, at the texts, and at the responses to the open-ended survey questions suggests that these questions (but also some others in the same tests) were perceived as ambiguous and 'tricky'. Moreover, many test takers found it difficult to retrieve the necessary information from the respective input texts.

21 Personal communication.

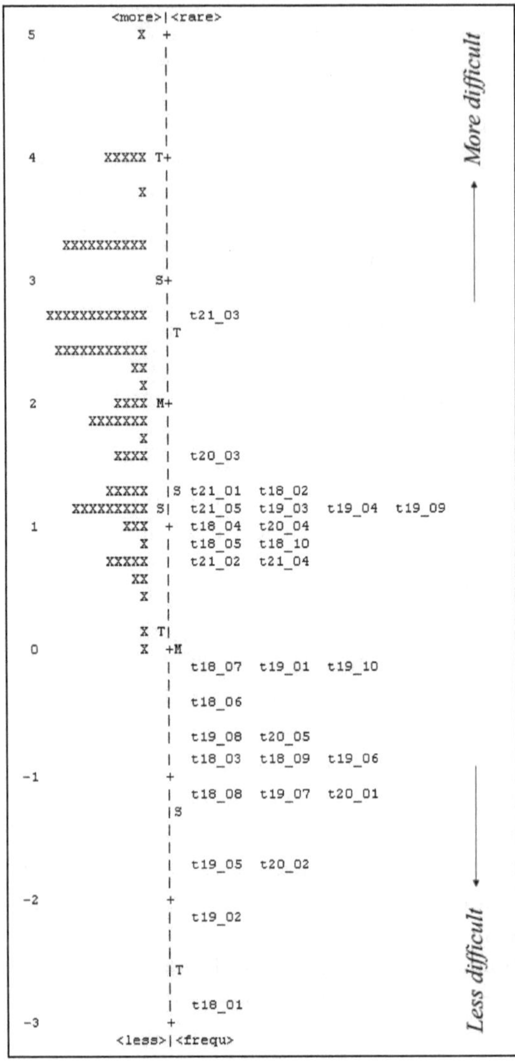

	N	*F.V.*	*S.D.*
t18.01	87	.99	.107
t18.02	87	.67	.474
t18.03	87	.93	.255
t18.04	85	.72	.453
t18.05	87	.72	.450
t18.06	86	.90	.308
t18.07	86	.87	.336
t18.08	86	.94	.235
t18.09	87	.93	.255
t18.10	87	.72	.450
t19.01	87	.86	.347
t19.02	87	.98	.151
t19.03	87	.68	.470
t19.04	84	.70	.460
t19.05	86	.97	.185
t19.06	84	.93	.259
t19.07	85	.94	.237
t19.08	86	.92	.275
t19.09	87	.69	.465
t19.10	86	.87	.336
t20.01	87	.94	.234
t20.02	87	.97	.184
t20.03	84	.60	.494
t20.04	84	.71	.454
t20.05	86	.92	.275
t21.01	85	.67	.473
t21.02	85	.75	.434
t21.03	85	.39	.490
t21.04	87	.76	.430
t21.05	85	.67	.473

Figure 5.19. Variable map and descriptive statistics for the items in test battery 2.

For example, if task 21.03 and the relevant portion of text 21 are taken into account, it can be seen that there can actually be ambiguity as to the choice of the correct option:

(131) 21.03 When talking about killer whales, Doug says that
(a) he will only film them from a safe distance.
(b) he has always been careful when diving with them.
(c) he tries to avoid any danger by facing them.
(d) he believes there are safe opportunities to film them. (task 21.03)

(132) I prefer to be face to face with the animals I'm filming. I haven't got in the water with killer whales yet, but I plan to. Of course, it's dangerous if you choose the wrong moment. They're big animals and can move fast, so I'd be stupid to film them searching for food! (extract from text 21)

It was, in particular, options (b) and (c) that 'distracted' the population, since 22 participants chose (b) and another 22 went for (c). The correct option (d) was chosen by 33 participants, and option (a) by 8 of them. The difficulty for this item is probably mainly caused by the tenses of the verbal groups used in the options. Specifically, the clause complex 'he has always been careful when diving with them' implies that Doug has already dived with whales, while he explicitly says that he has not yet done so. In 'he tries to avoid any danger by facing them' there is again an implication that he has already dived with whales, which is not true. Furthermore, the occurrence of the participant 'safe opportunities' in option (d) is likely to have caused problems because of being one of those expressions that some test takers perceived as 'not matching with any word in the text' and, therefore, difficult to recognise as overlapping with the relevant content of the text. In fact, the participant 'safe opportunities' refers back indirectly to 'it's dangerous if you choose the wrong moment' and to 'so I'd be stupid to film them searching for food', but is not used explicitly in the text. As far as task 20.03 is concerned, the correct option (c) was chosen by 50 test takers, (d) by 21, (b) by 11, and (a) by 2.

(133) 20.03 What does the writer say about himself?
(a) He loves going out and meeting people.
(b) He is very similar to his father.
(c) He enjoys being popular.
(d) He should go to bed earlier.

Option (c), in particular, makes reference to the following first extract from text 20, option (d) to the second extract, and option (b) to the third one:

(134) I like being liked and I love making everyone smile. (first extract from the first paragraph of text 20);

(135) There's nothing better than a night at home playing with the children. I rarely go to bed before midnight. Late evening is when fresh thoughts on cooking usually come to me, so I often write or plan my programmes then. When I eventually get to bed, I have no trouble sleeping! (second extract from the third paragraph of text 20);

(136) I grew up with music because my dad is the pianist, Chester Harriott – who's still playing, by the way. My working day is divided between television and writing cook books, though TV takes most of my time (third extract from the third paragraph of text 20).

'He enjoys being popular' is clearly related to 'I liked being liked' in option (c), but the different wordings in the text from the one in the question might have caused confusion to the test takers. 'I rarely go to bed before midnight' is the passage to which option (d) is related. If this statement is considered on its own and out of context, it might be actually interpreted as 'I should go to bed earlier', that is, as a speech act expressing a wish to change one's own habits, which might explain why some test takers went for this option. Finally, option (b) seems to be tricky because of its vagueness. Specifically, if the similarity to the father concerns the common love and practice of music, then father and son are similar, but if their job and their principal interests are made reference to, then the two men are not similar. As already suggested in 3.5.2 above for text 20, another possible cause of difficulty in test 20, but also in test 21, could be the 'information denseness' of the two input texts, and the fact that some candidates perceived the many events narrated in them as not related to each other.

As already said in 2.4, both test 18 and test 19 aim to assess the candidates' ability to "process a factual text, and to scan for specific information while disregarding redundant material". Furthermore, as seen in 3.2.1.2, text 19 is the most lexically dense text in the corpus, and the main difficulty encountered by the candidates whilst reading it was due to its peculiar text structure (it is a tourist leaflet) and to the presence of some low frequency lexical words (e.g. 'sightseeing', 'scheduled', 'reef', 'surroundings', 'brightly', 'coach'). On the other hand, although text 18 also takes the form of a leaflet and contains some words that were unknown to the test takers, it was generally perceived as easier to read than text 19 (see also the discussion below). According to the responses given to the open-ended questions, the main difficulty as to tasks 18 and 19 was again mainly concerned with the matching of the statements with the relevant text extracts (i.e. the identification of the relevant participants), and with the ambiguity of the statements. Specifically, the task in test 18 that was deemed most difficult is task 18.10, while candidates were divided in their opinion over the degree of difficulty of the tasks in test 19. From the statistical and psychometric point of view, by contrast, tasks 18.02, 19.03 and 19.04 are three of the most difficult items in the two tests. Task 19.03, for instance, consists in the following statement:

(137) 19.03 All travel between cities in Australia is by plane.

As said in 4.4.6.1 above, in order to find the information necessary to verify the truth of this and similar statements, test takers should identify the resources that encode reference to the relevant participants in the input text. The specific relevant information in text 19 is contained in the following extract:

(138) Coach from Sydney to Melbourne via Canberra and Albury.

As can be easily verified from this portion of text, the statement is false, in that at least the journeys between some cities are by coach. 27 candidates out of 86[22], however, answered by saying that the statement is true. These students might have answered wrongly either because of their lack of knowledge of the word 'coach', which, as said above, was written by some candidates, or because they did not focus on this part of the text. Task 19.04, on the other hand, was answered correctly by 59 candidates and wrongly by 25, while three

---

22 One candidate did not answer this question.

of them did not answer it. This task asks about the correctness of the following statement:

(139) 19.04 The cost covers accommodation and some meals.

The part of text that is necessary to answer it is the following:

(140) OUR PRICE INCLUDES: […] Nineteen nights' accommodation in good with full continental breakfast.

This statement is true, in that the semantic denotation of 'meals' includes breakfasts as well. In fact, in their responses to the open-ended questions some candidates expressed their doubts about whether breakfasts are to be considered as meals or not, and it was probably this lack of knowledge that caused them to fail the task.

As for the candidates' perception of the difficulty of the texts and tasks in test battery 2, a graphical as well as a numerical synopsis of their evaluations is provided by Figure 5.20[23]. As can be seen, according to the responses given to the closed-ended survey questions, both text 21 and its accompanying questions were on average judged as the most difficult ones in the battery, whilst text 18 was judged as the least difficult text to read. Furthermore, the tasks in test 20 and text 20 were judged as the second most difficult ones to perform and read respectively. By contrast, students found text 19 quite easy to read, and its accompanying questions were perceived as slightly more difficult than the text itself. These evaluations mainly accord with the actual performances on the tasks (see Figure 5.17 above), apart from the fact that test 18 turned out to be statistically less difficult than test 19.

23 For these data person reliability equals .82 and item reliability equals .98.

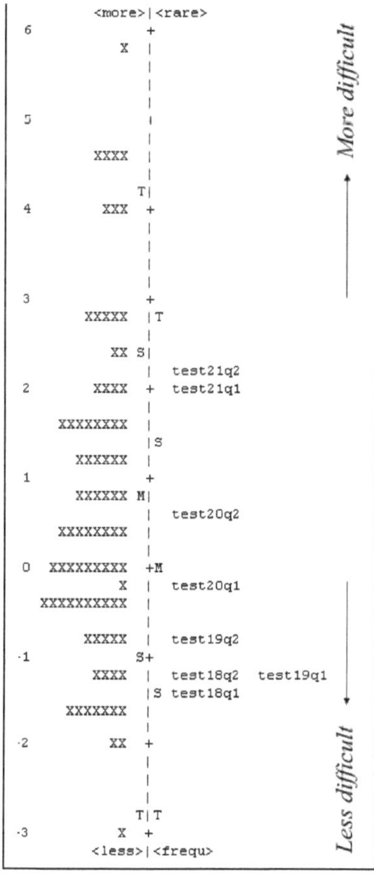

	N	*M*	*Mode*	*Med.*	*S.D.*
test18q1	85	1.88	2	2	.778
test18q2	85	1.94	2	2	.761
test19q1	86	1.93	2	2	.629
test19q2	86	2.08	2	2	.723
test20q1	85	2.22	2	2	.777
test20q2	84	2.49	3	3	.843
test21q1	85	2.92	3	3	.727
test21q2	86	2.94	3	3	.772

Figure 5.20. Variable map and descriptive statistics of the closed-ended survey questions about test battery 2.

Figure 5.21 shows the statistically significant correlations between the average performance on the tests in battery 2, the perceived text difficulty as well as the perceived task difficulty. It can be seen, in particular, that candidates tended to rate the tasks and the texts in a consistent way, in that the correlations between the responses to the two closed-ended survey questions were always significant. As for the correlations between performance and perception, on the other hand, the only significant correlations were between the average performance on test 20 and the evaluations of both text 20 and task 20[24]. As for this latter test, the more difficult they perceived its text and its questions, the lower the score they obtained.

	Performance	*Perceived text diffic.*	*Perceived task diffic.*
test18		√	√
test19		√	√
test20	√ (neg. with 2 and 3)	√	√
test21		√	√

Figure 5.21. Statistically significant correlations between the average performance on the tests in battery 2, the perceived text difficulty and the perceived task difficulty ('neg.' stands for negative correlation).

As a way of concluding this section devoted to test battery 2, a brief investigation into how the various groups of students differed in terms of their performance on the various tests and of their respective evaluations of them will be reported on. To this end, a series of ANOVA were carried out on the test takers' performance and on the responses to the evaluative questionnaire. The statistically significant differences found suggested that, even though all the students were supposed to be at level B1, the 'MLC1preint' students significantly outperformed the others on the battery overall (M=25.28), on test 20 (M=4.50), and on test 21 (M=3.50), while those that performed the worst belonged to the 'ENG' group (M=3.29 and 2.48). The engineering students also endorsed the highest average rate score on text 20, text 21 and text 18, and differed significantly from those of the other groups. The findings concerning 'test18q1', 'test20q1' and 'test21q1' are also confirmed by the Kruskal-Wallis Test run on each of the three variables.

24 The values of the correlation indices calculated between the evaluations of the texts, those of the tasks, and the performance on the tests in test batteries 2, 3, 4, and 5 are available but will not be reported on.

5.6 Test Battery 3 (Tests 11, 12, 22, and 23)

Test battery 3 consists of test 22 and test 23, both of which belong to the FCE sub-corpus (see 2.5 above), and of test 11 and test 12, two tests from the TOEFL sub-corpus (see 2.2 above). The two FCE tests were taken by the whole population depicted in Figure 5.22: 20 engineering students who had just finished attending a course at level B2 of the CEF (the 'ENG2' group); 14 communication science students whose level of English was also deemed to be at level B2 (i.e. the 'COSC2' group); 36 language majors who were attending a second-year English course (i.e. the 'MLC2' group); 19 language majors who were attending a first-year intermediate English course (i.e. the 'MLC1int' group); and 8 language majors who were attending a first-year advanced English course (the 'MLC1adv' group). Test 11 and test 12, by contrast, were not administered to the ENG2 group. Finally, since test 12 was also part of test battery 4 (see 5.7 below), it was administered to another population of students besides the one just described. The present section, therefore, will concern the analysis of the data collected by the administration of test 23, test 22 and test 11, while those about test 12 will be dealt with and discussed in section 5.7. The order in which the tests were taken is the one in which they are listed in table 5.22. The time limit for all the four tests in battery 3 was 60 minutes, whereas the time limit allowed for the completion of test 22 and test 23 was 30 minutes.

According to the descriptive statistics reported on in Figure 5.23, the test on which the population, excluding the engineering students, performed the best is test 11 (M% equals 85.4). It should be seen, in particular, that the Mode for this test equals 9 and the Median 8. This means that half of the test takers obtained a score of 8 or higher and that the most frequently obtained score was 9 (the Maximum). On test 22 and test 23, by contrast, the whole population obtained lower but similar average scores, and their 'M%' values are 76.8 and 78 respectively.

TEST BATTERY 3

Test 23. In case you lose your luggage
Test 22. A lot of advise
Test 11. The next artist
Test 12. *The final battle*

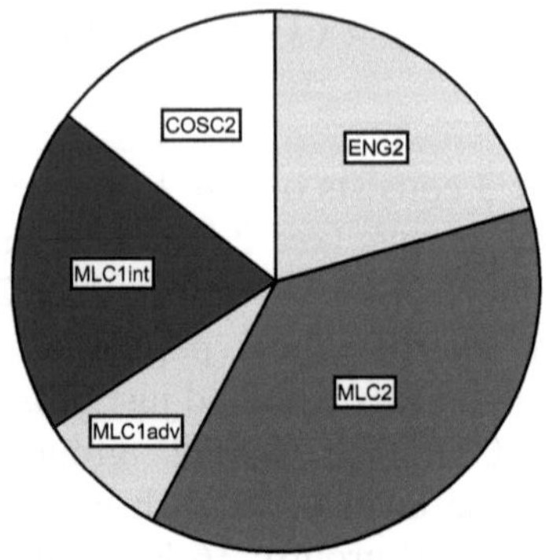

Figure 5.22. Test battery 3 and the population it was administered to.

	N	*Min*	*Max*	*M*	*M%*	*Mode*	*Median*	*S.D.*
test11	77	4	9	7.69	85.4	9	8	1.228
test22	97	2	7	5.38	76.8	6	6	1.278
test23	97	1	7	5.47	78	7	5	1.582

Figure 5.23. Descriptive statistics of tests 11, 22 and 23.

As far as test 11 is concerned, a closer look at the descriptive statistics about each one of its items in Figure 5.24 reveals that the most difficult items proved to be question 11.07 and question 11.09, while the easiest ones were questions 11.03, 11.01 and 11.06. As said in 2.2 above, question 11.03 elicits an answer to a 'stated detailed question', which is a sub-skill that everybody proved to have mastered. Students also proved to be equally good at answering a 'transition question' and to 'use context to determine meanings of difficult words', in that most of them answered question 11.01 and question 11.06 correctly. In particular, only 4 out of 76 students answered question 11.01 by choosing 'a survey of eighteenth-century art' as the topic that is likely to be discussed in the preceding paragraph. What might have misled this small group of candidates is the presence of the participant 'this survey of American artists' in the following clause in text 11: 'he is included in this survey of American artists because he was born in the United States' (see also 4.4.2.3 above).

	N	*Min*	*Max*	*F.V.*	*S.D.*
t11.01	76	0	1	.95	.225
t11.02	77	0	1	.94	.248
t11.03	76	1	1	1.00	.000
t11.04	77	0	1	.87	.338
t11.05	77	0	1	.83	.377
t11.06	77	0	1	.95	.223
t11.07	76	0	1	.67	.473
t11.08	77	0	1	.81	.399
t11.09	77	0	1	.71	.455

Figure 5.24. Descriptive statistics for the items in test 11.

According to the information about the TOEFL sub-corpus and given in 2.2 above, item 11.07 aims to assess the sub-skill of finding a 'definition from structural clues'. While 51 test takers managed to retrieve the correct definition of 'etchings' from the structural clues contained in the following extract from text 11, that is, 'an art form involving engraving', 25 of them did not.

(141) Whistler worked in various art forms, including etchings and lithographs. However, he is most famous for his paintings.

22 students chose 'a type of painting' (option 'a'), one went for 'the same as a lithograph' (option 'b'), and two for "an art form introduced by Whistler" (option 'd'). In fact, approximately 19% of test takers who responded to the second open-ended survey question concerning test 11 explicitly indicated the lack of knowledge of this word as the main cause of difficulty as to the completion of the tasks. Another 36% of the respondents said that they found the infrequent words the tasks asked about as particularly difficult. Finally, approximately 78% of the respondents indicated the use of some low frequency words, e.g. 'lithographs', 'etchings', as the only difficulty they encountered whilst reading text 11. In spite of the fact that only one candidate explicitly indicated question 11.09 as difficult and ambiguous, 13 students made the choice of option (b), one of them chose (a), and eight chose (d).

(142) 11.09. Which of the following is NOT true according to the passage?
(a) Whistler worked with a variety of art forms.
(b) Whistler's Mother is not the official name of his painting.
(c) Whistler is best known for his etchings.
(d) Whistler's Mother is painted in somber tones.

Option 'c', in particular, is not true, in that in text 11 the following statement is made instead:

(143) He is most famous for his paintings.

The statements contained in the other options, by contrast, overlap with those made in the text. In order to exclude option (b), for instance, test takers should have focused on the following portion of the text.

(144) He is most famous for his paintings, particularly Arrangement in Gray and Black No. 1: Portrait of the Artist's Mother or Whistler's Mother, as it is more commonly known.

On the basis of these data, it can be concluded that answering questions that ask about something that is 'not' true or false can be difficult and tricky, even though candidates might not perceive it as especially so.

In order to account for the difficulty of test 22 and of test 23 in more detail, a variable map[25] comprising all the items contained in them has been reproduced in Figure 5.25 along with some descriptive statistics. As stated in 2.5 above, all the items in test 22 are four-option multiple-choice questions, while each item in test 23 is part of a multiple-matching task. From Figure 5.25 it can be seen that the items that stand out as most difficult in test 22 are items 22.05, 22.02 and 22.03, whereas it is item 23.01, 23.05 and 23.03 that turned out to be the most difficult ones in test 23.

---

25 Person reliability and item reliability equal .41 and .91 respectively.

```
          <more>|<rare>
 4       .#####  +
                 |
                T|
                 |
                 |
                 |
                 |
                 |
 3   .#########  +
                 |
                 |
                S|
                 |
                 |T
      .########  |
                 |
 2               +
                 |
                 |
        ###### M|  t22.05   t23.01
                 |
                 |
          .####  |  t23.05
                 |S
 1               +
                 |
        ######   |
              .  |  t23.03
                S|  t22.02   t22.03
           ###   |
                 |  t22.04
                 |  t22.07
 0          ##   +M
                 |
            ##   |  t22.01   t23.02
                 |  t23.04
                T|
                 |
             #   |
                 |
-1               +
                 |S
                 |
                 |
                 |  t22.06   t23.06
                 |
                 |
                 |
-2               +
                 |
                 |
                 |T
                 |
                 |
                 |  t23.07
                 |
-3               +
          <less>|<frequ>
EACH '#' IS 2.
```

↑ More difficult

↓ Less difficult

	N	*F.V.*	*S.D.*
t22.01	97	.84	.373
t22.02	97	.74	.440
t22.03	97	.74	.440
t22.04	96	.78	.416
t22.05	97	.56	.499
t22.06	97	.94	.242
t22.07	97	.79	.407
t23.01	97	.55	.500
t23.02	97	.84	.373
t23.03	97	.71	.455
t23.04	97	.85	.363
t23.05	97	.62	.488
t23.06	97	.94	.242
t23.07	97	.98	.143

Figure 5.25. Variable map and descriptive statistics for the items in tests 22 and 23.

Item 22.05, in particular, assesses the detailed understanding of a specific expression used in text 22:

(145) 22.05 What is meant by 'you have got it out of your system' in line 34?
(a) You have planned your career sensibly.
(b) You are an experienced traveller.

(c) You have satisfied your wish to travel.
(d) You have learned to look after yourself.

54 students answered this question correctly by choosing option (c), 26 chose option (d), 15 option (a), and two option (b). In order to answer this question correctly test takers should focus on the following portion of text:

(146) Incidentally, if the travel bug is biting, now is the time to pack up and go. You can do temporary work for a while when you return, pick up where you left off and get the second job then. *[line 34]* Future potential employers will be relieved to see that you have got it out of your system, and are not likely to go off again.

As can be seen, the informal multi-word unit 'you have got it out of your system' can be fully understood if the informal idiom 'the travel bug (is biting)' is also correctly interpreted (see also 3.2.1.1 above), in that the pronoun 'it' refers back to the 'travel bug'. The unclear reference encoded by the pronoun 'it', therefore, also contributes to the difficulty of this item. In fact, in their responses to the first open-ended survey question, approximately 75% of the respondents said that the main cause of difficulty they encountered whilst reading the text were the infrequent vocabulary and some of the 'expressions' used in text 22, and some of them also made special reference to these two multi-word units. Moreover, about 7% of them responded by saying that they found this item especially difficult to answer and that they had to think carefully before responding. The lack of knowledge of these multi-word units, their opacity, and the inability to deduce their meanings from the context are, therefore, to be considered as the principal causes of failure on item 22.05.

As far as item 22.02 is concerned, 23 test takers chose option (b), one went for (c), another one for (d), and 72 for (a), i.e. the correct option:

(147) 22.02 According to the writer, why is the choice of your second job important?
(a) It will affect your future job prospects.
(b) It will last longer than your first job.
(c) It will be difficult to change if you don't like it.
(d) It should give you the opportunity to study.

The lack of success in answering this question can be partly explained by the fact that approximately 8% of the test takers explicitly reported the difficulty they had experienced in distinguishing between the notion of 'first job' and that of 'second job' as dealt with in the first four paragraphs of the text. In particular, according to the text, the latter is a job that fits "in with your ambitions now that you have learnt your way around, acquired some skills and have some idea of where you want to go" and, therefore, one that will affect a person's 'future job prospects'. These test and questionnaire data also support the first hypothesis formulated in 4.4.4 about the specific difficulty concerned with the recognition of the two main generic participants in text 22, namely 'first job' and 'second job'. The other difficulty envisaged in 4.4.4 had to do with the confusion that the use of the personal pronoun 'it' in text 22 might cause, in that this anaphoric device might encode reference to any one of the jobs (and to any other inanimate singular participant as well). Item 22.03 aims exactly to assess whether test takers can retrieve the identity of the participant encoded by this pronoun in a specific line of the text as follows:

(148) 22.03 'it' in line 26 refers to your
(a) first job.
(b) second job.
(c) application.
(d) career.

According to the data, some test takers did not manage to retrieve the correct referent: 21 of them chose option (c), two chose option (d), another two chose option (a), and the remaining 72 candidates went for the correct option (b). The main 'distracter', therefore, proved to be the participant 'your application', which presumes 'your letter of application for your next job' introduced in the following portion of text 22:

(149) This job will be studied very carefully when you send your letter of application for your next job. It should show evidence of serious career planning. Most important, *[line 26]* it should extend you, develop you and give you increasing responsibility.

The ambiguity as to the identity of the participant presumed by 'it' can be, first of all, explained by the short distance between the clause in which 'your letter of application' occurs and the two subsequent clauses in which the pronoun 'it' is used. Furthermore, if this portion of text is read expeditiously and locally, i.e. scanned (see 1.2 above), 'it' can equally well be interpreted as presuming the second job or the application. In fact, as stated in 2.5 above, test takers should read this text carefully, both globally and locally, to be able to answer this question.

As said in 2.5 above, the main aim of test 23 is to assess the candidates' ability to identify the main points in a text at paragraph level. According to the feedback obtained from the students, text 23 was not perceived as especially difficult to read, except for some terms which some students did not know. By contrast, the main difficulty the majority of them encountered in the tasks were due to the ambiguity of the summary sentences. Specifically, they complained about the fact that most of the sentences differed from each other only because of some subtle nuances. Sentences D and G, i.e. the correct responses to item 23.01 and item 23.05 respectively, were mentioned by some respondents as especially tricky:

(150) D No airline is free from mistakes.
G No airport can guarantee the safe handling of luggage.

In fact, these two statements basically differ because of the two generic participants that are at issue in them, namely 'airline' and 'airport', and the similarity between the two is likely to have misled the less careful candidates. In particular, 27 students out of 97 (27.8%) mistakenly chose G as the sentence that best summarises paragraph 1 (i.e. the paragraph associated with item 23.01), while 20 of them (20.6%) chose D as the most suitable summary sentence for paragraph 5 (i.e. the one associated with item 23.05). Another summary sentence that distracted another 11 candidates (11.3%) from the correct choice as to item 23.05 is sentence H:

(151) H Busy airports are likely to have more problems.

This sentence contains the participant 'airport' pre-modified by the Epithet 'busy'. By simply scanning and reading the text selectively for words in the text which match words in the question (as seen in 4.4.6.2 above), this sentence could be thought of as overlapping and, therefore, as presuming 'efficient transfer airports, such as Amsterdam, Copenhagen, Singapore and Zurich' introduced in paragraph 5. However, sentence H best summarises paragraph 4, in that 'busy airports' refers the reader back to "large transfer airports, known as 'hubs'" in the first clause complex in paragraph 4:

(152) These problems can become severe at large transfer airports, known as 'hubs', because of the large number of bags that are processed.

The test takers' evaluation of the three tests under investigation in this section accords with the students' average performance on the tests. As can be seen from Figure 5.26, both the text and the accompanying questions in test 22 were endorsed with the highest average rate scores, that is, they were evaluated as the most difficult ones. Both text 23 and task 23 were perceived as the second most difficult ones, whilst text and task 11 were evaluated as the least difficult in battery 3.

	N	*Min*	*Max*	*M*	*Mode*	*Median*	*S.D.*
test11q1	76	1	3	1.63	2	2	.608
test11q2	77	1	4	2.21	2	2	.749
test22q1	96	1	4	2.32	2	2	.657
test22q2	97	1	4	2.33	2	2	.718
test23q1	96	1	4	2.06	2	2	.678
test23q2	97	1	4	2.24	2	2	.674

Figure 5.26. Descriptive statistics of the closed-ended survey questions about test battery 3.

As was the case with most of the tests analysed so far, the perceived degree of difficulty of the three input texts under enquiry correlates with that of the corresponding tasks. By contrast, the average performance on each single test does not correlate with the evaluation of the corresponding text or with that of the accompanying tasks.

A series of ANOVA and of Kruskal-Wallis Tests carried out on the single items, on the tests as a whole, and on the responses to the closed-ended survey questions indicated that the five groups of test takers differed significantly from one another in their average performance on test 22, on test 23 and on the evaluation of text 22 (test22q1). The Engineering students (those belonging to group 'ENG2') obtained the best overall score on both test 22 (M=5.80) and test 23 (M=5.45), whilst the 'MLC1int' students obtained the lowest average score on both tests (M=4.63; M=4.95). The 'MLC1int' students were also the ones that endorsed the highest rate score on text 22 (M=2.58), whereas the 'MLC2' students were the group that assigned the lowest average score to it (M=2.46).

5.7 Test Battery 4 (from Test 12 to Test 16)

Test battery 4 is made up of the first three IELTS tests described in 2.3 above, tests 14, 15 and 16, and of tests 12 and 13, that is, two tests belonging to the TOEFL sub-corpus (see 2.2 above). The tests were administered to 75 candidates in the order in which they are listed in Figure 5.27. The population consisted of; 48 language majors who were attending a first-year advanced English course (i.e. the 'MLC1adv' group); 18 postgraduate language majors who were attending an English course targeted at their needs (i.e. the 'MLCpost2' group); and 9 second-year English language and literature majors who were also attending an English course ('LCM2'). As anticipated in 5.6 above, test 12 was not only administered to the population just described, but also to the students described in 5.6, with the exception of the engineering students. Test 12 was, therefore, taken by 19 'MLC1int' students, 56 'MLC1adv' students (viz. the 48 students just mentioned and the other 8 listed in 5.7 above), 36 'MLC2' students, 9 'LCM2' students, 14 'COSC2' students, and 18 'MLCpost2' students, for a total of 152 test takers. The time limit for the completion of the five tests in battery 4 was 65 minutes.

TEST BATTERY 4

Test 14. Use of University Grounds
Test 15. West Thames College
Test 16. Central Library
Test 13. One of the most mysterious things
Test 12: The final battle

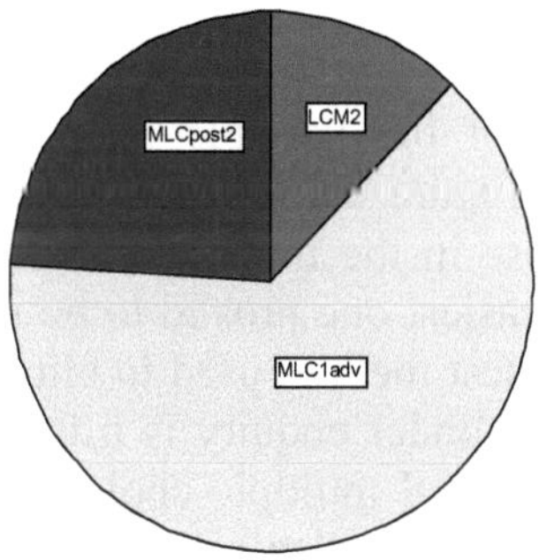

Figure 5.27. Test battery 4 and the population it was administered to.

As shown in Figure 5.28 below, the test of battery 4 which, on average, proved to be the most difficult is test 15, while the least difficult one is test 13. Tests 12 and 16, on the other hand, are characterised by similar average '<u>M</u>%' values. Since tests 15 and 16 are two parallel forms of the same (sub-)test, a paired t-test was run on them to verify whether the candidates' average performance was statistically significant. The output of the t-test indicated that the test takers' overall scores did not differ significantly, which suggests that the two tests are characterised by similar degrees of difficulty and that they measure the same sub-skills.

	N	*Min*	*Max*	*<u>M</u>*	*<u>M</u>%*	*Mode*	*Median*	*S.D.*
test12	152	3	11	8.26	75	10	9	2.044
test13	75	4	10	8.43	84.3	9	9	1.232
test14	75	1	5	4.01	80.2	4	4	.878
test15	75	1	7	4.72	67.4	5	5	1.331
test16	75	3	7	5.05	72	5	5	.971

Figure 5.28. Descriptive statistics of tests 12, 13, 14, 15, and 16.

Paragraph 5.7.1 in this section will be devoted to the detailed item analysis of tests 13, 14, 15, and 16, whereas paragraph 5.7.2 will deal with test 12, with the candidates' evaluation of each test, as well as with the differences between the candidates' performance on each test.

5.7.1 From Test 13 to Test 16

Inspection of the variable map[26] and of the descriptive statistics shown in Figure 5.29 below reveals that the most difficult item among those in the three IELTS tests and in test 13 is item 15.01, while the easiest one proved to be item 14.02. As said in 2.2 and 2.3 above, the test method used to elicit the candidates' responses in the IELTS tests under enquiry is true/false/not given, whereas the items in test 13 are all multiple-choice questions. As seen in 3.3.2 above, text 13 is the least lexically varied text in the corpus as far as the LV(re) index is concerned. Specifically, it was shown that it contains a large number of lexical word forms which belong to the same word families (e.g. 'migration', 'migrating', 'migrate'). In relation to this textual feature, it was noticed that question 13.02 is about the exact semantic difference between the words 'migration' and 'emigration', i.e. two cognate words that could be attributed to the same word family. From the data displayed in Figure 5.29, it can be seen, in particular, that this question proved to be the second most difficult one of test 13, which again suggests that candidates found it difficult to understand that, according to the text and if referred to animals, the distinguishing feature between the two concepts is the 'final destination', rather than the 'route taken'. In 3.3.2 it was also observed that a clue as to the choice of the correct option (d) in question 13.01 is the fact that this option contains only words that belong to frequently occurring word families. In fact, this item proved to be the easiest one in the battery. This suggests that if an option of a multiple-choice question contains a proportionally high number of words belonging to word families which occur frequently in the input text, it is likely to be perceived as the correct option.

26 For this data person reliability equals .34 and item reliability .91.

```
              <more>|<rare>
  4                 +
                    |
                    |
                    |
                    |  t15.01
                    |
               .#   |
                    |
                    |
                    |
  3                T+
                    |
                .   |
                    |
                    |
                    |T
             ####   |  t16.06
                   S|
                    |
                    |
  2       #######   +
                    |
                    |  t14.04
           ######   |
                   M|  t16.04
                    |
            #####   |
                    |S t15.05   t15.07
           ######   |  t13.08
                    |
  1         .####   +
                   S|
                #   |
                    |
                .   |
                    |  t16.03
                .   |  t13.02
                    |
                   T|  t16.01
                    |  t14.01   t15.04
  0                 +M
                    |  t13.06
                .   |  t13.03
                    |  t13.07
                    |  t15.02
                    |
                    |  t16.02   t13.09   t14.03
                .   |  t15.03
                    |
                    |  t13.10   t15.06
 -1                 +  t16.07   t14.05
                    |
                    |
                    |S t16.05   t13.05
                    |
                    |  t13.01   t13.04
                    |
                    |
                    |  t14.02
                    |
 -2                 +
              <less>|<frequ>
EACH '#' IS 2.
```

More difficult ⟶ / ⟵ Less difficult

	N	*F.V.*	*S.D.*
t13.01	74	.95	.228
t13.02	75	.75	.438
t13.03	75	.84	.369
t13.04	73	.95	.229
t13.05	75	.93	.251
t13.06	75	.83	.381
t13.07	75	.85	.356
t13.08	75	.59	.496
t13.09	75	.88	.327
t13.10	75	.91	.293
t14.01	75	.80	.403
t14.02	75	.96	.197
t14.03	75	.88	.327
t14.04	75	.45	.501
t14.05	75	.92	.273
t15.01	75	.13	.342
t15.02	75	.87	.342
t15.03	75	.89	.311
t15.04	75	.80	.403
t15.05	75	.56	.500
t15.06	75	.91	.293
t15.07	75	.56	.500
t16.01	75	.77	.421
t16.02	75	.88	.327
t16.03	75	.73	.445
t16.04	75	.49	.503
t16.05	75	.93	.251
t16.06	75	.32	.470
t16.07	75	.92	.273

Figure 5.29. Variable map and descriptive statistics for the items in tests 13, 14, 15, and 16.

As far as item 13.09 is concerned, it was suggested that its options are rather ambiguous, in that even though it is undeniably true that the author of the passage supports his/her argument through exemplification, he also compares and contrasts the behaviour of different animals. This ambiguity was also signalled by some respondents to the second open-ended survey question, who claimed that option (d) could also be considered as correct. This item, however, was answered correctly by 88% (its F.V. therefore equals .88) of the candidates, and its difficulty was, therefore, not proved by the data. More generally, approximately 45% of the examinees wrote that, in their opinion, most of the questions in test 13 are ambiguous, and about 23% of them said that they found question 13.08 especially tricky, since both option (b) and (c) in it could be correct. Question 13.08 is formulated as follows:

(153) 13.08. The author suggests that using 'instinct' to explain the ability of animals to return home during migration is
(a) absurd
(b) incomplete
(c) inaccurate
(d) reliable

In order to decide the correct answer to this question, test takers have to make reference to the following text extract:

(154) Just to say they have an "instinct" doesn't really explain the right conditions.

As can be seen, the participant 'instinct' is present in the text as well, which makes it lexically overlapping with the question. As said in 2.2. above, question 13.08 aims to test the ability to infer meanings; more specifically, it can be defined as being about the 'attitude' of the author towards the power of 'instinct' to explain the ability of animals to return home. It is claimed here, that the main textual information in support of option (b), i.e. the correct one, is given by the use of 'just' in 'just to say they have an instinct', in that this adverb seems to imply the correctness of 'instinct' and at the same time to call for other explanatory factors. Defining the author's attitude as 'inaccurate' would, by contrast, imply the incorrectness of instinct as an explanation of this ability, which is not the case. The possible causes of failure on this item – 27 candidates went for option (c) and 4 for option (d) – can be found in either the lack of knowledge of the specific connotations of the word 'inaccurate', or the test takers' inaccuracy whilst reading the text, or both.

Approximately 60% of the candidates reported that the main difficulty they encountered in completely mastering the contents of text 13 was due to their lack of knowledge of some infrequent and specific vocabulary used in it (e.g. 'eel', 'lemming', 'swallow', 'swift', 'house martin'). Question 13.05, in particular, was deemed ambiguous, in that it requires the specific knowledge of different but very similar species of birds. Question 13.05 and the portion of text it makes reference to are reproduced below.

(155) 13.05. According to the passage, swallows travelling from South Africa to England
(a) cover a relatively short distance
(b) often return to precisely the same point each year
(c) may not return for a number of years
(d) rebuild the same type of nest in similar locations each spring

(156) Migrating birds offer an even more amazing example of this ability. There are swifts and house martins that migrate from England to South Africa every year. They not only return to England the next spring, but many of them come back to nest in the very same house where they nested the year before. They fly the incredible distance of 6,000 miles, one way.

According to the *Longman Dictionary of Contemporary English*[27], a swallow is a "small black and white bird that comes to northern countries in the summer", whilst a swift is a "small brown bird that has pointed wings, flies very fast, and is similar to a swallow." In their responses to the open-ended survey questions, most of the students declared that they did not know these subtle semantic similarities and differences between the two species; nonetheless, most of them managed to answer the question correctly (the F.V. equals .93). In order to complete this discussion of the difficulty of test 13, a brief account of the students' performance on item 13.03 and item 13.04 will be given. Item 13.03 was answered correctly by 63 candidates who chose option (b), and incorrectly by 12 students who chose option (c). Item 13.04, on the other hand, was answered incorrectly just by 2 students who opted for (d) and by another one who went for (a). Despite the complexity pointed out in 4.4.5.3 above and related to the identification of the participants to be retrieved

27 *Longman Dictionary of Contemporary English* (2003). London: Longman.

through the interpretation of the conjunctive Adjunct 'too' and the anaphoric noun 'this ability', both items proved to be rather easy, even though the former was slightly more problematic.

Test 14 was dealt with specifically in 3.2.2, where it was pointed out that the lexical words 'university', 'grounds', and 'permit' occur especially frequently in the rather short input text. In the five statements in the task, on the other hand, it is 'university', 'permit', and 'hall' that occur frequently, which leads to the conclusion that these three words are of special relevance in the test as a whole. Furthermore, in 4.4.3 above it was suggested and linguistically substantiated that in text 14 it is not clear whether the identity of the participant encoded by the word 'permit' is the same throughout the text or not. This ambiguity, in particular, was deemed likely to cause processing difficulty. In fact, about 14% of the examinees explicitly said that they found it problematic to understand the exact differences, if any, between the permits and the 'parking options' available to students and members of staff. Approximately 30% of them, on the other hand, said that the major cause of difficulty they encountered was their lack of knowledge of the exact denotation of the words 'hall' and 'Warden of the Hall' ('warden' is actually not listed in any one of the word lists used in this study, whilst 'hall' occurs in the second word list). As for the difficulty of the tasks, approximately 25% of the respondents wrote that they found it difficult to retrieve the necessary information in the text, in that the wording of most of the statements is different from the corresponding necessary portion of text. To this some of them added that they could not easily decide the identity of the participants, and, in particular, what students and staff need and/or are allowed to do. Another 13% of the students found that most of the questions are ambiguous, and some of them mentioned questions 14.02, 14.01, 14.04, and 14.05 as especially difficult. As shown in Figure 5.29, according to the test administration data, the item in test 14 that proved to be most difficult is item 14.04. 34 candidates answered it correctly by selecting 'not given', 40 selected 'false', and 1 'true'. Item 14.01 and item 14.03 are the next two most difficult items in test 14, but their facility values are high, since the great majority of students answered them correctly. As already stated in 4.4.3 above, a characteristic that these three items share is the fact that in each one of them the participant ‘permit’ is at issue, which confirms the hypotheses formulated above about the ambiguity as to the identity of the participant(s).

An observation about the difficulty of test 14, test 15 and test 16 is that their most difficult items are generally those that are to be answered by choosing ‘not given’, that is, items 15.01, 16.06, 14.04, 15.05, and 16.03. In particular, item 15.01 is the most difficult item (its F.V. equals .13) in the battery, 16.06 is the second most difficult one (its F.V. equals .32), 14.04 is the third most difficult one of the four tests under enquiry (its F.V. equals .45), 15.05 is the fifth most difficult (its F.V. equals .56), and 16.03 is the eighth most difficult one (its F.V. equals .73). From this and in accordance with part of the students’ feedback, it can be concluded that the test method ‘true/false/not given’ can cause extra and specific difficulty. This also accords with what is claimed by Alderson (2000: 222), who maintains that:

> some tests reduce the possibility of guessing by including a third category such as ‘not given’, or ‘the text does not say’, but especially with items intending to test the ability to infer meaning, this can lead to considerable confusion.

An analysis of test 15 was carried out in 4.4.5.1 above, where it was hypothesised that the main difficulty in it is due to the fact that the input text as well as most of the statements contain phoric resources that code either relevance phoricity or both reminding and relevance phoricity. This hypothesis was partly confirmed by both the test and the questionnaire data. The students’ responses to the second open-ended survey question, in particular, revealed that about 50% of the test takers found it problematic to understand the meaning of the sentences and to ‘link them to the text’. Some of them also specified that the wording of the statements was often different from that in the text, which made it difficult to identify the relevant participants in the text. Other respondents found that all or some of the questions were ambiguous and, therefore, difficult to answer. Approximately 14% of the candidates mentioned item 15.01 as especially difficult, another 7% wrote that the major problems they encountered had to do with item 15.07, and, finally, three of them said that item 22.06 was particularly problematic. As far as the performance on item 15.01 is concerned, 58 test takers responded by choosing the option ‘false’, thus expressing their conviction that statement 15.01 is false, 7 said that the statement is ‘true’, and only 10 that it is ‘not given’, i.e. the correct answer. As for this specific item, some of the test takers said that they did not know the exact meaning of the word ‘merger’ used at the beginning of text 15, that is, the portion of text where the necessary information was to be looked for. The word ‘merger’ is actually not contained in any one of the four lists used for this study (see 3.3.1 above), and its low frequency is likely to have contributed to the poor performance on item 15.01. According to most respondents, however, the main problem is the ambiguity of the questions. 42 candidates answered item 15.07 by correctly choosing ‘true’ as the answer to this question, 24 of them went for ‘false’, and 9 for ‘not given’. In their responses to the second survey question some candidates expressed their doubts as to the correct choice for this item and said that, in their opinion, the answer could be either ‘true’ or ‘not given’, depending on how the statement and the text are interpreted. Some of them also wrote that their lack of knowledge of the meaning of ‘sixth form’ puzzled them at first, but it did not prevent them from answering the question. Item 15.05 turned out to be statistically as difficult as item 15.07, but no candidate mentioned this item as especially difficult in his/her response. In contrast, some candidates said that they found item 15.06 ambiguous, although 67 of them (91%) answered it correctly, and the exact reason they added to this was that they were not sure about whether ‘environment’ and ‘atmosphere’ should be considered as synonyms or not. Text 15 takes the form of a leaflet that presents and advertises West Thames College to prospective students. According to approximately 55% of the respondents to the first survey question, the main difficulty of text 15 is the presence of some infrequent lexical words, such as ‘merger’ and ‘guises’, and of the nominal groups ‘sixth form’ and ‘a typical student’, the exact meaning and reference of which was not clear to some students. 11% of the respondents, on the other hand, said that the text was difficult because of its ‘information density’, which is due, in their opinion, to the many dates, and names of colleges and of persons, and to all the figures and percentages in it. This last observation accords with and partly confirms the findings reported on in

3.2.1.2 and 3.4.2, namely that text 15 is the second densest text in the corpus if Ure's method is used to quantify lexical density, and the third densest one, if Halliday's method is applied.

Text 16 is a rather short notice about how the personal computers in a central library can be used. It is a text which test takers could encounter in an English speaking country 'on a daily basis' (see 2.3 above), and, according to the responses to the first survey question, it contained words unknown to approximately 22% of the candidates, such as 'spreadsheet' and 'loan'. 'Spreadsheet' actually does not belong to any one of the lists described in 3.3.1 above, whilst 'loan' belongs to the second word list. Furthermore, some of the students found the last paragraph particularly difficult to understand. Nearly all the respondents to the second open-ended survey question found most of the statements in the tasks ambiguous and sometimes tricky, since their correctness depends, in their opinion, on the reader/test taker's interpretation. Some other candidates' feedback was, on the other hand, more specific and focussed on the difficulty of items 16.06, 16.05, and 16.01. As can be seen from Figure 5.29, the items on which the students' performance was the poorest are item 16.06, item 16.04, and item 16.03. Item 16.06, in particular, is reproduced below together with the extract from text 16 containing the necessary information:

(157) 16.06 Library employees do not have detailed knowledge of computers.

(158) Please note that staff are not available to train people or give a lot of detailed instruction on how to use the programs. Prior knowledge is, therefore, necessary.

This item was answered correctly by the 24 candidates who chose 'not given' as the correct option, and wrongly by the 29 test takers who chose 'false' and the 22 who went for 'true'. According to some candidates' responses to the second open-ended survey question, it was the lack of knowledge of the semantic denotation of the adjective 'available' that caused them problems as to the choice of the correct option. In fact, the word 'available' is included in the third base word list described in 3.3.1 above (i.e. the 'academic' one), and should already be known by the population under investigation. The falsity of statement 16.06 could, on the other hand, be supported by the following argument: since employees are available to give 'some' (not a lot of) detailed instruction, they are also likely to have 'some' detailed knowledge of computers. However, explicit information about the actual state of their knowledge is not stated in the text. Item 16.04 and the part of text necessary to decide the correct answer are given below:

(159) 16.04 It is essential to reserve a computer three days in advance if you want to use one.

(160) Bookings may be made up to three days in advance.

37 test takers answered this item correctly by choosing 'false', while the remaining 38 chose 'true'. It can be claimed that the textual clues as to the correct completion of this task are the occurrence of the modal verb 'may', which clearly indicates possibility and not necessity, and the preposition 'up to'. As far as item 16.03 is concerned, 55 students answered it correctly by choosing 'not given' as the correct option and 20 chose 'false':

(161) 16.03 The information desk is closed at weekends.

(162) Bookings may be made in person at the information desk or by phoning 8673 8901 during normal office hours.

As can be seen by comparing the statement with the relevant part of the input text reported immediately after it, no information is given about the closure of the desk at weekends. The wrong choice made by some students might, however, be explained by the fact that in Italy public libraries (and, in particular, university libraries) are normally closed at weekends, and so are their information desks. In other words, these students are likely to have used their 'background knowledge', variously called and sub-categorised as 'scripts', 'frames', 'schemata', or 'knowledge of the world' in the literature (see 1.1 above and, for example, Clapham 1996: 16-32, Alderson 2000: 33-34; 44-46, and Bertuccelli Papi 2000: 181-202), in order to recover implicit information and process and interpret the text and this question. In this particular case, however, this knowledge turned out to be misleading for some candidates, independently of the group they belonged to (see also 5.7.2 below). An item that was deemed ambiguous by some candidates, but that did not turn out to be especially difficult, is item 16.01. It is reproduced below with the text portion containing the necessary information:

(163) 16.01 There are two computers and two printers available for public use at the library.

(164) 2 personal computers are available, for a fee of $5.00. There is also an ink jet printer attached to each terminal.

As far as statement 16.01 is concerned, the identity of both the participant 'computers' and that of the participant 'printers' can be easily retrieved from the input text, since the two of them are used in both the text and the question. However, whilst the numerative 'two' in the former nominal group clearly states the exact number of computers available, the number of ink jet printers is dubious. It can be argued, in particular, that the embedded clause 'attached to each terminal' does not clearly qualify the Head/Thing in terms of the number of printers at issue, and the nominal group 'an ink jet printer attached to each terminal' can be, therefore, defined as semantically opaque. It could, in fact, be equally interpreted as encoding reference to two printers each attached to different computers, or to just one printer attached to both computers.

5.7.2 Test 12 and the Perception of the Difficulty of the Tests in Battery 4

In 4.4.1 above it was suggested that the very presence of a large number of events and dates in test 12 can be regarded as its main difficulty, since the readers might get confused and find it problematic to distinguish between them. It was seen, in particular, that some of these events and dates are introduced as ranking structures (e.g. 'the final battle'), others as rank-shifted structures (e.g. 'the War of 1812' and also '1812'), and at least another one, i.e. 'the War of Independence', is introduced in question 12.01, albeit it can be referred back to the participant 'independence' via bridging. As the text unfolds some of these participants are presumed by means of ranking structures and often in thematic position, whilst others in rank-shifted position. 'The treaty', for instance, is mainly used as an object of a preposition, such as in 'the news that the Treaty of Ghent had been signed', or in 'negotiations for the treaty'. Some participants are, in other words, mainly foregrounded, others are backgrounded, and still others are characterised by both foregrounded and backgrounded occurrences throughout the test. Yet, they are all at issue in the tasks. The open feedback from approximately 15% of the test takers corroborates most of the suggestions made in 4.4.1, in that they explicitly mentioned their difficulty in distinguishing between all the many events and dates in the text. For about 7% of them, on the other hand, it was mainly the fact that the events are not narrated in chronological order that caused them processing difficulty. For another 40% it was the presence of some infrequent words, such as 'shanghaiing', 'ensuing', and 'casualties', that was the main cause of difficulty, and, finally, for about 4% of them the second paragraph was unclear.

As far as the tasks in test 12 are concerned, approximately 14% of the test takers deemed most of the questions ambiguous, and only by accurately reading and, if necessary, re-reading the text did they manage to answer them. Another 25% of them reported on their difficulty in retrieving the textual information necessary to answer the questions, with special reference to the information about the time, order, and place in which the events at issue took place. These and other test takers also claimed that they found questions 12.05, 12.08, 12.09, and 12.10 especially difficult to answer. With the exception of item 12.05, these evaluations correspond to their actual performance on the items. In fact, as can be learnt from Figure 5.30[28] below, the items that turned out to be most problematic are items 12.10, 12.08, and 12.09. Item 12.10, in particular, was answered correctly by the 59 respondents who chose option (c), while 69, the majority, chose option (b), 11 chose (d), and 9 chose (a). As stated in 2.2 above, this item assesses the ability to 'determine where to insert a piece of information', or, in more practical terms, the ability to fill in the

28 For these data person reliability equals .38 and item reliability equals .96.

following portion of text with the clause 'these U.S. troops were in fortified entrenchments outside the city':

(165) *10A* Early in January of 1815, some two weeks after the peace treaty had been signed, British troops in the southern part of the United States were unaware that the war had officially ended. *10B* Over 5,000 British troops attacked U.S. troops. *10C* During the ensuing battle, known as the Battle of New Orleans, the British suffered a huge number of casualties, around 2,000, and the Americans lost 71, all in a battle fought only because news of the peace treaty that had already been signed in Ghent had not yet reached the battlefield. *10D*

The high number of wrong answers to this question is likely to depend on the students' wrong and/or inaccurate interpretation of the identifying and cohesive resources in the missing clause, namely the Deictic 'these' and the Classifier 'U.S.' in the nominal group 'these U.S. troops'. It is argued, however, that this extra clause would sound somewhat 'unnatural' in the original text, in that it specifies a characteristic of the U.S. troops that could be more economically conveyed by means of an embedded clause such as 'the U.S. troops which were in fortified entrenchments outside the city'. In 2.2 above, item 12.08 was defined as an item that aims to test the ability of using context to determine the meaning of the 'difficult' word 'contested'. The majority of students (60%) actually managed to identify 'fought' as the best synonym of 'contested', while the rest chose either 'attacked', 'signed' or 'ended'. Item 12.09, on the other hand, was described as aiming to test the candidates' ability to 'answer implied detail questions' by asking them to decide the most likely date on which the Battle of New Orleans was fought as follows:

(166) 12.09 It can be determined from the passage that, of the following dates, the Battle of New Orleans was most probably fought
(a) on December 10, 1814
(b) on December 24, 1814
(c) on January 1, 1815
(d) on January 8, 1815

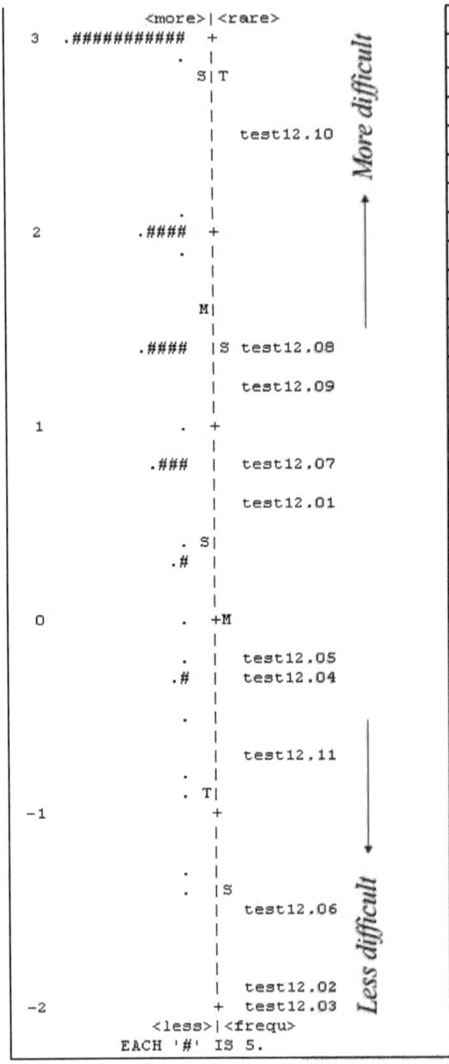

	N	*F.V.*	*S.D.*
t12.01	151	.71	.456
t12.02	152	.95	.224
t12.03	151	.95	.211
t12.04	152	.83	.378
t12.05	151	.81	.390
t12.06	152	.93	.260
t12.07	151	.68	.467
t12.08	144	.60	.491
t12.09	148	.62	.487
t12.10	148	.40	.491
t12.11	148	.87	.336

Figure 5.30. Variable map and descriptive statistics for the items in test 12.

According to the test administration data, 92 candidates went for the correct option (d), 24 went for (c), 17 for (b), and 15 for (a) (8 of them did not answer the question). In order to successfully answer this question, it can be claimed that test takers need to focus on the following two extracts:

(167) The War of 1812 was fought between Great Britain and the very young country of the United States only a relatively few years after the United States had won its independence from Britain. [...] The war lasted for a little more than two years, when a peace treaty was signed at Ghent, in Belgium, on the 24th of December, 1814. (extract 1 from test 12)

(168) Early in January of 1815, some two weeks after the peace treaty had been signed, British troops in the southern part of the United States were unaware that the war had officially ended. Over 5,000 British troops attacked U.S. troops. *10C* During the ensuing battle, known as the Battle of New Orleans, [...] (extract 2 from test 12)

In particular, in order to find out about the most likely date of occurrence of the battle, test takers ought to retrieve the identity of the participant 'the War of 1812' and that of 'the Treaty of Ghent', as well as to take the relevant time references 'a little more than two years' and 'some two weeks after' into account. As confirmed by the candidates' open feedback reported on above and by the test administration data, the retrieval of this type of information caused processing difficulty to the large majority of the population under investigation. As for item 12.05, it is interesting to notice that this 'stated detailed question'[29] about U.S. sailors mainly consists in the determination of the meaning of the verb 'shanghai' from the relevant contextual clues in text 12.

(169) 12.05 According to the passage, some U.S. sailors were ...
(a) taken forcibly to Shanghai
(b) made to go to Ghent
(c) forced to work on British ships
(d) responsible for causing the War of 1812

(170) The United States had declared war against Britain in June of 1812, mostly because of interference with U.S. shipping by the British and because of the shanghaiing of U.S. sailors for enforced service on British vessels. (extract from text 12)

---

29 See 2.2 above and Phillips (2001: 351-352).

Despite the fact that some candidates perceived this item as difficult, 81% of them answered it correctly, which suggests that they eventually managed to make up for their lack of knowledge of this term.

In order to conclude this section devoted to test battery 4, a brief investigation into the perceived difficulty of the various tests as well as into the performances of the various groups of test takers will be carried out. As can been seen from Figure 5.31, the text that was evaluated as the most difficult one is text 12 (M=2.35), followed by text 15 (M =2.19), text 13 (M=2.07), text 14 (M=1.83), and, finally, text 16 (M=1.58).

	N	*Min*	*Max*	*M*	*Mode*	*Median*	*S.D.*
test12q1	147	1	4	2.35	3	2	.766
test12q2	147	1	4	2.63	3	3	.769
test13q1	74	1	4	2.07	2	2	.816
test13q2	75	1	4	2.53	3	3	.827
test14q1	75	1	3	1.83	2	2	.665
test14q2	75	1	4	2.15	2	2	.766
test15q1	74	1	4	2.19	2	2	.734
test15q2	75	1	4	2.25	2	2	.660
test16q1	74	1	3	1.58	1	2	.619
test16q2	74	1	4	2.08	2	2	.807

Figure 5.31. Descriptive statistics of the closed-ended survey questions about test battery 4.

As far as the difficulty of the tasks is concerned, the questions in test 12 were, on average, perceived as the most difficult (M=2.63), while the tasks in test 16 as the easiest ones in the battery (M=2.08). The two extreme evaluations, therefore, were obtained by test 12 and test 16, which were deemed the most and the least difficult test respectively. These results, however, do not accord with the actual test results, in that, as discussed in 5.7.1 above, it was the tasks in test 15 that proved to be the most difficult ones, and those in test 13 that were, on average, the easiest to perform. In contrast, test 12 and test 16 are characterised by similar facility values, namely .75 and .72 respectively.

In Figure 5.32 the statistically significant correlations between the degree of perceived text difficulty, that of perceived task difficulty and the actual performance on the tasks are signalled.

	(1) Performance	*(2) Perceived text diff.*	*(3) Perceived task diffic.*
test12	√ (neg. with 2 and 3)	√	√
test13	√ (neg. with 2)	√	√
test14		√	√
test15		√	√
test16		√	√

Figure 5.32. Statistically significant correlations between the average performance on the tests in battery 4, the perception of text difficulty, and the perception of task difficulty ('neg.' stands for negative correlation).

As can be seen, the perceived difficulty of each text is significantly correlated with the perceived difficulty of the corresponding tasks. It can also be seen that the performance on test 12 is significantly (and negatively) correlated with both the perception of text difficulty and with that of task difficulty, that is, there is a direct relationship between the degree of perceived text and task difficulty and the actual performance on the test. Finally, the performance on test 13 is only correlated with the degree of perceived task difficulty.

In order to investigate into whether the various groups differed in their performance on the various tests and in their responses to the closed-ended survey questions, a series of ANOVA and/or Kruskal-Wallis Tests were carried out, taking into account the group they belonged to. These analyses indicated that the groups differed significantly in their performance on tests 12 and 14 and on their responses to questions 'test14q1' and 'test15q1'. Specifically, the group which outperformed the others significantly on test 12 is the 'MLCpost2' group ($\underline{M}$=8.94), whereas the worst performance was that of the 'MLC1int' group ($\underline{M}$=6.63). The 'MLC1adv' students, however, were also successful and obtained high average scores on the test and on most of its single items, and the 'LCM2' students were the ones that performed the best on item 12.09. As far as test 14 is concerned, the groups that turned out to be most successful on this test overall are the 'MLC1adv' group ($\underline{M}$=4.15) and the 'MLCpost2' group (M=4.00), whilst the 'LCM2' students' performance was poorer and differed significantly from those of these two groups (M=3.33). Finally, it is interesting to notice that the 'LCM2' group endorsed the highest average rate score on both text 14 and text 15 (in terms of their answers to questions 'test14q1' and 'test15q1'), while the 'MLCpost2' students' evaluations were significantly higher. This data suggests that the higher the students' proficiency, the lower the degree of perceived text difficulty.

5.8 Test Battery 5 (Tests 17, 24 and 25)

Test battery 5 contains test 17, that is, a test taken from the IELTS sub-corpus (see 2.3 above), and the two tests belonging to the CAE sub-corpus described in 2.6 above, i.e. tests 24 and 25. These tests were administered to 60 students, whose language proficiency was estimated at about level C1 of the CEF, in the order in which they are listed in Figure 5.33. The students belonged to the four groups also represented in Figure 5.33: 15 postgraduate language majors who were attending an English course targeted at their needs (i.e. the 'MLCpost2a' group); 11 language majors who were attending a first-year advanced English course (i.e. the 'MLC1adva' group); 20 second-year English language and literature majors who were attending an English course tailored at their needs (i.e. the 'LCM2a' group); and, finally, 14 third-year English language and literature majors who were also attending an English course (i.e. the 'LCM3' group). The time allowed to complete this battery was 60 minutes. As can be seen from the graph and the statistics shown in Figure 5.34 below, battery 5 as a whole proved to be rather difficult, in that 50% of the students obtained a score lower than 21, that is, below the Median, and no test taker managed to obtain the maximum score (31). However, six students (11.7%) obtained a score of 26, three of them (5%) obtained 27, and the overall 'M%' value equals 63.87 (that is, the average percentage of success was about 64%). In this respect, the open feedback received from some candidates suggests that one of the causes of their poor performance is the time limit, in that for them 60 minutes were insufficient to read the three texts and complete the tasks.

TEST BATTERY 5

Test 24. Emotional Intelligence
Test 25. Beginner takes all
Test 17. Green Wave

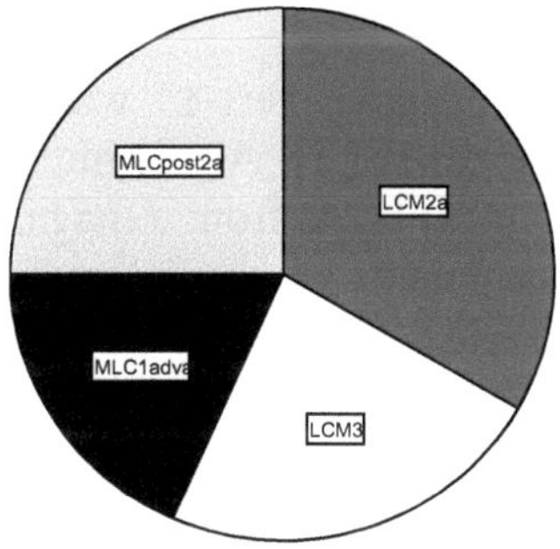

Figure 5.33. Test battery 5 and the population it was administered to.

According to table 5.35, the test that turned out to be the most difficult is test 17 (its 'M%' value equals 58), while the least difficult was test 25 (its 'M%' value equals 72.5). The performance on test 24 and test 17 was varied, which is suggested by the fact that there are three Modes for test 17 (7, 9 and 10), the Standard Deviation is high in both cases, and the scores obtained by the students on test 17 range from 2 to 11.

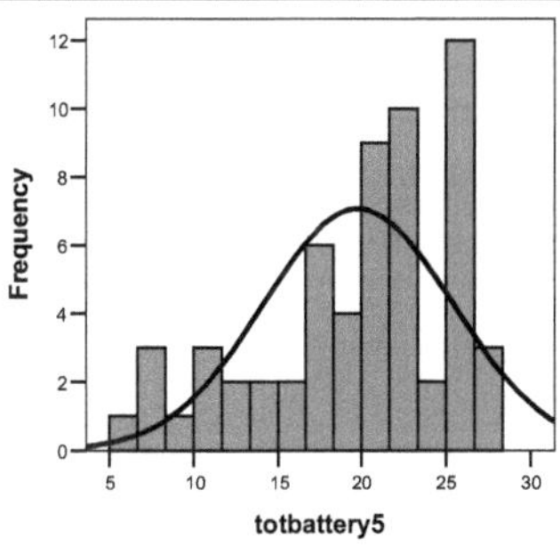

DESCRIPTIVE STATISTICS
N = 60
Mean = 19.8
Mode = 21
Median = 23
Standard Deviation = 5.62
Minimum = 6
Maximum = 27
M% = 63.87

Figure 5.34. Histogram displaying the distribution of the total score on 'totbattery5' for the whole population, and some descriptive statistics. The maximum score on battery 5 as a whole is 31.

	N	*Min*	*Max*	*M*	*M%*	*Mode*	*Median*	*S.D.*
test17	60	2	11	7.55	58	7, 9, 10	8	2.31
test24	60	2	12	7.9	65.8	9	8	2.333
test25	60	0	6	4.35	72.5	6	5	1.885

Figure 5.35. Descriptive statistics of test 17, test 24, and test 25.

As stated in 2.3 above, test 17 was chosen as representative of the first section of the IELTS Academic Reading module, and text 17, in particular, is a text taken from a magazine, journal, or newspaper that reports on the results of a survey about 'green consumerism' conducted in the mid 1990s in Great Britain. As seen in 3.5.2 above, text 17 is the least readable text in the corpus, which is mainly due to the large number of polysyllabic words in the text, e.g. 'environmentally', 'consumerism', and 'companies', and to the relatively small number of clause complexes. Furthermore, as seen in 3.1 above, text 17 is characterised by the highest STTR50 value, that is, by the fact of containing the highest number both of different grammatical and of lexical word-forms among all the texts in the corpus. Finally, as anticipated in 4.1.2, some students noticed that the exact identity of the generic participants 'green consumers' and 'green products' are not made clear until the end of the text, which caused them processing difficulty. By contrast, the identity of 'ethical spenders' is clearly stated when this participant is presented. Approximately 18% of the students said that they were distracted by the large number of figures, percentages, and other data in the text, and that these caused difficulty following the points made in the text whilst reading. Other test takers declared that they did not know the exact meaning of the multi-word unit 'go off the boil', but that they nonetheless managed to infer it from the co-text. Some other examinees made reference to the parts of the text in which the attitudes of politicians, companies, and the Government are expressed, and added that, in their opinion, they were not clearly stated. The following extracts from text 17 contain this information:

(171) Research in Britain has shown that 'green consumers' continue to flourish as a significant group amongst shoppers. This suggests that politicians who claim environmentalism is yesterday's issue may be seriously misjudging the public mood.

(172) The recently published report also predicts that the process will repeat itself with 'ethical' concerns, involving issues such as fair trade with the Third World and the social record of businesses. Companies will have to be more honest and open in response to this mood.

(173) However, concern for specific issues is lower than in 1990, suggesting that many consumers feel that Government and business have taken on the environmental agenda.

For a thorough discussion of this point, item t17.07 is also relevant, in that this multiple-choice question asks about the reasons for the politicians' attitude, and some students might have raised this issue because of the difficulty they encountered whilst trying to answer this question:

(174) 17.07 Politicians may have 'misjudged the public mood' because ...
(a) they are pre-occupied with the recession and financial problems.
(b) there is more widespread interest in the environment agenda than they anticipated.
(c) consumer spending has increased significantly as a result of 'green' pressure.
(d) shoppers are displeased with government policies on a range of issues.

The correct answer to this question is option (b), and it can be claimed that "the necessary information is not in one continuous text, but is scattered across more than one place"; its retrieval involves, therefore, the candidate's "ability to synthesise scattered information" (Buck, Tatsuoka and Kostin 1997: 441). The necessary information is, first of all, to be looked for at the beginning of the text (see the first extract above), in that one of the clues as to the choice of the correct option is given by the expression 'misjudged the public mood', which occurs in this part of the text and overlaps with the one used in the question. Furthermore, at the beginning of the text it is stated that green consumers 'continue to flourish', which can be related to 'more widespread interest' in option (b). Finally, the necessary information is also to be looked for in the last paragraph, where the participant 'environmental agenda' is used and, therefore, overlaps with 'environment agenda' in option (b). The process of retrieving all this information and synthesizing it turned out to be rather difficult for some examinees, and this is confirmed by the facility value for this item, .67. In particular, 40 candidates answered the question correctly, 9 chose option (a), 8 chose option (c), and 3 chose option (d).

As far as the other tasks in test 17 are concerned, an inspection of the variable map[30] and of the descriptive statistics shown in Figure 5.36 indicates that most of the items in this test were among the most difficult items in the battery overall, thus explaining the poor average performance on test 17. Specifically, the items that turned out to be most difficult are items t17.13, t17.12, t17.06, t17.04, and t17.10, while the least difficult ones are t17.09 and t17.11. As for t17.13, only 9% of the candidates managed to choose the correct word that fills in the gap, i.e. 'social record', while 29% went for 'social awareness', 13% for 'political beliefs', and 12% for 'financial constraints' (others went for other options). Item t17.12 was answered correctly (viz. 'armchair ethicals') by 19% of the candidates, while 29% chose 'environmental research', and 48% 'ethical spenders'. Finally, item t17.10 was answered correctly (viz. 'honesty and openness') by 48% of the candidates, whereas 31% chose 'environment research', and 22% 'social awareness'.

These results are confirmed by the responses to the second open-ended survey question, according to which approximately 31% of the test takers found the summary completion task (i.e., items t17.10-t17.13) especially demanding. Approximately 20% of them, on the other hand, responded by saying that they found the Yes/No/Not Given test method difficult (i.e., items t17.01-t17.06). This evaluation is also confirmed by the data, since the next two most difficult items in test 17, that is, t17.06 and t17.04, are statements the correct answer to which is 'not given'. In particular, these two items asked candidates to decide whether the following statements are true or false:

(175) 17.04 Consumers' green shopping habits are influenced by Mintel's findings.

(176) 17.06 Mintel undertakes market surveys on an annual basis.

If the text is read accurately, both globally and locally (see 1.2 above), it can be seen that this information is actually not stated. However, as suggested in 5.7.1 above, this test method can 'lead to considerable confusion', which is also proved by the poor performance on these items. Furthermore, the students were not familiar with this test method, and had received no training as to how to best

30 Person reliability equals .80 and item reliability equals .90.

tackle the task[31]. A more general comment from approximately 8% of the students is that they found it difficult to retrieve the necessary information for most of the questions, whilst about 10% deemed most of the items in test 17 ambiguous.

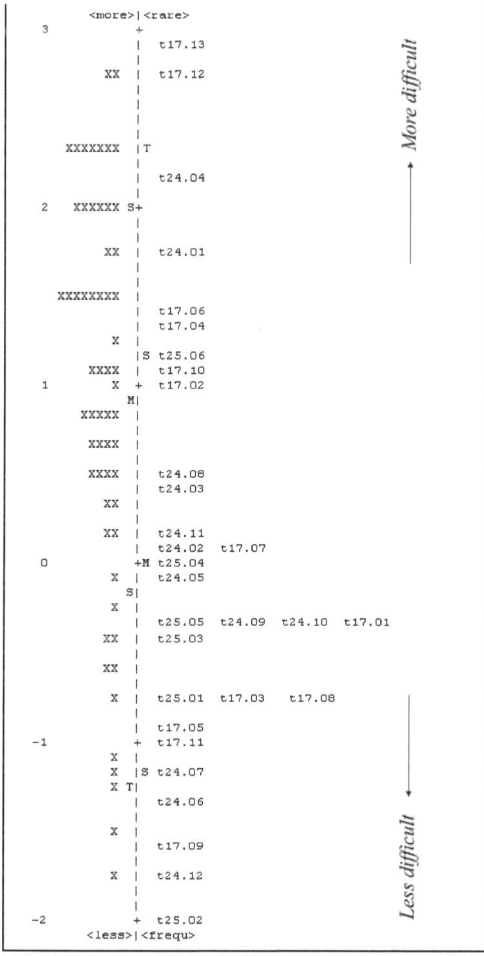

	N	*F.V.*	*S.D.*
t17.01	60	.73	.446
t17.02	60	.50	.504
t17.03	60	.80	.403
t17.04	60	.43	.500
t17.05	60	.82	.390
t17.06	59	.41	.495
t17.07	60	.67	.475
t17.08	60	.80	.403
t17.09	60	.88	.324
t17.10	58	.48	.504
t17.11	56	.84	.371
t17.12	58	.19	.395
t17.13	56	.09	.288
t24.01	60	.35	.481
t24.02	60	.67	.475
t24.03	60	.60	.494
t24.04	60	.27	.446
t24.05	60	.70	.462
t24.06	60	.87	.343
t24.07	60	.85	.360
t24.08	59	.59	.495
t24.09	60	.73	.446
t24.10	60	.73	.446
t24.11	59	.66	.477
t24.12	60	.90	.303
t25.01	60	.80	.403
t25.02	60	.92	.279
t25.03	60	.75	.437
t25.04	60	.68	.469
t25.05	60	.73	.446
t25.06	60	.47	.503

Figure 5.36. Variable map and descriptive statistics for the items in tests 17, 24, and 25.

31 Unfortunately, because of time constraints, it was not possible to train the students in this type of question and in this type of test, before they were asked to sit the test.

As said in 3.3.2 and 4.4.6.2 above, text 24 is a newspaper article and the second most 'academic' text in the corpus. According to the responses to the first open-ended question, approximately 62% of the test takers found the use of some infrequent words such as 'yard-stick', 'adroitness', 'bluntly', 'adept', and 'honing' as the main source of difficulty in the text. Approximately 5% of these respondents were more specific, and mentioned the academic words and the formal register of the text as mainly contributing to text difficulty. Approximately 14% of the respondents, on the other hand, complained about the fact that the four paragraphs deal with similar concepts and that the contents are not developed in a linear way, which makes it difficult to decide where statements are made. Furthermore, approximately 10% of the students were split in their opinion concerning the causes of text difficulty: some claimed that the topics the article deals with are difficult, others found that the text contains too much information and perceived it, therefore, as too 'dense', and still others that the text is too long to be read in a testing situation. Finally, as argued in 4.4.6.2 above, text 24 contains at least one clause complex that is structurally opaque and that is, therefore, likely to contribute to text complexity and to the difficulty of item t24.01 (i.e. the second most difficult item), namely:

(177) And the less straightforward the job, the more emotional intelligence matters

As for the difficulty of the other items, from Figure 5.36 it emerges that the most difficult item in test 24 turned out to be item t24.04. Specifically, 16 candidates (27%) answered it correctly by choosing paragraph D, 9 candidates chose paragraph A, 15 chose paragraph B, and 20 (34%) chose paragraph C. The item consisted of the following statement:

(178) In which section is the following mentioned?
24.04 the fact that emotional intelligence can be combined with other skills to improve people's ability to cope at work

And the following extract from text 24 contains the necessary information:

(179) In fact, studies that have measured people's emotional intelligence through the years show that most people grow more adept at handling their own emotions and impulses, at motivating themselves and at honing their empathy and social adroitness. There is an old-fashioned word for this growth in emotional intelligence: maturity. Not only can emotional intelligence be learnt, but individually we can add these skills to our tool kit for survival. This is especially relevant at a time when it seems a contradiction to put the words 'job' and 'stability' together.

It can be seen that the participant 'skills' occurs in both item 24.04 and in the input text, that is, there is lexical overlap between the two. However, whilst 'these skills' in the text presumes some sub-skills related to emotional intelligence (i.e., 'handling their own emotions and impulses', 'motivating themselves' and 'honing their empathy and social adroitness'), 'other skills' encodes relevance phoricity, in that it makes reference to skills that are different from emotional intelligence. These 'other skills', in particular, are part of 'our tool kit for survival', to which the sub-skills of emotional intelligence can be added. Besides these, another important point has to be made, namely that the participant 'skills' occurs in all paragraphs and not only in paragraph D, which decreases its potential for being the main clue to the correct answer of this item. All these considerations suggest that the high percentage of failure on this item is due to both the low reliability of the participant 'skills' as a key to the correct answer and the difficulty involved in processing the phoric relations described above. More generally, on the basis of the open feedback obtained from the candidates, it emerged that approximately 42% of the test takers did not find the statements particularly difficult to comprehend; what they deemed problematic, however, was linking them to the corresponding part of the text, that is retrieving the necessary information from it. Approximately 18% of them, on the other hand, found most of the statements ambiguous, in that, in their opinion, some statements can be matched with more than one paragraph. Finally, approximately 15% of the students wrote that their main difficulty was concerned with their lack of knowledge of some infrequent words and/or multi-word units used in the text and/or in the statements.

Test 25 was dealt with thoroughly in 4.4.2.2 above, where the cohesive markers that test takers ought to take into account for filling in the first five gaps correctly were discussed. As can be seen from Figure 5.36, the 6 items in test 25 turned out to be rather easy, and only item t25.06 has a rather low facility value (.47). As for this item, 28 candidates (46.7%) answered it correctly by choosing B as the extract that best fills in gap 6, 17 of them (28.3%) chose D, 5 (8.3%) chose C, 4 (6.7%) chose E, another 4 chose F, and, finally, 2 (3.3%) chose G. Extract D, i.e. the extra paragraph which does not fit into any of the gaps, proved therefore to be the main distracter for item t25.06. The main cohesive elements that provide a clue as to the correct choice of the missing text are underlined in the following extracts:

(180) (11) As they all agreed to this sum, it was decided that they should each 'pitch' to Evans. And so, one night in October, he sat in his study while four great film-makers rang, one after the other, to beg for the privilege of paying $3 million for an unfinished novel. Evans told me all this as we sat drinking coffee on a wooden verandah perched above the leafy garden of his home. He said that he had since turned down an offer to write the screenplay of *The Horse Whisperer*.

(181) (12) "'It's all been such a fairy tale so far, I don't want to spoil it. Writing at that level is a very tough business, and I don't want to become an employee of these people who I like and who have paid me so much money. I'd hate to find myself writing a draft or two and then have them say, "Thanks Nick, but now we'll bring in so-and-so." (6th gap)

(182) (13) He would be involved, he said, but at arm's length. The success of his novel had inevitably brought forth the offer of new backing for *Life and Limb* but he was no longer sure that he wanted to make it. "I think that I would be foolish not to write another novel,' he said."

In particular, the participants 'four great film-makers' and '$3 million' are introduced in the eleventh paragraph and presumed afterwards in the twelfth paragraph by means of 'these people' and 'so much money' respectively, and two inflected forms of the verb 'pay' are repeated, i.e. there is lexical repetition and therefore cohesion.

As far as the students' open feedback is concerned, approximately 40% of the respondents found the use of infrequent words and multi-word units as the main source of difficulty in the text (as seen in 3.3.2 above, text 25 actually contains a large number of words not present in any one of the four lists), while approximately 30% of them found it difficult to retrieve the identity of the participants at issue in a given portion of text and to distinguish between them. The students were also split in their opinion about the difficulty of the tasks. In particular, 40% of them found it difficult to recognise the cohesive markers necessary to reconstruct the text, and according to some of them, this was also due to the presence of the extra paragraph. Another 24% of the candidates, on the other hand, found that some gaps could be filled in by more than one extract, and some of them mentioned item t24.03, t24.04, t24.05, and t24.06 as the most difficult items in the test.

In order to complete this analysis of the difficulty of the tests in battery 5, some considerations about the perceived difficulty of the various texts and tasks will be made, and an investigation into the performances of the four groups of test takers will be carried out. As can be learnt from Figure 5.37, the text that was evaluated as the least difficult to read is text 17 (M=2.41), whilst text 25 was perceived as the most difficult one (M=2.78). In contrast, the tasks in test 24 were perceived as the most difficult ones (M=3.40), and those in test 17 as the easiest ones (M=3.02). These latter data, in particular, do not accord with the average performance on the three tests, in that it was test 17 that turned out to contain the most difficult items (see Figure 5.35). This discrepancy can be explained by two main reasons: firstly, the fact that the Yes/No/Not Given test method used in test 17 was not familiar to most test takers, and some of them might have, therefore, underestimated its difficulty; secondly, the fact that when the students were administered test 17, they were already fatigued and perhaps did not have enough time and stamina to think about the correct answers. The data displayed in Figure 5.38 below suggests that the test takers' performance on tests 17 and 25 correlate significantly with the evaluations of the difficulty of the respective input texts, but not with the evaluation of the tasks (the higher the evaluation of text difficulty, the lower the score obtained). It also suggests that for these two tests the evaluation of text difficulty is statistically correlated with the evaluation of task difficulty. By contrast, there are no significant correlations between the performance on test 24 and the evaluation of either its text or task difficulty. This latter consideration, in particular, indicates that test 24 was perceived as particularly difficult, but most test takers performed well on it.

	N	*Min*	*Max*	*M*	*Mode*	*Median*	*S.D.*
test17q1	51	1	4	2.41	2, 3	2	.876
test17q2	52	2	4	3.02	3	3	.671
test24q1	60	1	4	2.72	3	3	.739
test24q2	60	2	4	3.40	4	3	.643
test25q1	60	1	4	2.78	2	3	.904
test25q2	59	1	4	3.12	4	3	.892

Figure 5.37. Descriptive statistics of the closed-ended survey questions about test battery 5.

	(1) Performance	*(2) Perceived text diff.*	*(3) Perceived task diff.*
test17	√ (neg. with 2)	√	√
test24			
test25	√ (neg. with 2)	√	√

Figure 5.38. Statistically significant correlations between the average performance on the tests in battery 5, the perception of text difficulty, and the perception of task difficulty ('neg.' stands for negative correlation).

An investigation was also conducted into whether the various groups differed significantly in their performance on the various tests in battery 5 and in their responses to the closed-ended survey questions. To this end a series of ANOVA and/or Kruskal-Wallis Tests were carried out. The analyses indicated that the groups differed significantly in their evaluations of the tasks in test 24 and of those in test 25. Specifically, the 'MLCpost2a' students endorsed the highest average rate score on both, whilst the 'LCM3' endorsed the lowest rate score among the four groups. This suggests that in the case of battery 5 the more advanced students, i.e. the 'MLCpost2a' group, were more cautious about evaluating the difficulty of the tasks than the other groups.

5.9 Relationship between Test Takers' Performance, their Perception of Text/Task Difficulty and Results of the Quantitative Investigation

Some of the results discussed in the present chapter are represented synoptically in Figure 5.39 below. In particular, the average mean scores which test takers obtained on the tests, and the average evaluation of the difficulty of each text and task were converted to z-scores (see 3.6) and plotted in Figure 5.39. Before inspecting the graph, however, four things must be considered. Firstly, that the average score test takers obtained on a test, i.e. the 'Mean%' value, corresponds to the facility values (viz. the 'F.V.' values) of test 02, test 03, test 04, test 05, test 06, test 07, test 08, test 09, and test 10, and to the mean scores expressed as percentage (viz. the 'M%' values) of the other tests. Secondly, that the better the performance on a given item (i.e. the easier a given item), the longer the 'Mean%' bar on the right side of the graph. Thirdly, that the lower the rate score endorsed on a given text or task (i.e. the easier a given text or task are evaluated), the longer the 'textev' or 'taskev' bars on the left side of the graph. Fourthly, that these z-scores were calculated on the basis of the values of all 25 tests and not by considering each single battery.

As can be seen from Figure 5.39, the most difficult test in the corpus overall turned out to be test 06 ('F.V.'=.40), while text 6 was evaluated as the second least difficult text in the corpus ('test06q1'=1.47), and task 6 was endorsed an intermediate rate score ('test06q2'=2.35). In contrast, the easiest test turned out to be test 09 ('F.V.'=.90), task 09 was evaluated as the least difficult one in the corpus ('test09q2'=1.60), and text 09 was endorsed an intermediate rate score ('test09q1'=1.83). The two texts in the corpus that were evaluated as the most difficult ones are text 21 and text 25 ('test21q1'=2.92 and 'test25q1'=2.78 respectively), while the one perceived as the easiest one is text 5 ('test05q1'=1.31). On the other hand, the tasks in tests 24 and 25 were perceived as the two most difficult ones ('test24q2'=3.40 and 'test25q2'=3.12 respectively), whereas task 9 was perceived as the least difficult one ('test09q2'=1.60). If the DIALANG tests are not taken into account, test 17 stands out as the most difficult test in the corpus ('M%'=58; see also the discussion of test battery 5 in 5.8 above), and the tasks in test 17 as the third most difficult (in the corpus overall; 'test17q2'=3.02). Test 11, by contrast, is the easiest test ('M%'=85.4; see also the discussion of test battery 3 in 5.6 above), and text 16 the easiest text ('test16q1'=1.58).

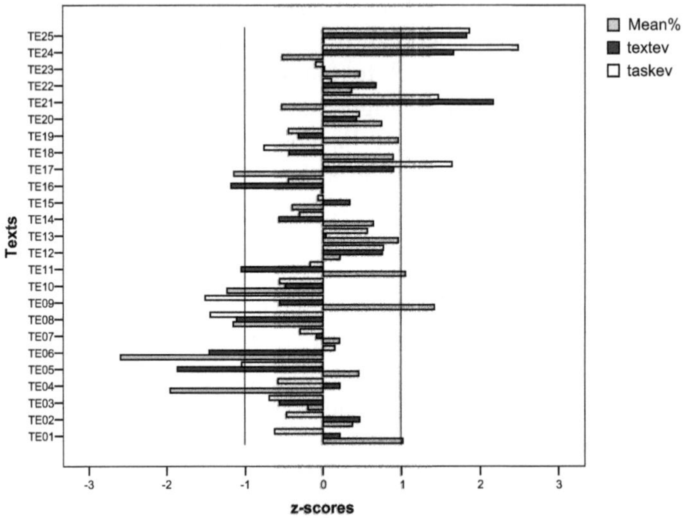

Figure 5.39. z-scores for the average performance on each test (expressed as percentage), for the evaluation of text difficulty and for the evaluation of task difficulty.

In order to explore the relationships, if any, between the average values of text and task evaluation for each test and the quantitative variables calculated in chapter 3 for each text in the corpus, a multiple regression analysis was run. The output from multiple regression, however, suggested that the values of these variables cannot predict the degree of perceived text and task difficulty statistically. To this end, therefore, the correlations between both the values of text and task evaluation and each quantitative variable were calculated by using Pearson Product Moment Correlation (r), Kendall, and Spearman's Rho. Subsequently, scatterplots were plotted for the pairs of variables that turned out to be significantly correlated. The linear

R-Square was also indicated, in that, according to Hatch and Lazaroton (1991: 440)

> this allows us to see how much of the variance (the variability of scores around the mean) in one measure can be accounted for by the other. To the degree that the two measures correlate, they share variance.

As can be seen from Figure 5.40, the correlation between the STTR50 values and the average rate scores endorsed on the texts is medium and positive. Furthermore, the STTR50 variable accounts for 15% of the variance in text evaluation. On the basis of these data, it can be suggested that the higher the values of the STTR50 index, the higher the average rate score endorsed on a given text, i.e. the more difficult the text was perceived.

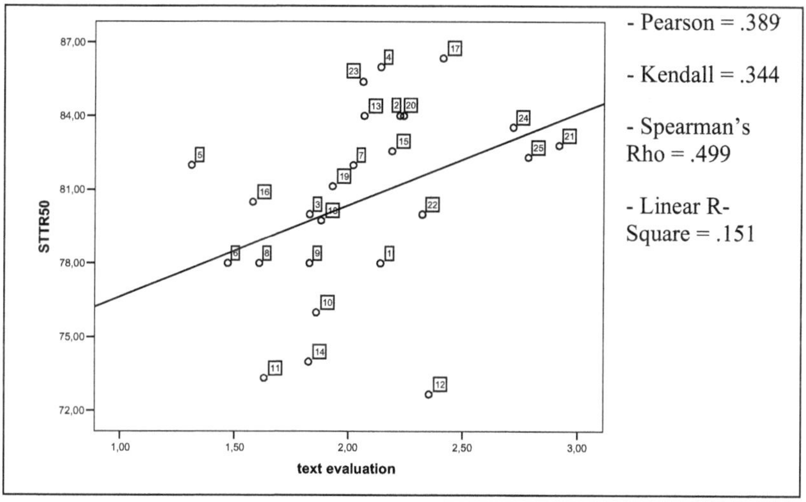

Figure 5.40. Scatterplot for the STTR50 index and text evaluation.

From Figure 5.41 it can be seen that the correlation between the LV(re) values and the average rate scores endorsed on the tasks is large and negative. Furthermore, the LV(re) variable accounts for about 28% of the variance in task evaluation. On the basis of these data, it can be suggested that the higher the values of the LV(re) index, the lower the average rate score endorsed on a given task, i.e. the less difficult the task was perceived.

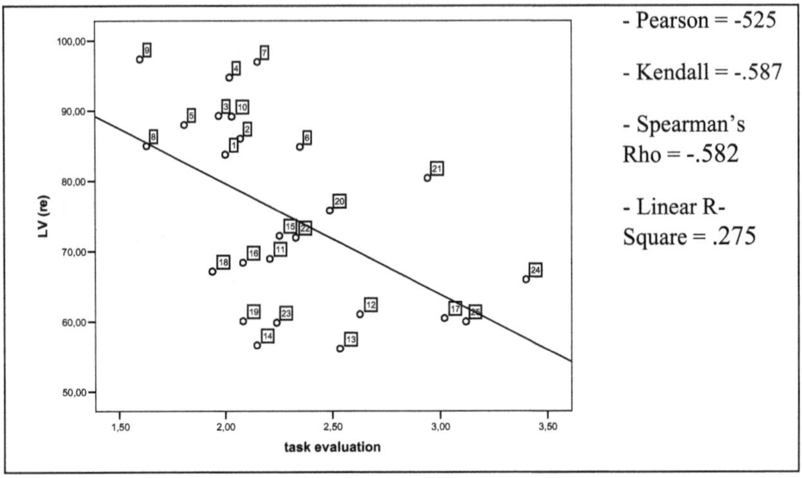

Figure 5.41. Scatterplot for the LV(re) index and task evaluation.

Figure 5.42 shows that there is medium and negative correlation between the GI values and the test scores obtained by the candidates. Figure 5.42 also shows that the GI variable accounts for 16% of the variance in the scores obtained. It can be suggested, therefore, that, according to these data, the higher the values of the GI index, the lower the average test scores obtained by the candidates.

Figure 5.43 shows that there is large and negative correlation between the H/P values and the test scores obtained by the candidates. The figure also shows that the H/P variable accounts for 30% of the variance in the scores obtained by the candidates. It can be suggested then that, according to these data, the higher the values of the H/P value, the lower the average test score obtained by the candidates.

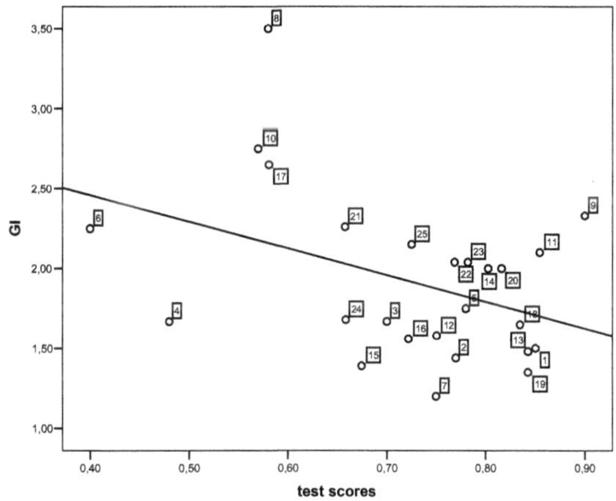

- Pearson = -399

- Kendall = -.211

- Spearman's Rho = -.312

- Linear R-Square = .159

Figure 5.42. Scatterplot for the GI index and the test scores.

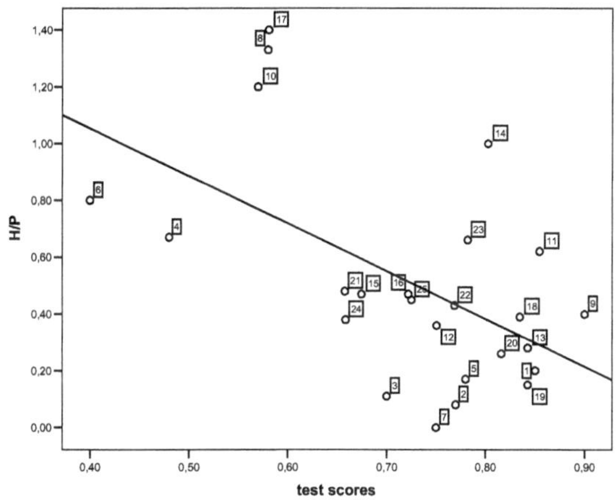

- Pearson = -551

- Kendall = -.324

- Spearman's Rho = -.476

- Linear R-Square = .303

Figure 5.43. Scatterplot for the H/P index and the test scores.

6. Conclusions

This study has investigated the text-inherent complexity and the receiver-oriented difficulty of a corpus of twenty-five reading comprehension tests. To this end, quantitative and qualitative research was carried out on the texts as well as on the tasks of the corpus and reported on in chapters 3 and 4. The thirteen calculations made use of in chapter 3 allowed for an objective measurement of various aspects of the complexity of the texts under investigation, while the qualitative analyses reported on in chapter 4 indicated some complex and marked types of cohesion in some of the tests of the corpus. The analyses described in chapter 5, on the other hand, drew on the data obtained from the administration of the tests and of the feedback questionnaires. They accounted for the difficulty of the tasks and the tests, as well as for the subjective judgement made by the test takers upon the difficulty of each text and the accompanying tasks. The data obtained by means of these different methodologies and from different sources was compared and 'triangulated' throughout chapters 3, 4 and 5, in order to increase the representativeness of this study (see 5.1 above).

As for the main results obtained by applying the quantitative calculations in chapter 3, it can be claimed, first of all, that although each calculation captured a specific aspect of the complexity of the texts, factor analysis and multidimensional scaling revealed the presence of two underlying macro-aspects: namely, one which is broadly concerned with the number of lexical words in a text and another broadly concerned with the length of its words, clauses, and sentences. Specifically, lexical density – calculated using both Ure's and Halliday's methods –, lexical variation, as well as the percentages of words that do not belong to any word list turned out to contribute to the quantification of the complexity due to the former macro-aspect. In contrast, readability formulae, the H/P ratio, the percentages of words belonging to the academic word list, and the type/token ratio turned out to contribute to the quantification of the latter macro-aspect. Grammatical intricacy is noteworthy, since the data proved that it contributes to the measurement of both macro-aspects: namely, it contributes negatively to the former and positively to the latter.

The analysis of the characteristics of the texts which scored noticeably high or low on the various calculations revealed more specific characteristics of these texts and of their complexity. Furthermore, the triangulation of these data with the indications of difficulty which emerged in chapter 4 and with the data obtained from the administration of the tests and the questionnaires – i.e. the data accounting for the difficulty of the corpus – made it possible to draw more reliable conclusions about the complexity of the texts, the difficulty of the tests, and their relationship.

The standardised type/token ratio, i.e. STTR50, is the ratio of the number of different words (types) to the number of running words (tokens) in a given text, and gives information about the degree of repetition in a text. Specifically, the higher the value of the index, the lower the amount of repetition in a text. According to the data reported on in 3.1, texts 17 and 4 score the two highest STTR50 values in the corpus, whereas texts 12, 11 and 14 score the lowest values. As shown in 5.4, 5.8 and 5.9 above, texts 17 and 4 were both judged to be difficult, namely the former was judged to be the fourth most difficult in the whole corpus, and the latter the second most difficult in the DIALANG sub-corpus. On the other hand, as seen in 5.6, 5.7 and 5.9, text 11 and text 14 were both evaluated as rather easy, but text 12 was evaluated as difficult. These evaluations, in particular, explain the general tendency described in 5.9, namely that there is (medium) positive correlation between the STTR50 values of the texts and the average rate scores endorsed on them.

Lexical density captures text complexity in terms of either the ratio of lexical items to words – i.e. Ure's method (LDU) – or the ratio of lexical items to ranking clauses – i.e. Halliday's method (LDH) – in a text. As seen in 3.2.1.2 and 3.4.2 above, the texts which score the highest lexical density values are texts 19 and 15, whereas those that score the lowest values are texts 8 and 3. Texts 19 and text 15 are two leaflets which advertise a trip to Australia and West Thames College respectively. In the former text function words tend to be omitted and a small number of ranking clauses to be used. In the latter, on the other hand, a large number of complex noun groups containing pre- and post-modifying elements are used, as well as a proportionally small number of ranking clauses. As shown in 5.5 above, the main difficulty of the tasks in test 19 concerned the retrieval of the information necessary to verify the truth of some of the statements, but the overall test proved to be easy. Moreover, both text 19 and its tasks were perceived as rather easy. Test 15, by contrast, turned out to be statistically rather difficult, but neither the text nor the tasks were evaluated as especially difficult. However, some test takers found that the 'information denseness' of the text was the main cause of difficulty, which can be interpreted as an effect of the high lexical density of text 15. Furthermore, part of the test takers' feedback confirmed the indications which emerged from the analyses reported on in 4.4.5.1, namely that the statements contained in the tasks are 'complex'. It also confirmed the suggestion that the retrieval of the identity of the relevant participants in the text can be difficult. As for the two lexically sparsest texts in the corpus, texts 8 and 3, neither their text nor their task was evaluated as especially difficult, but task 8 proved to be statistically rather difficult. As shown in 3.2.1.2 above, text 8 is likely to be an extract from a piece of fictional writing and is written in an intricate style. Its main difficulty, therefore, is likely to be due to its being grammatically intricate, rather than to its being lexically sparse. This supposition is confirmed by the fact that a large number of test takers found it difficult to infer the meanings implicitly conveyed by the text, and, in particular, found the last clause complex difficult to process.

As lexical density and grammatical intricacy are 'paired' properties, these last observations also provide a general conclusion about grammatical intricacy and an indication of how intricate texts can be actually difficult to process. As said in 3.4.2 above, in order to understand exactly what factors actually contribute to text complexity, the values of the indices LDU, LDH and GI should all be taken into account and interpreted critically. Along with these indices, the H/P index should also be considered, in that it measures the specific complexity caused by the use of hypotaxis. In fact, the correlation between GI and H/P is positive and high, which suggests that when a text is grammatically intricate it is likely to contain a large number of secondary clauses in hypotactic structures and vice versa. In contrast, as shown in 5.9 above, the correlation between the GI index and the average score obtained on the tests is negative, which is also the case of the correlation between the H/P index and the average score obtained. This suggests that the higher the values of the GI and the H/P indices, the lower the average test scores obtained by the candidates.

In this study lexical variation was calculated both as the ratio of the number of different lexical words to the total number of lexical words – i.e. LV (see 3.2.2 above) – and as the ratio of the number of word families in the text to the total number of lexical items – i.e. LV(re) (see 3.3.2 above). The LV(re) index, in particular, accounts for the relations between lexical words which belong to the same lemma and/or word family. By applying the more refined LV(re) index, in particular, it was possible to identify the lexically least varied texts in the corpus, – i.e. texts 13 and 14 – , as well as the lexically most varied ones, i.e. texts 9 and 7. Since text 13 aims to provide a rather didactic exposition of the concept of animal migration, its author had to repeat the concepts and to make reference to the same or similar participants throughout the text, which explains the paucity of lexical variety in this text. As seen in 4.4.5.3 and 5.7.1 above, test takers were successful on this test, evaluated its text and tasks as rather easy, and the main problems they raised had to do with the use of some hyponyms, specifically names of animals and species unknown to them. As for text 14, as seen in 4.4.3 and 5.7.1 above, test takers had problems interpreting the identity of the participant 'permit' correctly, and found it difficult to decide whether identity or similarity of reference was at issue in the test. These observations suggest, in particular, that the complexity of these lexically non-varied texts – and in these cases also expository and didactic, or informative texts – is due mainly to the presence of taxonomically related lexical words. The data also suggest that this aspect of textual complexity caused processing difficulty to the test takers. On the contrary, as seen in 3.2.2 above, the complexity of text 9 – the least varied text in the corpus and an extract from the review of a TV drama – is due to the fact that no key word or key concept can be identified on the basis of the frequency criterion. The reader has, therefore, to rely on other textual clues to successfully process the text. In particular, it is thematic progression that is likely to help them infer the information that is necessary to perform task 9. According to some of the test takers' feedback, however, processing the clause complex containing the necessary information was difficult, which suggests that the complexity of lexically varied texts can also cause difficulty. A general tendency reported on in 5.9 also suggests that the higher the values of the LV(re) index, the lower the average rate score endorsed on the tasks.

As seen in 3.3.1 above, in this study vocabulary difficulty was quantified using the information provided by two frequency word lists, namely the General Service List of English Words (GSL) and the Academic Word List (AWL). Since the criterion by which a word is assigned to a given word list is its frequency of occurrence in the English language (or rather in a representative corpus), the complexity of a text in terms of vocabulary is determined by the frequency of its words. As shown in 3.3.2 above, the texts characterised by the presence of the largest number of academic words are texts 1 and 24, whereas those that contain the highest number of words which do not belong to any word list are texts 7 and 9. According to the feedback derived from the questionnaire, the presence of academic and field-specific words in the former two texts caused processing difficulty, since they had to read them carefully to understand their exact meaning in the context and to successfully perform on the tests. The same holds true for the texts characterised by the presence of infrequent words and/or multi-word units, and it was especially so when their meaning could not be easily deduced from either their context or co-text, or when their semantic denotation or connotations were at issue in the tasks, e.g. tests 11, 12, 13, 22, and 24.

The readability formulae used in this study – i.e. FRE, FKGL, and GFI – assign an estimate of readability to a text on the basis of its word length – in terms of syllables – and of the average length of its sentences. The least readable texts in the corpus turned out to be texts 17, 1, 14, and 24, which were also all perceived as rather difficult, at least if compared to the other texts in their respective battery. In fact, many concurrent and related factors account for their perceived difficulty, and not just their low readability. For example, text 17 contains a large number of polysyllabic words which are also academic words. It also contains the highest number of secondary clauses in hypotactic structure, and it is the third most grammatically intricate text in the corpus.

As mentioned above, the quantitative measures of complexity that correlate best with the data derived from the tests and the questionnaires are STTR50, LV(re), GI, and H/P. They can, therefore, be considered as better predictors of the perceived difficulty of the texts and the tasks as well as of the actual difficulty of the tests than the other variables. Furthermore, as STTR50, GI and H/P belong to the second macro-aspect – i.e. the one concerned with the length of the words, clauses, and sentences in a text – it can be suggested that it is mainly the aspects of complexity associated with this macro-aspect that cause both the perceived and the actual difficulty of the texts and the tasks in the corpus.

Some of the qualitative analyses carried out in chapter 4 have already been made reference to in the foregoing concluding remarks, but there are others which revealed further and complementary aspects of the complexity inherent in the texts. In 4.4.1 above, it was pointed out that the complexity of test 12 was mainly caused by the many events and dates in the text, and, in particular, by the fact that some of them are foregrounded while others are realised mainly by rank-shifted structures and are, therefore, backgrounded. Yet, they are all at issue in the tasks. Part of the feedback corroborates these suggestions, in that the test takers found it difficult to retrieve the textual information about the time, order, and place in which the events took place. Since test takers are asked about this information in most of the tasks, they also found it difficult to answer the questions. Furthermore, as seen in 5.7.2 and 5.9 above, both the text and its tasks were evaluated as the most difficult among those in battery 4.

In 4.4.4 it was argued that one of the factors contributing to the complexity of text 22 is the presence of similar but distinct participants, namely ‘second job’, ‘first job’, ‘second temporary position’, and ‘permanent job’. The first part of the text, in particular, is devoted to the notion of ‘second job’, and reference is made to this participant repeatedly, also by means of the anaphoric noun ‘this job’ and the pronoun ‘it’. The unclear reference encoded by these anaphoric devices as well as the similarity of the participants caused confusion and, consequently, processing difficulty to the test takers.

This is specifically confirmed by the fact that they evaluated both the text and the tasks as the most difficult ones in battery 3 (see 5.6 and 5.9 above), and the test itself turned out to be the most difficult one in the battery.

Some aspects of the complexity of texts 2 and 25 were discussed in 4.4.2 above, where it was argued that the presence of both reported speech and narration in the texts contributes to their complexity and, according to the test takers' feedback, to the difficulty of their tasks. This textual feature, in particular, compels the author to use both first and third person pronouns to refer to the same participants, which causes confusion as to the reference encoded by these cohesive devices. Special tracking strategies, therefore, need to be deployed, and it is especially so in test 25, in which the test method used – i.e. gap-filling – increases the difficulty involved in the process of participant identification. In this case, in particular, careful attention should be paid to the lexically cohesive devices as well as to the reference devices used before and after each gap. Texts 2 and 25 were evaluated as difficult: specifically, text 25 was perceived as the second most difficult text in the corpus and text 2 as the most difficult text in battery 1. Their respective accompanying tasks were also found difficult, which confirms the processing difficulty caused by the characteristics of the texts and suggests that it is enhanced by the test method used. As seen in 4.4.5.2 above, in text 3 direct speech is also quoted. In this case, however, the difficulty of the text was due to the test takers' lack of knowledge of the context of situation, which prevented them from understanding the communicative intentions of the speaker, and, consequently, from easily performing the task.

As shown in 4.4.5.1 and 5.7.1, part of the difficulty of test 15 is due to the structural complexity of some of the nominal groups occurring in the input text as well as in the tasks, and in particular, to the fact that most of these groups contain a cascade of pre- and post-modifying relations. Furthermore, the nominal groups in the tasks also contain phoric resources which code either relevance phoricity and/or reminding phoricity or structural reference – i.e. esphora – and have to be referred to and compared with relevant and overlapping nominal groups in the text to correctly answer the questions. As discussed in 4.4.6.2 and 5.8 above, another complex structure which is likely to contribute to text complexity and to cause difficulty in processing text 24 is the following adverbial clause used to express proportions: 'and the less straightforward the job, the more emotional intelligence matters'. Specifically, it is the marked use of 'the' as opposed to its unmarked use as a Determiner that is likely to have created interpreting difficulty.

Figure 6.1 below is an attempt to provide a synopsis of the salient aspects which emerged from the investigation conducted into the text-inherent complexity and the test difficulty of each test of the corpus. The aspects have been grouped into four macro-aspects: 'Density', which comprises LDU, LDH, WLnotyp, and LV(re); 'Length', which comprises FKGL, H/P, WL3typ, STTR50, and GI; 'Cohesion'; and 'Difficulty', which comprises the candidates' performance on the tests and text and task evaluation. Reference is also made to the paragraphs in which these aspects are discussed (see also 3.6 above), and an attempt was made to grade their complexity/difficulty. Specifically, '1' stands for 'simple/easy', '2' stands for 'rather complex/difficult', and '3' stands for 'complex/difficult'. In the last column these four levels of complexity/difficulty have been averaged out (by dividing the sum by 4) in an attempt to provide some sort of synthesis of the investigation carried out.

From the statistical investigation carried out in chapter 5 some more findings emerged, the most relevant of which will be summarised briefly in the remaining part of these concluding remarks. Firstly, in 5.4 above the results of a series of statistical tests on the DIALANG sub-corpus suggested that the statistically different performance of the test takers on a given item did not always depend on the ability assessed by the item – either the ability of identifying the main idea of the text or inferencing ability – although this turned out to be the case of at least one pair of items, viz. tests 04 and 05. It was also suggested that part of the significant difference in the test takers' performance on two tests, viz. tests 07 and 08, was probably due to the different test methods used to elicit the responses, namely gap-filling and multiple-choice respectively. Two other factors which might have contributed to this last difference in performance are the lexical density of text 7 and the grammatical intricacy of text.

	Density	*Length*	*Cohesion*	*Difficulty*	*Total*
t01	1	3 (*WL3typ*: 3.3.2 *FKGL*: 3.5.2)	1	1 (3.5.2)	1.5
t02	1	1	3 (4.4.2)	2 (*textev*; *taskev*: 5.4; 4.4.2)	1.75
t03	2 (*LV(re)*: 3.3.2)	1	2 (4.4.5.2)	1 (4.4.5.2)	1.5
t04	3 (*LV(re)*: 3.3.2)	3 (*STTR50*: 3.1)	1	3 (*perfor*; *textev*; *taskev*: 5.4; 3.3.2)	2.5
t05	1	1	1	1	1
t06	1	1	1	3 (*perfor*; *taskev*: 5.4; 5.9)	1.5
t07	3 (*LV(re)*; *WLnotyp*: 3.3.2)	1	1	2 (*textev*; *taskev*: 5.4)	1.75
t08	1	3 (*GI*; *H/P*: 3.4.2; 3.2.1.2)	1	3 (*perfor*: 5.4; 3.2.1.2)	2
t09	3 (*LV(re)*; *WLnotyp*: 3.2.2; 3.3.2)	2 (*FKGL*: 3.5.2)	2 (3.2.2)	1 (3.2.2)	2
t10	2 (*LV(re)*; *WLnotyp*: 3.3.2)	3 (*GI*; *H/P*: 3.4.2)	1	3 (*perfor*; *taskev*: 5.4)	2.25
t11	1	1	1 (4.4.2.3)	1 (5.6)	1
t12	3 (*LDH*: 3.4.2)	3 (*FKGL*: 3.5.2)	3 (4.4.1)	3 (*perfor*; *textev*; *taskev*: 5.7.2)	3
t13	1 (*LV(re)*: 3.3.2)	1	2 (4.4.5.3)	1 (5.7.1; 5.7.2)	1.25
t14	1	3 (*FKGL*: 3.5.2; *H/P*: 3.4.2; *WL3typ*: 3.3.2)	3 (4.4.3)	2 (*perfor*: 5.7.1; *taskev*: 5.7.2)	2.25
t15	3 (*LDU*: 3.2.1.2; *LDH*: 3.4.2)	2 (*FKGL*: 3.5.2)	3 (4.4.5.1)	3 (*perfor*: 5.7.1)	2.75
t16	1	1	2 (5.7.1)	2 (*perfor*: 5.7.1; 5.9)	1.5
t17	1	3 (*STTR50*: 3.1; *WL3typ*: 3.3.2, *FKGL*: 3.5.2; *GI*; *H/P*: 3.4.2)	2 (4.1.2)	3 (*perfor*; *taskev*: 5.8)	2.25
t18	2 (*WLnotyp*: 3.3.2)	1	1 (4.4.6.1)	1 (5.5)	1.25
t19	3 (*LDU*: 3.2.1.2; *LDH*: 3.4.2)	1	1 (4.4.6.1)	1 (5.5)	1.5
t20	1	1 (*FKGL*: 3.5.2)	1	2 (*textev*; *taskev*: 5.5)	1.25
t21	1	1	1	3 (*perfor*; *textev*; *taskev*: 5.5)	1.5
t22	1	1	3 (4.4.4)	3 (*perfor*; *textev*; *taskev*: 5.6)	2

t23	1	3 (*STTR50*: 3.1)	2 (4.4.6.2)	2 (*perfor*; *textev*; *taskev*: 5.6)	2
t24	3 (*WLnotyp*: 3.3.2)	3 (*FKGL*: 3.5.2)	2 (4.4.6.2)	3 (*perfor*; *textev*; *taskev*: 5.8)	2.75
t25	1	1	3 (4.4.2.2)	2 (*perfor*; *textev*; *taskev*: 5.8)	1.75

Figure 6.1. Synoptic table summarising the aspects which contribute to the complexity and/or difficulty in each test of the corpus.

Finally, the statistical tests which were run on pairs of items in test 01 indicated that there were statistically significant differences between the performance on some of the items in test 01. In 5.5 above, further statistical tests were run to verify whether the performance on 'parallel' tests turned out to be statistically different or not. The paired t-test and the Wilcoxon S.R., in particular, indicated that, even though tests 20 and 21 in battery 2 are two parallel tests, the test takers' performance on them was statistically different. In contrast, the difference in the performance on tests 18 and 19, which are also parallel tests, did not turn out to be statistically significant. On the basis of these results it was argued that even though two tests are constructed following the same test specifications and aim to assess the same skills, they might prove to be different in terms of difficulty.

Further considerations of a more general type can be arrived at by inspecting Figure 6.2, in which the statistically significant correlations between the performance on the tests in the corpus – viz. either their facility values or their mean scores expressed as percentages –, the perception of text difficulty, and the perception of task difficulty are indicated together. It can be seen, first of all, that there is positive correlation between perceived text difficulty and perceived task difficulty in 22 out of 25 tests (88%). In fact, only in tests 02, 06, and 24 is there no such correlation. Another point which emerges is that there is negative correlation between text evaluation and the performance on the tests in 7 out of 25 tests, namely in tests 07, 08, 12, 13, 17, 20, and 25. In 28% of the tests, that is, the more difficult the text was evaluated, the worse the performance on it. Finally, it can be seen that there is negative correlation between task evaluation and the performance on the tests in 4 out of 25 tests, namely in tests 02, 07, 12, and 20. In 16% of the cases then the more difficult a task was evaluated, the worse the performance on it. These data suggest that, as far as this corpus of tests is concerned, the perception of the difficulty of a given input text is a slightly more reliable predictor of the actual difficulty of the tasks than the evaluation of the difficulty of the tasks. However, the correlation between text evaluation and performance is statistically significant only in a few cases, and the prediction is, therefore, not very reliable.

As seen in sections 5.4-5.8 above, the results of one-way between-groups analyses of variance (ANOVA) revealed that there were statistically significant differences in the average rate scores endorsed on some input texts and on some tasks across the various groups of test takers. Specifically, significant differences concerned the evaluation of texts 3, 4, 14, 15, 18, 20, 21, and 22 as well as of tasks 6, 24, and 25. In particular, the groups that performed best on these tests endorsed intermediate average scores on four texts – viz. texts 18, 21, 14, and 15 – which indicates that the most successful students tended not to choose extreme rate scores, and, therefore, to be cautious about evaluating text difficulty. They endorsed instead the highest rate scores on two texts – viz. texts 4 and 22 – and the lowest scores on two other texts – viz. texts 3 and 20. By contrast, the least successful groups of students endorsed the highest average rate scores on four texts – viz. texts 3, 14, 15, and 21 – that is, in 50% of the cases, intermediate scores on three other texts – viz. texts 4, 20, and 22 – and the lowest rate score on one text, viz. text 18. They tended, therefore, to evaluate the texts as more difficult than the most successful groups of students did. As far as the evaluation of task difficulty is concerned, the most successful groups endorsed the highest rate scores on tasks 24 and 25, and the lowest rate score on task 6. The least successful groups, in contrast, endorsed the highest rate score on task 6 and intermediate scores on texts 24 and 25. This suggests that the most successful test takers were aware of the difficulty of tasks 24 and 25 and expressed it by choosing high rate scores, whereas the least successful students did not perceive these tasks as especially difficult, even though they eventually performed worse on them than the other groups.

	(1) Performance	*(2) textev*	*(3) taskev*
test01		√	√
test02	√		√
test03		√	√
test04		√	√
test05		√	√
test06			
test07	√ (neg. with 2 and 3)	√	√
test08	√ (neg. with 2)	√	√
test09		√	√
test10		√	√
test11		√	√
test12	√ (neg. with 2 and 3)	√	√
test13	√ (neg. with 2)	√	√
test14		√	√
test15		√	√
test16		√	√
test17	√ (neg. with 2)	√	√
test18		√	√
test19		√	√
test20	√ (neg. with 2 and 3)	√	√
test21		√	√
test22		√	√
test23		√	√
test24			
test25	√ (neg. with 2)	√	√

Figure 6.2. Statistically significant correlations between the performance on the tests, the perception of text difficulty, and the perception of task difficulty ('neg.' stands for negative correlation)

We can thus conclude that there are general theoretical and practical implications of the results obtained in this study for research on text-inherent complexity and on test difficulty. The first of these is that in order to make choices as to the level of linguistic proficiency a given text is suited to – i.e. the need which this study originates from – the 'triangulation' of different methodologies should be attempted. In fact, as confirmed by this study, the comparison of different types of data can be revealing and lead to sound conclusions about both text complexity and test difficulty. In this way the characteristics of the texts, those of the readers, and those of the tasks are all, at least partly, taken into account, which should provide a more informed judgement on complexity and difficulty. As shown in this study, a text can in fact be inherently complex and cause processing difficulty, but the task can in the end prove to be easy for the test takers at a given level of linguistic proficiency. On the other hand, a less complex text can nonetheless cause processing difficulty, and the task/s based on it can turn out to be difficult for a given type of readers/test takers. Careful attention should be paid, therefore, to the interaction between these three factors, and attempts should be made to keep them under control. Another implication of this study is that both quantitative measures and qualitative analyses should be made use of, in that they can supply complementary indications of text complexity[1]. As shown in chapter 4, investigation into the textual relationships between the input text and the tasks can reveal further aspects of complexity and difficulty and should, therefore, also be attempted. Feedback questionnaires turned out to be important tools for analysis, since they made it possible to collect data on the perception and evaluation of difficulty: they should, therefore, also be used to carry out further research in this field.

1 Ideally it might become possible to elaborate methods for quantifying all the phenomena which initially are analysed qualitatively.

References

Alderson, Charles J. 2000. *Assessing Reading*. Cambridge: Cambridge University Press.

Alderson, Charles J. / Banerjee, Jayanti 2001. State of the Art Review: Language Testing and Assessment (part one). *Language Teaching* 34/4, 213-236.

Alderson, Charles J. / Banerjee, Jayanti 2002. State of the Art Review: Language Testing and Assessment (part two). *Language Teaching* 35/2, 79-113.

Alderson, Charles J. / Clapham, Caroline / Wall, Dianne 1995. *Language Test Construction and Evaluation*. Cambridge: Cambridge University Press.

Alderson, Charles J. / Cseresznyés, Mária 2003. *Into Europe: Reading and Use of English*. Budapest: Teleki László Foundation.

Anderson, Richard C. / Davison, Alice 1988. Conceptual and Empirical Bases of Readability Formulas. In Davison, Alice / Green, Georgia M. (eds), *Linguistic Complexity and Text Comprehension: Readability Issues Reconsidered*. Hillsdale, N.J.: Lawrence Earlbaum Associates, 23-53.

Bachman, Lyle F. / Palmer, Adrian S. 1996. *Language Testing in Practice*. Oxford: Oxford University Press.

Baicchi, Annalisa 2003. Relational Complexity of Titles and Texts: A Semiotic Taxonomy. In Merlini Barberesi, Lavinia (ed.), *Complexity in Language and Text*. Pisa: Edizioni PLUS – Università di Pisa, 319-341.

Bailey, Kathleen M. 1999. *Washback in Language Testing*. TOEFL Monograph Series. RM-99-04. <http://www.ets.org/Media/Research/pdf/RM-99-04.pdf>, last visited on 16th June 2008.

Baker, Rosemary 1997. *Classical Test Theory and Item Response Theory in Test Analysis*. Special Report No 2: Language Testing Update.

Bell, Allan 1991. *The Language of News Media*. Oxford: Blackwell.

Bertuccelli Papi, Marcella 2000. *Implicitness in Text and Discourse*. Pisa: Edizioni ETS.

Biber, Douglas 1988. *Variation across Speech and Writing*: Cambridge. Cambridge University Press.

Biber, Douglas / Conrad, Susan / Leech, Geoffrey 2002. *The Longman Student Grammar of Spoken and Written English*. London: Longman.

Biber, Douglas / Johansson, Stig / Leech, Geoffrey / Conrad, Susan / Finnegan, Edward 1999. *The Longman Grammar of Spoken and Written English*. London: Longman.

Botley, Simon / McEnery, Anthony M. 1999. Discourse Anaphora: The Need for Synthesis. In Botley, Simon / McEnery, Anthony M. (eds), *Computer-based and Computational Approaches to Discourse Anaphora*. Amsterdam/Philadelphia: Benjamins, 1-41.

Bradley, Diane C. / Garrett, Merrill E. 1983. Hemispheric Differences in the Recognition of Closed and Open Class Words. *Neuropsychologia, 21,* 155-159.

Brombin, Luisa 2002. *Appraisal in Popular Science Texts: a Corpus-based Study*. Undergraduate Degree Thesis. University of Padua.

Brombin, Luisa 2004. *Appraisal in Popular Science Texts*. In Taylor Torsello, Carol / Busà, Maria Grazia / Gesuato, Sara (eds), *Lingua Inglese e Mediazione Linguistica. Ricerca e Didattica con Supporto Telematico*. Padova: Unipress, 401-427.

Bruce, Bertram / Rubin, Andee 1988. Readability Formulas: Matching Tool and Task. In Davison, Alice / Green, Georgia M. (eds), *Linguistic Complexity and Text Comprehension: Readability Issues Reconsidered*. Hillsdale, N.J.: Lawrence Earlbaum Associates, 5-21.

Bruti, Silvia 2003. The Complexity of Phoric Relations: Anaphora vs. Cataphora. In Merlini Barberesi, Lavinia (ed.), *Complexity in Language and Text*. Pisa: Edizioni PLUS – Università di Pisa, 291-317.

Buck, Gary / Tatsuoka, Kumi / Kostin, Irene 1997. The Subskills of Reading: Rule-space Analysis of a Multiple-choice Test of Second Language Reading Comprehension. In *Language Learning* 47:3, September 1997, 423-466.

Butler, Chris 1985. *Computers in Linguistics*. New York: Basil Blackwell.
Candlin, C. 1984. Preface. In Alderson, J.C., Urquhart, A.H. (eds), *Reading in a Foreign Language*. London: Longman, pp. ix-xiii.
Carrell, P.L. 1983. Some Issues in Studying the Role of Schemata, or Background Knowledge, in Second Language Comprehension. *Reading in a Foreign Language* 1 (2), 81-92.
Castello, Erik 2002, *Tourist Information Texts: a Corpus Based Study of Four Related Genres*. Padova: Unipress.
Castello, Erik 2004. Calcolo della densità lessicale e dell'intricatezza grammaticale di corpora linguistici. In Taylor Torsello, Carol / Busà, Maria Grazia / Gesuato, Sara (eds), *Lingua inglese e mediazione linguistica. Ricerca e didattica con supporto telematico*. Unipress: Padova.
Chall, Jeanne S. / Dale, Edgar 1995. *Readability Revisited: The New Dale-Chall Formula*. Cambridge, Mass.: Brookline Books.
Clapham, Caroline 1996. *The Development of IELTS: A study of the Effect of Background Knowledge on Reading Comprehension*. Cambridge: Cambridge University Press.
Clark, Herbert H. / Haviland, Susan E. 1977. Comprehension and the Given-New Contract. In Freedle, Roy O. (ed.), *Discourse Production and Comprehension*. Norwood, N.J.: Ablex, 1-40.
Cobb, Tom 2002. Review of Paul Nation, *Learning Vocabulary in another Language*. Cambridge: Cambridge University Press. http://www.er.uqam.ca/nobel/r21270/cv/Nation_2001.htm, last visited on 16th June 2008.
Cobb, Tom 2006. *The Original Idea behind this Website: why & how to Use Frequency Lists to Learn Words*. <http://www.lextutor.ca/research/>, last visited on 16th June 2008.
Council of Europe 2001. *Common European Framework of Reference for Languages: Learning, teaching, assessment*. Cambridge: Cambridge University Press.
Council of Europe 2003. *Relating Language Examinations to the Common European Framework of Reference for Languages: Learning, Teaching, Assessment (CEF)*. Strasbourg: Language Policy Division.

Coxhead, Averil 2000. A New Academic Word List. *TESOL Quarterly 34, 2*, 213-238.

Davies, Alan 1978. Language Testing: Survey Articles 1 and 2. *Language Teaching and Linguistics Abstracts.* 11, 145-159, 215-231.

de Witt, Ray 1997. *How to Prepare for the IELTS.* London: The British Council.

Dörnyei, Zoltán 2003. *Questionnaires in Second Language Research: Construction, Administration, and Processing.* London: Lawrence Erlbaum Associates, Publishers.

Douglas, Dan 2000. *Assessing Languages for Specific Purposes.* Cambridge: Cambridge University Press.

Dressler, Wolfgang U. 1985. *Morphonology: the Dynamics of Derivation.* Ann Arbor: Karoma Publishers.

Dressler, Wolfgang U. / Mayerthaler, Willi / Panagl, Oswald / Wurzel, Wolfgang U. 1987. *Leitmotifs in Natural Morphology.* Amsterdam: Benjamins.

Eggins, Suzanne 2004 [1994]. *An Introduction to Systemic Functional Linguistics.* London and New York: Continuum.

Ellis, Richard / Hopkins, Konrad 1985. *How To Succeed in Written Work and Study.* London and Glasgow: Collins.

Enright, Mary K. / Grabe, William / Koda, Keiko / Mosenthal, Peter Mulcahy-Ernt, Patricia / Shedl, Mary 2000. TOEFL *2000 Reading Framework: A Working Paper.* TOEFL Monograph Series, MS-17, April 2000.

ETS, 2005. *Test and Score Data Summary: 2004-05 Test Year Data – Test of English as a Foreign Language.* Princeton, NJ, USA: Educational Testing. <http://www.ets.org/toefl/>, last visited on 16th June 2008.

Flesch, Rudolf 1948. A New Readability Yardstick. *Journal of Applied Psychology,* 32, 221-233, June.

Flesch, Rudolf 2002. *How to Write Plain English.* <http://www.mang.canterbury.ac.nz/courseinfo/academicWriting/Flesch.htm>, last visited 31st July 2002.

Francis, Gill 1986. *Anaphoric Nouns.* Birmingham University: English Language Research.

Francis, Gill 1994. Labelling Discourse: an Aspect of Nominal Group Lexical Cohesion. In Coulthard, M. (ed.), *Advances in Written Text Analysis*. London: Routledge.

Francis, W. Nelson / Kucera, Henry 1982. *Frequency Analysis of English Usage*. Boston: Houghton Mufflin Company.

Freedle, Roy O. 1997. The Relevance of Multiple-Choice Reading Test Data in Studying Expository Passage Comprehension: The Saga of 15 Year Effort Towards an Experimental/Correlation Merger. *Discourse Processes*, 23, 399-440.

Fries, Peter 1983. On the Status of Theme in English: Arguments from Discourse. In Petöfi, János S. / Sözer, Emel (eds), *Micro and Macro Connexity of Texts*. Hamburg: Helmut Buske, 116-152.

Garrod, Simon C. / Sanford, Anthony J. 1990. Referential Processes in Reading: Focusing on Roles and Individuals. In Balota David A. / Flores D'Arcais Giovanni B. / Rayner Keith (eds), *Comprehension Processes in Reading*. Hillsdale, New Jersey: Lawrence Erlbaum Associates, 465-485.

Gervasi, Vincenzo / Ambriola, Vincenzo 2003. Quantitative Assessment of Textual Complexity. In Merlini Barberesi, Lavinia (ed.), *Complexity in Language and Text*. Pisa: Edizioni PLUS – Università di Pisa, 199-230.

Ghadessy, Mohsen 1995. Thematic Development and its Relationship to Registers and Genres. In Ghadessy, Mohsen, (ed.), *Thematic Development in English Text*. London: Pinter, 129-146.

Gibson, Timothy R. 1993. *Towards a Discourse Theory of Abstracts and Abstracting*. Two volumes. Monographs in Systemic Linguistics, Number 5. Nottingham: Department of English Studies, University of Nottingham.

Givón, Talmy 1990. *Syntax: A Functional-Typological Introduction*. Volume II. Amsterdam: Benjamins.

Givón, Talmy 1995. Coherence in Text vs. Coherence in Mind. In Gernsbacher, Morton Ann / Givón, Talmy (eds), *Coherence in spontaneous text*. Amsterdam/Philadelphia: Benjamins, 59-115.

Gleason, Henry Allan, Jr., 1973. Contrastive Analysis in Discourse Structure. In Makkai Adam / Lockwood David (eds), *Readings in Stratificational Linguistics*. University, Al: Alabama University Press, 258-276.

Gordon, Barry / Caramazza, Alfonso 1985. Lexical Access and Frequency Sensitivity: Frequency Saturation and Open/Closed Class Equivalence. *Cognition, 21*, 95-115. Available on-line at: <http://www.wjh.harvard.edu/~caram/PDFs/1985_Gordon_Caramazza.pdf>, last visited on 16th June 2008.

Green, Rita 2004. *IRT for Language Testers*, Worksheet 1.

Grimes, Joseph E. 1975. *The Thread of Discourse.* Berlin, New York and Amsterdam: Mouton.

Gunning, Robert 1952. *The Technique of Clear Writing*. McGraw Hill.

Gutwinski, Waldemar 1976. *Cohesion in Literary Texts: a Study of some Grammatical and Lexical Features of English Discourse.* The Hague: Mouton (=*Janua Linguarum Series Minor* 204).

Halliday, Michael A.K. / Hasan, Ruqaiya 1989 [1985]. *Language, Context, and Text: Aspects of a Language in a Social-Semiotic Perspective.* Oxford: Oxford University Press.

Halliday, Michael A.K / Hasan, Ruqaiya 1976. *Cohesion in English.* London: Longman.

Halliday, Michael A.K 1976. *Halliday: System and Function in Language*. In Kress Gunther (ed.). London: Oxford University Press.

Halliday, Michael A.K 1978. *Language as Social Semiotic: the Social Interpretation of Language and Meaning*. London: Edward Arnold.

Halliday, Michael A.K 1987. Spoken and Written Modes of Meaning. In Horowitz, Rosalind / Samuels, Jay S. (eds), *Comprehending Oral and Written Language*. New York: Academic Press, 55-82.

Halliday, Michael A.K 1989 [1985]. *Spoken and Written Language*. Oxford: Oxford University Press.

Halliday, Michael A.K 1989. Part A. In Halliday Michael / Hasan Ruqaiya, 3-49.

Halliday, Michael A.K 1997. Sulla Grammatica dell'Inglese Scientifico. In Taylor Torsello, Carol (ed), *Grammatica: Studi Interlinguistici*. Padova: Unipress.

Halliday, Michael A.K 2004. *An Introduction to Functional Grammar* (Third Edition), revised by Matthiessen, Christian M.I.M. London: Arnold.

Halliday, Michael A.K / Martin, Jim R. 1993, *Writing Science. Literacy and Discursive Power*. London: Palmer.

Hammersley, Martin / Atkinson. Paul 1983. *Ethnography: Principles and Practice.* London: Routledge.

Hasan, Ruqaiya 1987. The Grammarian's Dream: Lexis as Most Delicate Grammar. In Halliday, Michael A.K / Fawcett, Robin P. (eds), *New Developments in Systemic Functional Linguistics, Volume 1: Theory and Description.* London: Frances Pinter, 184-211.

Hasan, Ruqaiya 1989. Part B. In Halliday Michael A.K / Hasan Ruqaiya, 52-118.

Hasan, Ruqaiya 1996. *Ways of Saying, Ways of Meaning: Selected Papers by Ruqaiya Hasan.* In Cloran, Carmen / Butt, David / Williams Geoffrey (eds). London and New York: Cassel.

Hatch, Evelyn / Lazaroton, Anne 1991. *The Research Manual: Design and Statistics for Applied Linguistics.* Boston, Massachusetts: Heinle & Heinle Publishers.

Heatley, Alex / Nation, Paul I.S. / Coxhead, Averil 2002. *RANGE and FREQUENCY programs.* <http://www.victoria.ac.nz/lals/staff/paul-nation/nation.aspx>, last visited on 16th June 2008.

Henning, Grant 1987. *A Guide to Language Testing: Development, Evaluation, Research.* Cambridge Mass.: Newbury House.

Johnson, K. 1998. *Readability.* <http://www.timetabler.com/reading.html#1>, last visited on 16th June 2008.

Linnarud, Moira 1975. *Lexis in Free Production* (rapport – arbetsrapport).

Linnarud, Moira 1977. *Some Aspects of Style in the Source and the Target Language* (rapport – arbetsrapport).

Linnarud, Moira 1986. Lexis in Composition: A Performance Analysis of Swedish Learners' Written English. *Lund Studies in English* 74. Malmö: CWK Gleerup.

Mahnke, Kathleen M. / Duffy, Carolyn B. 1996 [1993]. *The Heinemann ELT TOEFL: Preparation Course.* Hampshire and London: McMillan Heinemann.

Malinowski, Bronisław 1923. The Problem of Meaning in Primitive Languages. Supplement I to Ogden, Charles K. / Richards, Ivor A., *The Meaning of Meaning*, 296-336. New York: Harcourt Brace & World.

Martin, Jim R. 1992. *English Text: System and Structure*. Amsterdam and Philadelphia: John Benjamins.

Martin, Jim R. / Matthiessen, Christian M.I.M. 1991. Systemic Typology and Topology. In Christie Frances (ed.), *Literacy in Social Processes*. Darwin: Centre of Studies of Language in Education, Northern Territory University, Darwin, 345-383.

Martin, Jim R. / Rose, David 2003. *Working with Discourse: Meaning beyond the clause*. London and New York: Continuum.

Masi, Silvia 2003. The Complexity of Polyphonic Discourse. In Merlini Barberesi, Lavinia (ed.), *Complexity in Language and Text*. Pisa: Edizioni PLUS – Università di Pisa, 401-432.

Masi, Silvia 2003. The Literature on Complexity. In Merlini Barberesi, Lavinia (ed.), *Complexity in Language and Text*. Pisa: Edizioni PLUS – Università di Pisa, 117-145.

Matthiessen, Christian M.I.M. 1995. *Lexicogrammatical Cartography: English Systems*. Tokyo: International Language Sciences Publishers.

McEnery, Tony / Wilson, Andrew 1996. *Corpus Linguistics*. Edinburgh: Edinburgh University Press.

Merlini Barbaresi, Lavinia 2003, Towards a Theory of Text complexity. In Merlini Barberesi, Lavinia (ed.), *Complexity in Language and Text*. Pisa: Edizioni PLUS – Università di Pisa, 23-66.

Merlini Barberesi, Lavinia 1988. *Markedeness in English Discourse*. Parma: Edizioni Zara.

Montin, Enrica 2002. *The Grammar of Scientific English: a Corpus-based Analysis of Lexical Density and Grammatical Intricacy*. Undergraduate Degree Thesis. University of Padua.

Morley, John 1998. *Truth to Tell: Form and Function in Newspaper Headlines*. Bologna: Clueb.

Nation, Paul I.S. 1990. *Teaching and Learning Vocabulary*. New York: Newbury Press.

Nation, Paul I.S. 2001. Using Small Corpora to Investigate Learner Needs: Two Vocabulary Research Tools. In Ghadessy, Mohsen

/ Henry, Alex / Roseberry, Robert L. (eds), *Small Corpus Studies and ELT: Theory and Practice*. Amsterdam/Philadelphia: Benjamins, 31-45.

Nation, Paul I.S. 2006. *RANGE BNC*. <http://www.victoria.ac.nz/lals/staff/paul-nation/nation.aspx>, last visited on 16th June 2008.

Nation, Paul I.S. / Waring, Robert 1997. Vocabulary Size, Text Coverage and Word Lists. In Schmitt Norbert / McCarthy M. (eds), *Vocabulary: Description, Acquisition and Pedagogy*. Cambridge: Cambridge University Press, 6-19. <http://www.lextutor.ca/research/nation_waring_97.html>, last visited on 16th June 2008.

Pallant, Julie 2001. *SPSS Survival Manual*. London and New York: Open University Press.

Partington, Alan 1998. *Patterns and Meanings: Using Corpora for English Language Research and Teaching*. Amsterdam, Philadelphia: John Benjamins.

Peirce, Charles S. 1965. *Collected Papers*. In Hartshorne, C. / Weiss, P. (eds). Cambridge Ma.: Harvard University Press.

Perfetti, Charles A. 1969. Lexical Density and Phrase Structure Depth as Variables in Sentence Retention. *Journal of Verbal Learning and Verbal Behavior*, 8, 719-724.

Phillips, Deborah 2001. *Longman Complete Course for the TOEFL TEST: Preparation for the Computer and Paper Tests*. New York: Longman.

Quirk, Randolph / Greenbaum, Sidney / Leech, Geoffrey / Svartvik, Jan 1985. *A Comprehensive Grammar of the English Language*. London and New York: Longman.

Schmitt, Norbert 2000. *Vocabulary in Language Teaching*. Cambridge: Cambridge University Press.

Scott, Michael 2006. *Introduction to Oxford WordSmith Tools 4.0*. Oxford: Oxford University Press.

Sinclair, John 1991. *Corpus, Concordance, Collocation*. Oxford: Oxford University Press.

Stubbs, Mike 1996. *Text and Corpus Analysis: Computer-Assisted Studies of Language and Culture*. Blackwell: Oxford.

Taylor Torsello, Carol 1984. *English in Discourse: A course for Language Specialists (1)*. Padova: CLEUP.

Taylor Torsello, Carol 1992. *English in Discourse: A Course for Language Specialists (2)*. Padova: Cleup.

Taylor Torsello, Carol 1996. Theme as the Interpreter's Path Indicator through the Unfolding Text. In *The Interpreter's Newsletter*, 7. Trieste: Edizioni Lint Trieste, 113-149.

Taylor Torsello, Carol 2002. The Economy of Text as Information Exchange. In Iamartino Giovanni / Bignami Marialuisa / Pagetti Carlo (eds), *The Economy Principle in English: Linguistic, Literary, and Cultural Perspectives*. Milan: Unicopli, 16-41.

Taylor Torsello, Carol 2003 Participant Identification in Newspaper Articles. In Merlini Barberesi Lavinia (ed.), *Complexity in Language and Text*. Pisa: Edizioni PLUS – Università di Pisa, 233-251.

Taylor Torsello, Carol / Baldry, Anthony 2005. SFL in Text-Based, Web-enhanced Language Study. In Hasan, Ruqaiya / Matthiessen, Christian M.I.M / Webster, Jonathan (eds), *Continuing Discourse on Language: A Functional Perspective*. London: Equinox, 311-342.

Taylor, Wilson L. 1953. Cloze Procedure: a New Tool for Measuring Readability. *Journalism Quarterly* 30, 415-453.

Thorndike, Robert L. 1982. Educational Measurement – Theory and Practice. In Spearitt, Donald (ed.), *The Improvement of Measurement in Education and Psychology*. Hawthorn, Victoria: The Australian Council for Educational Research.

Tuzzi, Arjuna 2001. Esempi di Utilizzo della Statistica nell'Analisi dei Testi. In Taylor Torsello, Carol / Brunetti, Giuseppe / Penello, Nicoletta (eds), *Corpora Testuali per Ricerca, Traduzione e Apprendimento Linguistico*. Padova: Unipress, 257-269.

University of Cambridge Local Examinations Syndicate (UCLES), 2000 [1998]. *Cambridge First Certificate in English: Examination Papers from the University of Cambridge Local Examinations Syndicate*. Cambridge: Cambridge University Press.

University of Cambridge Local Examinations Syndicate (UCLES), 2001 [2000]. *CAMBRIDGE IELTS 2 – Examination Papers from the University of Cambridge Local Examinations Syndicate*. Cambridge: Cambridge University Press.

University of Cambridge Local Examinations Syndicate (UCLES), 2001a [1997]. *Cambridge Preliminary English Test: Examination Papers from the University of Cambridge Local Examinations Syndicate*. Cambridge: Cambridge University Press.

University of Cambridge ESOL Examinations (Cambridge ESOL), 2008. *PET Handbook for Teachers,* <http://www.cambridgeesol.org/assets/pdf/resources/teacher/pet_handbook.pdf>, last visited on 16th June 2008.

University of Cambridge ESOL Examinations (Cambridge ESOL), 2008a. Excerpts adapted from "Sample paper R&W PET.pdf", downloaded on 25th April 2008 from <http://www.candidates.cambridgeesol.org/cs/digitalAssets/105746_pet_sample_papers.zip>.

University of Cambridge ESOL Examinations (Cambridge ESOL), 2008b. *FCE Handbook for Teachers,* <http://www.cambridgeesol.org/assets/pdf/resources/teacher/fce_handbook.pdf>, last visited on 16th June 2008.

University of Cambridge ESOL Examinations (Cambridge ESOL), 2008c. Excerpts adapted from "Sample Paper R CAE.pdf ", downloaded on 25th April 2008 from <http://www.candidates.cambridgeesol.org/cs/digitalAssets/106850_cae_sample_papers.zip>.

University of Cambridge ESOL Examinations (Cambridge ESOL), 2008d. *CAE Handbook*, <http://www.cambridgeesol.org/assets/pdf/resources/teacher/cae_handbook.pdf>, last visited on 16th June 2008.

University of Cambridge ESOL Examinations (Cambridge ESOL), The British Council, IDP: IELTS Australia, 2007. *IELTS – English for International Opportunity. Handbook 2007.* Cambridge: UCLES. <http://www.ielts.org/pdf/IELTS_Handbook_2007.pdf>, last visited on 16th June 2008.

Ure, Jean 1971. Lexical Density and Register Differentiation. In Perren, George E. / Trim, John L.M. (eds), *Applications of Linguistics: Selected Papers of the Second International Congress of Applied Linguistics*. Cambridge: Cambridge University Press, 443-452.

Urquhart, A.H. 1987. Comprehensions and Interpretations. *Reading in a Foreign Language*. 3, 387-409.

Urquhart, Sandy / Weir, Cyril J. 1998. *Reading in a Second Language: Process, Product and Practice*. Cambridge: Cambridge University Press.

Ventola, Eija 1995. Packing and Unpacking Information in Academic Texts. In Ventola, Eija / Mauranen, Anna (eds), *Academic Writing: Intercultural and Textual Issues*. Amsterdam: John Benjamins, 153-194.

Weir, Cyril J. / Huizhong, Yang / Yan, Jin 2000. *An Empirical Investigation of the Componentiality of L2 Reading in English for Academic Purposes*. Cambridge: Cambridge University Press.

West, Michael 1953. *A General Service List of English Words*. London: Longman.

Widdowson, Henry G. 1979. *Explorations in Applied Linguistics*. Oxford: Oxford University Press.

Appendix

Test 01[1]

Here are four texts (1, 2, 3, 4) which talk about an eye examination. Choose the best title ('a', 'b', 'c', 'd') for each text.

(1) Since eye problems are related to physical health and hereditary factors, problems currently or previously experienced by you or your family will be identified and taken into account. *What is the best title for text 1?*
(2) This checks your all round vision and can indicate neurological problems. *What is the best title for text 2?*
(3) This determines if your eyes are coordinated and working together properly. *What is the best title for text 3?*
(4) Examination of the various parts of the eye can give an overall picture of the condition of your eyes and your general health. *What is the best title for text 4?*

(a) Interior and exterior of eyes.
(b) Muscle balance analysis.
(c) Visual field screening.
(d) Full history and symptoms.

Test 02

Our hands met. I looked him straight in the face. The smile slowly creased up into lines of thought. He waved his left hand impatiently to silence the loquacious Proprietor who had begun the parrot formula he had repeated at least fifteen times so far. 'I have the honour, sir, to introduce …'
'That's right,' said the Minister not to anyone in particular, but to some mechanism of memory inside his head. 'You are … no, not Nwege, … it's Odili, isn't it?'
'Yes, sir'
Odili, the great,' said the Minister boyishly. 'Where have you been all this time?'

Who wrote this?
(a) The Proprietor.
(b) The Minister.
(c) Nwege.
(d) Odili.

---

1 Tests 01 to 10 were taken from the online test DIALANG. <http://www.dialang.org/english/index.htm>, last visited on 16th June 2008.

Test 03

Complete the task by filling the gap(s). Click on the box to make a list of options appear. Choose your answer by clicking on it. Which sentence best fills the gap(s) in the text? Click on the box to make a list of options appear. Choose your answer by clicking on it.

"I do think this course should be fun," he said. "And people should get out of it what they want. Some come here wanting to be serious and even a bit competitive about their work. And that's fine. Others are very hesitant and find that a bit of humour can relax them. ... I've taught high court judges and dustmen and the mix is part of the fun."

(a) They tend to learn more.
(b) You get all sorts on these courses.
(c) I prefer the serious student.
(d) We only accept professional people.

Test 04

Read the text, and choose one of the options below, then click on the button using the mouse.

"People seem to think that ecotourism is about indulging in trips to remote or fragile areas. If you really care about these environments you wouldn't go near them. Real 'ecotourism' is when people spend their free time putting life back into places which have already suffered at the hands of man – in other words not trampling over the rare and beautiful but getting kicks out making good our past mistakes."

This sentence is a summary of the text:

"Ecotourism should involve the improvement of damaged environments ... the exploitation of natural areas."

Which is the best group of words for the gap (…) in the sentence and completes the summary of the text
(a) As well as
(b) In spite of
(c) Instead of
(d) Due to

Test 05

Read the text, and choose one of the options below, then click on the button using the mouse.

"The new digital cameras are great fun and very easy to use. They let you review your pictures the moment you take them, so you can re-shoot right away if you're not satisfied. But remember, a digital camera is just a computer "XXXX". It's not a replacement four your ordinary camera."

What is the meaning of the missing word (XXXX) in the text?
(a) "something that is poor quality"
(b) "an item that is not essential, something extra"
(c) "something expensive but good value for money"
(d) "a fashion which always remains popular"

Test 06

"With your travel agency, the holiday you book is the holiday you get. If you arrive and find we've failed to live up to our promises, let us know what the problem is within one day of your arrival. We'll spend 24 hours doing everything possible to sort the problem out. In the unlikely event that we can't resolve your problem and make you happy within 24 hours, we'll fly you home and give you your money back."

Read the text and decide how you would describe it.

It is …
(a) A commitment
(b) An appeal
(c) A warning
(d) A vow

Test 07

"To the Western visitor, Bangkok is not just another city, it's another way of life. (a) Wooden houses built on stilts over the water stand in the shadows of gleaming tower blocks. (b) The air is filled with different smells. (c) Exotic temples glimmer in the sunlight. (d) Their cool interiors offer a welcome refuge from the heat and noise of the city."

There is one sentence missing from the text. The missing sentence is: *"Tradition and development rub shoulders here as nowhere else."* Read the text and decide where the sentence fits best.

(a)
(b)
(c)
(d)

Test 08

"She had gone alone, but the children were to go to the station to meet her. And loving the station as they did, it was only natural that they should be there a good hour before there was any chance of Mother's train arriving, even if the train were punctual, which was most unlikely."

What can you say about their mother's train?
(a) It would probably be early.
(b) It would probably be on time.
(c) It would probably be late.
(d) It had been cancelled.

Test 09

"Suzanne Somers is always going to struggle to lose the image of the superficial character she played in a recent TV drama. Try as she might, she is not particularly convincing here as an ace journalist who sets out to uncover the mystery behind the death of seven people in a night-club. Ed Begley and Michael Nouri offer satisfactory support, but the direction is strictly routine, and the same goes for the script."

What is the writer's opinion of the film?
(a) Some of the acting is excellent.
(b) The film is well-directed.
(c) The leading performance is weak.
(d) The script is well-written.

Test 10

"Injuries can have serious repercussions for the victim. About three million people a year suffer personal injuries in an accident yet only a quarter of them ever consider making a claim. The majority of the victims do not realise they have a potential claim, or indeed they think a claim will take time and be costly to settle. Or the victims feel the legal system for compensation is stacked against them by being complicated and burdensome."

What is the main idea of the text?
(a) The legal system for compensations is far too difficult to understand.
(b) Claims for compensations are very expensive and time consuming.
(c) Anybody injured in an accident can suffer personal injuries for a long time.
(d) Most people involved in accidents could claim for compensation and possibly win.

Test 11[2]

The next artist in this survey of American artists is James Whistler; he is included in this survey of American artists because he was born in the United States, although the *[line 2]* majority of his artwork was completed in Europe. Whistler was born in Massachusetts in 1834, but nine years later his father moved the family to St. Petersburg, Russia, to work on the construction of a railroad. The family returned to the United States in 1849. Two years later Whistler entered the U.S. military academy at West Point, but he was unable to graduate. At the age of twenty-one, Whistler went to Europe to study art despite familial *[line 7]* objections, and he remained in Europe until his death.

Whistler worked in various art forms, including *[line 8]* etchings and lithographs. However, he is most famous for his paintings, particularly Arrangement in Gray and Black No. 1: Portrait of the Artist's Mother or Whistler's Mother, as it is more commonly known. This painting shows a side view of Whistler's mother, dressed in black and posing against a gray wall. The *[line 11]* asymmetrical nature of the portrait, with his mother seated off-center, is highly characteristic of Whistler's work.

11.01 The paragraph preceding this passage most likely discusses
(a) a survey of eighteenth-century art
(b) a different American artist √
(c) Whistler's other famous paintings
(d) European artists

11.02 Which of the following best describes the information in the passage?
(a) Several artists are presented.
(b) One artist's life and works are described. √
(c) Various paintings are contrasted.
(d) Whistler's family life is outlined.

11.03 Whistler is considered an American artist because
(a) he was born in America. √
(b) he spent most of his life in America.
(c) he served in the U.S. military.
(d) he created most of his famous art in America.

11.04 The word "majority" in line 2 is closest in meaning to
(a) seniority
(b) maturity

---

2 Tests 11 and 12 are excerpts from the "Longman Complete Course for the TOEFL Test, Preparation for the Computer and Paper Tests", pp. 345-346 and 351-352, by Deborah Phillips. Copyright © 2001 by Addison-Wesley Longman, Inc. Reprinted by permission of Pearson Education, Inc.

(c) large pieces
(d) high percentage √

11.05 It is implied in the passage that Whistler's family was
(a) unable to find any work at all in Russia
(b) highly supportive of his desire to pursue art
(c) working class √
(d) military

11.06 The word "objections" in line 7 is closest in meaning to
(a) protests √
(b) goals
(c) agreements
(d) battles

11.07 In line 8, the "etchings" are
(a) a type of painting
(b) the same as a lithograph
(c) an art form introduced by Whistler
(d) an art form involving engraving √

11.08 The word "asymmetrical" in line 11 is closest in meaning to
(a) proportionate
(b) uneven √
(c) balanced
(d) lyrical

11.09 Which of the following is NOT true according to the passage?
(a) Whistler worked with a variety of art forms.
(b) Whistler's Mother is not the official name of his painting.
(c) Whistler is best known for his etchings. √
(d) Whistler's Mother is painted in somber tones.

Test 12

The final battle of the War of 1812 was the Battle of New Orleans. This battle gave a clear demonstration of the need for effective communication during wartime; it also showed the disastrous results that can come to pass when communication is inadequate.
The War of 1812 was fought between Great Britain and the very young country of the United States only a relatively few years after the United States had won its independence from Britain. The United States had declared war against Britain in June of 1812, mostly because of interference with U.S. shipping by the British and because of the shanghaiing of U.S. sailors for enforced service on British vessels.

The war lasted for a little more than two years, when a peace treaty was signed at Ghent, in Belgium, on the 24th of December, 1814.
Unfortunately, the news that the Treaty of Ghent had been signed and that the war was officially over was not communicated in a timely manner over the wide distance to where the war was being contested. Negotiations for the treaty and the actual signing of the treaty took place in Europe, and news of the treaty had to be carried across the Atlantic to the war front by ship. A totally unnecessary loss of time was incurred as a result of the amount of time that it took to inform the combatants of the treaty.
10A Early in January of 1815, some two weeks after the peace treaty had been signed, British troops in the southern part of the United States were unaware that the war had officially ended. *10B* Over 5,000 British troops attacked U.S. troops. *10C* During the ensuing battle, known as the Battle of New Orleans, the British suffered a huge number of casualties, around 2,000, and the Americans lost 71, all in a battle fought only because news of the peace treaty that had already been signed in Ghent had not yet reached the battlefield. *10D*

12.01 The main idea of this passage is that …
(a) the War of Independence was unnecessary
(b) the War of 1812 was unnecessary
(c) the Treaty of Ghent was unnecessary
(d) the Battle of New Orleans was unnecessary √

12.02 Look at the word "it" in paragraph 1. Click on the word or phrase that it refers to.

[12.02a[3] Look at the underlined word "it" in paragraph 1. Write the word or phrase "it" refers to.]

12.03 Look at the expression "come to pass" in paragraph 1. This expression could best be replaced by …
(a) happen √
(b) overthrow
(c) self-destruct
(d) circumvent

12.04 According to the passage, when did the United States win its independence from Britain?
(a) Shortly before the War of 1812 √
(b) During the War of 1812
(c) Just after the War of 1812
(d) Long after the War of 1812

---

3 The questions in square brackets are the adapted versions of the original questions. These changes were made in order to successfully administer the tests on-line by means of the software QuestionMark Perception, version 3.3.1.

12.05 According to the passage, some U.S. sailors were …
(a) taken forcibly to Shanghai
(b) made to go to Ghent
(c) forced to work on British ships √
(d) responsible for causing the War of 1812

12.06 Click on the sentence in paragraph 2 that indicates when the War of 1812 officially ended.

[12.06a Which sentence in paragraph 2 indicates when the War of 1812 officially ended?
(a) Sentence 1
(b) Sentence 2
(c) Sentence 3 √]

12.07 It is NOT stated in the passage that Ghent was …
(a) where negotiations took place
(b) the site of the final battle √
(c) where the treaty was signed
(d) far from the battlefield

12.08 Look at the word contested in paragraph 3. Click on the word or phrase in paragraph 4 that is closest in meaning to contested.

[12.08a Look at the word "ontested" in paragraph 3. Write the word or phrase in paragraph 4 that is closest in meaning to "ontested".]

12.09 It can be determined from the passage that, of the following dates, the Battle of New Orleans was most probably fought
(a) on December 10, 1814
(b) on December 24, 1814
(c) on January 1, 1815
(d) on January 8, 1815 √

12.10 The following sentence could be added to paragraph 4.
These U.S. troops were in fortified entrenchments outside the city.
Where would it best fit into the paragraph? Click on the square (.) to add the sentence,
to the paragraph.

[12.10a The following sentence could be added to paragraph 4
These U.S. troops were in fortified entrenchments outside the city.
Where would it best fit into the paragraph?
(a) 10A
(b) 10B

(c) 10C √
(d) 10D]

12.11 Click on the paragraph that describes the battle that took place after the signing of the treaty.

[12.11a Which paragraph describes the battle that took place after the signing of the treaty?
(a) PARAGRAPH 1
(b) PARAGRAPH 2
(c) PARAGRAPH 3
(d) PARAGRAPH 4 √]

Test 13[4]

One of the most mysterious things in nature is the ability of certain creatures to find their way home, sometimes from great distances. Birds are not the only ones who can do this. Bees, eels, and salmon are able to return to a particular place after long journeys, too.

Most migrations take place between breeding grounds and regions where animals feed. For some animals, such as the lemming, the move is a one-way trip. Some scientists call this movement emigration, because these animals never return to their homes. For other animals, such as birds, the migration includes a return trip home. Birds move in periodic migrations, or at regular times during their lives, and often to the same place year after year.

Many experiments have been made with birds in an attempt to find out what guides them on their way home. In one case, seven swallows were taken 400 miles from home. When they were set free, five of them returned to their nests. In another case, a certain kind of sea bird was taken from its nest off the Welsh coast to Venice by plane. When it was released, it made its way home to its nest, a distance of 930 miles if it flew in a straight line.

[Line 17] Migrating birds offer an even more amazing example of this ability. There are swifts and house martins that migrate from England to South Africa every year. They not only return to England the next spring, but many of them come back to nest in the very same house where they nested the year before. They fly the incredible distance of 6,000 miles, one way.

Certain types of butterflies migrate, too, and find their way home over long distances. In the tropics one can sometimes see *[Line 23]* great mass flights of butterflies all flying steadily in one direction. They may go a thousand miles and more and then return again in another season.

---

4 Test taken from "Heinemann TOEFL Preparation Course", pp. 311-312, by M.K. Mahnke, Carolyn Duffy & Jacqeline Flamm, 1996, Copyright ©. Reproduced with permission of Macmillan Publishers Ltd.

Despite all the efforts that have been made to explain how these creatures find their way home, we still have no sure explanation. Since many of the birds fly over great bodies of water, we can't explain it by saying they use landmarks to guide them. Just to say they have an "instinct" doesn't really explain the right conditions. The reason they do it may be to obtain food or to reproduce under the right conditions. *[Line 31]* But the signals and guideposts they use on their flights are still a mystery to man.

13.01. What does this passage mainly discuss?
(a) How migratory creatures find their way home √
(b) Popular places for migration
(c) Why animals migrate
(d) The emigration patterns of butterflies

13.02. According to paragraph 2, the main difference between animal emigration and animal migration is
(a) the breeding places
(b) the route taken
(c) the time spent
(d) the final destination √

13.03. In paragraph 5 the author mentions butterflies in order to
(a) include the tropics in the discussion of migration
(b) give another example of migratory creatures √
(c) show that these small flying creatures don't get lost
(d) show that butterflies are sturdy creatures

13.04. In line 17 the phrase "his ability" refers to
(a) the ability to fly without stopping
(b) the ability to build nests year after year
(c) the ability to find their way home √
(d) the ability to fly in a straight line

13.05. According to the passage, swallows travelling from South Africa to England
(a) cover a relatively short distance
(b) often return to precisely the same point each year √
(c) may not return for a number of years
(d) rebuild the same type of nest in similar
locations each spring

13.06. The word "ass" in line 23 is closest in meaning to
(a) visible
(b) voluntary
(c) rapid
(d) large √

13.07. The word "ignals" in line 31 is closest in meaning to
(a) indications √
(b) instincts
(c) radar
(d) maps

13.08. The author suggests that using "nstinct" to explain the ability of animals to return home during migration is
(a) absurd
(b) incomplete √
(c) inaccurate
(d) reliable

13.09. Which of the following devices does the author use to present the information in the passage?
(a) Support through exemplification √
(b) Narration
(c) Cause and effect
(d) Comparison and contrast

13.10. The attitude of the author toward the ability of certain creatures to find their way home may best be expressed by which of the following?
(a) Confused
(b) Sympathetic
(c) Admiring √
(d) Indifferent

Test 14[5]

Look at the information on the following page about the use of vehicles in the University grounds.
In boxes 14.01-14.05 on your answer sheet write
TRUE (T) if the statement is true
FALSE (F) if the statement is false
NOT GIVEN (NG) if the information is not given in the passage
Example
The campus roads are not open to general members of the public. Answer: TRUE.

14.01 University employees do not need to pay for their parking permits. (F)
14.02 Parking in Halls of Residence is handled by the Wardens of the Halls. (T)

---

5 Tests 14, 15, 16, and 17 were taken from: University of Cambridge Local Examinations Syndicate, "Cambridge IELTS 2 Student's Book with Answers", 2001, © UCLES, reprinted with permission of the publisher (Cambridge University Press) and UCLES, pp. 98-99, 102-103, 116-117, 83-86.

14.03 Having a University permit does not allow staff to park at Halls. (T)
14.04 Parking permits cost £20 a year. (NG)
14.05 Students living in Hall do not need permission to park in Hall car parks. (F)

USE OF UNIVERSITY GROUNDS BY VEHICULAR TRAFFIC
The University grounds are private.
The University authorities only allow authorised members of the University, visitors and drivers of vehicles servicing the University to enter the grounds.
Members of staff who have paid the requisite fee and display the appropriate permit may bring a vehicle into the grounds. A University permit does not entitle them to park in Hall car parks however, unless authorised by the Warden of the Hall concerned.
Students may not bring vehicles into the grounds during the working day unless they have been given special permission by the Security Officer and have paid for and are displaying an appropriate entry permit. Students living in Halls of Residence must obtain permission from the Warden to keep a motor vehicle at their residence.
Students are reminded that if they park a motor vehicle on University premises without a valid permit, they will be fined £20.

Test 15

Look at the introduction to West Thames College on the following page and at the statements (Questions 15.01-15.07) below.
In boxes 15.01-15.07 on your answer sheet write
TRUE (T) if the statement is true
FALSE (F) if the statement is false
NOT GIVEN (NG) if the information is not given in the passage

15.01 Chiswick Polytechnic was closed at the same time West Thames College was opened. (NG)
15.02 Most of the students at the college come from outside the local area. (T)
15.03 The college changed its name to West Thames College in 1993. (T)
15.04 There are currently 6000 students over the age of 19 attending the college. (F)
15.05 Students under the age of 16 cannot attend any of the courses offered by the college. (NG)
15.06 The college offers a more mature environment in which to learn than a school. (T)
15.07 There are fewer subjects to study in the sixth form of a school than at the college. (T)

WEST THAMES COLLEGE BACKGROUND INFORMATION FOR CANDIDATES
West Thames College (initially known as Hounslow Borough College) came into existence in 1976 following the merger of Isleworth Polytechnic with part of Chiswick Polytechnic. Both parent colleges, in various guises, enjoyed a long tradition of service to the community dating back to the 1890s.

The college is located at London Road, Isleworth, on a site occupied by the Victorian house of the Pears family, Spring Grove House. An earlier house of the same name on this site had been the home of Sir Joseph Banks, the botanist who named Botany Bay with Captain Cook in 1770. Later he founded Kew Gardens.

Situated at the heart of West London, West Thames College is ideally placed to serve the training and education needs of local industry and local people. But its influence reaches much further than the immediate locality.

Under its former name, Hounslow Borough College, it had already established a regional, national and international reputation for excellence. In fact, about eight per cent of its students come from continental Europe and further afield, whilst a further 52 per cent are from outside the immediate area. Since 1 April 1993, when it became independent of the local authority and adopted its new title, West Thames College has continued to build on that first class reputation.

These days there is no such thing as a typical student. More than half of West Thames College's 6000 students are over 19 years old. Some of these will be attending college part-time under their employers' training schemes. Others will want to learn new skills purely out of interest, or out of a desire to improve their promotion chances, or they may want a change in career.

The college is also very popular with 16-18 year olds, who see it as a practical alternative to a further two years at school. They want to study in the more adult atmosphere the college provides. They can choose from a far wider range of subjects than it would be practical for a sixth form to offer. If they want to go straight into employment they can still study at college to gain qualifications relevant to the job, either on a day-release basis or through Network or the Modern Apprenticeship Scheme.

Test 16

Read the passage about personal computers on the following page and look at the statements below (Questions 16.01-16.07). In boxes 16.01-16.07 on your answer sheet write

TRUE (T) if the statement is true

FALSE (F) if the statement is false

NOT GIVEN (NG) if the information is not given in the passage

16.01 There are two computers and two printers available for public use at the library. (T)

16.02 You can buy floppy disks at the information desk. (F)

16.03 The information desk is closed at weekends. (NG)

16.04 It is essential to reserve a computer three days in advance if you want to use one. (F)

16.05 If you are more than a quarter of an hour late, you could lose your reservation for the computer. (T)

16.06 Library employees do not have detailed knowledge of computers. (NG)

16.07 The library runs courses for people who want to learn about computers. (T)

CENTRAL LIBRARY

PERSONAL COMPUTERS AVAILABLE FOR PUBLIC TO USE

- 2 personal computers are available, for a fee of $5.00. There is also an ink jet printer attached to each terminal. The library has a number of commercially available programs for word processing and spreadsheets.
- A4 paper can be bought from the desk if you wish to print your work. Alternatively you can bring your own paper. If you wish to store information however, you will need to bring your own floppy disk.

Bookings
Because of high demand, a maximum of one hour's use per person per day is permitted. Bookings may be made up to three days in advance. Bookings may be made in person at the information desk or by phoning 8673 8901 during normal office hours. If for some reason you cannot keep your appointment, please telephone. If the library is not notified and you are 15 minutes late, your time can be given to someone else. Please sign in the visitors' book at the information desk when you first arrive to use the computer.
Please note that staff are not available to train people or give a lot of detailed instruction on how to use the programs. Prior knowledge is, therefore, necessary. However, tutorial groups are available for some of the programs and classes are offered on a regular basis. Please see the loans desk for more information about our computer courses.

Test 17

You should spend about 20 minutes on Questions 17.01-17.13 which are based on Reading Passage 1 below.

GREEN WAVE WASHES OVER MAINSTREAM SHOPPING (1)
Research in Britain has shown that 'green consumers' continue to flourish as a significant group amongst shoppers. This suggests that politicians who claim environmentalism is yesterday's issue may be seriously misjudging the public mood.

A report from Mintel, the market research organisation, says that despite recession and financial pressures, more people than ever want to buy environmentally friendly products and a 'green wave' has swept through consumerism, taking in people previously untouched by environmental concerns. The recently published report also predicts that the process will repeat itself with 'ethical' concerns, involving issues such as fair trade with the Third World and the social record of businesses. Companies will have to be more honest and open in response to this mood.

Mintel's survey, based on nearly 1,000 consumers, found that the proportion of those who look for green products and are prepared to pay more for them has climbed from 53 per cent in 1990 to around 60 per cent in 1994. On average, they will pay 13 per cent more for such products, although this percentage is higher among women, managerial and professional groups and those aged 35 to 44.

Between 1990 and 1994 the proportion of consumers claiming to be unaware of or unconcerned about green issues fell from 18 to 10 per cent but the number of green spenders among older people and manual workers has risen substantially. Regions such as Scotland have also caught up with the south of England in their environmental concerns. According to Mintel, the image of green consumerism as associated in the past with the more eccentric members of society has virtually disappeared. The consumer research manager for Mintel, Angela Hughes, said it had become firmly established as a mainstream market. She explained that as far as the average person is concerned environmentalism has not 'gone off the boil'. In fact, it has spread across a much wider range of consumer groups, ages and occupations.

Mintel's 1994 survey found that 13 per cent of consumers are 'very dark green', nearly always buying environmentally friendly products, 28 per cent are 'dark green', trying 'as far as possible' to buy such products, and 21 per cent are 'pale green' – tending to buy green products if they see them. Another 26 per cent are 'armchair greens'; they said they care about environmental issues but their concern does not affect their spending habits. Only 10 per cent say they do not care about green issues.

Four in ten people are 'ethical spenders', buying goods which do not, for example, involve dealings with oppressive regimes. This figure is the same as in 1990, although the number of 'armchair ethicals' has risen from 28 to 35 per cent and only 22 per cent say they are unconcerned now, against 30 per cent in 1990. Hughes claims that in the twenty-first century, consumers will be encouraged to think more about the entire history of the products and services they buy, including the policies of the companies that provide them and that this will require a greater degree of honesty with consumers.

Among green consumers, animal testing is the top issue – 48 per cent said they would be deterred from buying a product if it had been tested on animals followed by concerns regarding irresponsible selling, the ozone layer, river and sea pollution, forest destruction, recycling and factory farming. However, concern for specific issues is lower than in 1990, suggesting that many consumers feel that Government and business have taken on the environmental agenda.

Questions 17.01-17.06
Do the following statements agree with the claims of the writer of Reading Passage 1? In boxes 17.01-17.06 on your answer sheet write
YES (Y) if the statement agrees with the claims of the writer
NO (N) if the statement contradicts the claims of the writer
NOT GIVEN (NG) if it is impossible to say what the writer thinks about this

17.01 The research findings report commercial rather than political trends. (Y)
17.02 Being financially better off has made shoppers more sensitive to buying 'green'. (N)
17.03 The majority of shoppers are prepared to pay more for the benefit of the environment according to the research findings. (Y)

17.04 Consumers' green shopping habits are influenced by Mintel's findings. (NG)
17.05 Mintel have limited their investigation to professional and managerial groups. (N)
17.06 Mintel undertakes market surveys on an annual basis. (NG)

Questions 17.07-17.09
Choose the appropriate letters A-D and write them in boxes 17.07-17.09 on your answer sheet.

17.07 Politicians may have 'misjudged the public mood' because ...
(a) they are pre-occupied with the recession and financial problems.
(b) there is more widespread interest in the environment agenda than they anticipated. √
(c) consumer spending has increased significantly as a result of 'green' pressure.
(d) shoppers are displeased with government policies on a range of issues.

17.08 What is Mintel?
(a) an environmentalist group
(b) a business survey organisation √
(c) an academic research team
(d) a political organisation

17.09 A consumer expressing concern for environmental issues without actively supporting such principles is ...
(a) an 'ethical spender'.
(b) a 'very dark green' spender.
(c) an 'armchair green'. √
(d) a 'pale green' spender.

Questions 17.10-17.13

Complete the summary using words from the box below. Write your answers in boxes 17.10-17.13 on your answer sheet.
NB There are more answers than spaces, so you will not use them all.

The Mintel report suggests that in future companies will be forced to practise greater ... (17.10) ... in their dealings because of the increased awareness amongst ... (17.11) ... of ethical issues. This prediction is supported by the growth in the number of ... (17.12) ... identified in the most recent survey published. As a consequence, it is felt that companies will have to think more carefully about their ... (17.13) ... .

environmental research
honesty and openness (17.10)
ethical spenders
politicians
social awareness

social record (17.13)
armchair ethicals (17.12)
environmentalists consumers
consumers (17.11)
political beliefs
financial constraints

Test 18[6]

Part 3
Questions 18.01-18.10
Look at the sentences below about a journey to the Arctic on board a ship.
Read the text on paper to decide if each sentence is correct or incorrect.
If it is correct, mark A on your answer sheet.
If it is not correct, mark B on your answer sheet.

18.01 This trip is for people who like peace and quiet. (A)
18.02 Many different activities are organised on board. (B)
18.03 The voyage begins in Scotland. (A)
18.04 The ship follows a fixed route. (B)
18.05 There are different types of accommodation. (B)
18.06 Passengers serve themselves in the dining room. (A)
18.07 Whales can be seen in the morning near Tromsø. (A)
18.08 There are some examples of traditional buildings in Tromsø. (A)
18.09 The ship stays overnight in Tromsø. (B)
18.10 Bear Island used to be a busy fishing centre. (A)

EXPLORING THE ARCTIC
The Arctic is one of the few places in the world untouched by pollution where you can see nature at its wildest and most beautiful. Join our ship the *Northern Star* from 2 to 18 July, for a 17-day voyage to the Arctic. During the voyage you are able to relax and get away from it all. There are no parties or film-shows to attend, quizzes to enter, or entertainers to watch. However, we do have specialists on board who are willing to answer any of your questions about the Arctic and who will talk about the animals and birds that you see on the trip.
After setting off from Scotland, we go north along the coast of Norway to Bear Island. Along the way you'll see thousands of seabirds and wonderful scenery, with rivers of ice and huge cliffs. You will have the chance to see reindeer, polar bears, and other Arctic animals. Although we have a timetable, experience has shown that

6 Tests 18, 20, 21 are excerpts adapted from "Sample paper R&W PET.pdf", © University of Cambridge ESOL Examinations, downloaded from: <http://www.candidates.cambridgeesol.org/cs/digitalAssets/105746_pet_sample_papers.zip>, pp. 16, 17 and 25.

we may have to change our direction a little, depending on the weather and on which animals appear.

The *Northern Star* is a very special ship and our past voyages have been very popular. Our cabins all have the same excellent facilities, which include a private bathroom and refrigerator. Our chefs are happy to prepare any food for people on special diets. Choose just what you want to eat from the wide variety available from the dining room buffet. There is a library, shop, clinic and plenty of space for relaxation. If you need some exercise, why not go jogging every morning around the decks, or do some swimming in the indoor pool.

Prices include economy class air travel and 16 nights on board the Northern Star, all meals and excursions and all lectures.

Day 1: Board the Northern Star.

Days 2-7: We sail slowly north along the coast of Norway, stopping at places of interest.

Day 8: Tromsø. You need to get up at sunrise to see the whales as we sail towards Tromsø. Visit Tromsø to see the Arctic Museum, the cathedral and the beautiful old wooden houses. In the evening we sail away along the west coast to Bird Island, which is excellent for bird-watching.

Days 9-10: Bear Island. We arrive here in the early evening and stay overnight. Bear Island once had an active fishing industry, but today little of this remains. We will explore the island, looking out for Arctic flowers.

Days 11-16: Spitsbergen. A place of mountains and rivers of ice, it is home to a large variety of animals.

Day 17: Leave the ship in Spitsbergen and fly to London from Tromsø.

Test 19[7]

PART 3

Questions 19.01-19.10

Look at the statements below about a tour of Australia.

Read the text on the next page to decide if each statement is correct or incorrect.

If it is correct, mark A on your answer sheet.

If it is not correct, mark B on your answer sheet.

19.01 If you start your holiday on April 1st, you will return on April 19th. (B)
19.02 Return flights are from Melbourne. (A)
19.03 All travel between cities in Australia is by plane. (B)
19.04 The cost covers accommodation and some meals. (A)
19.05 You can make a reservation now for one of next year's tours. (A)
19.06 You have to pay the full price on the day you book. (B)

---

7 Tests 19 was taken from: University of Cambridge Local Examinations Syndicate, "Preliminary English Test 2 Student's Book", 2001, © UCLES, reprinted with permission of the publisher (Cambridge University Press) and UCLES, pp. 10-11.

19.07 You can only take this holiday in the Spring or Autumn. (B)
19.08 If you want to go on a city sightseeing tour, you will have to pay extra. (B)
19.09 The trip to Ayers Rock ends with a flight. (A)
19.10 Booking is possible seven days a week. (B)

GLOBEWISE: WE'LL SHOW YOU THE VERY BEST OF AUSTRALIA
On a fully guided 22-day tour
You can pay just £90 now to be sure of our place on this successful and popular touring holiday next year. There are departures right through the year. We make sure you see the very best of everything which Australia has to offer.
PERTH
Wonderful long, golden beaches, superb restaurants serving fabulous food in delightful surroundings, lush green parks and the beautiful Swan River. City sightseeing tour included.
ALICE SPRINGS .
Fly over the outback to famous Alice Springs in the heart of Aboriginal country. Full sightseeing tour. See Flying Doctor base and 'School of the Air'.
AYERS ROCK
Drive to Yulara National Park. Visit the mysterious Olgas and Ayers Rock with its caves and Aboriginal rock paintings. Fly on to the lively seaside town of Cairns and relax in the sun.
GREAT BARRIER REEF
We've included a full day's boat trip on the famous reef with the chance to see amazing, brightly coloured fish and other sea creatures.
SYDNEY
You'll love the excitement and beauty of Australia's biggest city – we've included a sightseeing tour and a visit to the famous Opera House. You can also book a day trip to the Blue Mountains.
MELBOURNE
We've included a city sightseeing tour – or you can visit the Penguin Parade.
We promise that by the end of the trip you'll be wanting to return!
FOR ONLY £1985
OUR PRICE INCLUDES

- Scheduled flights by Australia's national airline, Qantas, from London or Manchester to Perth, returning from Melbourne.
- After arriving in Perth, Air Australia flights between Perth/Alice Springs/ Cairns/Sydney.
- Coach from Sydney to Melbourne via Canberra and Albury.
- All airport transfers in Australia.
- Nineteen nights' accommodation in good grade hotels with full continental breakfast
- Day trip to Ayers Rock, with a full day Barrier Reef boat trip, a visit to an Australian sheep station and city sightseeing tours in Perth, Alice Springs, Sydney, Canberra and Melbourne.

- All state and local taxes. Hotel baggage handling. Experienced Globewise Tour Manager at all stages of the trip.

Sold out for this year booking now for next year
PHONE US ON 01303 692154
QUOTING REFERENCE GW/398 FOR OUR NEW FULL COLOUR BROCHURE
OUR OFFICES ARE OPEN: Monday to Friday 9 a.m. – 8 p.m. Saturday 9 a.m. – 4 p.m.
BROCHURE REQUESTS ONLY: Sunday 10 a.m. – 2 p.m.

Test 20

Questions 20.01-20.05
Read the text and questions below.
For each question mark the correct letter (a), (b), (c), or (d) on your answer sheet.

AINSLEY HARRIOTT
I've always been a bit of an entertainer and played the funny man. I was a part-time comedian for years, so I learned how to stand in front of audiences. It made me sure of myself. I like being liked and I love making everyone smile.
I've lived in London all my life and have just moved to a larger house with my wife Clare and our two children, Jimmy and Madeline. We spend a lot of time just singing and dancing around the house. I grew up with music because my dad is the pianist, Chester Harriott – who's still playing, by the way. My working day is divided between television and writing cook books, though TV takes most of my time. I spend about five days a fortnight working on the cooking programmes I appear in. I eat all sorts of things at home but I only buy quality food. When I'm cooking, I experiment with whatever is in the fridge – it's good practice for my TV series.
I'm a football fan and enjoy going to matches, but I'm a home-loving person really. I don't like going to the pub but we do go out to eat about twice a month. There's nothing better than a night at home playing with the children. I rarely go to bed before midnight. Late evening is when fresh thoughts on cooking usually come to me, so I often write or plan my programmes then. When I eventually get to bed, I have no trouble sleeping!

20.01 What is the writer's main purpose in writing the text?
(a) To describe how he lives. √
(b) To say what makes him laugh.
(c) To talk about his cooking ideas.
(d) To explain how he started in TV.

20.02 What would a reader learn about Ainsley from the text?
(a) He is a very good musician.
(b) He likes lo plan the family meals.
(c) He is nervous about performing on stage.
(d) He enjoys spending time with his family. √

20.03 What does the writer say about himself?
(a) He loves going out and meeting people.
(b) He is very similar to his father.
(c) He enjoys being popular. √
(d) He should go to bed earlier.

20.04 What does he say about his working life?
(a) He would like lo appear less on TV.
(b) He gets his best ideas at certain times. √
(c) He prefers being a comedian.
(d) He should practise cooking more.

20.05 Which of the following is the best description of the writer?
(a) The popular TV comedian who enjoys cooking, watching football, and having a busy social life.
(b) The TV cook who loves making people laugh, watching football and, above all, having a happy family life. √
(c) The singing TV cook who likes making jokes, playing with his children, and having an early night.
(d) The cook and comedian who takes great care about the way he cooks his food and enjoys listening to music more than anything.

Test 21

Questions 21.01-21-05
Read the text and questions below.
For each question, mark the correct letter (a), (b), (c) or (d) on your answer sheet.

DOUG ALLAN, WILDLIFE CAMERAMAN
I recently spent two years in the Arctic filming the series Blue Planet. I love being in an environment that hasn't changed for 20,000 years. Of course it's freezing, but it must be a healthy place because you never catch colds.
When I'm filming, I like to really feel how lonely the environment is. Filming underwater involves cutting through thick ice and diving in tied to a line. The person at the other end has to be ready to pull you out fast if necessary.
Originally I was a research diver for the British Antarctic Survey project, but for me science lacked excitement. I'd always enjoyed photography, and whenever camera teams passed through, they encouraged me to watch and learn. I was then able to move into filming in 1985 and have concentrated on Arctic and Antarctic wildlife ever since.
I prefer to be face to face with the animals I'm filming. I haven't got in the water with killer whales yet, but I plan to. Of course, it's dangerous if you choose the wrong moment. They're big animals and can move fast, so I'd be stupid to film them searching for food!

I've never had problems with polar bears, although once I was frightened when one tried to get into my tent. Polar bears are bold, clever and dangerous. But I made this one see I wasn't about to attack it – I'm sure it realised I wouldn't hurt it.
When I come back home from my trips, I work in the mornings and spend the afternoons swimming to keep fit. Now I'm fifty, filming is harder. The challenge for me is to continue to deliver high-quality work.

21.01 In this text, Doug Allan is describing ...
(a) the challenges of the environment he works in. √
(b) the career opportunities in TV camera world.
(c) the difficulties of having to work alone.
(d) the beautiful scenery of the Arctic.

21.02 What does Doug say about his early career?
(a) He wasted the years he spent as a scientist.
(b) It was a good chance to learn about filming. √
(c) He was bored by working only in the Antarctic.
(d) It taught him how to become a skilled diver.

21.03 When talking about killer whales, Doug says that ...
(a) he will only film them from a safe distance.
(b) he has always been careful when diving with them.
(c) he tries to avoid any danger by facing them.
(d) he believes there are safe opportunities to film them. √

21.04 How does Doug describe his experience with a polar bear?
(a) The animal was much more afraid than Doug was.
(b) Doug felt nervous that the bear might come back.
(c) The bear seemed to know Doug wasn't a danger. √
(d) Doug realised he was wrong to trust polar bears.

21.05 What might Doug say about his job as a cameraman?
(a) Now I've reached fifty, I realise the dangers involved in getting the best pictures are too great.
(b) I'm glad I gave up working as a scientist because I'm now in a position to be able to retire early.
(c) Although I find the work more difficult now I'm not as young, I still have other film projects I'm keen to do. √
(d) I've made wildlife films in every country in the world, but I want to keep filming until I get too old.

Test 22[8]

You are going to read an extract from a newspaper article. For Questions 22.01-22.07, choose the answer ('a', 'b', 'c' or 'd') which you think fits best according to the text.
Mark your answers on the separate answer sheet.

A lot of advice is available for college leavers heading for their first job. In this article we consider the move to a second job. We are not concerned with those looking for a second temporary position while hunting for a permanent job. Nor are we concerned with those leaving an unsatisfactory job within the first few weeks. Instead, we will be dealing with those of you taking a real step on the career ladder, choosing a job to fit in with your ambitions now that you have learnt your way around, acquired some skills and have some idea of where you want to go.
What sort of job should you look for? Much depends on your long-term aim. You need to ask yourself whether you want to specialise in a particular field, work your way up to higher levels of responsibility or out of your current employment into a broader field.
Whatever you decide, you should choose your second job very carefully. You should be aiming to stay in it for two to three years.
This job will be studied very carefully when you send your letter of application for your next job. It should show evidence of serious career planning. Most important, *[line 26]* it should extend you, develop you and give you increasing responsibility.
Incidentally, if the travel bug is biting, now is the time to pack up and go. You can do temporary work for a while when you return, pick up where you left off and get the second job then. *[line 34]* Future potential employers will be relieved to see that you have got it out of your system, and are not likely to go off again.
Juliette Davidson spent her first year after leaving St. Aldate's College working for three solicitors. It was the perfect first job in that 'OK ... they were very supportive people. I was gently introduced to the work, learnt my way round an office and improved my word processing skills. However, there was no scope for advancement. One day I gave in my notice, bought an air ticket and travelled for a year.'
Juliette now works as a Personal Assistant to Brenda Cleverdon, the Chief Executive of Business in the Community. 'In two and a half years I have become more able and my job has really grown,' she says. 'Right from the beginning my boss was very keen to develop me. My job title is the same as it was when I started but the duties have changed. From mainly typing and telephone work, I have progressed to doing most of the correspondence and budgets. I also have to deal with a variety of queries, coming from chairmen of large companies to people wanting to know how to start their own business. Brenda involves me in all her work but also gives me specific projects to do and events to organise.'

---

8 Tests 22 and 23 were taken from: University of Cambridge Local Examinations Syndicate, "Cambridge First Certificate in English 4, Student's Book", 2001, © UCLES, reprinted with permission of the publisher (Cambridge University Press) and UCLES, pp. 6-7, 4-5.

22.01 Who is intended to benefit from the advice given in the article?
(a) students who have just finished their studies
(b) people who are unhappy with their current job
(c) those who are interested in establishing a career √
(d) people who change jobs regularly

22.02 According to the writer, why is the choice of your second job important?
(a) It will affect your future job prospects. √
(b) It will last longer than your first job.
(c) It will be difficult to change if you don't like it.
(d) It should give you the opportunity to study.

22.03 'it' in line 26 refers to your
(a) first job.
(b) second job. √
(c) application.
(d) career.

22.04 If you have a desire to travel, when does the writer suggest that you do it?
(a) straight after you have left college
(b) when you are unable to find a permanent job
(c) after you have done some temporary work
(d) between the first and second job √

22.05 What is meant by 'you have got it out of your system' in line 34?
(a) You have planned your career sensibly.
(b) You are an experienced traveller.
(c) You have satisfied your wish to travel. √
(d) You have learned to look after yourself.

22.06 How did Juliette Davidson benefit from the experience of her first job?
(a) It was a good introduction to working in an office. √
(b) She met a variety of interesting people.
(c) It enabled her to earn enough money to travel.
(d) She learnt how to use a word processor.

22.07 In what way is Juliette's current job better than her first job?
(a) She has a more impressive job title.
(b) She now knows how to start her own business.
(c) She has been able to extend her skills. √
(d) She is more involved in the community.

Test 23

Part 1

You are going to read a newspaper article about air travel. Choose the most suitable summary sentence from the list A-I for each part (23.01-23.07) of the article. There is one extra summary sentence which you do not need to use. There is an example at the beginning (O).

Mark your answers on the separate answer sheet.

A It is important to report missing luggage.
B Airlines make special efforts to avoid errors.
C Airlines sometimes blame travellers for lost luggage.
D No airline is free from mistakes.
E Plan ahead when travelling.
F Unexpected events can result in luggage going missing.
G No airport can guarantee the safe handling of luggage.
H Busy airports are likely to have more problems.
I Losing your luggage can be very inconvenient.

IN CASE YOU LOSE YOUR LUGGAGE

Alex McWhirter looks at airlines that lose bags.

0 (I)

There is nothing more disappointing than arriving at an airport overseas to discover that your baggage has been left behind. At best you will have to put up with wearing the clothes you stand up in for hours or days, until the airline reunites you with your luggage. At worst, you may be in a different climate zone, thousands of miles from home and forced to wear wholly unsuitable clothes.

23.01 (D)

Although airlines rarely reveal how many cases they lose, it is a fact of life that sooner or later regular travellers will be parted from their luggage. Even the best airlines slip up from time to time, and it is impossible for any carrier to guarantee that a passenger's checked luggage will go on the same flight, particularly when a journey calls for one or more changes of aircraft.

23.02 (B)

The system works like this. Airlines insist on exaggerated check-in times (which require passengers to report to the airport at a given time before departure) designed to allow sufficient time for baggage to pass through the airport and be loaded on to the plane. Minimum connecting times (MCTs) are the shortest time it takes to transfer between two flights. These, too, are exaggerated to allow for baggage transfers.

23.03 (F)
In normal circumstances the system works well. But extra security checks at airports and problems with air traffic combine to cause delayed flights. All this can cause the baggage system to fail. Then there is the possibility of human error, or an accident in which the destination label is torn off.

23.04 (H)
These problems can become severe at large transfer airports, known as 'hubs', because of the large number of bags that are processed. Last year, for example, London's Heathrow airport handled more than 41 million passengers, of whom nine million were changing planes. British Airways alone handled two million transfer passengers at Heathrow, with most making the one-mile transfer between Terminal 1 (for Domestic and European flights) and Terminal 4 (for long-distance flights).

23.05 (G)
Even efficient transfer airports, such as Amsterdam, Copenhagen, Singapore and Zurich have their bad days. The risk of baggage being lost when changing planes is higher than average at certain airports. Even the United States has problems – Miami airport is well known for luggage going missing when transatlantic passengers make immediate connections for destinations in Latin America.

23.06 (E)
You should choose direct flights whenever possible and check in well before the official time. If a change of plane is unavoidable, or makes your flight less expensive, then try to fly the same airline throughout. Try to allow more connecting time by taking an earlier flight to the transfer airport, and make sure you label your luggage inside and out with your home and holiday addresses. Don't forget to include the flight numbers.

23.07 (A)
If, after all this, your luggage still goes missing, you must contact the appropriate airline official in the baggage hall and complete a *property irregularity report* (PIR). This must be done before leaving the airport.

Test 24[9]

Answer questions 24.01-24.12 by referring to the newspaper article about emotional intelligence on page 3. Indicate your answers on the separate answer sheet.
For questions 24.01-24.12, answer by choosing from the sections of the article (A-D). Some of the choices may be required more than once.

9 Tests 24 and 25 are excerpts adapted from “"Sample Paper R CAE.pdf”, © University of Cambridge ESOL Examinations, downloaded from: <http://www.candidates.cambridgeesol.org/cs/digitalAssets/106850_cae_sample_papers.zip >, pp. 10 and 11.

In which section is the following mentioned?
24.01 the significance of emotional intelligence in work that is challenging (C)
24.02 increased accuracy in the way emotional intelligence is described (B)
24.03 the means by which we are assessed at work having become more comprehensive (A)
24.04 the fact that emotional intelligence can be combined with other skills to improve people's ability to cope at work (D)
24.05 areas in which emotional intelligence cannot be expected to offer solutions (D)
24.06 people having succeeded despite inadequacies in emotional intelligence (C)
24.07 the assumption that people have the academic skills to perform their jobs well (A)
24.08 the negative effect that a lack of emotional intelligence can have on a person's other skills (C)
24.09 the means of predicting who will excel in the workplace (A)
24.10 the reason why organisations promote emotional intelligence in the workplace (C)
24.11 misconceptions about what emotional intelligence involves (B)
24.12 the kind of staff relations that ensure an organisation has an advantage over its rivals (C)

EMOTIONAL INTELLIGENCE – THE KEY TO SUCCESS

Daniel Goleman examines the 'people skills' that are essential for a place at the top of your profession

(A) The rules for work are changing. We are being judged by a new yardstick – not just by how clever we are, or by our training or expertise, but also by how well we handle ourselves and each other. This yardstick is increasingly used in choosing who will be hired and who will not, who will be passed over and who will not. The new rules can be used to indicate who is likely to become a star performer and who is most prone to mediocrity. And, no matter what field we work in currently, they measure the traits that are crucial to our marketability for future jobs. These rules have little to do with what we were told at school was important. The ability to do well in examination is largely irrelevant to this standard. The new measure takes it for granted that we all have enough intellectual ability and technical know-how to do our jobs. It focuses instead on social skills and personal qualities, such as initiative and empathy, adaptability and persuasiveness – the 'people skills' that make up what is now commonly referred to as emotional intelligence.

(B) In a time when few guarantees of job security have led to the very concept of a 'job' being rapidly replaced by 'portable skills', personal qualities begin to play an important role in the workplace. Talked about loosely for decades under a variety of names, from 'character' and 'personality' to 'soft skills', there is, at last, a more precise understanding of these human talents as well as a new name for them. 'Emotional intelligence' is generally defined as the ability to monitor and regulate one's own and others' feelings, and to use feelings to guide thought and action. In our work-life it comprises five basic elements: self-awareness, self-regulation, motivation, empathy and adeptness in social relationships. There is a common assumption that it may demand not 'being nice', but rather, for example, bluntly

confronting someone with the uncomfortable truth. Nor does emotional intelligence mean giving free rein to feelings – 'letting it all hang out'. Rather, it means managing feelings so that they are expressed appropriately and effectively, enabling people to work together smoothly towards their common goal.

(C) More and more businesses are seeing that encouraging emotional intelligence skills is a vital component of management philosophy. And the less straightforward the job, the more emotional intelligence matters – if only because a deficiency in these abilities can hinder the use of whatever technical expertise or intellect a person may have. There are many examples of people who have risen to the top notwithstanding flaws in emotional intelligence, but as work becomes more complex and collaborative, companies where people work together best have a competitive edge. In the new workplace, with its emphasis on teamwork and a strong customer orientation, the crucial set of emotional competencies is becoming increasingly essential for excellence in every job and in every part of the world.

(D) Whereas one's IQ undergoes few changes, emotional intelligence continues to develop as we go through life and learn from our experiences; our competence in it can keep growing. In fact, studies that have measured people's emotional intelligence through the years show that most people grow more adept at handling their own emotions and impulses, at motivating themselves and at honing their empathy and social adroitness. There is an old-fashioned word for this growth in emotional intelligence: maturity. Not only can emotional intelligence be learnt, but individually we can add these skills to our tool kit for survival. This is especially relevant at a time when it seems a contradiction to put the words 'job' and 'stability' together. Emotional intelligence is no magic formula for uncompetitive organisations, no guarantee of more market share or a healthier bottom line. But if the human ingredient is ignored, then nothing else works as well as it might.

Test 25

For questions 25.01-25.06, you must choose which of the extracts A-G fit into the numbered gaps in the following magazine article. There is one extra paragraph which does not fit in any of the gaps. Indicate your answers on the separate answer sheet.

BEGINNER TAKES ALL

Even before it was published, The Horse Whisperer was the hottest book of the year. A first novel by British screenwriter Nicholas Evans, it has earned its author record-breaking sums. He talks here about his inspiration and his triumph

The first months of the year were not kind to Nicholas Evans, screenwriter, producer and aspiring director. The year began badly when Life and Limb, a film project he had been working on for months, fell through 'almost overnight'. His disappointment mingled with stomach-churning worry: it had been two years since he had earned any money and the promise of that film had been the only buffer between him and an increasingly irate bank manager.

25.01 (F)
Although he was acting very much on impulse, the seeds for the story had been with him for some time, sown by a farrier he met on Dartmoor while staying with a friend. The farrier had told him the story of a docile horse that had turned, no one knew why, into a fiend. Its owners were desperate until they heard of a gypsy who, simply by talking to the animal, transformed its temperament in a matter of hours. Such men, the farrier said, were known as 'horse whisperers'.

25.02 (A)
'It was a funny time,' he says now. 'I was observing people, but essentially I was alone and I really felt as though my life was falling apart. I'd tried for ten years to make a go of it as a film-maker, and here I was, hugely in debt and wondering how I was going to feed the children, and thinking maybe it was all just folly.'

25.03 (E)
When pushed, he ventured that Evans might get $30,000 as an advance on the book. 'I had in mind how much I needed to pay off a bit of the overdraft and keep us going, and it was more than that. I'd spent seven months on The Horse Whisperer, and there were at least another two to go. $30,000 was a really difficult figure. I was also advised to write a 12-page synopsis of the remainder of the book.'

25.04 (G)
The events that followed have become publishing history. Within a week – a week of hotly contested auctions – the novel had been sold to Transworld Publications in the UK for $550.000 and to Delacorte in the US for $3.15 million, both record-breaking advances for a first novel.

25.05 (C)
As they all agreed to this sum, it was decided that they should each 'pitch' to Evans. And so, one night in October, he sat in his study while four great film-makers rang, one after the other, to beg for the privilege of paying $3 million for an unfinished novel. Evans told me all this as we sat drinking coffee on a wooden verandah perched above the leafy garden of his home. He said that he had since turned down an offer to write the screenplay of *The Horse Whisperer*.

25.06 (B)
He would be involved, he said, but at arm's length. The success of his novel had inevitably brought forth the offer of new backing for *Life and Limb* but he was no longer sure that he wanted to make it. 'I think that I would be foolish not to write another novel,' he said.

(A) Evans' imagination was captured. He began researching the subject with a view to writing a screenplay – he was, after all, a film-maker. But disillusionment with the film world following the demise of Life and Limb prompted him to write the story as a book. And so throughout the spring he drove across the US, stopping at ranches and learning about horses and the men who work with them.

(B) 'It's all been such a fairy tale so far, I don't want to spoil it. Writing at that level is a very tough business, and I don't want to become an employee of these people who I like and who have paid me so much money. I'd hate to find myself writing a draft or two and then have them say, "Thanks Nick, but now we'll bring in so-and-so".'

(C) 'We couldn't believe it; we sat there with our jaws gaping. We'd never sent the manuscript to New York, we still don't know how it got there,' Evans says. Nor did they send it to Hollywood, but within that same week the major studios were fighting over it. 'My agent in the UK wisely involved an agent over there and when he phoned us to say, " think we can get $3 million outright," we laughed in disbelief.'

(D) As a screenwriter, he had yearned for the freedom of novelists and, when he had it, found himself 'in the middle of this immense and terrifying plain without the support of screenplay rules to guide me.' But he carries us smoothly through. Even so, he remains baffled as to why the story has captured imaginations in the mind-blowing way that it has. (extra paragraph)

(E) He thought that again towards the end of August, by which time he had returned home and written the first half of the book. 'At that point the bank manager was getting really very heavy with us, and I needed to know whether it was worth going on. I plucked up the courage to show it to a friend who was a literary agent; he read it and said it was "fine".'

(F) A wise man, finding himself in Evans' position, would have got a job. He could have gone back to being a television executive, or begun a television project that had been on hold. Instead, he made a decision that most people, Evans included, would consider insane. He bought a ticket to America and set off for three months to research his first novel.

(G) In October, together with the first two hundred pages of the novel, this was sent to seven UK publishers on the eve of their departure for the annual spending spree at the internationally renowned Frankfurt Book Fair. Within days his agent was on the telephone to report that he had just turned down the first offer of $75,000. 'I said, "You what?" And he said, "It's OK. I just sense something is happening".

Index

Linguistic Insights

Studies in Language and Communication

This series aims to promote specialist language studies in the fields of linguistic theory and applied linguistics, by publishing volumes that focus on specific aspects of language use in one or several languages and provide valuable insights into language and communication research. A cross-disciplinary approach is favoured and most European languages are accepted.

The series includes two types of books:

- **Monographs** – featuring in-depth studies on special aspects of language theory, language analysis or language teaching.
- **Collected papers** – assembling papers from workshops, conferences or symposia.

Each volume of the series is subjected to a double peer-reviewing process.

Vol. 1 Maurizio Gotti & Marina Dossena (eds)
Modality in Specialized Texts. Selected Papers of the 1st CERLIS Conference.
421 pages. 2001. ISBN 3-906767-10-8. US-ISBN 0-8204-5340-4

Vol. 2 Giuseppina Cortese & Philip Riley (eds)
Domain-specific English. Textual Practices across Communities and Classrooms.
420 pages. 2002. ISBN 3-906768-98-8. US-ISBN 0-8204-5884-8

Vol. 3 Maurizio Gotti, Dorothee Heller & Marina Dossena (eds)
Conflict and Negotiation in Specialized Texts. Selected Papers of the 2nd CERLIS Conference.
470 pages. 2002. ISBN 3-906769-12-7. US-ISBN 0-8204-5887-2

Editorial address:

Prof. Maurizio Gotti — Università di Bergamo, Facoltà di Lingue e Letterature Straniere, Via Salvecchio 19, 24129 Bergamo, Italy
Fax: 0039 035 2052789, E-Mail: m.gotti@unibg.it

Vol. 4 Maurizio Gotti, Marina Dossena, Richard Dury, Roberta Facchinetti & Maria Lima
Variation in Central Modals. A Repertoire of Forms and Types of Usage in Middle English and Early Modern English.
364 pages. 2002. ISBN 3-906769-84-4. US-ISBN 0-8204-5898-8

Vol. 5 Stefania Nuccorini (ed.)
Phrases and Phraseology. Data and Descriptions.
187 pages. 2002. ISBN 3-906770-08-7. US-ISBN 0-8204-5933-X

Vol. 6 Vijay Bhatia, Christopher N. Candlin & Maurizio Gotti (eds)
Legal Discourse in Multilingual and Multicultural Contexts. Arbitration Texts in Europe.
385 pages. 2003. ISBN 3-906770-85-0. US-ISBN 0-8204-6254-3

Vol. 7 Marina Dossena & Charles Jones (eds)
Insights into Late Modern English. 2nd edition.
378 pages. 2003, 2007.
ISBN 978-3-03911-257-9. US-ISBN 978-0-8204-8927-8

Vol. 8 Maurizio Gotti
Specialized Discourse. Linguistic Features and Changing Conventions.
351 pages. 2003, 2005.
ISBN 3-03910-606-6. US-ISBN 0-8204-7000-7

Vol. 9 Alan Partington, John Morley & Louann Haarman (eds)
Corpora and Discourse.
420 pages. 2004. ISBN 3-03910-026-2. US-ISBN 0-8204-6262-4

Vol. 10 Martina Möllering
The Acquisition of German Modal Particles. A Corpus-Based Approach.
290 pages. 2004. ISBN 3-03910-043-2. US-ISBN 0-8204-6273-X

Vol. 11 David Hart (ed.)
English Modality in Context. Diachronic Perspectives.
261 pages. 2003. ISBN 3-03910-046-7. US-ISBN 0-8204-6852-5

Vol. 12 Wendy Swanson
Modes of Co-reference as an Indicator of Genre.
430 pages. 2003. ISBN 3-03910-052-1. US-ISBN 0-8204-6855-X

Vol. 13 Gina Poncini
Discursive Strategies in Multicultural Business Meetings.
2nd edition. 338 pages. 2004, 2007.
ISBN 978-3-03911-296-8. US-ISBN 978-0-8204-8937-7

Vol. 14 Christopher N. Candlin & Maurizio Gotti (eds)
Intercultural Aspects of Specialized Communication.
2nd edition. 369 pages. 2004, 2007.
ISBN 978-3-03911-258-6. US-ISBN 978-0-8204-8926-1

Vol. 15 Gabriella Del Lungo Camiciotti & Elena Tognini Bonelli (eds)
Academic Discourse. New Insights into Evaluation.
234 pages. 2004. ISBN 3-03910-353-9. US-ISBN 0-8204-7016-3

Vol. 16 Marina Dossena & Roger Lass (eds)
Methods and Data in English Historical Dialectology.
405 pages. 2004. ISBN 3-03910-362-8. US-ISBN 0-8204-7018-X

Vol. 17 Judy Noguchi
The Science Review Article. An Opportune Genre in the Construction of Science.
274 pages. 2006. ISBN 3-03910-426-8. US-ISBN 0-8204-7034-1

Vol. 18 Giuseppina Cortese & Anna Duszak (eds)
Identity, Community, Discourse. English in Intercultural Settings.
495 pages. 2005. ISBN 3-03910-632-5. US-ISBN 0-8204-7163-1

Vol. 19 Anna Trosborg & Poul Erik Flyvholm Jørgensen (eds)
Business Discourse. Texts and Contexts.
250 pages. 2005. ISBN 3-03910-606-6. US-ISBN 0-8204-7000-7

Vol. 20 Christopher Williams
Tradition and Change in Legal English. Verbal Constructions in Prescriptive Texts.
2nd revised edition. 216 pages. 2005, 2007. ISBN 978-3-03911-444-3.

Vol. 21 Katarzyna Dziubalska-Kołaczyk & Joanna Przedlacka (eds)
English Pronunciation Models: A Changing Scene.
2nd printing. 476 pages. 2005, 2008. ISBN 978-3-03911-682-9.

Vol. 22 Christián Abello-Contesse, Rubén Chacón-Beltrán, M. Dolores López-Jiménez & M. Mar Torreblanca-López (eds)
Age in L2 Acquisition and Teaching.
214 pages. 2006. ISBN 3-03910-668-6. US-ISBN 0-8204-7174-7

Vol. 23 Vijay K. Bhatia, Maurizio Gotti, Jan Engberg & Dorothee Heller (eds)
Vagueness in Normative Texts.
474 pages. 2005. ISBN 3-03910-653-8. US-ISBN 0-8204-7169-0

Vol. 24 Paul Gillaerts & Maurizio Gotti (eds)
Genre Variation in Business Letters. 2nd printing.
407 pages. 2008. ISBN 978-3-03911-681-2.

Vol. 25 Ana María Hornero, María José Luzón & Silvia Murillo (eds)
Corpus Linguistics. Applications for the Study of English.
2nd printing. 526 pages. 2006, 2008. ISBN 978-3-03911-726-0

Vol. 26 J. Lachlan Mackenzie &
María de los Ángeles Gómez-González (eds)
Studies in Functional Discourse Grammar.
259 pages. 2005. ISBN 3-03910-696-1 / US-ISBN 0-8204-7558-0

Vol. 27 Debbie G.E. Ho
Classroom Talk. Exploring the Sociocultural Structure
of Formal ESL Learning. 2nd edition.
254 pages. 2006, 2007. ISBN 978-3-03911-434-4

Vol. 28 Javier Pérez-Guerra, Dolores González-Álvarez,
Jorge L. Bueno-Alonso & Esperanza Rama-Martínez (eds)
'Of Varying Language and Opposing Creed':
New Insights into Late Modern English.
455 pages. 2007. ISBN 978-3-03910-788-9

Vol. 29 Francesca Bargiela-Chiappini & Maurizio Gotti (eds)
Asian Business Discourse(s).
350 pages. 2005. ISBN 3-03910-804-2 / US-ISBN 0-8204-7574-2

Vol. 30 Nicholas Brownlees (ed.)
News Discourse in Early Modern Britain.
Selected Papers of CHINED 2004.
300 pages. 2006. ISBN 3-03910-805-0 / US-ISBN 0-8204-8025-8

Vol. 31 Roberta Facchinetti & Matti Rissanen (eds)
Corpus-based Studies of Diachronic English.
300 pages. 2006. ISBN 3-03910-851-4 / US-ISBN 0-8204-8040-1

Vol. 32 Marina Dossena & Susan M. Fitzmaurice (eds)
Business and Official Correspondence:
Historical Investigations.
209 pages. 2006. ISBN 3-03910-880-8 / US-ISBN 0-8204-8352-4

Vol. 33 Giuliana Garzone & Srikant Sarangi (eds)
Discourse, Ideology and Specialized Communication.
494 pages. 2007. ISBN 978-3-03910-888-6

Vol. 34 Giuliana Garzone & Cornelia Ilie (eds)
The Use of English in Institutional and Business Settings.
An Intercultural Perspective.
372 pages. 2007. ISBN 978-3-03910-889-3

Vol. 35 Vijay K. Bhatia & Maurizio Gotti (eds)
Explorations in Specialized Genres.
316 pages. 2006. ISBN 3-03910-995-2 / US-ISBN 0-8204-8372-9

Vol. 36 Heribert Picht (ed.)
Modern Approaches to Terminological Theories and Applications.
432 pages. 2006. ISBN 3-03911-156-6 / US-ISBN 0-8204-8380-X

Vol. 37 Anne Wagner & Sophie Cacciaguidi-Fahy (eds)
Legal Language and the Search for Clarity / Le langage juridique et la quête de clarté. Practice and Tools / Pratiques et instruments.
487 pages. 2006. ISBN 3-03911-169-8 / US-ISBN 0-8388-5

Vol. 38 Juan Carlos Palmer-Silveira, Miguel F. Ruiz-Garrido & Inmaculada Fortanet-Gómez (eds)
Intercultural and International Business Communication: Theory, Research and Teaching.
2nd edition. 343 pages. 2006, 2008. ISBN 978-3-03911-680-5

Vol. 39 Christiane Dalton-Puffer, Dieter Kastovsky, Nikolaus Ritt & Herbert Schendl (eds)
Syntax, Style and Grammatical Norms: English from 1500–2000.
250 pages. 2006. ISBN 3-03911-181-7 / US-ISBN 0-8204-8394-X

Vol. 40 Marina Dossena & Irma Taavitsainen (eds)
Diachronic Perspectives on Domain-Specific English.
280 pages. 2006. ISBN 3-03910-176-0 / US-ISBN 0-8204-8391-5

Vol. 41 John Flowerdew & Maurizio Gotti (eds)
Studies in Specialized Discourse.
293 pages. 2006. ISBN 3-03911-178-7 / US-ISBN 0-8204-8393-1

Vol. 42 Ken Hyland & Marina Bondi (eds)
Academic Discourse Across Disciplines.
320 pages. 2006. ISBN 3-03911-183-3 / US-ISBN 0-8204-8396-6

Vol. 43 Paul Gillaerts & Philip Shaw (eds)
The Map and the Landscape: Norms and Practices in Genre.
256 pages. 2006. ISBN 3-03911-182-5 / US-ISBN 0-8204-8395-4

Vol. 44 Maurizio Gotti & Davide Giannoni (eds)
New Trends in Specialized Discourse Analysis.
301 pages. 2006. ISBN 3-03911-184-1 / US-ISBN 0-8204-8381-8

Vol. 45 Maurizio Gotti & Françoise Salager-Meyer (eds)
Advances in Medical Discourse Analysis:
Oral and Written Contexts.
492 pages. 2006. ISBN 3-03911-185-X / US-ISBN 0-8204-8382-6

Vol. 46 Maurizio Gotti & Susan Šarcević (eds)
Insights into Specialized Translation.
396 pages. 2006. ISBN 3-03911-186-8 / US-ISBN 0-8204-8383-4

Vol. 47 Khurshid Ahmad & Margaret Rogers (eds)
Evidence-based LSP: Translation, Text and Terminology.
584 pages. 2007. ISBN 978-3-03911-187-9

Vol. 48 Hao Sun & Dániel Z. Kádár (eds)
It's the Dragon's Turn: Chinese Institutional Discourses.
262 pages. 2008. ISBN 978-3-03911-175-6

Vol. 49 Cristina Suárez-Gómez
Relativization in Early English (950-1250):
the Position of Relative Clauses.
149 pages. 2006. ISBN 3-03911-203-1 / US-ISBN 0-8204-8904-2

Vol. 50 Maria Vittoria Calvi & Luisa Chierichetti (eds)
Nuevas tendencias en el discurso de especialidad.
319 pages. 2006. ISBN 978-3-03911-261-6

Vol. 51 Mari Carmen Campoy & María José Luzón (eds)
Spoken Corpora in Applied Linguistics.
274 pages. 2008. ISBN 978-3-03911-275-3

Vol. 52 Konrad Ehlich & Dorothee Heller (Hrsg.)
Die Wissenschaft und ihre Sprachen.
323 pages. 2006. ISBN 978-3-03911-272-2

Vol. 53 Jingyu Zhang
The Semantic Salience Hierarchy Model:
The L2 Acquisition of Psych Predicates
273 pages. 2007. ISBN 978-3-03911-300-2

Vol. 54 Norman Fairclough, Giuseppina Cortese & Patrizia Ardizzone (eds)
Discourse and Contemporary Social Change.
555 pages. 2007. ISBN 978-3-03911-276-0

Vol. 55 Jan Engberg, Marianne Grove Ditlevsen, Peter Kastberg &
Martin Stegu (eds)
New Directions in LSP Teaching.
331 pages. 2007. ISBN 978-3-03911-433-7

Vol. 56 Dorothee Heller & Konrad Ehlich (Hrsg.)
Studien zur Rechtskommunikation.
322 pages. 2007. ISBN 978-3-03911-436-8

Vol. 57 Forthcoming.

Vol. 58 Carmen Frehner
Email – SMS – MMS
294 pages. 2008. ISBN 978-3-03911-451-1

Vol. 59 Isabel Balteiro
The Directionality of Conversion in English:
A Dia-Synchronic Study.
276 pages. 2007. ISBN 978-3-03911-241-8

Vol. 60 Maria Milagros Del Saz Rubio
English Discourse Markers of Reformulation.
237 pages. 2007. ISBN 978-3-03911-196-1

Vol. 61 Forthcoming.

Vol. 62 Sandrine Onillon
Pratiques et représentations de l'écrit.
458 pages. 2008. ISBN 978-3-03911-464-1

Vol. 63 Hugo Bowles & Paul Seedhouse (eds)
Conversation Analysis and Language for Specific Purposes.
334 pages. 2007. ISBN 978-3-03911-469-6

Vol. 64 Vijay K. Bhatia, Christopher N. Candlin &
Paola Evangelisti Allori (eds)
Language, Culture and the Law.
The Formulation of Legal Concepts across Systems and Cultures.
342 pages. 2008. ISBN 978-3-03911-470-2

Vol. 65–67 Forthcoming.

Vol. 68 Eva Alcón-Soler (ed.)
Learning How to Request in an Instructed
Language Learning Context.
260 pages. 2008. ISBN 978-3-03911-601-0

Vol. 69 Domenico Pezzini
The Translation of Religious Texts in the Middle Ages.
428 pages. 2008. ISBN 978-3-03911-600-3

Vol. 70 Tomoko Tode
Effects of frequency in classroom second language learning. Quasi-experiment and stimulated-recall analysis.
195 pages. 2008. ISBN 978-3-03911-602-7

Vol. 71 Forthcoming.

Vol. 72 Cynthia J. Kellett Bidoli & Elana Ochse (eds)
English in International Deaf Communication.
444 pages. 2008. ISBN 978-3-03911-610-2

Vol. 73 Joan C. Beal, Carmela Nocera & Massimo Sturiale (eds)
Perspectives on Prescriptivism.
269 pages. 2008. ISBN 978-3-03911-632-4

Vol. 74 Carol Taylor Torsello, Katherine Ackerley & Erik Castello (eds)
Corpora for University Language Teachers.
308 pages. 2008. ISBN 978-3-03911-639-3

Vol. 75 Forthcoming.

Vol. 76 Marina Dossena & Ingrid Tieken-Boon van Ostade (eds)
Studies in Late Modern English Correspondence. Methodology and Data.
291 pages. 2008. ISBN 978-3-03911-658-4

Vol. 77-79 Forthcoming.

Vol. 80 Martin Solly, Michelangelo Conoscenti & Sandra Campagna (eds)
Verbal/Visual Narrative Texts in Higher Education.
384 pages. 2008. ISBN 978-3-03911-672-0

Vol. 81 Meiko Matsumoto
From Simple Verbs to Periphrastic Expressions: The Historical Development of Composite Predicates, Phrasal Verbs, and Related Constructions in English.
235 pages. 2008. ISBN 978-3-03911-675-1

Vol. 82-83 Forthcoming.

Vol. 84 Maria Grazia Guido
English as a Lingua Franca in Cross-cultural Immigration Domains.
285 pages. 2008. ISBN 978-3-03911-689-8

Vol. 85 Erik Castello
Text Complexity and Reading Comprehension Tests.
352 pages. 2008. ISBN 978-3-03911-717-8